AF783196

NEUROMETHODS

Series Editor
Wolfgang Walz
University of Saskatchewan
Saskatoon, Canada

For further volumes:
http://www.springer.com/series/7651

Neuromethods publishes cutting-edge methods and protocols in all areas of neuroscience as well as translational neurological and mental research. Each volume in the series offers tested laboratory protocols, step-by-step methods for reproducible lab experiments and addresses methodological controversies and pitfalls in order to aid neuroscientists in experimentation. *Neuromethods* focuses on traditional and emerging topics with wide-ranging implications to brain function, such as electrophysiology, neuroimaging, behavioral analysis, genomics, neurodegeneration, translational research and clinical trials. *Neuromethods* provides investigators and trainees with highly useful compendiums of key strategies and approaches for successful research in animal and human brain function including translational "bench to bedside" approaches to mental and neurological diseases.

Brain Metallomics

Edited by

Andreas M. Grabrucker

Department of Biological Sciences, University of Limerick, Limerick, Ireland

Humana Press

Editor
Andreas M. Grabrucker
Department of Biological Sciences
University of Limerick
Limerick, Ireland

ISSN 0893-2336 ISSN 1940-6045 (electronic)
Neuromethods
ISBN 978-1-0716-5467-5 ISBN 978-1-0716-5468-2 (eBook)
https://doi.org/10.1007/978-1-0716-5468-2

© The Editor(s) (if applicable) and The Author(s), under exclusive license to Springer Science+Business Media, LLC, part of Springer Nature 2026

This work is subject to copyright. All rights are solely and exclusively licensed by the Publisher, whether the whole or part of the material is concerned, specifically the rights of translation, reprinting, reuse of illustrations, recitation, broadcasting, reproduction on microfilms or in any other physical way, and transmission or information storage and retrieval, electronic adaptation, computer software, or by similar or dissimilar methodology now known or hereafter developed.
The use of general descriptive names, registered names, trademarks, service marks, etc. in this publication does not imply, even in the absence of a specific statement, that such names are exempt from the relevant protective laws and regulations and therefore free for general use.
The publisher, the authors and the editors are safe to assume that the advice and information in this book are believed to be true and accurate at the date of publication. Neither the publisher nor the authors or the editors give a warranty, expressed or implied, with respect to the material contained herein or for any errors or omissions that may have been made. The publisher remains neutral with regard to jurisdictional claims in published maps and institutional affiliations.

This Humana imprint is published by the registered company Springer Science+Business Media, LLC, part of Springer Nature.
The registered company address is: 1 New York Plaza, New York, NY 10004, U.S.A.

If disposing of this product, please recycle the paper.

Preface

Neuroscience has long focused on the chemical and electrical signals orchestrating brain function. While the bulk metals sodium (Na) and potassium (K) have prominent roles in the generation of electric potential and action potential, and the importance of calcium (Ca) in the synaptic release of neurotransmitters has been well studied, recent advances have illuminated the pivotal roles of trace and ultra-trace metals in neural processes. Metals such as iron (Fe), zinc (Zn), copper (Cu), and manganese (Mn) are not merely trace elements in the brain; they are essential players in a wide range of neurological activities, from roles in energy metabolism, neurotransmitter synthesis, and cell signaling in neurons and glial cells, to the maintenance of synaptic plasticity. The emerging field of neurometallomics—an intersection of neuroscience, biochemistry, and metal biology—may offer a comprehensive understanding of how these metals contribute to normal brain development, function, aging, and neurological disorders.

This book, *Brain Metallomics*, aims to summarize the diverse roles metals play in the brain, encompassing their distribution, regulatory mechanisms, and involvement in disease. It focuses on methods to measure neurometals with various technologies. It is a compendium of current knowledge, from the molecular underpinnings of metal ion homeostasis and detection to the latest research on how metal dysregulation can lead to brain pathologies.

As the first volume dedicated entirely to neurometallomics, this book is designed to serve as a valuable resource for researchers, clinicians, and students alike. Each chapter not only delves into specific aspects of metal biology in the brain but also offers practical insights into the sophisticated methods used to study metal distribution and function. It discusses potential therapeutic approaches to modulate metal homeostasis in disease, providing a roadmap for future research and clinical practice.

The concept of neurometallomics reflects a paradigm shift in our understanding of brain biology—one that acknowledges the intricate interplay between metal ions and neuronal activity. It integrates metals into the functional properties of proteins (neurometalloproteomics) but should inspire to go beyond and research the links between brain metabolomics, transcriptomics, lipidomics, etc., and metallomics, enabling a more holistic view of how the brain works. This book is a testament to the collaborative spirit of the field, bringing together contributions from leading experts. By uniting their insights, this book seeks to inspire further research and discussion on the critical roles that metals play in the brain, fostering a sense of community among its readers.

This work will deepen the reader's understanding of neurometallomics and spark new ideas and innovations in the study of brain health and disease. As the field evolves, we envision that neurometallomics will serve as a foundational resource and a springboard for future discoveries.

Limerick, Ireland *Andreas M. Grabrucker*

Contents

Contributors

Asma Ahmadani • *Department of Nutrition and Health, College of Medicine & Health Sciences, United Arab Emirates University, Al Ain, United Arab Emirates*

Agostinho Almeida • *LAQV, REQUIMTE, Department of Chemical Sciences, Faculty of Pharmacy of the University of Porto, Rua Jorge Viterbo Ferreira 228,, Porto, Portugal*

Amita Attlee • *Department of Nutrition and Health, College of Medicine & Health Sciences, United Arab Emirates University, Al Ain, United Arab Emirates*

Dorota Bederska-Łojewska • *Department of Neurobiology, Maj Institute of Pharmacology, Polish Academy of Sciences, Kraków, Poland*

Shayon Bhattacharya • *Department of Biological Sciences, SSPC, The Research Ireland Centre for Pharmaceuticals, Bernal Institute, University of Limerick, Limerick, Ireland; Health Research Institute, University of Limerick, Limerick, Ireland*

Anna Błażewicz • *Department of Pathobiochemistry and Interdisciplinary Applications of Ion Chromatography, Medical University of Lublin, Lublin, Poland*

Asuncion Carmona • *University of Bordeaux, CNRS, LP2I Bordeaux, UMR 5797, Chemical Imaging and Speciation in Neuroscience, Gradignan, France*

Maria Cristina D'Adamo • *Department of Medicine & Surgery, LUM University "GIUSEPPE DEGENNARO", Casamassima, BA, Italy*

Mónica Fernández-Monreal • *University of Bordeaux, CNRS, INSERM, Bordeaux Imaging Center, BIC, Bordeaux, France*

Andreas M. Grabrucker • *Department of Biological Sciences, University of Limerick, Limerick, Ireland; Bernal Institute, University of Limerick, Limerick, Ireland; Health Research Institute (HRI), University of Limerick, Limerick, Ireland*

Midori Kato-Negishi • *Department of Bio-Analytical Chemistry, Research Institute of Pharmaceutical Sciences, Faculty of Pharmacy, Musashino University, Tokyo, Japan*

Masahiro Kawahara • *Department of Bio-Analytical Chemistry, Research Institute of Pharmaceutical Sciences, Faculty of Pharmacy, Musashino University, Tokyo, Japan*

Ines Kelkoul • *University of Bordeaux, CNRS, LP2I Bordeaux, UMR 5797, Chemical Imaging and Speciation in Neuroscience, Gradignan, France*

Michał Kiełbus • *Department of Biochemistry and Molecular Biology, Medical University of Lublin, Lublin, Poland*

Lara Lobo • *Department of Physical and Analytical Chemistry, Faculty of Chemistry, University of Oviedo, Oviedo, Spain*

Nicola M. Lowe • *Centre for Global Development, University of Lancashire, Preston, Lancashire, UK*

Marta Marina-Latorre • *Department of Physical and Analytical Chemistry, Faculty of Chemistry, University of Oviedo, Oviedo, Spain*

Rukhsana Nawaz • *Department of Clinical Psychology, College of Medicine & Health Sciences, United Arab Emirates University, Al Ain, United Arab Emirates*

Richard Ortega • *University of Bordeaux, CNRS, LP2I Bordeaux, UMR 5797, Chemical Imaging and Speciation in Neuroscience, Gradignan, France*

ROSARIO PEREIRO • *Department of Physical and Analytical Chemistry, Faculty of Chemistry, University of Oviedo, Oviedo, Spain*
MAURO PESSIA • *Department of Physiology, College of Medicine and Health Sciences, United Arab Emirates University, Al Ain, United Arab Emirates; Department of Physiology & Biochemistry, Faculty of Medicine and Surgery, University of Malta, Msida, Malta*
NOÉMIE PIED • *University of Bordeaux, CNRS, INSERM, Bordeaux Imaging Center, BIC, Bordeaux, France*
SANDEEP B. SUBRAMANYA • *Department of Physiology, College of Medicine & Health Sciences, United Arab Emirates University, Al Ain, United Arab Emirates*
BERNADETA SZEWCZYK • *Department of Neurobiology, Maj Institute of Pharmacology, Polish Academy of Sciences, Kraków, Poland*
KEN-ICHIRO TANAKA • *Department of Bio-Analytical Chemistry, Research Institute of Pharmaceutical Sciences, Faculty of Pharmacy, Musashino University, Tokyo, Japan*
RICHARD B. THOMPSON • *Department of Biochemistry and Molecular Biology, University of Maryland School of Medicine, Baltimore, MD, USA*
JUSTYNA TUREK • *Department of Neurobiology, Maj Institute of Pharmacology, Polish Academy of Sciences, Kraków, Poland*
JULIA WOJNICKA • *Department of Pathobiochemistry and Interdisciplinary Applications of Ion Chromatography, Medical University of Lublin, Lublin, Poland*

Check for updates

Chapter 1

Foundations of Brain Metallomics

Andreas M. Grabrucker

Abstract

Metallomics is an emerging field that focuses on the comprehensive study of the roles, functions, and dynamics of metal ions in biological systems. In the context of the brain, metallomics seeks to unravel the complex interactions between metal ions and neural processes. This chapter explores the foundations of brain metallomics, highlighting its history, significance, methodologies, and the intricate interplay between metals and neurological functions. Metallomics holds significant potential in the prevention and treatment of brain disorders. By mapping the metal ion distributions and understanding their biochemical pathways, metallomics can identify imbalances or dysregulations linked to neurological conditions such as Alzheimer's, Parkinson's, and multiple sclerosis. This knowledge facilitates the development of targeted therapies that restore proper metal ion homeostasis, potentially halting or reversing disease progression. Furthermore, metallomics can aid in the early diagnosis of brain disorders by detecting specific metal ion biomarkers, enabling timely intervention and improved patient outcomes. In the future, linking the metallome to the proteome, transcriptome, epigenome, microbiome, etc., may reveal individual signatures of specific pathologies that can be used to optimize disease prediction, prevention, and intervention strategies.

Keywords Trace metals, Neuroscience, Metallome, Zinc, Copper, Metal detection, Nervous system

1 Introduction: The History of Brain Metallomics

Brain metallomics, the study of metal ions and their roles in brain function, is a relatively new field with roots in the broader disciplines of biochemistry, neurobiology, and analytical chemistry. The recognition of the importance of metals in biological systems dates back to the early 18th century, when the role of iron in blood and its necessity for life were first acknowledged [1]. However, the specific interest in metals within the brain began to take shape in the mid-20th century. Initial studies focused on the essentiality of trace elements such as iron, copper, and zinc for overall health, but the direct link to brain function was not made clear for some time.

In the 1950s and 1960s, pioneering research unveiled the presence and significance of these metals in the brain. For instance, the

Andreas M. Grabrucker (ed.), *Brain Metallomics*, Neuromethods, vol. 237,
https://doi.org/10.1007/978-1-0716-5468-2_1,
© The Author(s), under exclusive license to Springer Science+Business Media, LLC, part of Springer Nature 2026

discovery that copper was a critical component of enzymes such as cytochrome c oxidase and superoxide dismutase highlighted the potential roles of metal ions in neural metabolism and antioxidant defense [2, 3]. The field of neurochemistry, which emerged prominently in the 1960s, provided a framework for studying the brain's chemical composition, including its metal content. Techniques such as atomic absorption spectroscopy (AAS) and colorimetric assays were employed to quantify metal ions in neural tissues (e.g., see [4]). Though less sensitive and specific than modern techniques, these methods laid the groundwork for understanding the distribution and concentration of metals in the brain. One significant early finding was the association between altered metal levels and neurological diseases. For example, studies in the 1960s and 1970s reported abnormal iron and copper concentrations in the brains of individuals with Alzheimer's, Parkinson's, and Wilson's diseases [5–8]. These observations suggested that metal dyshomeostasis could be linked to neurodegeneration.

The 1980s and 1990s marked a period of rapid technological advancement that greatly enhanced the study of brain metallomics. Inductively coupled plasma mass spectrometry (ICP-MS) emerged as a powerful tool for detecting trace metal concentrations with high sensitivity and precision. Coupled with laser ablation techniques, ICP-MS enabled spatially resolved analysis of metal distribution in brain tissues. Another significant development was the advent of X-ray fluorescence (XRF) imaging and synchrotron radiation-based techniques. These methods provided high-resolution maps of metal ion distribution at the cellular and subcellular levels, offering unprecedented insights into the localization of metals within the brain. During this period, the application of nuclear magnetic resonance (NMR) spectroscopy and magnetic resonance imaging (MRI) also advanced. Paramagnetic metals such as manganese and gadolinium were used as contrast agents [9], allowing researchers to visualize brain structures and functions of metal content. These imaging techniques enabled noninvasive in vivo studies of metal ion dynamics.

The early 2000s saw the emergence of metalloproteomics [10], a subdiscipline focused on identifying and characterizing metal-binding proteins. This approach integrated proteomics with metallomics, utilizing advanced mass spectrometry techniques to study metalloproteins in complex biological samples. Research in metalloproteomics revealed the critical roles of metal ions in modulating the activity of numerous brain proteins involved in neurotransmission, signal transduction, and neuroprotection [11, 12]. For instance, zinc has been shown to play a vital role in synaptic plasticity and memory formation by interacting with neurotransmitter receptors and signaling pathways [13–16].

In recent years, brain metallomics has continued to evolve with the development of more sophisticated analytical techniques and interdisciplinary approaches. High-resolution imaging methods, such as secondary ion mass spectrometry (SIMS) and nanoscale secondary ion mass spectrometry (NanoSIMS), have enabled even more precise localization of metal ions at the nanoscale. The application of genomics and bioinformatics has also enriched metallomics research. Understanding the genetic regulation of metal transporters, chaperones, and metalloproteins provides insights into the molecular mechanisms underlying metal homeostasis and its dysregulation in neurological diseases. Current research focuses on elucidating the complex interactions between metal ions, proteins, and other biomolecules in the brain [17]. Studies are increasingly investigating the role of metal ions in neuroinflammation, oxidative stress, and mitochondrial dysfunction, which are critical processes in the pathogenesis of several brain disorders [18, 19].

2 The Significance of Metal Ions in the Brain

The history of brain metallomics research reflects a dynamic, interdisciplinary journey from early observations of the necessity of metals in biology to the sophisticated, multifaceted analyses of today. Advancements in analytical techniques, from atomic absorption spectroscopy to high-resolution imaging and mass spectrometry, have significantly enhanced our understanding of the roles of metal ions in the brain. As research continues to uncover the intricate balance of metals in neural function and disease, brain metallomics holds promise for novel diagnostic and therapeutic approaches to address neurological disorders.

2.1 Metal Homeostasis and Brain Function

Metal ions, such as iron (Fe), copper (Cu), zinc (Zn), and manganese (Mn), are essential for numerous biological processes, including enzyme function, neurotransmission, and oxidative stress regulation. These metals are vital for maintaining brain health and cognitive function [20]. For instance, iron is a key component of hemoglobin and is crucial for oxygen transport and mitochondrial respiration [21]. Copper and zinc play pivotal roles in synaptic signaling and neuroprotection [22, 23], while manganese regulates neurotransmitter synthesis and antioxidant defense mechanisms [24].

The brain maintains a delicate balance of metal ion concentrations through tightly regulated transport, storage, and excretion mechanisms. Disruptions in metal homeostasis can lead to severe neurological disorders. For example, excessive accumulation of iron has been associated with neurodegenerative diseases such as Alzheimer's and Parkinson's. Conversely, copper deficiency can impair cognitive development and cause neurodegeneration [25]. Therefore, understanding the regulation of metal ions in the brain is critical for elucidating the pathophysiology of various neurological conditions.

For example, Alzheimer's disease is characterized by abnormal accumulation of amyloid-beta plaques and tau tangles, both of which can interact with metal ions such as copper and zinc, potentially exacerbating oxidative damage and neurotoxicity [26, 27]. Parkinson's disease, on the other hand, is associated with elevated iron levels in the substantia nigra, contributing to dopaminergic neuron degeneration through oxidative stress and mitochondrial dysfunction [28]. In turn, zinc deficiency has been linked as a risk factor for neurodevelopmental disorders such as Autism Spectrum Disorders (ASD) [29]. It is, therefore, essential to understand how brain metal levels are regulated and how each metal contributes to healthy brain function. However, it should be mentioned that agonistic and antagonistic relationships between them influence metal concentrations. Instead of a specific metal, it may be more appropriate to consider the metal profile as a whole.

The uptake of essential metals into the human brain is a highly regulated process critical for maintaining normal neurological function. Essential metals such as iron, zinc, copper, and manganese are transported across the blood–brain barrier (BBB), a selectively permeable membrane that protects the brain from harmful substances, while allowing essential nutrients to enter. The BBB is formed by endothelial cells tightly joined by tight junctions, supported by pericytes and astrocytes. Essential metals must cross this barrier to enter the brain via specific transport systems [30, 31].

For example, iron uptake primarily occurs in the brain through the transferrin receptor-mediated pathway. Iron circulates in the blood bound to transferrin, a glycoprotein. The transferrin-iron complex binds to transferrin receptors on the endothelial cells of the BBB, initiating receptor-mediated endocytosis. Once inside the cell, the iron is released from transferrin and transported across the cell membrane into the brain, likely through divalent metal transporter 1 (DMT1) [32].

Zinc is the second most prevalent trace metal in the brain. Zinc uptake into the brain is mediated by a family of zinc transporters, including ZIP (Zrt-, Irt-like Protein) and ZnT (Zinc Transporter) proteins. ZIP transporters facilitate zinc influx into cells and cytosol from organelles, while ZnT transporters primarily mediate zinc efflux or sequestration into organelles [33]. At the BBB, ZIP transporters are believed to play a significant role in transporting zinc into brain endothelial cells, while ZnTs subsequently release it into the brain parenchyma.

Copper transport across the BBB involves copper transporter 1 (CTR1), a high-affinity copper transporter. CTR1 facilitates the uptake of copper into endothelial cells of the BBB. Inside the cells, copper is chaperoned by proteins such as ATOX1 to various cellular compartments or across the BBB into the brain, where it is incorporated into enzymes [34]. In contrast, manganese enters the brain through both active and passive transport mechanisms. DMT1, which also transports iron, facilitates manganese uptake at the BBB. Additionally, manganese can cross the BBB via a transferrin-independent mechanism, potentially through the divalent metal cation transporter system [35, 36].

The brain tightly regulates the uptake and homeostasis of essential metals to prevent toxicity, as both deficiency and excess can be detrimental. Specialized proteins, such as metal-binding/buffering proteins like metallothioneins and ferritin, sequester metals for cell signaling or to buffer excess metals and prevent oxidative stress and neurotoxicity [37]. Additionally, the expression of metal transporters at the BBB can be regulated by various factors, including metal levels, hormones, and inflammatory signals [38, 39].

2.2 Metal Neurochemistry

Research in metalloproteins aims to elucidate their metal-binding sites, affinity, and kinetics, providing insights into their functional roles in the brain. Advanced proteomic techniques, such as tandem mass spectrometry and bioinformatics tools, identify and quantify metalloproteins in neural tissues, revealing their involvement in health and disease. Investigating the metallomic profile of brain diseases can lead to the identification of novel biomarkers and therapeutic targets.

Important metal-binding proteins in the brain are critical in maintaining metal homeostasis, preventing toxicity, and facilitating essential biochemical processes. Some of the frequently researched metal-binding proteins in the brain include Ferritin, which stores and sequesters excess iron, protecting cells from oxidative damage and releases iron in a controlled manner for cellular processes; Metallothioneins (MTs) that bind zinc, copper, and other divalent metals for cell signaling purposes and metal storage, Amyloid Precursor Protein (APP) that besides its role in Alzheimer's disease, can bind metals such as copper, zinc, and iron, influencing their homeostasis, Prion Protein (PrP) that binds copper and may play a role in copper metabolism defense (misfolding of PrP is linked to prion diseases, where metal-binding properties are disrupted), and Superoxide Dismutase 1 (SOD1), an antioxidant enzyme that requires copper and zinc for its activity. It converts superoxide radicals into hydrogen peroxide and oxygen, protecting cells from oxidative damage.

However, it should be noted that about 10% of the human genome encodes zinc-binding proteins, among them over 300 enzymes, many of which are expressed in the brain [40]. With additional proteins binding copper and iron, it is evident that metal ions are likely part of every cellular process in the brain.

While metals are indispensable for proper brain function, their neurochemistry is a double-edged sword, with both deficiency and excess posing significant risks to neuronal health. Therefore, the brain has evolved mechanisms to tightly regulate metal concentrations, ensuring they are available for essential biochemical processes while minimizing the risk of toxicity. Disruption in metal homeostasis, whether due to genetic mutations, environmental exposure, or aging, can lead to the accumulation or lack of metals in specific brain regions, contributing to abnormal neurochemistry.

Metals such as iron, zinc, copper, and manganese are crucial for numerous biochemical processes within the brain, including proper protein folding, enzyme function, and cell signaling. Iron, e.g., is a critical cofactor in several enzymes involved in neurotransmitter synthesis and mitochondrial function. Iron is essential for the activity of tyrosine hydroxylase, which catalyzes the first step in the synthesis of the neurotransmitter dopamine [41]. Iron is also a component of cytochrome c oxidase, an enzyme in the mitochondrial electron transport chain that is vital for ATP production [42].

Copper is involved in the activity of several critical enzymes, including dopamine β-hydroxylase, which converts dopamine to norepinephrine, another important neurotransmitter [43]. Copper is also released at synapses, where it acts as a neuromodulator [22]. Manganese is required for the activity of several enzymes, including glutamine synthetase, which helps detoxify ammonia in the brain by converting glutamate to glutamine. It is also a cofactor for Mn-SOD, an enzyme critical for protecting mitochondria from oxidative damage [44]. However, excessive manganese exposure, often through environmental sources, can lead to manganism, a neurotoxic condition with symptoms similar to Parkinson's disease due to the metal's accumulation in the basal ganglia and its impact on dopaminergic neurons [45].

Zinc, finally, is a highly versatile metal in the brain, playing roles in enzyme function, neurotransmitter regulation, and signal transduction [46, 47]. It is a cofactor for several enzymes, including carbonic anhydrase and matrix metalloproteinases, which are involved in maintaining pH balance and extracellular matrix remodeling, respectively [48, 49]. Zinc also modulates neurotransmission by interacting with ion channels and receptors, such as the N-methyl-D-aspartate (NMDA) receptor [50] and the GABA receptor [51], as well as synaptic proteins such as SHANK3 [52], which are involved in synaptic plasticity and memory formation. Disruption in zinc homeostasis has been associated with several brain disorders [53, 54].

In addition, toxic metals such as lead (Pb) and mercury (Hg) have been linked to brain disorders such as ASD [19]. Their overload produces pathologies such as inflammation, while, at the cellular level, these metals interfere with the function of essential metals. However, it is noteworthy that other non-essential metals are detected in healthy individuals. For example, the average human body (70 kg), including brain tissue, contains 0.68 g rubidium, 0.32 g strontium, 60 mg aluminum, 40 mg cerium, 20 mg tin, 20 mg titanium, 15 mg nickel, 7 mg lithium, 5 mg germanium, 2 mg antimony, 2 mg silver, 1.5 mg niobium, 1 mg zirconium, 0.7 mg gallium, 0.7 mg tellurium,

0.6 mg yttrium, among others. While these metals may not have a biological function, they may contribute to the overall balance of essential metals and the establishment of a specific metal profile. However, due to their ultra-trace levels, research on these metals is in its infancy. Nevertheless, understanding the balance of metal ions in the brain is critical for developing therapeutic strategies for many, if not all, brain pathologies.

3 Analytical Techniques in Brain Metallomics

Advancements in analytical techniques have propelled the field of brain metallomics. These methods enable precise quantification and localization of metal ions in neural tissues. Key techniques that will be discussed in detail in the following chapters include:

3.1 Mass Spectrometry (MS)

MS-based methods, such as inductively coupled plasma mass spectrometry (ICP-MS), allow for sensitive detection and quantification of metal ions in biological samples. Coupled with laser ablation (LA-ICP-MS), this technique can provide spatial distribution maps of metals within brain tissues.

3.2 X-Ray Fluorescence (XRF)

XRF imaging is a powerful tool for visualizing the distribution of metal ions in the brain at high resolution. Synchrotron-based XRF can offer insights into the metal content at the cellular and subcellular levels.

3.3 Nuclear Magnetic Resonance (NMR)

NMR spectroscopy can be used to study metal–protein interactions and the structural dynamics of metalloproteins in the brain. Paramagnetic metals such as manganese can also be used as contrast agents in magnetic resonance imaging (MRI) to assess brain function.

3.4 Electron Microscopy

Techniques such as transmission electron microscopy (TEM) and energy-dispersive X-ray spectroscopy (EDX) enable the visualization of metal nanoparticles and their interactions with neural structures.

3.5 Fluorescence Microscopy

Fluorophores that detect metals in biological tissue are specialized fluorescent molecules that bind selectively to metal ions, enabling the visualization and quantification of these metals within cells and tissues. These fluorophores emit light upon excitation, and their fluorescence properties—such as intensity, wavelength, or lifetime—change in response to the binding of specific metal ions, such as zinc, copper, or calcium. Such changes allow researchers to map the distribution of metals in biological systems using fluorescent microscopy.

4 State-of-the-Art Brain Metallomics—from Cellular to Organ Level

Currently, the metals sodium (Na), potassium (K), magnesium (Mg), calcium (Ca) (bulk metals), iron (Fe), zinc (Zn) (trace metals), copper (Cu), manganese (Mn), Molybdenum (Mo), and cobalt [Co, as cobalamin (Vitamin B12)] (ultra-trace metals) are considered essential for humans. Chromium (Cr) may potentially join this list in the future [20]. While these metals do not constitute the full metallome, as many other metals can be found in brain tissue, their presence is vital to the physiological processes taking place in the brain. In contrast, the presence of certain toxic metals, the overload of essential metals, and the deficiency of essential metals are linked to brain pathologies. However, while an average metallome of the brain may be measured, it is evident that different brain regions, different brain cell types such as neurons and glial cells, and even different subcellular compartments of brain cells have their unique metallome. It is, therefore, not surprising that detailed metallome mapping at subcellular and cellular levels across the whole brain is still missing. However, several studies have shed some light on the metallome of certain brain regions (Fig. 1).

For example, lower iron levels have been detected in the Cortex compared to the Hippocampus. In contrast, the substantia nigra and globus pallidus seem particularly rich in iron. On the other hand, zinc levels are very high in the Hippocampus. Other brain regions with high zinc content include the Cortex, the olfactory bulb, and the amygdala.

Copper levels are high in the Hippocampus and lower but noteworthy in the Cortex, Cerebellum, and Basal Ganglia. The Basal Ganglia, however, are high in manganese levels. Brain regions with moderate manganese levels include the Cerebellum and Pituitary gland [26, 27].

The bulk metals seem more evenly distributed across the brain, with calcium and magnesium being detectable in all regions but at higher levels in the hippocampus, cortex, cerebellum, and thalamus. However, the precise metallome of each brain region can vary depending on factors such as age, diet, environmental exposure, and disease state.

Interestingly, selective enrichment of specific metals in nervous tissue is not restricted to humans. A study analyzing the metal content of various animals, including the Common Earthworm, Desert Locust, Turkish Snail, Atlantic Herring, mouse, rat, and pig, found that already in the earthworm head ganglion, zinc is more concentrated than in other nervous tissues [55].

On a cellular level, metallomes may differ depending on the type of cells and their metabolic and functional status. For example, the subcellular distribution of metals in neurons is highly organized and plays a critical role in various cellular functions, including

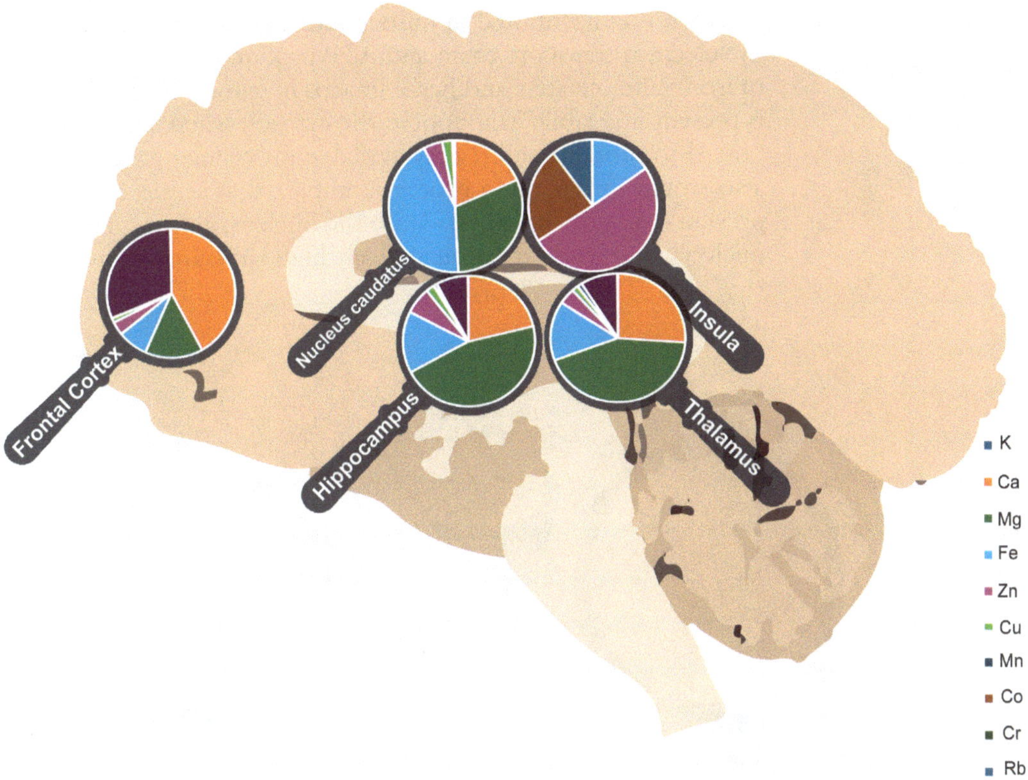

in ug/g dry weight	hippocampus	frontal cortex	thalamus	nucleus caudatus	insula
K	17000 +/- 2400	N/A	15000 +/- 2100	16800 +/- 1000	N/A
Ca	307 +/- 39	153.9 +/- 47.5	380 +/- 140	321 +/- 40	N/A
Mg	658 +/- 104.5	53.1 +/- 6.3	630+/-54.8	535 +/- 132.3	N/A
Fe	219 +/- 36.9	27.4 +/- 8.7	217 +/- 63.2	746 +/- 96.5	2.94 +/- 0.1
Zn	95 +/- 37.7	12.2 +/- 3.48	58 +/- 9.0	81 +/- 20.8	9.60 +/- 0.37
Cu	29 +/- 11.2	4 +/- 1.7	21+/- 4.3	42 +/- 15.9	N/A
Mn	1.54 +/- 0.29	0.149 +/- 0.022	1.84 +/- 0.26	2.48 +/- 0,26	N/A
Co	10.28 +/- 1.63	N/A	N/A	6.07 +/- 0.99	4.44 +/- 0.14
Cr	0.1 +/- 0.08	0.15 +/- 0.05	0.14 +/- 0.09	N/A	N/A
Rb	2.11 +/- 0.33	N/A	16.4 +/- 3.4	2.27 +/- 0.13	1.97 +/- 0.18
Al	112 +/- 56	113 +/- 54	129 +/- 76	N/A	N/A
Ni	0.27 +/- 0.01	0.26 +/- 0.02	0.19 +/- 0.11	N/A	N/A
Pb	09 +/- 0.7	0.42 +/- 0.28	0.71 +/- 0.42	N/A	N/A
Cd	0.022 +/- 0.005	0.022 +/- 0.016	0.05 +/- 0.04	N/A	N/A

Fig. 1 Trace metal concentration in different brain regions. Within the brain, trace metals are unequally concentrated in different brain regions. The figure shows the concentrations in μg/g dry weight brain tissue in the human hippocampus, frontal cortex, thalamus, caudate nucleus, and insula for the essential metals K, Ca, Mg, Fe, Zn, Cu, Mn, and Co, and the non-essential metals Cr, Rb, Al, Ni, Pb, Cd (based on [73]). Concentrations were measured using atomic absorption spectroscopy, particle-induced X-ray emission (PIXE), or inductively coupled plasma mass spectrometry. This metal homeostasis can be challenged by environmental factors, the presence of pathologies such as aggregates of metal-binding proteins, and aging. Little is known about gender effects, the influence of brain lateralization, circadian fluctuations, activity-dependent changes, hormonal influences, and related factors

neurotransmission, enzyme activity, and protection against oxidative stress (Fig. 2).

Iron is concentrated in mitochondria, where it is essential for the electron transport chain and ATP production as a component of iron-sulfur clusters and heme groups in cytochromes [56]. Iron is present in a labile iron pool in the cytosol, where iron is loosely bound to proteins and readily available for cellular processes. It is also stored in ferritin, a protein complex that sequesters iron to prevent oxidative damage [57]. In the nucleus, and particularly the nucleoli, iron levels are comparatively high [58]. Iron is involved in

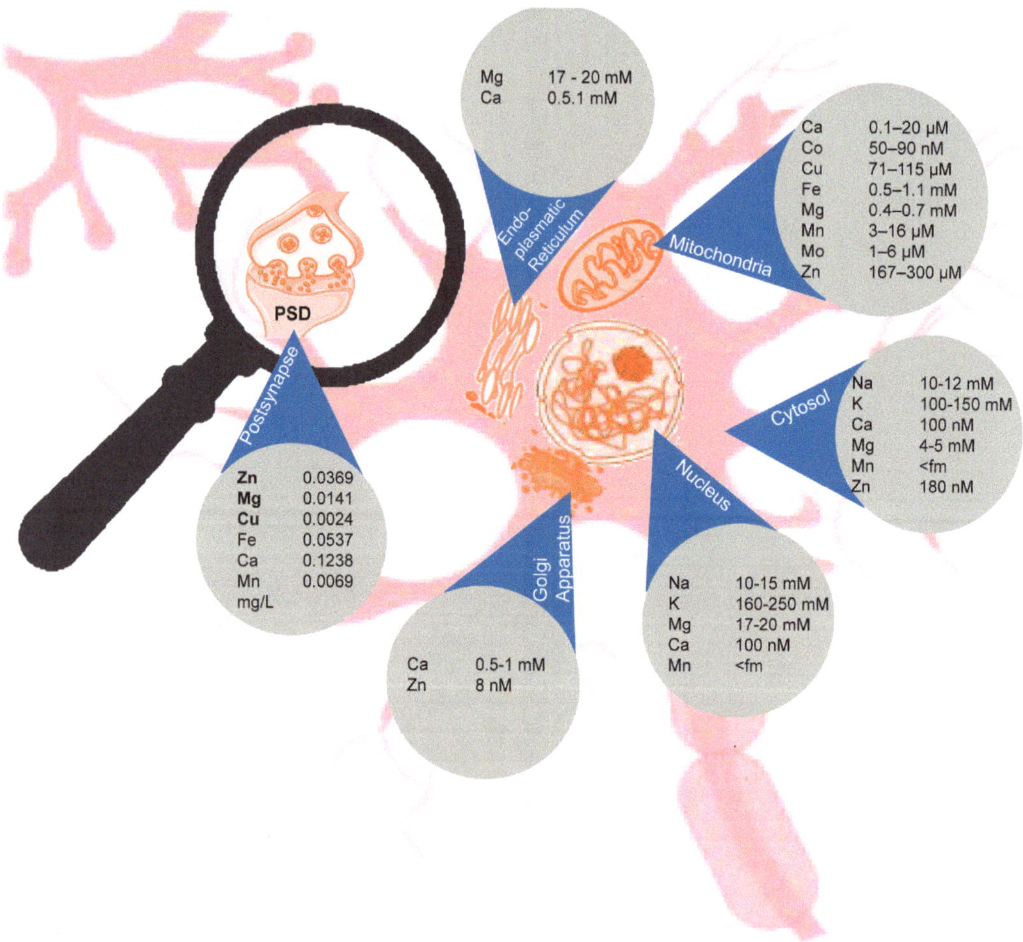

Fig. 2 Trace metal concentration on the subcellular level. Metal concentrations shown are based on several studies investigating metal levels in subcellular compartments with different methodologies (based on [61, 74–76]). However, it is important to note that the levels indicated for some metals are incomplete, as data on their absolute concentrations are missing for specific organelles; that there are few to no cell-type-specific analyses available; and that few studies investigate metal levels in cells within living tissue. Thus, metal levels may be affected by growth media conditions used for in vitro cultivation. Cell cycle stage, age, and metabolic activity are usually not considered

DNA synthesis and repair, serving as a cofactor for ribonucleotide reductase, an enzyme essential for DNA replication.

In contrast, free zinc is selectively stored in synaptic vesicles of glutamatergic neurons, particularly in the hippocampus. It is released during neurotransmission and acts as a neuromodulator by interacting, among others, with receptors like NMDA and AMPA at the postsynapse and intracellular postsynaptic proteins [59–61]. In the nucleus, zinc levels are high due to the role zinc plays in gene expression and DNA replication as a structural component of zinc finger transcription factors. Besides, zinc contributes to mitochondrial function and protects against apoptosis by inhibiting caspases, which drive programmed cell death [62]. While the aforementioned subcellular compartments are high in zinc, the cytosol is very low in free zinc (in the picomolar to femtomolar range) because zinc is stored bound to proteins such as metallothioneins, which buffer zinc levels and release it for cell signaling in a controlled manner [63].

Copper is crucial to the function of cytochrome c oxidase, a key enzyme in the electron transport chain that is vital for energy production. Thus, higher levels of copper can be detected in mitochondria. However, these need to be tightly controlled by cells [64]. Copper is involved in the processing and delivery of copper-dependent enzymes, including those needed for neurotransmitter synthesis. Thus, it is also required in the Golgi Apparatus. However, copper is also distributed in the cytoplasm and associated with chaperone proteins that regulate its transport and prevent toxicity [65].

Manganese is critical for the function of manganese superoxide dismutase (MnSOD), an enzyme that protects mitochondria from oxidative damage by neutralizing superoxide radicals. Thus, manganese is found in mitochondria [66]. Manganese is also involved in the function of enzymes that glycosylate proteins, a process necessary for proper protein folding and trafficking that occurs in the Endoplasmic Reticulum (ER). Besides, Manganese is a cofactor for various enzymes, including those involved in neurotransmitter synthesis and metabolism, such as glutamine synthetase, and thus needs to be present in certain quantities in the cytosol [67].

Calcium is stored in the ER, which is crucial in regulating intracellular calcium signaling. Calcium release from the ER is involved in processes such as synaptic plasticity and neurotransmitter release. Mitochondria also store calcium, which helps regulate energy production and apoptosis [68]. However, excessive calcium uptake by mitochondria can trigger cell death [69]. At synapses, calcium levels fluctuate activity-dependently, as channels and pumps in the plasma membrane regulate the influx and efflux of calcium ions to maintain neuronal excitability and trigger synaptic transmission. In particular, calcium is essential for triggering vesicle fusion with the presynaptic membrane [70].

Magnesium levels are higher in the mitochondria compared to the cytosol, where the metal is essential for ATP production, as it stabilizes ATP molecules and acts as a cofactor for ATPases involved in energy metabolism [71]. In the nucleus, magnesium stabilizes DNA and is involved in DNA replication, repair, and RNA synthesis [72].

However, precise data on the metallomes of cells and their subcellular compartments that report absolute concentrations of metals are still largely lacking, and the data present in the literature sometimes vary greatly based on the methodologies used for assessment.

The subcellular distribution of metals in neuronal cells, including the example of neurons mentioned above, is tightly regulated by a network of metal transporters, chaperones, and storage proteins, ensuring their availability for essential biochemical processes. It is, therefore, important to consider that while large-scale changes on a brain region level or whole organ level may be linked to brain disorders, dysregulation of metal homeostasis at the subcellular level, which may not be readily detectable on a larger scale, can already have a significant impact on cell function, and thus contribute to brain pathologies.

5 Conclusions

Brain metallomics represents a rapidly advancing field that bridges neurobiology, chemistry, and analytical science. The intricate balance of metal ions in the brain is crucial for maintaining neuronal health and function. Through state-of-the-art analytical techniques and metalloproteomic studies, researchers are uncovering the complex roles of metals in the brain, offering new perspectives on the molecular underpinnings of neurological disorders. Continued advancements in brain metallomics hold promise for developing novel diagnostic and therapeutic approaches to combat neurodegenerative diseases and improve cognitive health.

Acknowledgments

The Bernal Institute, Health Science Institute, and Department of Biological Sciences of the University of Limerick, Ireland, supported this research.

Conflict of Interest The author has no conflicts of interest to declare that are relevant to the content of this chapter.

References

1. Beutler E (2002) History of iron in medicine. Blood Cells Mol Dis 29(3):297–308
2. Kaila VR, Johansson MP, Sundholm D, Laakkonen L, Wiström M (2009) The chemistry of the CuB site in cytochrome c oxidase and the importance of its unique His-Tyr bond. Biochim Biophys Acta 1787(4):221–233
3. McCord JM, Fridovich I (1969) Superoxide dismutase. An enzymic function for erythrocuprein (hemocuprein). J Biol Chem 244:6049–6055
4. Maske H (1955) Über den topchemischen Nachweis von Zink im Ammonshorn verschiedener Säugetiere. Naturwissenschaften 42:424
5. Foley PB, Hare DJ, Double KL (2022) A brief history of brain iron accumulation in Parkinson disease and related disorders. J Neural Transm (Vienna) 129(5–6):505–520
6. Hallgren B, Sourander P (1960) The non-haemin iron in the cerebral cortex of Alzheimer's disease. J Neurochem 5:307–310
7. Hallgren B, Sourander P (1958) The effect of age on the non-haemin iron in the human brain. J Neurochem 3:41–51
8. Teschke R, Eickhoff A (2024) Wilson disease: copper-mediated cuproptosis, iron-related ferroptosis, and clinical highlights, with comprehensive and critical analysis update. Int J Mol Sci 25(9):4753
9. Pan D, Schmieder AH, Wickline SA, Lanza GM (2011) Manganese-based MRI contrast agents: past, present and future. Tetrahedron 67(44):8431–8444
10. Barnett JP, Scanlan DJ, Blindauer CA (2012) Protein fractionation and detection for metalloproteomics: challenges and approaches. Anal Bioanal Chem 402:3311–3322
11. Yoo J, Han J, Lim MH (2023) Transition metal ions and neurotransmitters: coordination chemistry and implications for neurodegeneration. RSC Chem Biol 4(8):548–563
12. Burdette SC, Lippard SJ (2003) Meeting of the minds: metalloneurochemistry. Proc Natl Acad Sci USA 100(7):3605–3610
13. Sabouri S, Rostamirad M, Dempski RE (2024) Unlocking the brain's zinc code: implications for cognitive function and disease. Front Biophys 2:1406868
14. Zhang C, Dischler A, Glover K, Qin Y (2022) Neuronal signalling of zinc: from detection and modulation to function. Open Biol 12(9):220188
15. Chen B, Yu P, Chan WN, Xie F, Zhang Y, Liang L, Leung KT, Lo KW, Yu J, Tse GMK, Kang W, To KF (2024) Cellular zinc metabolism and zinc signaling: from biological functions to diseases and therapeutic targets. Signal Transduct Target Ther 9(1):6
16. Nakashima AS, Dyck RH (2009) Zinc and cortical plasticity. Brain Res Rev 59(2):347–373
17. Stanton JE, Malijauskaite S, McGourty K, Grabrucker AM (2021) The metallome as a link between the "Omes" in autism spectrum disorders. Front Mol Neurosci 14:695873
18. Wang L, Yin YL, Liu XZ, Shen P, Zheng YG, Lan XR, Lu CB, Wang JZ (2020) Current understanding of metal ions in the pathogenesis of Alzheimer's disease. Transl Neurodegener 9:10
19. Błażewicz A, Grabrucker AM (2022) Metal profiles in autism spectrum disorders: a crosstalk between toxic and essential metals. Int J Mol Sci 24(1):308
20. Grabrucker AM (2023) The role of essential metals in human nutrition. Nova Science Publishers, Hauppauge, NY, USA
21. Abbaspour N, Hurrell R, Kelishadi R (2014) Review on iron and its importance for human health. J Res Med Sci 19(2):164–174
22. D'Ambrosi N, Rossi L (2015) Copper at synapse: release, binding and modulation of neurotransmission. Neurochem Int 90:36–45
23. Krall RF, Tzounopoulos T, Aizenman E (2021) The function and regulation of zinc in the brain. Neuroscience 457:235–258
24. Soares ATG, Silva AC, Tinkov AA, Khan H, Santamaría A, Skalnaya MG, Skalny AV, Tsatsakis A, Bowman AB, Aschner M, Ávila DS (2020) The impact of manganese on neurotransmitter systems. J Trace Elem Med Biol 61:126554
25. Pfaender S, Grabrucker AM (2014) Characterization of biometal profiles in neurological disorders. Metallomics 6(5):960–977
26. DeBenedictis CA, Vilella A, Grabrucker AM (2019) The role of trace metals in Alzheimer's disease. In: Wisniewski T (Ed.)Eds. Alzheimer's disease. Codon Publications, Brisbane (AU), Ch. 6. https://www.ncbi.nlm.nih.gov/books/NBK552144/doi:10.15586/alzheimersdisease.2019.ch6
27. Vilella A, Daini E, DeBenedictis CA, Grabrucker AM (2020) Targeting metal homeostasis as a therapeutic strategy for Alzheimer's disease. In: Huang X (Ed.)Eds. Alzheimer's disease: drug discovery. Exon Publications, Brisbane (AU), Ch. 5. https://www.ncbi.nlm.nih.gov/books/NBK566119/doi:10.36255/exonpublications.alzheimersdisease.2020.ch5
28. Zeng W, Cai J, Zhang L et al. (2024) Iron deposition in Parkinson's disease: a mini-review. Cell Mol Neurobiol 44:26

29. Ross MM, Hernandez-Espinosa DR, Aizenman E (2023) Neurodevelopmental consequences of dietary zinc deficiency: a status report. Biol Trace Elem Res 201(12):5616–5639
30. Bradbury MWB (1992) Trace metal transport at the blood-brain barrier. In: Bradbury MWB (Ed.)Eds. Physiology and pharmacology of the blood-brain barrier. Handbook of experimental pharmacology, vol 103. Springer, Berlin, Heidelberg
31. Smith QR, Rabin O, Chikhale EG (1997) Delivery of metals to brain and the role of the blood-brain barrier. In: Connor JR (Ed.)Eds. Metals and oxidative damage in neurological disorders. Springer, Boston, MA
32. Mills E, Dong XP, Wang F, Xu H (2010) Mechanisms of brain iron transport: insight into neurodegeneration and CNS disorders. Future Med Chem 2(1):51–64
33. Kambe T, Taylor KM, Fu D (2021) Zinc transporters and their functional integration in mammalian cells. J Biol Chem 296:100320
34. An Y, Li S, Huang X, Chen X, Shan H, Zhang M (2022) The role of copper homeostasis in brain disease. Int J Mol Sci 23(22):13850
35. Liu Q, Barker S, Knutson MD (2021) Iron and manganese transport in mammalian systems. Biochim Biophys Acta, Mol Cell Res 1868(1):118890
36. Balachandran RC, Mukhopadhyay S, McBride D, Veevers J, Harrison FE, Aschner M, Haynes EN, Bowman AB (2020) Brain manganese and the balance between essential roles and neurotoxicity. J Biol Chem 295(19):6312–6329
37. Yang R, Roshani D, Gao B, Li P, Shang N (2024) Metallothionein: a comprehensive review of its classification, structure, biological functions, and applications. Antioxidants 13(7):825
38. Bird AJ (2015) Cellular sensing and transport of metal ions: implications in micronutrient homeostasis. J Nutr Biochem 26(11):1103–1115
39. Liu T, Bowen RL, Wilson AC, Atwood CS (2022) Estropause, sex hormones and metal homeostasis in the mouse brain. Front Neurol 13:841822
40. Andreini C, Banci L, Bertini I, Rosato A (2006) Counting the zinc-proteins encoded in the human genome. J Proteome Res 5(1):196–201
41. Frantom PA, Seravalli J, Ragsdale SW, Fitzpatrick PF (2006) Reduction and oxidation of the active site iron in tyrosine hydroxylase: kinetics and specificity. Biochemistry 45(7):2372–2379
42. Lyons JA, Aragão D, Slattery O, Pisliakov AV, Soulimane T, Caffrey M (2012) Structural insights into electron transfer in caa3-type cytochrome oxidase. Nature 487(7408):514–518
43. Lutsenko S, Washington-Hughes C, Ralle M, Schmidt K (2019) Copper and the brain noradrenergic system. J Biol Inorg Chem 24(8):1179–1188
44. Aschner JL, Aschner M (2005) Nutritional aspect of manganese homeostasis. Mol Asp Med 26:353–362
45. Kwakye GF, Paoliello MM, Mukhopadhyay S, Bowman AB, Aschner M (2015) Manganese-induced Parkinsonism and Parkinson's disease: shared and distinguishable features. Int J Environ Res Public Health 12(7):7519–7540
46. Bitanihirwe BK, Cunningham MG (2009) Zinc: the brain's dark horse. Synapse 63(11):1029–1049
47. Grabrucker AM (2016) Zinc in the developing brain. In: Moran VH, Lowe NM (eds) Nutrition and the Developing Brain, 1st edn. CRC Press
48. Tu C, Foster L, Alvarado A, McKenna R, Silverman DN, Frost SC (2012) Role of zinc in catalytic activity of carbonic anhydrase IX. Arch Biochem Biophys 521(1–2):90–94
49. Birkedal-Hansen H, Moore WG, Bodden MK, Windsor LJ, Birkedal-Hansen B, DeCarlo A, Engler JA (1993) Matrix metalloproteinases: a review. Crit Rev Oral Biol Med 4(2):197–250
50. Krall RF, Moutal A, Phillips MB, Asraf H, Johnson JW, Khanna R, Hershfinkel M, Aizenman E, Tzounopoulos T (2020) Synaptic zinc inhibition of NMDA receptors depends on the association of GluN2A with the zinc transporter ZnT1. Sci Adv 6:eabb1515
51. Hosie AM, Dunne EL, Harvey RJ, Smart TG (2003) Zinc-mediated inhibition of GABA(A) receptors: discrete binding sites underlie subtype specificity. Nat Neurosci 6(4):362–369
52. Arons MH, Lee K, Thynne CJ, Kim SA, Schob C, Kindler S, Montgomery JM, Garner CC (2016) Shank3 is part of a zinc-sensitive signaling system that regulates excitatory synaptic strength. J Neurosci 36(35):9124–9134
53. Fan YG, Wu TY, Zhao LX, Jia RJ, Ren H, Hou WJ, Wang ZY (2024) From zinc homeostasis to disease progression: Unveiling the neurodegenerative puzzle. Pharmacol Res 2024 199:107039
54. Grabrucker S, Grabrucker AM (2017) Chapter 9—Zinc in Autism. In: White AR, Aschner M, Costa LG, Bush AI (eds) Biometals in neurodegenerative diseases. Academic Press, pp 153–173
55. DeBenedictis CA, Raab A, Ducie E, Howley S, Feldmann J, Grabrucker AM (2020) Concentrations of essential trace metals in the brain of animal species-a comparative study. Brain Sci 10(7):460
56. Cheng R, Dhorajia VV, Kim J, Kim Y (2022) Mitochondrial iron metabolism and

neurodegenerative diseases. Neurotoxicology 88:88–101
57. Lv H, Shang P (2018) The significance, trafficking and determination of labile iron in cytosol, mitochondria and lysosomes. Metallomics 10(7):899–916
58. Reinert A, Morawski M, Seeger J, Arendt T, Reinert T (2019) Iron concentrations in neurons and glial cells with estimates on ferritin concentrations. BMC Neurosci 20(1):25
59. Frederickson CJ, Bush AI (2001) Synaptically released zinc: physiological functions and pathological effects. BioMetals 14(3–4):353–366
60. Bender PTR, McCollum M, Boyd-Pratt H, Mendelson BZ, Anderson CT (2023) Synaptic zinc potentiates AMPA receptor function in mouse auditory cortex. Cell Rep 42(8):112932
61. Jan HH, Chen IT, Tsai YY, Chang YC (2002) Structural role of zinc ions bound to postsynaptic densities. J Neurochem 83(3):525–534
62. Lu Q, Haragopal H, Slepchenko KG, Stork C, Li YV (2016) Intracellular zinc distribution in mitochondria, ER and the Golgi apparatus. Int J Physiol Pathophysiol Pharmacol 8(1):35–43
63. Maret W (2017) Zinc in cellular regulation: the nature and significance of "Zinc Signals". Int J Mol Sci 18(11):2285
64. Borchard S, Bork F, Rieder T, Eberhagen C, Popper B, Lichtmannegger J, Schmitt S, Adamski J, Klingenspor M, Weiss KH, Zischka H (2018) The exceptional sensitivity of brain mitochondria to copper. Toxicol In Vitro 51:11–22
65. Polishchuk R, Lutsenko S (2013) Golgi in copper homeostasis: a view from the membrane trafficking field. Histochem Cell Biol 140(3):285–295
66. Harischandra DS, Ghaisas S, Zenitsky G, Jin H, Kanthasamy A, Anantharam V, Kanthasamy AG (2019) Manganese-induced neurotoxicity: new insights into the triad of protein misfolding, mitochondrial impairment, and neuroinflammation. Front Neurosci 13:654
67. Horning KJ, Caito SW, Tipps KG, Bowman AB, Aschner M (2015) Manganese is essential for neuronal health. Annu Rev Nutr 35:71–108
68. Verkhratsky A (2004) Endoplasmic reticulum calcium signaling in nerve cells. Biol Res 37(4):693–699
69. Duchen MR (2000) Mitochondria and calcium: from cell signalling to cell death. J Physiol 529(Pt 1):57–68
70. Südhof TC (2012) Calcium control of neurotransmitter release. Cold Spring Harb Perspect Biol 4(1):a011353
71. Killilea DW, Killilea AN (2022) Mineral requirements for mitochondrial function: a connection to redox balance and cellular differentiation. Free Radic Biol Med 182:182–191
72. de Baaij JH, Hoenderop JG, Bindels RJ (2015) Magnesium in man: implications for health and disease. Physiol Rev 95(1):1–46
73. Grochowski C, Blicharska E, Krukow P, Jonak K, Maciejewski M, Szczepanek D, Jonak K, Flieger J, Maciejewski R (2019) Analysis of trace elements in human brain: its aim, methods, and concentration levels. Front Chem 7:115
74. Killilea DW, Killilea AN (2022) Mineral requirements for mitochondrial function: a connection to redox balance and cellular differentiation. Free Radic Biol Med 182:182–191
75. Carmona A, Devès G, Ortega R (2008) Quantitative micro-analysis of metal ions in subcellular compartments of cultured dopaminergic cells by combination of three ion beam techniques. Anal Bioanal Chem 390(6):1585–1594
76. Bischof H, Burgstaller S, Waldeck-Weiermair M, Rauter T, Schinagl M, Ramadani-Muja J, Graier WF, Malli R (2019) Live-cell imaging of physiologically relevant metal ions using genetically encoded FRET-based probes. Cells 8(5):492

Check for updates

Chapter 2

Introduction to Brain Metallomics and Its Relevance to Neuroscience

Nicola M. Lowe and Andreas M. Grabrucker

Abstract

Metallomics is an emerging interdisciplinary field that focuses on the comprehensive analysis of the roles, uptake, transport, storage, and function of metal ions in biological systems. This introductory chapter provides an overview of essential metals in the brain, emphasizing their significance within the broader context of neuroscience, including how metal dyshomeostasis contributes to the pathogenesis of neurological disorders, including Alzheimer's disease, Parkinson's disease, and Wilson's disease. By highlighting the intersection of metal biology with neurochemistry and neuropathology, this chapter underscores the relevance of metallomics in advancing our understanding of brain function and in identifying novel diagnostic and therapeutic targets for neurological diseases.

Keywords Trace metals, Neuroscience, Metallome, Zinc, Copper, Metal detection, Nervous system

1 Introduction: Overview of Essential Metal Ions in the Brain

Of the 20 elements known to be essential for human health, ten are metals [1]. Metal ions are essential for the proper functioning of the brain, influencing a wide array of cellular processes, including neurotransmission, enzyme catalytic function, oxidative metabolism, and synaptic plasticity. Sadly, we still understand the role of metals much less than that of proteins, even though metal–protein interactions appear vital for numerous brain functions, and their imbalance might contribute to neurological disorders and disease.

The metals that have been identified as having an essential role in human physiology and biochemistry are listed in Table 1, along with their category in the periodic table, generic properties and the concentration (normal range for healthy adults) of each element in the blood plasma (or serum) and cerebrospinal fluid (CSF).

Like most organisms on Earth, our bodies contain only a small portion of all metals in the universe. An overwhelming 99.98% of our bodies consists of oxygen (O), carbon (C), hydrogen (H),

Andreas M. Grabrucker (ed.), *Brain Metallomics*, Neuromethods, vol. 237,
https://doi.org/10.1007/978-1-0716-5468-2_2,

© The Author(s), under exclusive license to Springer Science+Business Media, LLC, part of Springer Nature 2026

Table 1

Properties of the essential metal ions and the normal concentration range in human plasma and cerebrospinal fluid (CSF)

Essential metals	Category/chemical properties	Normal range in plasma	Normal range in CSF	References
Sodium	Alkali/highly reactive	135–147 mmol/L	135–157 mmol/L	Plasma: [2]; CSF: [3, 4]
Potassium		3.5–5.0 mmol/L	2.6–3.0 mmol/L	Plasma: [2]; CFS: [3]
Calcium	Alkaline Earth/reactive	2.2–2.58 mmol/L	1.0–1.5 mmol/L	Plasma: [2]; CFS: [3]
Magnesium		0.8–1.2 mmol/L	0.9–1.4 mmol/L	Plasma: [2]; CFS: [3, 5]
Manganese	Transition metals/exist in multiple oxidation states (except Zinc)	5–20 nmol/L	15–27 nmol/L	Plasma: [6]; CFS: [7]
Iron		11–32 μmol/L	0.2–1.1 μmol/L	Plasma: [2]; CFS: [8]
Zinc		11.5–18.5 μmol/L	0.48–0.54 μmol/L	Plasma: [2]; CFS [3, 9]
Copper		11–22 μmol/L	1.7–5.0 μmol/L	Plasma: [2]; CFS: [3,10]
Molybdenum		5–20 nmol/L	1–3 nmol/L	Plasma: [11]; CFS: [3]
Cobalt		0.02–0.06 nmol/L	0.17–3.2 nmol/L	Plasma: [2]; CSF: [12]

nitrogen (N), phosphorus (P), sulfur (S), chlorine (Cl), and the four metals calcium (Ca), potassium (K), sodium (Na), and magnesium (Mg) [13]. Hence, Ca, K, Na, and Mg are classified as common elements or common metals. The remaining metals are present in much smaller quantities and are thus categorized as trace and ultra-trace metals. The distinction between these categories has become less clear, particularly as technological advances enable the detection of ever-lower metal concentrations. Typically, metals like iron (Fe), zinc (Zn), copper (Cu), manganese (Mn), molybdenum (Mo), and cobalt (Co) are identified as trace metals, as their average concentrations in a 70 kg human body are 4.2 g (Fe), 2.3 g (Zn), 0.072 g (Cu), 0.012 g (Mn), 0.005 g (Mo), and 0.003 g (Co). These metals are crucial because they perform biological functions essential to survival. Therefore, it is unsurprising that these metals rank among the most concentrated elements in human tissues. Other metals are also measurable, with most of them falling into the ultra-trace range, including gallium (Ga), silver (Ag), niobium (Nb), zirconium (Zr), yttrium (Y), gold (Au), scandium (Sc), tantalum (Ta), vanadium (V), and tungsten (W). While it is often assumed that these metals are neutral within the human body and their presence stems from incidental exposure, this assumption lacks scientific support. Even ultra-trace metals can influence metal balance and may serve a physiological purpose indirectly. Regrettably, the effects of many metals classified as "non-essential" on the human brain remain unclear. In contrast, metals like mercury (Hg) and lead (Pb) are known to be toxic even in very low concentrations [14]. In this chapter, we focus on the essential metals, providing an overview of their functional roles within the CNS and the consequences of dysregulation.

2 Metal Homeostasis

The concentration of metals in human tissues is tightly regulated to maintain them within a physiological range optimal for their specific function. Exceeding this range will result in symptoms of toxicity, while falling below this range will also result in suboptimal function due to a deficiency of the metal. The dietary recommendations for sufficiency while preventing toxicity are specific for each metal.

The first site of homeostatic regulation following oral exposure (from food or a supplement) is in the intestinal lumen. Because they exist in ionic form, metal ions cannot diffuse across lipid membranes passively. Specialized transport systems regulate their uptake into enterocytes in the gastrointestinal tract and subsequent release into circulation, where the metal ions are transported to the target tissue bound to blood proteins or associated with amino acids. In the CNS, the blood–brain barrier provides an additional important site for homeostatic regulation. Dysregulation of these essential metal ions has been implicated in the pathophysiology of neurological disorders, including Alzheimer's and Parkinson's disease [15].

2.1 The "Common" Metals

Sodium and Potassium

An adult male typically has an average sodium content of 92 g, with about half (46 g) existing in the extracellular fluid (ECF) at a concentration of 135–145 mmol/L. Potassium is also a positively charged ion (K^+). An adult typically has around 105 g of potassium in their body. It is the most abundant intracellular cation, with only a small amount found in extracellular fluid [16]. Dietary recommendations for adults are 2.4 g/day of sodium (upper limit) and 3.51 g/day of Potassium (reference nutrient intake (RNI)) [17, 18].

Sodium absorption occurs primarily in the ileum and colon. Transport from the lumen into the enterocytes is facilitated by the sodium/hydrogen exchanger (NHE) protein family, diffusion through sodium channels, and cotransport with sugars and phosphates [19]. Potassium uptake from the lumen of the small intestine into the portal system is thought to be a passive process through tight junctions, driven by both concentration and electrochemical gradients. Some transcellular absorption can occur, primarily in the colon, via active (ATPase) or cotransporter mechanisms [20]. Once in the circulation, transfer of sodium and potassium across the BBB occurs via active, transcellular mechanisms, the tight junctions of the BBB preventing their passive diffusion. The sodium concentration in CSF is similar to that of blood plasma, whereas the potassium concentration is slightly lower than blood plasma (Table 1).

The sodium and potassium gradients across neuronal membranes, upheld by the Na^+/K^+-ATPase, are essential for creating resting membrane potentials and facilitating action potential (AP) propagation. Sodium is a positively charged ion (Na+). The sodium-potassium pump, Na^+/K^+-ATPase, transports sodium out of and potassium into the cell against their concentration gradients, using ATP as the energy source, thereby preserving a concentration difference between the ECF and intracellular fluid. This distribution is vital for generating electrical impulses in nerve cells. A flux of sodium and potassium ions triggered by voltage-gated sodium channels (VGSCs) results in an electrical impulse (AP), which neurons utilize to transmit information. Currently, nine different sodium channel isoforms have been identified in the nervous system. Recent studies indicate that VGSCs play a critical role in ensuring proper electrophysiological functions in neurons [21]. Following an AP, the Na^+/K^+ pump restores the distribution of Na and K ions to their original levels, allowing the neuron to fire another AP when needed.

Calcium

Calcium (Ca2+), a divalent cation, is the most prevalent metal in the body, making up about 1.9% of total body weight. The majority of calcium is found in the form of hydroxyapatite, a key component of teeth and bones, with only a small amount (approximately 1%) located in cells and extracellular fluids. Absorption of dietary calcium takes place in the small intestine via passive diffusion through tight junctions (in the jejunum and ileum, where concentrations are high) and in the duodenum by active transport, carrier-mediated, and ion exchange mechanisms, regulated by calcitriol (vitamin D). Dietary recommendations (RNI) for adults are 7 g Ca/day [18].

Concentration in the CSF is lower than plasma, and transfer across the BBB occurs via tightly regulated transcellular, carrier-mediated processes. In the brain, calcium plays a crucial role in the release of neurotransmitters [22] stored in presynaptic vesicles at synapses. An AP triggers the opening of voltage-gated calcium channels at the presynapse, causing calcium to flow into the cell from the extracellular space. Calcium then interacts with proteins that regulate vesicle exocytosis. This brief rise in intracellular presynaptic calcium results in vesicle fusion with the synaptic cell membrane, facilitating the release of neurotransmitters into the synaptic cleft. Calcium also plays a crucial role as a second messenger in many brain processes. Changes in intracellular calcium concentrations, both temporally and spatially defined, regulate various cellular functions. These fluctuations result from several signaling pathways, including P38-MAPK, IP3-Ca2+, Rya-Ca2+ signaling, as well as the activation of L-type, N-type, and T-type calcium channels, cAMP-PKA signaling, and Wnt-Ca2+ signaling [23]. Generally, these pathways operate on the principle that a stimulus, such as ligand binding to a receptor, triggers receptor activation, leading—through various steps—to a temporary rise in cytosolic calcium levels. Proteins with calcium-binding sites in the cytoplasm bind Ca^{2+}, resulting in their activation. Calmodulin proteins are notable examples among these calcium-binding proteins. This activation can produce various outcomes, including the downstream phosphorylation and subsequent activation of other proteins, which ultimately elicit the cell's biological response.

Magnesium

Magnesium ranks as the fourth most abundant metal in the human body, following calcium, potassium, and sodium. After potassium, it is the second most concentrated intracellular cation in human cells [24]. An adult's total magnesium levels typically range from 21 g to 28 g [25], and the dietary recommendation (RNI) for adult men and women is 300 mg/day and 270 mg/day, respectively [18].

Absorption from the gut lumen occurs in the proximal sections of the small intestine by diffusion through the gap junctions along electrochemical and concentration gradients. When the nutrient concentration in the diet is low, active transport in the distal small intestine and colon can occur via active transcellular transport. The concentration of Mg in the plasma is higher than in the CSF; thus, transport across the BBB is regulated but not fully understood [26].

Magnesium protects the brain from excessive excitation that can lead to neuronal cell death (excitotoxicity) [27]. It does this by inhibiting glutamatergic excitatory transmission through blocking calcium channels in the N-methyl-D-aspartate (NMDA) receptor, which plays a crucial role in synaptic transmission, learning, and memory. Magnesium deficiency may occur due to consuming a diet low in magnesium-rich foods (leafy green vegetables, nuts, legumes, and whole grains) and high in processed foods. Secondary deficiency may be associated with chronic diseases that reduce absorption (such as Crohn's disease) or increase excretion (such as diabetes). The neurological consequences of magnesium deficiency impact both the central and peripheral nervous systems, resulting in muscle cramps and tremors. Severe deficiency may result in seizures [28].

2.2 The "Trace Metals"

Iron

Iron is the most prevalent trace metal in the human body. Approximately 65% of the iron is located in red blood cells, where it forms part of hemoglobin and plays a crucial role in transporting oxygen

[29], which is also its major function in the brain. Iron is also a cofactor for the synthesis of neurotransmitters and myelin [30,31]. Dietary recommendations for adults are 8.7 mg/day for men and women over 50 years. For women aged 19–50 years, it is 14.8 mg/day [18]. Iron deficiency is the most common nutrient deficiency, affecting 25%–30% of the global population. Primary deficiency is caused by diets low in animal proteins, which are rich sources of bioavailable iron (heme iron), and diets high in phytate, which inhibit iron absorption.

Absorption from the gut lumen occurs in the proximal small intestine. Heme iron is transferred into the enterocytes intact via a carrier protein. Once inside the cell, the iron (Fe2+) is released into the intracellular pool. Non-heme iron is absorbed in the ferrous form (Fe2+) and transported into enterocytes via divalent metal transporter 1 (DMT1). Iron originating from both heme and non-heme dietary sources is exported across the basolateral membrane into the circulation via ferroportin, which is regulated by the hormone hepcidin. Before entering the circulation, the iron is oxidized to the Fe3+ form by hephaestin and bound primarily to transferrin in the blood plasma.

The concentration of iron in the CSF is ten-fold lower than in blood plasma; thus, regulation across the BBB is tightly controlled, unsurprisingly given the metal ion's highly oxidative nature. Transferrin-bound iron binds to transferrin receptor 1 on the endothelial cells of the BBB and is taken into the cell by endocytosis. Export into the CSF is poorly understood but involves metal-binding proteins, including ferroportin, hephaestin, and zinc-iron (ZIP) transporters. Research indicates that neuroinflammatory conditions, such as those in multiple sclerosis, are sustained by macrophage and microglial regulation of brain iron homeostasis. In active lesions of multiple sclerosis, myelin is degraded, and its remnants, including iron, are ingested by these cells. Iron builds up in the extracellular space as brain cells age, leading to oxidative stress in certain brain regions [32, 33]. This type of microglial degeneration has been identified in both aged and Alzheimer's brain tissue [34, 24].

Dietary iron deficiency in children significantly impacts cognitive development, including motor skills, learning, and memory. Deficiency may also result in myelination deficits, leading to slower nerve impulse conduction [30, 31].

Zinc

Zinc (Zn^{2+}) is the second most abundant trace metal in the body, with an average content of 2–3 g. It exists in two forms: as "free" (aqueous) Zn^{2+} or bound to proteins. Free Zn^{2+} serves as a signaling ion both intracellularly and extracellularly. Zinc is crucial for many physiological processes involved in brain development and function. Specifically, it contributes to neural tube closure, neurogenesis, neuronal migration, myelination, and neurotransmission [35]. In the CNS, it functions as a neurotransmitter released by "zincergic" neurons [36]. For adults, dietary recommendations (RNI) are 9.5 mg/day and 7.0 mg/day, respectively [18]. Similar to iron, inadequate zinc intake is a global public health concern, affecting an estimated 17%–30% of the global population, particularly those in low and middle-income countries, where access to affordable and bioavailable sources of zinc-rich food is limited (animal-source foods) [37, 38]. Zinc absorption occurs in the jejunum and ileum. Zinc uptake into enterocytes is facilitated by ZIP4 and DMT1, and from the enterocyte into the circulation by ZnT1, with metallothionein playing a role in regulating the transfer of zinc into the blood by sequestering zinc when zinc intake levels are high. The zinc concentration in the plasma is 11.5–18.5 μmol/L, where most of the zinc is bound to albumin and alpha macroglobulin. In the CSF, the concentration is considerably lower, at 0.08–0.15 nmol/L, indicating tight regulation of zinc transfer across the BBB, which involves both ZIP (uptake) and ZnT (output) transporters [39, 40].

Vesicular zinc plays a role in synaptic signaling and neuromodulation [41]. For example, zinc can affect neurotransmission by modulating neurotransmitter receptors, ion channels, cell adhesion proteins, and pre- and postsynaptic calcium uptake [35]. Additionally, zinc regulates dynamic processes within the postsynaptic density (PSD) [42]. Collectively, these functions contribute to critical mechanisms, such as long-term potentiation (LTP), which is believed to underlie learning and memory. Besides, zinc facilitates neurogenesis during both development and adulthood by influencing stem cell apoptosis, proliferation, and neuronal differentiation [43]. Zinc also serves as a key regulator of the extracellular matrix (ECM). By regulating ECM proteins through a putative zinc–metalloprotease–BDNF axis, it may help establish proper neuronal connectivity [44].

The consequences of zinc deficiency are difficult to identify because overt clinical symptoms do not appear until deficiency becomes severe, and the effects of marginal deficiency are highly generalized and challenging to definitively characterize [45]. However, in patients with Acrodermatitis Enteropathica, a genetic disorder affecting the gene coding for the ZIP4 transporter, resulting in a failure of zinc absorption from the gut, neurological symptoms of a severe zinc deficiency can be observed, including developmental delays, motor dysfunction, seizures, hypotonia, and cognitive impairments [46].

Copper

Copper is essential for multiple human enzymes, including those that play key roles in the CNS, such as energy metabolism (cytochrome c oxidase), neurotransmitter production (e.g., Dopamine β-hydroxylase), and antioxidant defense (e.g., superoxide dismutase) [47].

Dietary recommendations (RNI) for copper intake for adults are 1.2 mg/day [18], with absorption taking place primarily from the lumen of the duodenum and proximal jejunum via transport proteins CTR1 (copper transporter 1) and DMT1 (divalent metal transporter 1). Once inside the enterocytes, copper is bound to chaperone proteins and then transferred to a copper-transporting ATPase (ATP7A) for transfer across the basolateral membrane into the circulation.

The concentration of copper in the CSF is much lower than that of the blood plasma (Table 1). Transport across the BBB is tightly regulated and involves CTR1 for uptake into neurons and endothelial cells of the BBB [48], and ATP7A, the copper-transporting ATPase, for efflux [49]. Similar to zinc, severe primary copper deficiency is rare, and the impact of a deficiency or dysregulation of copper metabolism on neurofunction is best observed in patients with Menkes disease, a genetic condition caused by mutations in the ATP7A gene impacting the absorption of copper from the gut and transport into the CSF, resulting in severe neurodegeneration and seizures due to a secondary copper deficiency. Conversely, Wilson's disease is caused by a mutation in the ATP7B gene, which encodes the copper transporter responsible for the efflux of copper primarily from the liver, but also from the brain and kidneys. Thus, copper accumulates in tissues including the liver and brain, resulting in tremors, psychiatric symptoms, and involuntary muscle contractions [50].

Manganese

The human body contains about 10–20 mg of manganese, with 25%–40% residing in the bones. Manganese serves as a cofactor for several enzymes, such as hydrolases, integrins, and transferases. One notable manganese-binding enzyme is glutamine synthetase (GS), which binds approximately 80% of the manganese found in the brain [51].

An adequate intake of manganese for adults is 2.3 mg/day for men, 1.8 mg/day for women [52]. It is absorbed primarily in the duodenum and proximal jejunum, and, similar to iron and zinc, uptake into the gut enterocytes and endothelial cells of the BBB is facilitated by DMT1 and ZIP14. [53, 54] In addition, MnT1 (manganese transporter 1) plays a role in the regulation of Mn uptake from the gut when intake is high [6]. MnT1 may also play a role in manganese dysregulation, facilitating the uptake of manganese to toxic levels in neurons [55].

Manganism results from excessive manganese exposure and causes neurodegeneration due to manganese accumulation in the brain's dopaminergic areas. The symptoms resemble those of Parkinson's disease, including stiffness in limbs, tremors, and balance difficulties, as well as psychological symptoms like mania and emotional instability [56]. While manganism manifests effects from acute exposure, long-term exposure to low levels may increase the risk of developing Parkinson's disease [57]. The mechanisms of neurotoxicity include inflammation, oxidative stress, and apoptosis.

Molybdenum

In humans, molybdenum acts as a cofactor for a limited number of enzymes. The three enzymes considered vital for health are sulfite oxidase, xanthine oxidase, and aldehyde oxidase. Every mammalian molybdoenzyme includes molybdopterin, an organic compound incorporating functional molybdenum [58]. The adequate intake of molybdenum for adults is 200 μg/day [59]. The concentration of molybdenum in human plasma is significantly higher than that of the CSF (Table 1), reflecting tight regulation; however, absorption from the gut and transport across the BBB are not well characterized, but thought to involve a sodium-dependent active transport system [60].

Individuals suffering from genetic disorders that prevent the utilization of molybdenum, and thus, the breakdown of sulfites show signs of brain damage and seizures [61].

Cobalt

Cobalt's sole recognized biological role is as a metal element in vitamin B12, which is also referred to as cobalamin (cyanocobalamin). Vitamin B12 is essential for myelination in the CNS [62]. Insufficient vitamin B12 can lead to the accumulation of methylmalonyl-CoA, potentially causing demyelination of neurons as it is incorporated into the myelin sheath, resulting in metabolic acidosis. There are no specific dietary recommendations for cobalt. It is absorbed in the gut as a component of B12, bound to intrinsic factor in the ileum, and transferred to transcobalamin II (TCII) in the blood and delivered to the BBB, where it binds with a membrane-bound receptor (CD320) specific for vitamin B12 and internalized by endocytosis, and released into the CSF [63].

3 Metallomics for Neuroscience

Historically, the overload or deficiency of a particular metal ion was often investigated in isolation, and pathological effects were attributed to the observed changes. For example, calcium deficiencies can result in neurological signs and symptoms like confusion, memory loss, delirium, depression, and hallucinations [64]. Migraine sufferers have been found to have reduced magnesium levels in both blood plasma and cerebrospinal fluid (CSF) [65]. Additionally, low magnesium has been associated with depressive and anxiety disorders [66]. Mild zinc deficiency impacts the CNS, with various brain disorders linked to it, including depression, Alzheimer's disease, and autism spectrum disorders [67]. Molybdenum excess can cause hallucinations and convulsions [68], while a deficiency in vitamin B12 (cobalamin) may lead to spinal cord demyelination, resulting in a variety of neuropsychiatric and neurological symptoms [69].

However, the dysregulation of a specific metal ion likely leads to a characteristic biometal profile [70], where simultaneous alterations in other metal ions accompany a certain alteration in one biometal. For instance, in Alzheimer's disease, higher levels of zinc and manganese have been found in senile plaques, but lower levels of copper and zinc in plasma. In autism spectrum disorders, lower levels of calcium, iron, magnesium, manganese, and zinc, but higher levels of aluminum, cadmium, mercury, and lead have been reported in hair samples. In epilepsy, lower plasma zinc and higher plasma copper levels were measured; in ADHD, lower plasma zinc and higher plasma manganese and lead levels are reported. Mood disorders are linked to lower levels of zinc, calcium, and iron in the blood, but higher lead levels. In schizophrenia, low zinc and calcium levels, but high copper and cadmium levels, were observed in hair analyses [70].

Therefore, metallomics approaches in neuroscience are an important future direction because they address a crucial—but historically underexplored—aspect of brain function: the role of specific metal profiles in neural biology. Understanding metal profiles in disease can help map how, where, and why changes occur, providing insights into disease mechanisms and possible intervention points. Metal signatures or profiles in blood, CSF, or brain tissue could become early indicators of neurodegenerative diseases, possibly even before clinical symptoms appear. Understanding how metal profiles are established by metal transporters, metalloproteins, and metal-related pathways could lead to new drug targets—e.g., drugs that restore metal balance, chelate toxic metals, or modulate metalloprotein function.

Neuroscience often struggles to connect molecular chemistry with brain physiology. Metallomics naturally bridges this gap because metals interact directly with biomolecules (proteins, lipids, DNA), influencing their structure and function. Thus, metallomics offers a systems-level understanding of how metals contribute to brain health and disease, a knowledge gap that remains largely unaddressed in neuroscience today. However, integrating with other omics data will be key to effectively utilizing metallomics data.

4 Omics Integration of Brain Metallomics

Given the complexity of metal ion regulation and its implications for neurological health and disease, omics-based approaches provide comprehensive, high-throughput tools for systematically analyzing the metallome—the complete set of metal species within a biological system.

Several powerful analytical platforms are employed for quantitative and spatial profiling of metal ions, such as inductively coupled plasma mass spectrometry (ICP-MS) for total metal content analysis in brain tissues, cerebrospinal fluid, and isolated cells. Additionally, spatially resolved techniques are critical for understanding metal distribution within specific brain regions, like laser ablation ICP-MS (LA-ICP-MS), which combines laser ablation with ICP-MS for spatial mapping of metals in tissue sections at micrometer resolution, enabling correlation with histological features. X-ray fluorescence microscopy (XFM) is also utilized for nondestructive imaging of metal ions in biological specimens at subcellular resolution, along with secondary ion mass spectrometry (SIMS) and NanoSIMS for high spatial resolution. Furthermore, X-ray absorption spectroscopy (XAS), including X-ray absorption near edge structure (XANES) and extended X-ray absorption fine structure (EXAFS), provides insights into the oxidation states and coordination environments of metal centers in biological samples, thereby adding another layer of information to metallomics data. To fully characterize the metal-related molecular landscape of the brain, metallomics is increasingly integrated with other omics platforms:

Quantitative proteomics identifies changes in metal-binding proteins and metallochaperones under physiological or pathological conditions. RNA sequencing (transcriptomics) can uncover regulatory networks that control metal homeostasis, including metal transporters and metal-responsive transcription factors. Analysis of low-molecular-weight metabolites (metabolomics) may reveal metal-dependent biochemical pathways and markers of oxidative stress.

For example, multiomics-based metallomics has advanced our understanding of several neurodegenerative diseases. In Alzheimer's disease, combined imaging and proteomics approaches have revealed aberrant Zn^{2+} and Cu^{2+} homeostasis associated with amyloid plaques [71]. In Parkinson's disease, metallomic analyses highlight iron dysregulation within the substantia nigra [72, 73]. Moreover, omics techniques have illuminated the roles of manganese toxicity in manganism and copper transport defects in Wilson's and Menkes diseases [74, 75].

Technological innovations such as multimodal imaging, high-throughput single-cell metallomics, and machine-learning-driven data integration are poised to further enhance the resolution and interpretability of brain metallomic studies. These tools will not only deepen our understanding of metal biology in the brain but also help discover novel biomarkers and therapeutic targets for neurological disorders.

References

1. Jomova K, Makova M, Alomar SY, Alwasel SH, Nepovimova E, Kuca K, Rhodes CJ, Valko M (2022) Essential metals in health and disease. Chem Biol Interact 367:110173
2. Taber CW, Thomas CL (1993) Taber's cyclopedic medical dictionary. F.A. Davis. Chicago, Philadelphia
3. Mandal R, Guo AC, Chaudhary KK, Liu P, Yallou FS, Dong E, Aziat F, Wishart DS (2012) Multi-platform characterization of the human cerebrospinal fluid metabolome: a comprehensive and quantitative update. Genome Med 4(4):38
4. Urbanits S (2004) Cellular diagnostics of CSF. EJIFCC 15(3):51–53
5. Haensch C (2010) Cerebrospinal fluid magnesium level in different neurological disorders. Neurosci Med 1(2):60–63
6. Baj J, Flieger W, Barbachowska A, Kowalska B, Flieger M, Forma A, Teresiński G, Portincasa P, Buszewicz G, Radzikowska-Büchner E, Flieger J (2023) Consequences of disturbing manganese homeostasis. Int J Mol Sci 24(19): 14959.
7. Melø TM, Larsen C, White LR, Aasly J, Sjøbakk TE, Flaten TP, Sonnewald U, Syversen

T (2003) Manganese, copper, and zinc in cerebrospinal fluid from patients with multiple sclerosis. Biol Trace Elem Res 93(1–3):1–8
8. Ficiarà E, Munir Z, Boschi S, Caligiuri ME, Guiot C (2021) Alteration of iron concentration in Alzheimer's disease as a possible diagnostic biomarker unveiling ferroptosis. Int J Mol Sci 22(9):4479
9. Agarwal RP, Henkin RI (1982) Zinc and copper in human cerebrospinal fluid. Biol Trace Elem Res 4(2–3):117–124
10. Molina JA, Jiménez-Jiménez FJ, Aguilar MV, Meseguer I, Mateos-Vega CJ, González-Muñoz MJ, de Bustos F, Porta J, Ortí-Pareja M, Zurdo M, Barrios E, Martínez-Para MC (1998) Cerebrospinal fluid levels of transition metals in patients with Alzheimer's disease. J Neural Transm 105(4-5):479–488
11. Oskarsson A, Kippler M (2023) Molybdenum—a scoping review for Nordic Nutrition Recommendations. Food Nutr Res 67:10326
12. Rakow A, Kowski A, Treskatsch S, von Baehr V, Weynandt CL, Tafelski S, Klotz E, Duda GN, Perka C, Huesker K, Schoon J (2025) Metal concentrations in blood and cerebrospinal fluid of patients with arthroplasty implants. JAMA Netw Open 8(3):e252281
13. Emsley J (1998) The elements, 3rd edn. Clarendon Press, Oxford
14. Tchounwou PB, Yedjou CG, Patlolla AK, Sutton DJ (2012) Heavy metal toxicity and the environment. Exp Suppl 101:133–164
15. Doroszkiewicz J, Farhan JA, Mroczko J, Winkel I, Perkowski M, Mroczko B (2023) Common and trace metals in Alzheimer's and Parkinson's Diseases. Int J Mol Sci 24(21):15721
16. Hinderling PH (2016) The pharmacokinetics of potassium in humans is unusual. J Clin Pharmacol 56:1212–1220
17. SACN: Scientific Advisory Committee on Nutrition (2015) Dietary reference values for food energy and nutrients for the United Kingdom. UK Department of Health and Social Care. Salt reduction targets for 2024 (updated guidelines). Public Health England Wellington House, Waterloo Road London SE1 8UG, 133-155
18. SACN: Scientific Advisory Committee on Nutrition (2016) RNI for K Government Dietary Recommendations: Government recommendations for energy and nutrients for males and females aged 1–18 years and 19+ years. Scientific Advisory Committee on Nutrition. Public Health England Wellington House, Waterloo Road London SE1 8UG, 133-155
19. Zachos NC, Tse M, Donowitz M (2005) Molecular physiology of intestinal Na+/H+ exchange. Annu Rev Physiol 67:411–443
20. McDonough AA, Youn JH (2017) Potassium homeostasis: the knowns, the unknowns, and the health benefits. Physiology 32(2):100–111
21. Wang J, Ou SW, Wang YJ (2017) Distribution and function of voltage-gated sodium channels in the nervous system. Channels 11(6):534–554
22. Südhof TC (2012) Calcium control of neurotransmitter release. Cold Spring Harb Perspect Biol 4(1):a011353
23. Fields RD, Lee PR, Cohen JE (2005) Temporal integration of intracellular Ca2+ signaling networks in regulating gene expression by action potentials. Cell Calcium 37(5):433–442
24. Romani A, Scarpa A (1992) Regulation of cell magnesium. Arch Biochem Biophys 298(1):1–12
25. Saris NE, Mervaala E, Karppanen H, Khawaja JA, Lewenstam A (2000) Magnesium. An update on physiological, clinical and analytical aspects. Clin Chim Acta 294(1-2):1–26
26. de Baaij JH, Hoenderop JG, Bindels RJ (2015) Magnesium in man: implications for health and disease. Physiol Rev 95(1):1–46
27. Kirkland AE, Sarlo GL, Holton KF (2018) The role of magnesium in neurological disorders. Nutrients 10(6):730
28. Xue W, You J, Su Y, Wang Q (2019) The effect of magnesium deficiency on neurological disorders: a narrative review article. Iran J Public Health 48(3):379–387
29. Gell DA (2018) Structure and function of haemoglobins. Blood Cells Mol Dis 70:13–42
30. Todorich B, Pasquini JM, Garcia CI, Paez PM, Connor JR (2009) Oligodendrocytes and myelination: the role of iron. GLIA 57(5):467–478
31. Lozoff B, Beard J, Connor J, Barbara F, Georgieff M, Schallert T (2006) Long-lasting neural and behavioral effects of iron deficiency in infancy. Nutr Rev 64(5 Pt 2):S34–S43
32. Craelius W, Migdal MW, Luessenhop CP, Sugar A, Mihalakis I (1982) Iron deposits surrounding multiple sclerosis plaques. Arch Pathol Lab Med 106(8):397–399
33. Bagnato F, Hametner S, Yao B, van Gelderen P, Merkle H, Cantor FK, Lassmann H, Duyn JH (2011) Tracking iron in multiple sclerosis: a combined imaging and histopathological study at 7 Tesla. Brain 134(Pt 12):3602–3615
34. Lopes KO, Sparks DL, Streit WJ (2008) Microglial dystrophy in the aged and Alzheimer's disease brain is associated with ferritin immunoreactivity. GLIA 56(10):1048–1060
35. Grabrucker AM (2016) Zinc in the developing brain. In: Moran VH, Lowe N (eds) Nutr Dev Brain. CRC Press, Boca Raton, FL, pp 143–168
36. Sindreu C, Storm DR (2011) Modulation of neuronal signal transduction and memory formation by synaptic zinc. Front Behav Neurosci 5:68
37. Wessells KR, Brown KH (2012) Estimating the global prevalence of zinc deficiency: results based on zinc availability in national food supplies and the prevalence of stunting. PLoS One 7(11):e50568

38. Brown KH, Rivera JA, Bhutta Z, Gibson RS, King JC, Lönnerdal B, Ruel MT, Sandtröm B, Wasantwisut E, Hotz C (2004) International Zinc Nutrition Consultative Group (IZiNCG) technical document #1. Assessment of the risk of zinc deficiency in populations and options for its control. Food Nutr Bull 25(1 Suppl 2):S99–S203
39. Krall RF, Tzounopoulos T, Aizenman E (2021) The function and regulation of zinc in the brain. Neuroscience 457:235–258
40. Colvin RA, Davis N, Nipper RW, Carter PA (2000) Zinc transport in the brain: routes of zinc influx and efflux in neurons. J Nutr 130(5S Suppl):1484S–1487SS
41. Gower-Winter SD, Levenson CW (2012) Zinc in the central nervous system: from molecules to behavior. Biofactors 38(3):186–193
42. Grabrucker AM (2014) A role for synaptic zinc in ProSAP/Shank PSD scaffold malformation in autism spectrum disorders. Dev Neurobiol 74(2):136–146
43. Gower-Winter SD, Corniola RS, Morgan TJ, Levenson CW (2013) Zinc deficiency regulates hippocampal gene expression and impairs neuronal differentiation. Nutr Neurosci 16(4):174–182
44. Koh JY, Lim JS, Byun HR, Yoo MH (2014) Abnormalities in the zinc-metalloprotease-BDNF axis may contribute to megalencephaly and cortical hyperconnectivity in young autism spectrum disorder patients. Mol Brain 7:64
45. King JC (2011) Zinc: an essential but elusive nutrient. Am J Clin Nutr 94(2):679S–684SS
46. Cui W, Wang N, Shi M, Xu X, Jiang H (2025) Acrodermatitis enteropathica in the pediatric population: a literature review of real-world studies. Front Nutr 12:1590075
47. Scheiber IF, Mercer JF, Dringen R (2014) Metabolism and functions of copper in brain. Prog Neurobiol 116:33–57
48. An Y, Li S, Huang X, Chen X, Shan H, Zhang M (2022) The role of copper homeostasis in brain disease. Int J Mol Sci 23(22):13850
49. Sarkar B (2000) Copper transport and its defect in Wilson disease: characterization of the copper-binding domain of Wilson disease ATPase. J Inorg Biochem 79(1-4):187–191
50. Ortiz JF, Morillo Cox Á, Tambo W, Eskander N, Wirth M, Valdez M, Niño M (2020) Neurological manifestations of Wilson's disease: pathophysiology and localisation of each component. Cureus 12(11):e11509
51. Crossgrove J, Zheng W (2004) Manganese toxicity upon overexposure. NMR Biomed 17(8):544–553
52. European Food Safety Authority (EFSA) (2013) Scientific opinion on dietary reference values for manganese. EFSA J 11(2):3349
53. Roth J, Ponzoni S, Aschner M (2013) Manganese homeostasis and transport. Met Ions Life Sci 12:169–201
54. Aschner M, Aschner JL (1991) Manganese neurotoxicity: cellular effects and blood-brain barrier transport. Neurosci Biobehav Rev 15(3):333–340
55. Chen P, Chakraborty S, Mukhopadhyay S, Lee E, Paoliello MM, Bowman AB, Aschner M (2015) Manganese homeostasis in the nervous system. J Neurochem 134(4):601–610
56. Avila DS, Puntel RL, Aschner M (2013) Manganese in health and disease. Met Ions Life Sci 13:199–227
57. Lucchini RG, Martin CJ, Doney BC (2009) From manganism to manganese-induced parkinsonism: a conceptual model based on the evolution of exposure. NeuroMolecular Med 11(4):311–321
58. Rajagopalan KV (1988) Molybdenum: an essential trace element in human nutrition. Annu Rev Nutr 8:401–427
59. DHSC: Department of Health and Social Care, UK (2011) Dietary reference values for food energy and nutrients for the UK. The Stationery Office
60. Adamus JP, Ruszczyńska A, Wyczałkowska-Tomasik A (2024) Molybdenum's role as an essential element in enzymes catabolizing redox reactions: a review. Biomolecules 14(7):869
61. Atwal PS, Scaglia F (2016) Molybdenum cofactor deficiency. Mol Genet Metab 117(1):1–4
62. Stabler SP (2020) Vitamin B12. In: Marriott BP, Birt DF, Stallings VA, Yates AA (eds) Present knowledge in nutrition. 11th edn. Elsevier, Washington, DC, pp 257–271
63. Alruwaili M, Basri R, AlRuwaili R, Albarrak AM, Ali NH (2023) Neurological implications of vitamin B12 deficiency in diet: a systematic review and meta-analysis. Healthcare 11(7):958
64. Schafer AL, Shoback DM (2000) Hypocalcemia: diagnosis and treatment. In: Feingold KR, Anawalt B, Boyce A et al (eds) Endotext [Internet]. MDText.com, Inc., South Dartmouth, MA
65. Dolati S, Rikhtegar R, Mehdizadeh A, Yousefi M (2020) The role of magnesium in pathophysiology and migraine treatment. Biol Trace Elem Res 196(2):375–383
66. Botturi A, Ciappolino V, Delvecchio G, Boscutti A, Viscardi B, Brambilla P (2020) The role and the effect of magnesium in mental disorders: a systematic review. Nutrients 12(6):1661

67. Sauer AK, Hagmeyer S, Grabrucker AM (2016) 'Zinc deficiency'. In: Erkekoglu P, Kocer-Gumusel B (eds) Nutritional deficiency. IntechOpen, London
68. Novotny JA (2011) Molybdenum nutriture in humans. J Evid Compl Alt Med 163:164–168
69. Nawaz A, Khattak NN, Khan MS, Nangyal H, Sabri S, Shakir M (1010) Deficiency of vitamin B12 and its relation with neurological disorders: a critical review. JoBAZ 81:10
70. Pfaender S, Grabrucker AM (2014) Characterization of biometal profiles in neurological disorders. Metallomics 6(5):960–977
71. Abelein A (2023) Metal binding of Alzheimer's amyloid-β and its effect on peptide self-assembly. Acc Chem Res 56(19):2653–2663
72. Andersen JK (2004) Iron dysregulation and Parkinson's disease. J Alzheimers Dis 6(6 Suppl):S47–S52
73. Brooks J, Everett J, Lermyte F, Tjendana Tjhin V, Sadler PJ, Telling N, Collingwood JF (2020) Analysis of neuronal iron deposits in Parkinson's disease brain tissue by synchrotron X-ray spectromicroscopy. J Trace Elem Med Biol 62:126555
74. Zhang S, Wu L, Zhang J, Wang X, Yang X, Xin Y, Chen L, Li J, Niu P (2023) Multi-omics analysis reveals Mn exposure affects ferroptosis pathway in zebrafish brain. Ecotoxicol Environ Saf 253:114616
75. Kulkarni PP, She YM, Smith SD, Roberts EA, Sarkar B (2006) Proteomics of metal transport and metal-associated diseases. Chemistry 12(9):2410–2422

Check for updates

Chapter 3

Studying Metal Ions in Biology: Some Practical Aspects

Richard B. Thompson

Abstract

Metal ions are now established as playing numerous essential roles throughout the kingdoms of biology, and elucidating these roles has become vital to answering important biological questions and solving many practical problems in medicine, biotechnology, and agriculture. This chapter aims to highlight several practical issues in the study of metal effects in biology and suggest means by which they may be addressed. These include cleanliness, speciation, metal ion buffers, and kinetics.

Keywords Metal ions, Zinc, Metal ion buffers, Speciation, Metal ion sensors

1 Introduction

It has now been more than a century since people began to recognize the importance of metal ions in biology, and overall, the study of metals in biology has been a success story. As our knowledge accumulated and improved methods were developed, our overall understanding has improved commensurately, with important practical (mainly biomedical) [1]. Our collaborators and we were a small part of this story. However, throughout our careers, we tried several approaches and consequently made numerous mistakes when answering questions about the levels and functions of metal ions in biology, particularly when using optical methods. In some cases, we learned from our mistakes and, with much help from some outstanding chemists (see "Acknowledgments"), evolved reliable practices for measuring and studying metal ions in biological contexts. Particularly for those entering the field of metallobiochemistry from other disciplines, this chapter attempts to summarize some of that experience, hoping that our successors will benefit from our experience (and our mistakes).

Andreas M. Grabrucker (ed.), *Brain Metallomics*, Neuromethods, vol. 237,
https://doi.org/10.1007/978-1-0716-5468-2_3,
© The Author(s), under exclusive license to Springer Science+Business Media, LLC, part of Springer Nature 2026

2 Cleanliness is Next to Godliness: Purity of Reagents

Nowadays, we take highly purified chemical reagents and solvents for granted, but this was far from the case until the American Chemical Society began promulgating standards for reagent purity decades ago, and manufacturers began providing highly purified reagents with specified impurity levels. As we study different metal ions at lower and lower levels, the purity of water, buffer salts, biomolecules, and other components, particularly with respect to metal-ion contaminants, becomes more important and more challenging to assess and achieve. For example, if we consider measuring free Zn^{2+} ions at nanomolar to picomolar levels in a typical physiological buffer containing 125 mM NaCl, what level of Zn impurity is likely present in the sodium chloride? The ACS reagent grade NaCl is specified to have ≤5 ppm "heavy metals as Pb," which means that the lead (and maybe lead alone) in the NaCl was measured by atomic absorbance, inductively coupled plasma mass spectrometry (ICP-MS), or ICP-optical emission spectroscopy (ICP-OES), and found to be less than 5 ppm by weight. Thus, the 125 mmol of "reagent grade" NaCl in that liter of buffer contains up to 175 nanomoles of lead in some form. What this does not tell you is how much zinc is there, nor other metal ions that are chemically similar enough to mimic Zn^{2+}, such as Cd^{2+}, Hg^{2+}, or Cu^{2+}; indeed, the chemical similarity of the first two to Zn^{2+} accounts in large part for their toxicity. If we are trying to measure the effects of zinc ions at nanomolar levels, the potential for interference is evident. Furthermore, this issue also arises for all other components of the buffer (or growth medium or other aqueous medium).

Fortunately, there are some things one can do to partially alleviate the problem, some of which are additionally easy and cheap. First, one can use a chromatographic medium called Chelex-100 (Bio-Rad), which has iminodiacetic acid moieties attached to a hydrophilic chromatographic stationary phase. One passes the water or other aqueous buffer through a small chromatography column containing Chelex, and the multivalent metal ions are tightly bound to the iminodiacetate moieties, effectively filtered out with high efficiency.

The water one works with is also an issue: the water from the "deionized water" tap in the laboratory is much better than "tap water" but still unsatisfactory. Fortunately, many biochemical and particularly molecular biology (and NMR and analytical chemistry) labs require much purer water than just "deionized"; one source that we have found satisfactory has been a Milli-Q purification system (Sigma-Aldrich), which has a conductivity sensor on the output to verify performance; typically, the water has a resistance of 18 megohm per cm, an index of the lack of dissolved ions. Other highly purified water has occasionally been found satisfactory.

One issue with using resins like Chelex is that they perforce bind ions such as Ca^{2+} and Mg^{2+}, which may be constituents of growth media or solutions such as artificial cerebrospinal fluid (ACSF), and thus treating such media containing calcium and magnesium with Chelex drastically changes their composition. This is unacceptable, but fortunately, we found an approach that minimizes the problem. In the case of ACSF, we found that highly purified inorganic salts used in semiconductor production were available: for instance, $CaCl_2$ is available in a Puratronic™ grade from ThermoFisher (99.9965%, metals basis). Adding the calcium salt (and magnesium salts of comparable purity) after Chelex treatment of the medium prevents stripping of these salts from the medium while minimizing contamination by other metals. A similar approach may be used to formulate other media containing divalent cations.

3 Further Cleanliness

Laboratory glassware is typically made of borosilicate glass (e.g., Pyrex, Kimax), but the "starting materials" with which it is made, including silica, boric oxide, sodium, or potassium oxide, and alumina, are mainly mined. Thus, they are hardly pure, and impurities can leach into aqueous solutions, contaminating them. Better is "fused quartz" or simply "quartz," since it does not have the other ingredients, but it is more expensive, and one seldom finds quartzware apart from cuvettes. Best of all is "synthetic fused silica" (Suprasil™) made by flame hydrolysis of tetrachlorosilane, so there are no (mined) mineral impurities to speak of and optics made with this material have superior transmission in the ultraviolet and infrared (and much reduced fluorescence) [2]. For most purposes, reliable cleanliness of silicaware can be achieved by overnight immersion in moderately dilute (10% or more) reagent-grade nitric or sulfuric acid, followed by copious rinses in the Milli-Q-purified water mentioned above. Any metal contamination in the acid will be soluble in water and removed by rinsing.

Of course, much glassware in the laboratory has been replaced by (disposable) plasticware. A lack of repeated use requiring varying degrees of cleanliness seems attractive, but these plastic items are often made using molding processes that employ chemicals called "mold release agents": the mold is sprayed or coated with the agent to prevent the plastic from sticking to the mold material. Unfortunately, one popular class of release agents is zinc salts of fatty acids, such as zinc stearate, and small amounts may remain on otherwise brand-new plasticware. Fortunately, the mold release agent can be easily rinsed off in dilute acid before using the plasticware. Many workers use plasticware made of polytetrafluoroethylene (PTFE; the best-known trade name is Teflon™), which may not require mold release agents. Another potential source of contamination is plastic disposable pipette tips: the 200 μL size is often colored yellow, and the 1 mL size is colored blue. The yellow and blue pigments are sources of metal contamination; this can be avoided by using so-called "natural" pipette tips that omit these pigments.

Other ubiquitous items in chemical laboratories that can be a source of metal (especially zinc ion) contamination are the disposable paper wipes, such as KimWipes™ (zinc oxide is sometimes included as part of the paper formulation), often used, as well as disposable gloves that are powdered to facilitate putting them on. We have successfully used polyethylene gloves without powder. Our friends who do trace element research sometimes use laminar flow hoods like those used in tissue culture work, and may work in "clean room" conditions, wearing garment covers to minimize airborne particulate contamination.

4 Speciation

When using biological specimens and media that are not possible to free from contamination, it may be asked how it is possible to study the effects of low metal ion concentrations on biological processes without having them obscured by metal ion contamination. The short answer is that it generally isn't, at least without a great deal of trouble. Fortunately, in biology, one seldom considers (or measures the effect(s) of) a metal ion by itself, as if it were in a vacuum. Instead, in general, we are discussing complex aqueous solutions, with numerous other solutes dissolved in them, and in the case of cells and tissues, multiple compartments and phases that may be separated from one another by semipermeable membranes. From our standpoint, the central issue is "speciation," which refers to the fact that in biological media of all types, including cells themselves, there is an enormous variety of ligands of many chemical types that bind metal ions with varying degrees of affinity, selectivity, and persistence [3]. Particularly for divalent and trivalent metal cations, it is typically the case that a large proportion of any given metal ion is tightly bound to biological ligands, and only a small fraction is relatively available to participate in reactions of any kind. Thus, in blood serum that contains typically up to 15 micromolar total zinc ion, most is bound to relatively abundant ligands like serum albumin or glutamic acid, which are present in nearly millimolar concentrations and have affinities in the micromolar range [4], so that of the total zinc, just under nanomolar concentrations are weakly enough bound to participate in other reactions [5]. The corollary to this is that for any reaction in which the zinc or other metal ion participates, an essential step is exchanging one or more ligands bound to the metal ion with ligand(s) on the target molecule, such as a protein. A closely related issue is that when studying the effect of zinc ions, for instance, on a biochemical process or reaction, how does one prepare a buffer containing the zinc ion and know what fraction of the zinc ion one has added is actually available to participate in or influence the reaction?

Also, from a chemical standpoint, what is meant by describing a metal ion as being "free" or "available" to participate, for instance, in a reaction? In aqueous solutions in biology, the multivalent metal ions always interact electrostatically at least weakly or transiently with the ubiquitous, polar water molecules; strictly speaking, they are never free. However, since the water is mainly bound weakly and transiently, the metal ion readily exchanges the water for stronger ligands that perforce bind for extended periods. The affinities of metal ions for numerous ligands have been measured and compiled, along with the metal ions' preferences for the geometrical arrangement of ligands surrounding them [6–8]. It is beyond our scope to describe in detail how the electronic structure of the metal ion determines the affinity and preferred geometry of ligands binding to it, nor the details of calculating speciation in mixtures of ligands and metal ions, but a few points below may prove of use.

5 Metal Ion Buffers

If we consider the binding of a fairly high-affinity ligand to a metal ion as simply a reversible equilibrium reaction, it should be evident that under a certain set of conditions of pH, temperature, and ionic strength, a very small proportion of the metal ion will remain unbound (at least transiently) and be available to react or bind some other ligand. Moreover, if we add a small concentration of metal ions to the mixture, the vast majority of what is added will in turn be bound, and the actual concentration of unbound metal ions will change only slightly. This behavior is, of course, very similar to that of a pH buffer, in which the addition of acid or base to an aqueous solution (where protons are added to or removed from the system) has much less effect on the solution's pH than if the buffer were not present. Thus, a high-affinity ligand may act as a "metal ion buffer" and minimize the effects of added metal ions (perhaps as a contaminant) or added ligands that bind the metal on the "free" or available metal ion concentration. We note an important difference between pH buffers and metal ion buffers: with pH, there is an abundant, ready supply of protons and hydroxide ions in water to protonate or deprotonate the buffer; the same is not often true of metal ions or ligand(s). Nevertheless, by choosing a reasonably high-affinity ligand for a given metal ion, the buffering effects can be quite satisfactory, and solutions with low free metal ion concentrations can be reliably prepared.

An example may serve to illustrate the point: if we prepare a solution of 2 mM nitriloacetic acid (NTA) at pH 7.0, 25°C and add

0.7 micromolar total zinc ion, the unbound zinc ion will comprise only 1.06 picomolar. Moreover, if a contaminant adds an additional 1.5 micromolar of zinc, the free concentration increases only to 4.26 picomolar, so if the phenomenon we are studying is only sensitive to nanomolar concentrations of zinc ions, the effect of the contaminant will be negligible [9]. Similarly, adding a strongly binding ligand from some source will bind up some of the free metal ions, but the equilibrium with the metal-ion buffer ligand will shift slightly, releasing some metal ions to counterbalance and minimize the effects of the added ligand.

Computational methods have been developed [10] and incorporated into free software (such as MINEQL+ www.mineql.com) that enables one to calculate the speciation of metal ions in aqueous solutions as a function of the pH, ionic strength, temperature, and the concentrations and identities of the ligands; using this software, we and others have published formulations of metal ion buffers and buffered media for particular applications [11–13].

We would add that metal ion buffers are particularly useful for accurate calibration of metal ion sensor response, particularly of optical sensors such as fluorescence or bioluminescence-based sensors [9, 14]. For example, fluorescence-based sensors are small- or macro-molecules that exhibit a concentration-dependent change in their intensity, or (preferably) wavelengths or lifetime upon binding the metal ion of interest; the advantages of the latter two changes over simple intensity are well-known [15, 16]. It is generally valid to assume that adding a small amount of the sensor molecule does not substantially alter the free concentration or speciation of the analyte metal ion, but this may not hold in very small compartments, such as organelles. Often, these sensors are calibrated in situ with a two-point calibration: they are exposed to a high concentration (well above K_D) and a low concentration (well below K_D) of the metal ion, and the free metal ion concentration is calculated from the apparent fractional saturation of an intervening value. Better are multipoint calibration curves using a set of suitable metal ion buffers spanning well above and below K_D, since no presumption of the K_D value under the experimental conditions need be made, and the calibration is less vulnerable to inaccuracies in either the low or high metal ion concentration measurements.

Zinc, such as Ca and Mg, only exists as a +2 ion in the aqueous solutions commonly found in biology, so we need not worry about any change in the ionization state of any of these metals. By comparison, the ubiquitous Fe, Cu, as well as Mn, Co, and Cr, exhibit multiple ionization states, and indeed their employment in biological systems is often to serve in (essential) oxidation/reduction reactions. Their propensity to exist in a particular ionization state is determined strongly by the presence and strength of other oxidants and reductants (especially atmospheric oxygen), and also by the number, strength, and orientation of their ligands. For instance, Fe^{2+} bound in the heme prosthetic groups of hemoglobin and cytochrome c is dramatically less prone to oxidation than the aquo Fe^{2+} ion in aerobic solution at neutral pH. Thus, for the latter group of metal ions, maintaining control of the ionization state by the levels of reductants and oxidants is essential to elucidating their biological effects.

6 Kinetics as an Issue

Another important issue in considering biological activities is the kinetics not only of reactions involving breaking and forming covalent bonds in a particular process but also of metal ions adsorbing to and desorbing from specific ligands [10]. We can consider a simple binding reaction of a metal ion to (for instance) a discrete site on a protein or other ligand as having an apparent rate K_{OBS} that is a function of an association rate constant K_{ON} (on-rate), a dissociation rate constant K_{OFF} (off-rate), and the sum of the concentrations (activities) of the metal ion [M] and the binding site [P]:

$$K_{OBS} = K_{ON}([M][P]) + K_{OFF}$$

Thus, the time for the reaction to proceed to 50% completion is

$$t_{1/2} = 0.693 / K_{OBS}$$

For a diffusion-controlled reaction, K_{ON} is as fast as $10^8 M^{-1} s^{-1}$ in water at room temperature, such that the reaction may proceed to equilibrium in milliseconds at micromolar concentrations of the reactants. For an enzyme that requires a bound metal ion to catalyze its reaction, it would seem of value not only that the metal ion binding affinity should be high (K_D nanomolar or tighter) but also that the K_{OFF} rate should be slow, such that when the ambient metal ion concentration is low, the enzyme does not rapidly lose its metal ion and become inactive. An example is human erythrocyte carbonic anhydrase II, a zinc-containing enzyme. The zinc affinity is high at 4 picomolar, and the K_{OFF} is suitably slow at $1.4 \times 10^{-6} s^{-1}$ [17]. The desirably slow off rate (concentration-independent) has an important consequence: the on rate must also be relatively slow for the affinity to be so tight at realistic concentrations of protein or metal ion, in this case, only about $3.5 \times 10^5 M^{-1} s^{-1}$ [18]. Thus, at nanomolar concentrations of protein and metal ion, in the absence of other factors, the average $t_{1/2}$ for binding the metal and thereby making the apoenzyme active can be tens of minutes. This comes into play when we consider expression of the protein and binding the metal ion to make it active: first, there would be a substantial lag between completing expression of the apoenzyme polypeptide and when it becomes active, which may be metabolically undesirable; and second, since binding the metal confers significant additional thermodynamic stability, it probably is desirable for this reason as well that metalation occur in vivo as quickly as possible. It was found that small molecules such as 2,6-dipicolinic acid [19] and glutamate both significantly catalyze the metal-binding reaction, reducing the equilibration time in vitro to minutes instead of hours. It is not known what catalyst(s) may act in vivo. However, the message from this example is more far-reaching: that the kinetics of metal ion binding reactions should generally be considered when evaluating their biological effects, particularly at low concentrations.

In summary, we have briefly recounted some important factors that must always be considered when studying the roles and effects of active metal ions in biological experiments. Central among these are cleanliness, metal ion speciation and buffers, and kinetics. Indeed, other factors can and do come into play, but the foregoing are worth remembering from the beginning.

Acknowledgments

We are grateful to Bright Focus Foundation, the National Science Foundation, the Office of Naval Research, and National Institutes of Health for their support; and outstanding scientists including Thomas Baldwin, Harold Bright, David Christianson, Carol Fierke, Chris Frederickson, Joseph Lakowicz, Imre Lengyel, Emanuel Margoliash, Jim Moffett, and Rod Zika, for many helpful discussions, as well as their patience and insight as we inched up the learning curve.

Conflict of Interest Richard Thompson is a principal of Pokegama Technologies, Inc., a manufacturer of (among other things) metal ion buffers.

References

1. Bertini I, Sigel A, Sigel H (eds) (2001) Handbook on metalloproteins. Marcel Dekker, New York
2. Thompson RB, Scarlata S (2017) Fluorescence made easier: Fluorescence techniques for the novice. Episode 1: The Basics. In: Geddes CD (ed) Reviews in fluorescence 2016. Reviews in Fluorescence. Springer, pp 1–8. https://doi.org/10.1007/978-3-319-48260-6_1
3. Stumm W, Morgan JJ (1996) Aquatic chemistry: chemical equilibria and rates in natural waters, 3rd edn. Wiley-Interscience, New York
4. Blindauer CA, Harvey I, Bunyan KE, Stewart AJ, Sleep D, Harrison DJ, Berezenko S, Sadler PJ (2009) Structure, properties, and engineering of the major zinc binding site on human albumin. J Biol Chem 284(34):23116–23124. https://doi.org/10.1074/jbc.M109.003459
5. Maares M, Hackler J, Haupt A, Heller RA, Bachmann M, Diegmann J, Moghaddam A, Schomburg L, Haase H (2022) Free zinc as a predictive marker for COVID-19 mortality risk. Nutrients 14(7). https://doi.org/10.3390/nu14071407
6. Dawson RMC, Elliott DC, Elliott WH, Jones KM (eds) (1989) Data for biochemical research, 3rd edn. Oxford University Press, Oxford
7. Vasak M, Kagi JHR (1994) Metallothioneins. In: King RB (eds) Encyclopedia of inorganic chemistry, vol 4. Wiley, New York, pp 2229–2241
8. Cotton FA, Wilkinson G (1988) Advanced inorganic chemistry, 5th edn. Wiley-Interscience, New York
9. Thompson RB, Fierke CA (2017) Measuring and imaging metal ions with fluorescence-based

biosensors: speciation, selectivity, kinetics, and other issues. Methods Enzymol 589:281–299
10. Hering JG, Morel FMM (1988) Kinetics of trace metal complexation: role of alkaline earth metals. Environ Sci Technol 22:1469–1478
11. McCranor BJ, Szmacinski H, Zeng HH, Stoddard AK, Hurst T, Fierke CA, Lakowicz JR, Thompson RB (2014) Fluorescence lifetime imaging of physiological free Cu(II) levels in live cells with a Cu(II)-selective carbonic anhydrase-based biosensor. Metallomics 6(5):1034–1042. https://doi.org/10.1039/c3mt00305a
12. Pao P-J, Emri E, Abdirahman SB, Soorma T, Zeng H-H, Hauck SM, Thompson RB, Lengyel I (2018) The effects of zinc supplementation on primary human retinal pigment epithelium. J Trace Elem Med Biol 49:184–191. https://doi.org/10.1016/j.jtemb.2018.02.028
13. Bozym R, Hurst TK, Westerberg N, Stoddard A, Fierke CA, Frederickson CJ, Thompson RB (2008) Determination of zinc using carbonic anhydrase-based fluorescence biosensors. Methods Enzymol 450:287–309. https://doi.org/10.1016/s0076-6879(08)03414-9
14. Matveeva EG, Stoddard AK, Zeng H-H, Franke G, Bourne L, Fierke CA, Thompson RB (2022) Ratiometric zinc biosensor based on bioluminescence resonance energy transfer: trace metal ion determination with tunable response. Int J Mol Sci 23(23):14936
15. Grynkiewicz G, Poenie M, Tsien RY (1985) A new generation of calcium indicators with greatly improved fluorescence properties. J Biol Chem 260(6):3440–3450
16. Thompson RB, Patchan MW (1995) Lifetime-based fluorescence energy transfer biosensing of zinc. Anal Biochem 227:123–128
17. Huang C-c, Lesburg CA, Kiefer LL, Fierke CA, Christianson DW (1996) Reversal of the hydrogen bond to zinc ligand histidine-119 dramatically diminishes catalysis and enhances metal equilibration kinetics in carbonic anhydrase II. Biochemistry 35(11):3439–3446
18. Kiefer LL, Paterno SA, Fierke CA (1995) Hydrogen bond network in the metal binding site of carbonic anhydrase enhances zinc affinity and catalytic efficiency. J Am Chem Soc 117:6831–6837
19. Pocker Y, Fong CTO (1983) Inactivation of bovine carbonic anhydrase by dipicolinate: kinetic studies and mechanistic implications. Biochemistry 22:813–818

Chapter 4

Metal Meets Amyloid: Computational Approaches to Binding, Aggregation, and Neurotoxicity

Shayon Bhattacharya

Abstract

Metal ions such as Cu^{2+}, Zn^{2+}, and Fe^{3+} are deeply implicated in the pathogenesis of neurodegenerative diseases such as Alzheimer's and Parkinson's disease, due to their ability to modulate the aggregation, toxicity, and redox cycling of amyloidogenic peptides, including amyloid-β (Aβ), α-synuclein (α-Syn), and tau. Yet, the transient, heterogeneous, and polymorphic nature of metal-bound intermediates makes them challenging to characterize experimentally. Computational modeling has become indispensable in uncovering atomistic insights into these complex systems. This chapter provides a didactic yet comprehensive overview of state-of-the-art molecular modeling strategies—from quantum mechanical methods like density functional theory (DFT) to long-timescale molecular dynamics (MD) simulations—that enable the study of metal–amyloid interactions across length and time scales. I highlight approaches for building and validating metal–peptide coordination models, tuning force fields, sampling conformational dynamics, and interpreting biologically relevant outcomes. Case studies, including the simulation of Cu^{2+}-associated Aβ monomeric fragments and α-Syn oligomers at hard and soft biointerfaces, demonstrate how computational workflows complement experiments such as cyclic voltammetry, atomic force microscopy and nuclear magnetic resonance spectroscopy. By addressing the limitations of experimental resolution, these tools also offer forward-looking avenues for therapeutic targeting, chelator design, and redox-sensitive modulation of toxic aggregates.

Keywords Amyloid-β, α-Synuclein, Tau, Metal ions, Molecular dynamics, Density functional theory, Cu^{2+}, Zn^{2+}, Fe^{3+}, Protein aggregation, Neurodegeneration, Redox chemistry, Computational modelling

1 Introduction

Metal dyshomeostasis and protein misfolding are two converging hallmarks of neurodegenerative diseases, notably Alzheimer's disease (AD), Parkinson's disease (PD), and related tauopathies [1]. Central to these disorders is the accumulation of amyloidogenic proteins such as amyloid-β (Aβ), α-synuclein (α-Syn), and tau into toxic oligomers and, consequently, fibrillar aggregates. Transition metals such as copper (Cu^{2+}), zinc (Zn^{2+}), and iron (Fe^{2+}/Fe^{3+}) are not only found enriched in pathological inclusions such as senile plaques and Lewy bodies but have also been implicated in altering the structural and functional dynamics of these proteins [2, 3]. These interactions

Andreas M. Grabrucker (ed.), *Brain Metallomics*, Neuromethods, vol. 237,
https://doi.org/10.1007/978-1-0716-5468-2_4,
© The Author(s), under exclusive license to Springer Science+Business Media, LLC, part of Springer Nature 2026

contribute to aggregation kinetics, generate reactive oxygen species (ROS), and can even lead to oxidative protein modifications—making the protein–metal interface a key molecular trigger in disease progression [4, 5] (Fig. 1).

Our recent computational modeling studies have significantly advanced our understanding of amyloidogenic peptides and their interactions with biological surfaces and chemical effectors. A key focus has been the modulation of aggregation by small molecules and physiological interfaces. For instance, combined atomic force microscopy (AFM) and molecular dynamics (MD) studies demonstrated that levodopa alters the morphology of Aβ42 and α-synuclein fibrils, suggesting a chemical effector-driven disassembly mechanism at nanoscale resolution [6] (Fig. 2A). In a complementary vein, direct visualization of Aβ42 oligomer proliferation on fibril surfaces using liquid AFM, supported by atomic-scale simulations, confirmed the role of nonperiodic catalytic sites in accelerating secondary nucleation [7] (Fig. 2B).

Several studies have explored the structural dynamics of α-Syn helical tetramers [8–10] (which were found to be aggregation-resistant) and their stability at membrane interfaces [11]. Our computational models have uncovered how surface charge and curvature influence tetramer conformations, with implications for selective therapeutic stabilization of aggregation-resistant forms [12–15] (Fig. 2C). Extensions of these models have revealed how intramolecular allosteric coupling between terminal and central hydrophobic domains modulates helix–helix association propensity in partially folded monomers—a possible early trigger in amyloid formation [16, 17] (Fig. 2D).

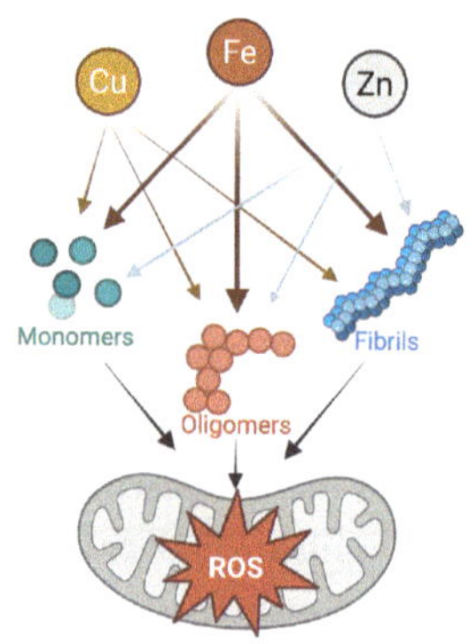

Fig. 1 Metal-mediated modulation of amyloid aggregation and oxidative stress. Schematic representation of how biologically relevant metal ions (Cu, Fe, Zn) influence amyloid peptide aggregation pathways. Metal coordination can alter monomer conformations, promote oligomer formation, modulate fibril structure, and affect aggregation kinetics. Redox-active metals (particularly Cu and Fe) may catalyze the generation of reactive oxygen species (ROS), contributing to mitochondrial dysfunction and neurotoxicity. The interplay between metal binding, aggregation state, and oxidative stress represents a central mechanistic axis in metal-associated neurodegenerative pathology

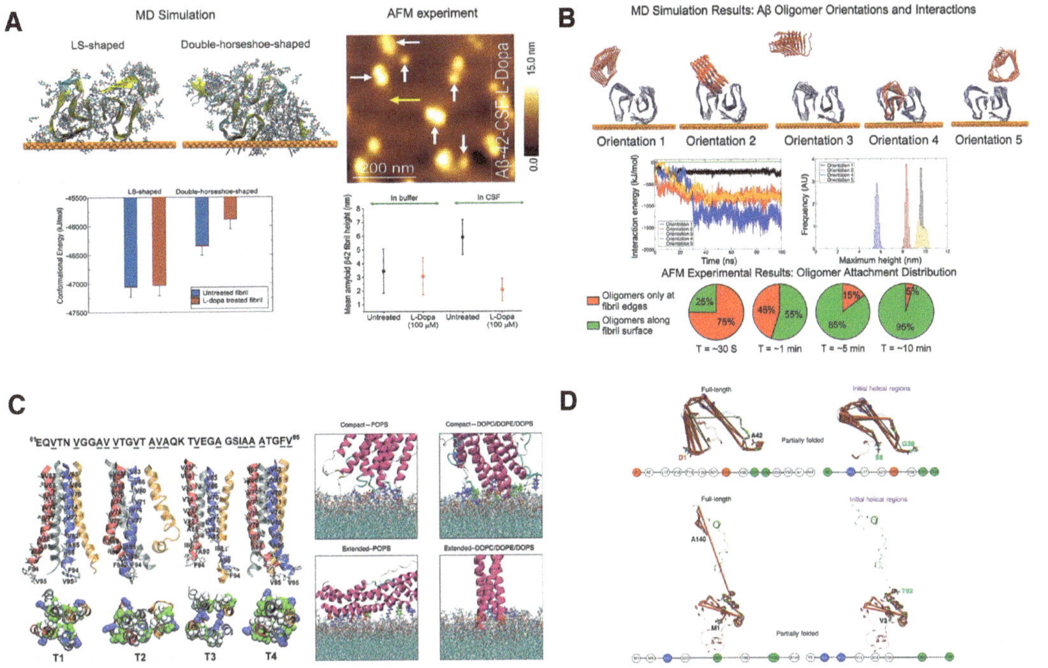

Fig. 2 Computational insights into amyloid structure, surface interactions, and aggregation mechanisms. (**A**) Combined molecular dynamics (MD) simulations and atomic force microscopy (AFM) experiments reveal that small molecules modulate amyloid morphology. Representative LS-shaped and double-horseshoe Aβ42 fibril conformations are shown alongside AFM imaging of fibrillar assemblies. Conformational free-energy analysis and height distributions demonstrate that levodopa alters fibril morphology and reduces fibril stability in both buffer and cerebrospinal fluid (CSF), supporting a chemical effector–driven disassembly mechanism at nanoscale resolution. (**B**) MD simulations and liquid-phase AFM visualization of Aβ42 oligomer–fibril interactions. Multiple oligomer orientations on fibril surfaces were characterized computationally, revealing heterogeneous binding modes and interaction energetics. Time-resolved AFM analysis confirms preferential oligomer proliferation at nonperiodic catalytic sites along fibril surfaces, supporting a secondary nucleation mechanism mediated by structurally distinct surface regions. (**C**) Atomistic models of amyloid tetramer conformations at lipid membrane interfaces. Simulations demonstrate that surface charge, lipid composition (POPS vs. DOPC/DOPE/DOPS), and curvature modulate tetramer compactness and orientation. These findings highlight how membrane physicochemical properties influence aggregation-prone versus aggregation-resistant conformational states. (**D**) Allosteric coupling in partially folded amyloid monomers. Structural mapping reveals intramolecular communication between the terminal and central hydrophobic domains, modulating the helix–helix association propensity. Differential coupling between full-length and truncated variants suggests that early conformational fluctuations may act as triggers for oligomerization and downstream fibril formation

We have shown that the capacity of terminal regions to stabilize helical states across Aβ and α-Syn to be an emerging design principle to halt toxic aggregation [17]. Other modeling efforts have charted general design rules for controlling aggregation in amyloidogenic peptides by predicting the relative stability of helical vs. disordered conformers [18], and through retaining stereochemistry and chiral integrity of amino acids [19].

In the tau family of peptides, our MD simulations have captured distinct effects of disease-causing mutations such as P301L and

Δ K280 on hexamer stability and elongation propensity, offering a molecular basis for fibril polymorphism and differential toxicity in frontotemporal dementia [20]. These tau models highlight the power of MD to discriminate between nucleation and elongation mechanisms linked to specific residues. At the fibrillar level, MD simulations of the Non-Amyloid-b Component (NAC) region of α-Syn have linked specific sequence and environmental features to strain-dependent toxicity and thermodynamic stability [21].

These approaches are underpinned by protocols combining free energy calculations, conformational subspace mapping, and benchmarking against available experimental observables [22]. Recently, we have comprehensively reviewed the growing role of atomistic and multiscale simulations in resolving early oligomeric species and surface-mediated aggregation pathways, which remain experimentally elusive [23]. Lastly, label-free interface-resolved AFM studies of Aβ 40/42 aggregation, supplemented by MD simulations, quantified interface-specific interaction energies and identified lateral fibril alignment as a distinct feature of Aβ42, which may inform new biofunctional design strategies [24].

Beyond their established pathological roles, amyloidogenic proteins such as Aβ, α-synuclein, and tau exhibit highly dynamic behaviors that are profoundly influenced by their biochemical environment. Computational modeling has helped decipher these structural transitions at atomic resolution— particularly under conditions that mimic biological surfaces, mutations, and lipid interactions. However, a critical modulator of these aggregation pathways is often underappreciated in early-stage modeling: the presence of metal ions. In a recent study, we combined liquid-phase AFM with molecular dynamics simulations under cerebrospinal fluid (CSF)-mimicking conditions at a gold–water interface to reveal Ca^{2+}-mediated coaggregation between Aβ40 and Aβ42 isoforms [25]. Transition metals such as Cu^{2+}, Zn^{2+}, and Fe^{3+} can coordinate with amyloidogenic peptides, influencing their conformational ensembles, charge states, and nucleation kinetics in ways that remain difficult to probe experimentally. The next sections explore how metals shape the aggregation landscapes of Aβ , α-synuclein, and tau, and how computational approaches can help map these complex, metal-mediated mechanisms.

1.1 Amyloid-β and Metal Ion Interactions

Among the most widely studied protein–metal interactions are those involving Aβ peptides. Aβ42, the longer and more aggregation-prone is oform, exhibits metal-binding primarily at its N-terminal region (residues 1–16), which contains key donor residues such as histidines (H6, H13, H14), aspartates (D1, D7), and glutamates (E3, E11) [26]. Copper can coordinate in multiple geometries; most notably in 3N1O or 4N arrangements, with subtle variations depending on pH, buffer conditions, and redox state [27–29]. Zinc, being redox-inactive, generally forms tetrahedral geometries but may also engage in interpeptide bridges that stabilize amorphous aggregates [30, 31].

Interestingly, while both Cu^{2+} and Zn^{2+} are enriched in amyloid plaques, they exert opposing effects on aggregation. For instance, Zn^{2+} at sub-stoichiometric concentrations promotes oligomer formation but may inhibit fibril elongation [32]. Cu^{2+}, on the other hand, can accelerate early aggregation stages and also catalyze ROS production via Fenton-like reactions, resulting in oxidative stress and neuronal damage [33, 34]. Fe^{3+} also exhibits redox cycling and has been implicated in oxidative crosslinking of Aβ and ferroptosis pathways in AD [35].

Despite extensive biochemical and spectroscopic studies, significant gaps remain in our understanding of Aβ–metal interactions, especially at the level of transient oligomeric intermediates, which are believed to be the most neurotoxic species. Structural heterogeneity, dynamic conformations, and the polymorphic nature of Aβ oligomers pose challenges for conventional experimental approaches such as NMR and X-ray crystallography. As a result, much of the atomic-level detail has been inferred from computational methods such as density functional theory (DFT), classical molecular dynamics (MD), and hybrid QM/MM approaches [36, 37].

1.2 α-Synuclein and Metal-Dependent Aggregation

α-Synuclein (α-Syn), the principal component of Lewy bodies in PD, is another intrinsically disordered protein (IDP) that exhibits metal-responsive aggregation. The N-terminal region of α-Syn contains multiple histidines (notably His50) and carboxylates (Asp121, Glu123) that serve as potential metal-binding sites [38, 39]. In particular, Cu^{2+} binding to α-Syn has been reported to accelerate oligomer formation and stabilize atypical structures at higher Cu^{2+} concentrations— precursors to toxic pore-forming aggregates [40]. In our own work, we demonstrated via DFT-MD combined with liquid-based atomic force microscopy (AFM) that α-Syn aggregates into "annular oligomers" upon incubation with Cu^{2+} at the gold–water interface, with structural features distinct from aggregates formed in metal-free conditions [41]. MD simulations revealed concentration-dependent packing and thermodynamic preferences for Cu^{2+}-stabilized oligomers.

Zn^{2+}, in contrast, has a less pronounced effect on α-Syn aggregation, but its presence can modulate α-Syn interactions with lipid membranes and potentially influence endocytosis pathways [42]. Fe^{3+} may also promote α-Syn aggregation and has been recently implicated in mediating dopaminergic neuron death via iron-catalyzed oxidative stress, including iron dysregulation outside the brain through gut-first PD [43].

Here again, experimental resolution of metal-bound oligomeric intermediates is limited, especially in the physiological buffer milieu. Computational modeling becomes indispensable to resolve questions around binding stoichiometry, coordination geometry, and allosteric

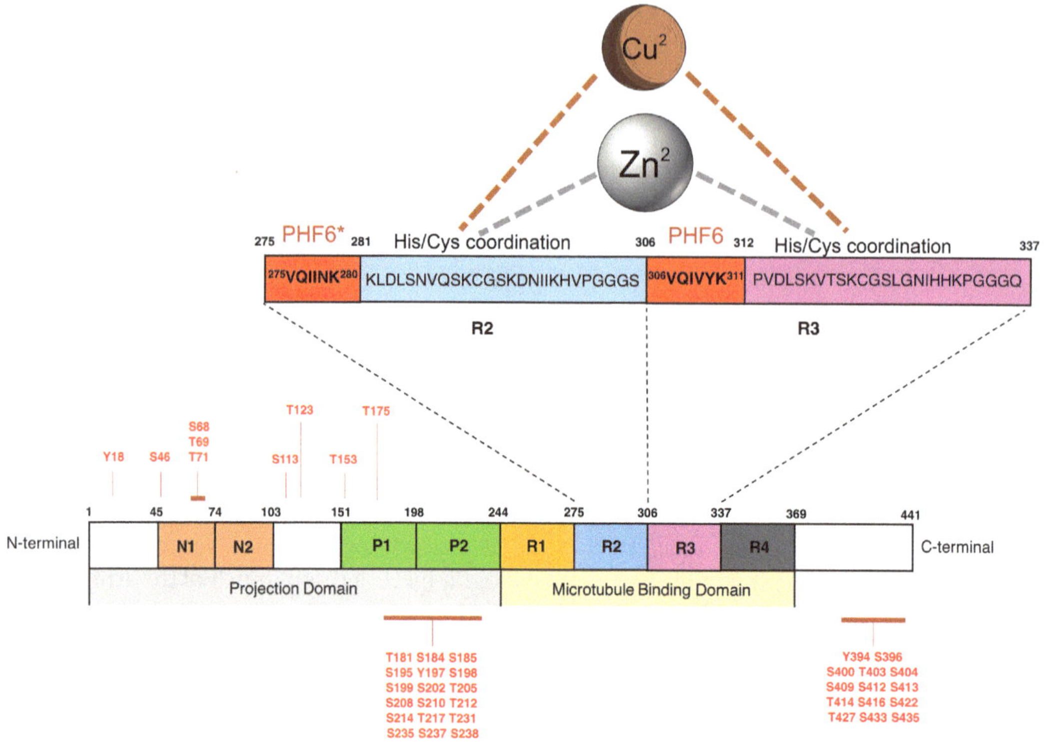

Fig. 3 Metal-binding regions and hyperphosphorylation hotspots in tau protein. Schematic representation of full-length human tau highlighting its domain organization, including the N-terminal projection domain (N1–N2), proline-rich regions (P1–P2), and the microtubule-binding domain (MBD; R1–R4 repeats). The hexapeptide motifs PHF6* (^{275}VQIINK^{280}) in R2 and PHF6 (^{306}VQIVYK^{311}) in R3, which nucleate β-sheet aggregation, are indicated. Cu^{2+} and Zn²⁺ ions have been reported to coordinate primarily within His- and Cys-containing segments of the R2 and R3 repeat domains, modulating local structure and aggregation propensity

effects induced by metal interactions on aggregation-prone sequences like the NAC domain (residues 61–95).

1.3 Tau and Metal Ion Pathology

Tau, a microtubule-associated protein that aggregates into neurofibrillary tangles in AD and other tauopathies, also displays metal sensitivity. While tau lacks classical metal-binding motifs like those in Aβ or α-Syn, it has been shown to bind Cu^{2+} and Zn^{2+} at histidine- and cysteine-containing regions in its repeat domains (especially R2 and R3) in the microtubule binding domain (MBD) [44, 45] (Fig. 3). Metal binding to tau can promote conformational changes, disulfide crosslinking, and oxidative modifications that increase aggregation propensity [46].

Notably, Cu^{2+}-induced oxidation of Cys291 and Cys322 in tau leads to dityrosine and disulfide crosslinking, forming soluble tau oligomers that are thought to be seeding-competent and neurotoxic [47, 48]. Additionally, Fe^{3+} binding has been shown to induce tau hyperphosphorylation through kinase activation and

ROS production, linking metal overload to both tau aggregation and post-translational modification cascades [49].

Despite increasing recognition of metal–tau interactions, high-resolution structures of metal-bound tau aggregates are rare. This makes molecular simulations vital for investigating how local metal binding might influence long-range conformational behavior, aggregation pathways, and microtubule interactions.

1.4 Computational Modeling: Bridging the Gaps

Given the experimental limitations across all three protein systems—Aβ, α-Syn, and tau— computational approaches provide a unique opportunity to characterize metal-binding at atomic resolution. DFT calculations help predict coordination geometries, binding energies, redox states, and electronic properties of metal–peptide complexes. Classical MD simulations offer insights into the effect of metal ions on aggregation pathways, secondary structure transitions, and interfacial behavior.

A number of DFT studies have explored Cu^{2+} and Zn^{2+} coordination in amyloidogenic peptides using functionals such as B3LYP or MPWB1K with LANL2DZ or def2-TZVP basis sets, revealing preferences for 3N1O geometries and Asp/Glu over Tyr as primary oxygen donors [36, 37, 50]. For α-Syn, hybrid QM/MM and MD simulations have been used to examine binding at the His50 and C-terminal Asp/Glu sites, capturing conformational transitions [51–53]. Tau-metal interactions remain relatively underexplored computationally, though some efforts have begun using MD simulation to probe metal ion-induced misfolding in the R3 repeats [54].

Multiscale simulations combining DFT, MD, and enhanced sampling techniques (e.g., metadynamics, REST2) are particularly promising (Fig. 4). These allow one to probe rare events such as metal-induced conformational transitions, nucleation of aggregation seeds, or binding to surfaces such as membranes or inorganic substrates. Our own studies have leveraged this approach at the gold–water interface for α-Syn [6, 7, 24, 55, 56], and future work may extend to lipid membrane surfaces or exosomal boundaries.

Moreover, integrating modeling results with experimental observables—such as AFM morphology, Raman shifts, or EPR spectra—can provide converging evidence and increase the predictive power of simulations. With the growing availability of structural templates from the protein data bank (PDB), AlphaFold [57] models, and cryo-EM datasets, the field is now poised for large-scale modeling efforts across multiple neurodegenerative proteins and metal species.

1.5 Scope of this Chapter

This chapter provides a detailed methodology for modeling metal–amyloid interactions using a combination of classical MD, DFT, and hybrid QM/MM approaches. It outlines:

1. Software tools and metal parameterization strategies.

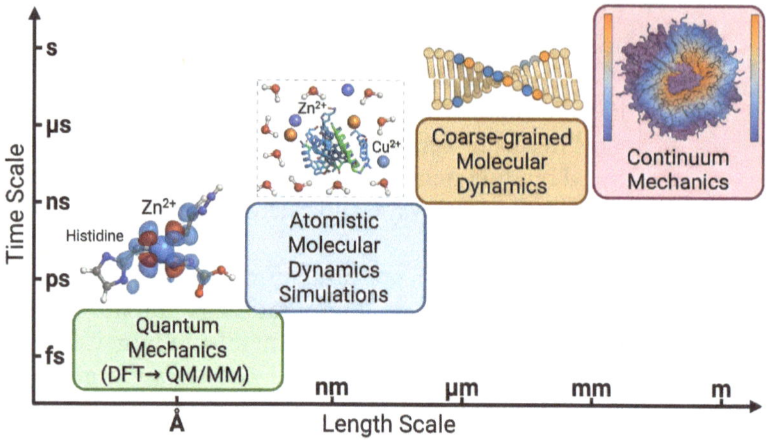

Fig. 4 Multiscale computational framework for modeling metal–amyloid interactions. Schematic representation of the length and time scales accessible to different computational techniques used in studying amyloid–metal systems. Quantum mechanical approaches (DFT and QM/MM) provide electronic-level resolution of metal coordination geometry and redox properties at the Å–fs scale. Atomistic molecular dynamics (MD) simulations capture peptide conformational dynamics, aggregation, and metal coordination stability at the nm–ns/μs scale. Coarse-grained MD enables modeling of oligomerization and surface-mediated assembly processes at extended length and time scales, while continuum models describe mesoscale aggregation and macroscopic structural behavior. The integration of these approaches enables a comprehensive understanding of metal-mediated amyloid aggregation mechanisms across scales

2. Preparation of monomeric and oligomeric structures with metals.
3. Surface and interfacial modeling (e.g., gold–water interface).
4. Advanced sampling techniques and redox state modeling.
5. Case studies from Aβ and α-Syn systems, including examples from our own published work.

Emphasis is placed on practical execution, reproducibility, and how computational modeling can answer biological questions that remain unresolved experimentally. A curated list of known metal–amyloid complex structures from the PDB is also provided, serving as a launchpad for further investigations.

2 Materials

The modeling of amyloid–metal ion interactions requires a combination of quantum chemical precision and large-scale dynamic simulation. The first step involves DFT and hybrid QM/MM approaches to resolve the local coordination geometry and electronic environment of metal-bound states. These quantum-level insights inform the parametrization and setup of molecular dynamics (MD) simulations, enabling the exploration of longer-time and length-scale processes such as peptide folding, aggregation, and surface interactions. This section provides a detailed description of the structural inputs, computational tools, simulation engines, and parameters used in this multiscale framework, with emphasis on reproducibility and adaptation across Aβ, α-synuclein, and tau systems.

2.1 Structural Inputs and Experimental Templates

Atomistic models of the amyloidogenic peptides are obtained from either experimentally resolved structures or predictive modeling databases. For Aβ42, NMR-derived structures such as PDB ID 1IYT [58] and 1Z0Q [59] offer starting conformations that capture the α-helical propensity of the N-terminal and central hydrophobic domains. In the case of α-synuclein, structures bound to membrane-mimicking environments (PDB IDs 1XQ8 [60], 2KKW [61]) provide partially folded starting states. Tau repeat domains have limited structural resolution but are usually modeled based on cryo-EM-resolved filament cores (PDB ID 6VHA [62]) or generated de novo using AlphaFold [57] models of full-length isoforms.

When modeling metal interactions, initial complex geometries can be extracted from structures like PDB ID 1ZE9 [63], which represents a Zn^{2+}-bound Aβ(1–16) peptide. In cases where experimental geometries are unavailable, metal-peptide binding poses are constructed using homology modeling [27], or relaxed via DFT optimization [64].

2.2 Quantum Chemical Modeling: DFT and QM/MM

Density functional theory [65] (DFT) is a computational quantum mechanical modeling method used to investigate the electronic structure (or nuclear structure) of many-body systems, in particular atoms, molecules, and the condensed phases. DFT calculations are essential to predict the preferred metal coordination geometry, ligand environment, and charge distribution of metal ions in contact with amyloid peptides. For this purpose, we usually employ Gaussian 16 [66] or ORCA 5.0 [67], which offer flexibility in treating transition metals with appropriate basis sets and pseudopotentials. A hybrid QM/MM employing the ONIOM [68] method has also been used for studying the interplay between transition-metal ions and amyloid monomers and oligomers [37].

Metal-bound peptide fragments (typically 5–15 residues) have also been optimized using hybrid functionals such as B3LYP [69], PBE0 [70], or MPWB1K [71], combined with basis sets like LANL2DZ [72] for metals and 6–31 G(d) or def2-TZVP for the peptide [27]. Solvation effects could be included using implicit models such as PCM [73] or COSMO [74]. The results yield detailed insights into:

(1) Preferred coordination spheres (e.g., 3N1O vs. 4 N).

(2) Identification of dominant ligating residues (His, Glu, Asp, Tyr).

(3) Effect of pH on metal-binding modes.

(4) Spin states and redox potential of Cu^{2+} and Fe^{3+} [75].

In larger systems, QM/MM approaches implemented via ChemShell [76] or CP2K [77] could be used to preserve the quantum treatment of the metal-binding site, while embedding it within a full peptide or oligomeric environment. These hybrid schemes proved essential in modeling metal binding at fibril interfaces or protein–surface contact regions.

2.3 Molecular Dynamics Simulations: System Preparation and Parametrization

Following quantum-level refinement of metal coordination, full peptide and peptide-metal systems could be simulated using classical MD. The systems are popularly prepared using CHARMM-GUI [78], tleap (AMBER [79]), or GROMACS [80] topology builders, and solvated in cubic or dodecahedral boxes with a minimum of 10 Å water buffer around the solute. Ions are added to neutralize the system and mimic physiological conditions (typically, 0.15 M NaCl). Protein topologies are assigned using a variety of force fields optimized for intrinsically disordered proteins (IDPs), such as:

- CHARMM36m [81], for balanced treatment of folded/disordered transitions.
- CHARMM22* [82], which has shown good performance on Aβ monomers.
- AMBER ff03w [83] and ff14SB [84], often paired with dispersion-corrected water models like TIP4P-D [85] and TIP4P/2005 [86].

Water models were carefully selected to match the target force field and to ensure consistency in temperature-pressure coupling and hydrogen bonding dynamics.

2.4 Force Field Considerations for Metal Ions

A critical consideration in MD modeling of metalloproteins is the representation of the metal center. For nonbonded models (i.e., metals treated as point charges with Lennard-Jones interactions), we use parameter sets such as:

- 12-6-4 LJ parameters [87] developed by Li and Merz for divalent ions.
- Cu^{2+} and Zn^{2+} parameters validated in coordination with histidine-rich environments [88, 89].

In bonded models, tools such as MCPB.py [88] (AMBER) or CGenFF [90, 91] (CHARMM) are used to derive bonded interaction parameters for metal–ligand complexes based on quantum mechanical force constants. This approach allowed inclusion of bond/angle terms for His–Cu, Asp–Zn, or Tyr–Cu bridges with more accurate geometry retention. The choice of metal representation strongly influences simulation stability, coordination retention, and ligand-exchange sampling, and must be validated through test simulations or energy decomposition analysis.

2.5 Software Packages and Computational Resources

Simulations are performed using:

- GROMACS package for high-speed production runs on GPU clusters.
- AMBER 22 for systems requiring flexible ion treatment and advanced restraint schemes.
- CP2K or ORCA/ChemShell for hybrid QM/MM calculations.

Trajectory analysis could be performed using Python libraries such as MDAnalysis [92] and MDTraj [93], or packages such as VMD [94], PyMOL [95], and cpptraj [96], with specific workflows for:

- Tracking metal coordination stability.
- Root mean square fluctuations (RMSF) and radius of gyration (Rg).
- Solvent-accessible surface area (SASA).
- H-bond and salt bridge formation.
- Redox active site mapping from spin density profiles.

Long simulations (up to 1 μs per system) are usually run on national HPC resources, with typical hardware including NVIDIA GPUs [79] and ≥128 GB RAM per node for DFT-level jobs [97].

2.6 Impact of Force Field Selection on Amyloid–Metal Simulations

The outcome of MD simulations, especially in metal-coordinated systems, can vary significantly depending on the choice of force field. Force fields tuned for folded globular proteins (e.g., older AMBER99sb [98]) tend to overstabilize secondary structure and fail to capture the flexibility of IDPs like Aβ or α-Syn. CHARMM36m and CHARMM22*, in contrast, have been shown to better represent the disordered ensemble of Aβ42 and the transient helicity of α-synuclein, which is critical for capturing metal binding-induced conformational changes [99].

Moreover, not all force fields treat polarizability equally, and fixed-charge models may struggle to replicate the subtle electronic redistribution induced by transition metals. In such cases, combining DFT-based ESP charges [100] with nonbonded ion models, or using polarizable force fields like Drude [101] or AMOEBA [102], may offer better fidelity, albeit at higher computational cost. Ultimately, careful benchmarking—either against experimental data (e.g., NMR chemical shifts, AFM morphologies) or quantum mechanical energetics—is essential for validating the choice of force field in any given system.

3 Methods

This section presents a step-by-step methodology for modeling amyloid–metal ion interactions using multiscale computational techniques. The approach begins with the construction of metal-coordinated peptide systems based on experimentally informed or DFT-optimized geometries, followed by molecular dynamics (MD) simulations to investigate structural dynamics, aggregation propensity, and surface effects. The methods described are applicable to amyloid-β (Aβ), α -synuclein (α-Syn), and tau peptides in monomeric and oligomeric states.

3.1 System Preparation

Selection of Initial Structures

Initial peptide structures are obtained from the Protein Data Bank (PDB) or AlphaFold2 predictions, as described in Sect. 2.1. When metal-bound structures are unavailable, representative fragments containing putative metal-binding residues (e.g., His13, His14, Asp1 in Aβ; His50, Asp121 in α-Syn) are selected and manually coordinated with Cu^{2+}, Zn^{2+}, or Fe^{3+} ions using Avogadro [103] or PyMOL [95, 104]. Protonation states of titratable residues are usually assigned using PROPKA or H++ servers at physiological pH [105, 106].

DFT Optimization of Coordination Geometries

DFT calculations are not only used to refine metal-binding geometries but also to extract atomic partial charges and to accurately derive force field parameters for subsequent MD simulations. As mentioned in Sect. 2.2, these calculations are usually carried out using Gaussian 16 and ORCA 5.0. Implicit solvent models (PCM or COSMO) are used to approximate aqueous conditions. Electrostatic potential (ESP) charges could be derived using the restrained electrostatic potential [100] (RESP) protocol or CHELPG [107] (CHarges from Electrostatic Potentials using a Grid-based method) fitting to use in MD force fields.

3.2 Classical Molecular Dynamics Simulations

System Assembly

The peptide–metal complexes are embedded in standard cubic/rectangular water boxes using GROMACS or CHARMM-GUI. Solvation is done with TIP3P or TIP4P-D water models, and Na^+ and Cl^- ions are added to neutralize charge and mimic 150 mM ionic strength. Box size ensured at least 10 Å buffer from peptide to box edge.

Force Field Assignment

Peptides were parameterized using CHARMM36m or AMBER ff14SB force fields. For nonbonded metal ion models, 12-6-4 Lennard-Jones parameters are employed as indicated earlier in the section. Bonded metal–peptide models are usually constructed using standard docking methods [108] or via the MCPB.py (AMBER) using DFT-derived metal-ligand geometries and force constants.

Final topologies include bonded terms for coordination with His, Asp, and Glu sidechains.

Equilibration and Production Runs

Energy minimization of the systems is followed by 500 ps of NVT (canonical ensemble) equilibration [109] and 1 ns of NPT (isothermal-isobaric ensemble) equilibration [110] (Berendsen barostat [111]). Production MD is usually run at 310 K for 1 μs per system or longer to necessitate adequate sampling with 2 fs timestep and LINCS [112] constraint on hydrogen bonds. For α-Syn and Aβ oligomers, MD simulations may need to be extended up to 5 μs per replica depending on when and where the system properties converge [113].

3.3 Enhanced Sampling and Surface Systems

Replica Exchange and REST2

To accelerate sampling of disordered states and capture rare metal exchange events, advanced/enhanced sampling methods [114] such as replica exchange molecular dynamics [115] (REMD) and replica exchange with solute scaling [116] (REST2) are employed via PLUMED, a plugin for use with GROMACS.

Surface-Guided Aggregation

To model aggregation at hard bio-nano interfaces such as gold–water interfaces, Au(111) slabs are built using CHARMM-GUI Nanomaterial Modeler [117]. The metal surface is restrained with harmonic potentials while peptides are allowed to adsorb freely. The Synhaivska et al. study serves as a reference for copper-mediated α -Syn aggregation at interfaces [41].

To model aggregation at soft biointerfaces, such as electrified liquid/liquid interfaces, explicit biphasic simulation boxes are constructed that contain aqueous and organic phases separated by a dynamic interface [118, 119]. A weak harmonic potential is applied to the peptide center of mass to gently bias it toward the interface, mimicking the effect of an applied Galvani potential while preserving natural conformational flexibility and interfacial reorientation. The recent Holzinger et al. study illustrates copper-mediated Aβ peptide modulation at the electrified water/DCE interface [120].

3.4 Trajectory Analysis

Structural Metrics

MDAnalysis, Gromacs Tools and cpptraj are excellent programs to extract root mean square deviations (RMSD), radius of gyration (Rg), secondary structure content (DSSP [121]) and solvent accessible surface area (SASA)—a few of the many different structural metrics [18] that could be used to assess the shapes, sizes, and conformational changes due to metal binding. Metal coordination numbers are tracked via distance cutoffs and angular criteria.

Energetic and Electronic Properties

Energetic characterization of metal–amyloid interactions can be performed at multiple levels, depending on the desired resolution and computational cost. For structurally sampled states obtained from MD trajectories, binding free energies are most commonly estimated using end-point post-processing approaches such as MM/PBSA (Molecular Mechanics/Poisson–Boltzmann Surface Area) and MM/GBSA (Generalized Born Surface Area) [122, 123]. These methods decompose the binding free energy into molecular mechanics terms (electrostatics and van der Waals interactions) and implicit solvation contributions, enabling per-residue energy decomposition and mechanistic interpretation of electrostatic versus hydrophobic driving forces.

For systems simulated with CHARMM-based force fields, implicit solvent models such as GBMV [124] (Generalized Born using Molecular Volume) can be used to compute single-point conformational energies from representative MD snapshots or ensemble-averaged structures. In aggregation studies, energies are often normalized per monomer to enable size-independent comparisons across oligomeric states. These approaches provide a thermodynamic ranking of metal-bound versus metal-free assemblies while retaining structural context from MD sampling.

For more rigorous free-energy estimates, particularly when comparing metal-coordination states (e.g., Cu^{2+} vs Cu^{+}, or different coordination motifs), alchemical free-energy methods [125] such as thermodynamic integration (TI), free-energy perturbation (FEP), or λ-dynamics may be employed. These techniques allow transformation between defined chemical states within the same simulation framework and can capture subtle coordination-dependent stability differences that are not accessible via endpoint methods.

Electronic properties can be evaluated from representative MD-derived structures using QM or QM/MM calculations. Predictive modeling of metal coordination environments, although demonstrated in molecular electronic systems [126], provides transferable principles for tuning metal–peptide binding affinity and redox reactivity. These include metal–ligand charge-transfer, spin densities, and coordination-geometry validation. While frontier orbital energies (HOMO–LUMO) may provide qualitative insight into redox reactivity in isolated snapshots, they are not direct measures of binding affinity and must be interpreted cautiously. Redox potential shifts (e.g., Cu^{2+}/Cu^{+}) can instead be estimated using thermodynamic cycles (Born–Haber-type approaches [127]) or continuum-solvation-corrected QM calculations [128] on MD-derived coordination geometries.

Together, this hierarchy of energetic analyses from MM/PBSA and GBMV ensemble post-processing to alchemical free-energy calculations and QM-based electronic characterization provides a comprehensive framework for dissecting the thermodynamic and redox determinants of metal–amyloid interactions. Additional practical considerations, common pitfalls, and troubleshooting recommendations are provided in (*see* **Note 1-10**).

4 Notes

1. **Choosing Metal Ions**: Different diseases are associated with different metal profiles. Cu^{2+} and Zn^{2+} dominate in Alzheimer's disease plaques, while Fe^{3+} is enriched in Parkinsonian Lewy bodies. Simulations should reflect biologically relevant metal concentrations and ratios.
2. **Titration States of Histidines**: Use pKa prediction tools like PROPKA [105] to decide on the protonation of His residues (especially δ vs ε tautomeric states), as metal binding often depends on specific tautomeric forms.
3. **DFT Model Size**: Fragment selection is crucial. Use only the relevant binding pocket (e.g., 3–6 residues) for DFT to reduce computational cost, and validate geometry using both PCM and explicit solvent snapshots.
4. **Force Field Compatibility**: Ensure metal-ion parameters match the water model and force field. For example, the 12-6-4 Cu^{2+} model is optimized for TIP3P; for TIP4P-D, parameters may need to be adjusted.
5. **Simulation Box Size**: For systems containing multiple metal-bound peptides, increase the water box size or use semi-isotropic pressure coupling to prevent artefactual aggregation.
6. **Metal Exchange Events**: These are rare and require enhanced sampling (e.g., metadynamics, REST2) or targeted biasing (e.g., distance restraints between competing ligands and metals).
7. **Surface Simulations**: Be mindful of charge neutrality when modeling metal surfaces and their interaction with peptides. Use counterions or restrain dipoles at the interface.
8. **Interpreting Spin Densities**: Redox-sensitive systems (e.g., Cu^{2+}/Cu^{+}) can show delocalized unpaired spin density. Visualize these using isosurfaces and correlate them with experimental EPR spectra, if available.

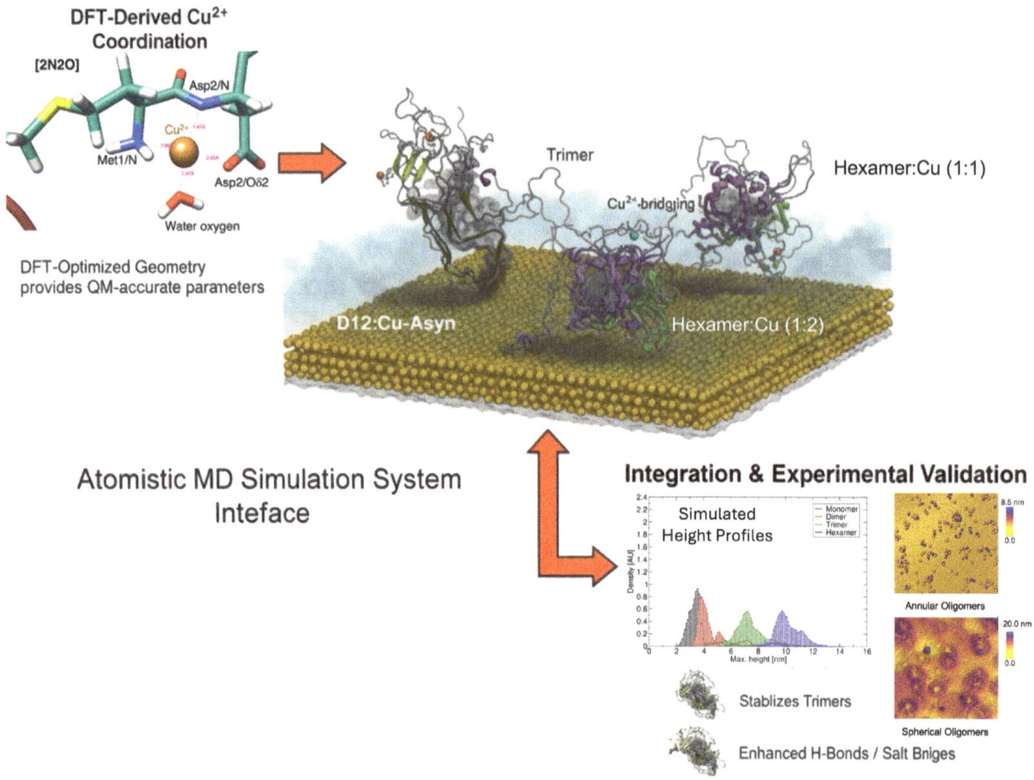

Fig. 5 Multiscale modeling workflow for Cu^{2+}-mediated α-synuclein aggregation at the Au(111)–water interface. Density functional theory (DFT) calculations were used to determine the high-affinity Cu^{2+} coordination geometry at the α-syn N-terminus (Met1, Asp2, and a coordinated water molecule), providing quantum-mechanically derived structural parameters for atomistic simulations. These DFT-derived models were embedded into explicit-solvent molecular dynamics (MD) simulations of α-syn monomers and oligomers interacting with an Au(111) surface. The simulations reveal Cu^{2+}-dependent stabilization of specific oligomeric states, including trimers and hexamers, and Cu^{2+}-bridging interactions that promote compact surface-associated assemblies. Simulated height distributions and interfacial conformations were directly compared with liquid-phase AFM measurements, showing agreement in oligomer dimensions and morphologies (annular and spherical species). This integrative DFT–MD–AFM framework provides mechanistic insight into metal-ion-mediated modulation of α-syn aggregation at biologically relevant interfaces

9. **Reproducibility**: Always archive DFT-optimized coordinates, parameter files, and key MD input scripts for reproducibility and transparency.
10. **Validation**: Cross-reference simulation results with available NMR, AFM, or MS data whenever possible. Experimental benchmarks anchor credibility for modeling predictions.

5 Case Study 1: Multiscale Modeling of Cu^{2+}-Mediated α-Synuclein Aggregation at a Gold–Water Interface

To illustrate the integration of quantum chemical calculations with atomistic molecular dynamics (MD), I highlight our multiscale investigation of Cu^{2+}-mediated aggregation of α-synuclein (α-Syn) at a gold–water interface work [41]. This computational study was performed in direct connection with liquid-phase AFM experiments, enabling a quantitative comparison between simulated structural metrics and experimentally measured height distributions (Fig. 5).

5.1 Experimental Motivation and Structural Context

Liquid AFM experiments revealed that Cu^{2+} modulates α-Syn aggregation behavior on Au(111), promoting compact oligomeric assemblies with altered surface morphology. These observations raised mechanistic questions:

1. How does Cu^{2+} coordination influence protofibrillar stability?
2. Does copper stabilize specific oligomer sizes?
3. How does surface confinement affect metal-mediated assembly?

To address these questions, a structurally consistent protofibrillar model of α-Syn in the Greek-key fold was used as the starting point for computational analysis.

5.2 Multiscale System Construction: DFT–MD Integration

DFT-Derived Cu^{2+} Coordination Model

The high-affinity Cu^{2+}-binding site at the N-terminus of α-Syn was first characterized using density functional theory (DFT). The optimized coordination geometry included:

(1) NH_2 terminal nitrogen (Met1).

(2) Deprotonated amide nitrogen of Asp2.

(3) Carboxylate oxygen of Asp2.

(4) Oxygen from a coordinated water molecule.

This 2N2O coordination motif was used to constrain the Cu^{2+}–ligand distances and angles in subsequent MD simulations. Partial charges for Cu^{2+}-coordinated constructs were derived using the RESP protocol based on quantum mechanical electrostatic potential calculations.

This step ensured that the Cu^{2+} coordination sphere was grounded in quantum mechanical accuracy before being propagated into classical MD simulations.

Atomistic MD System Setup

The atomistic simulations were constructed using:

(1) Greek-key protofibrillar models of α-Syn monomer, dimer, trimer, and hexamer.
(2) An explicit Au(111) slab (15×15 nm, two-layer thickness ~0.3 nm).
(3) Periodic boundary conditions.
(4) Explicit water (CHARMM-modified TIP3P).
(5) Neutralizing ions.

The fibril axis was initially oriented parallel to the Au(111) surface, with the NAC region positioned proximal to the interface to create a defined starting adsorption geometry.

Cu^{2+}-coordinated systems were built with controlled Cu^{2+}:monomer ratios of 1:1 and 1:2 to investigate concentration-dependent effects.

Gold atoms were fixed in crystallographic positions to maintain a stable surface representation.

Production simulations were carried out in the NPT ensemble using GROMACS with:

(1) 2 fs timestep.
(2) PME electrostatics.
(3) LINCS and SETTLE constraints.
(4) Parrinello-Rahman barostat.
(5) 200 ns production trajectories.

5.3 Structural and Energetic Analyses

A comprehensive structural analysis framework was applied:

(i) **Adsorption and Height Profiles**
Maximum height above Au(111) was computed and directly compared with AFM height distributions. The simulated monomer heights (~3–4 nm) showed quantitative agreement with experimental measurements.

(ii) **Shape Metrics (Asphericity)**
Protein shape was quantified via the asphericity parameter (Δ), derived from the principal moments of gyration. Oligomers generally exhibited spherical behavior ($\Delta \approx 0$), whereas monomers adopted slightly rod-like conformations.

(iii) **Salt-Bridge Stability**
The E46–K80 salt bridge was tracked as a key stabilizer of the Greek-key fold. Trimeric assemblies displayed the most persistent salt-bridge stability across the simulation timescale.

(iv) **Hydrogen-Bonding Networks**
Intermolecular and intramolecular hydrogen bond counts revealed enhanced stabilization in Cu^{2+}-coordinated oligomers, particularly for trimers and hexamers.

(v) **Conformational Energetics**
Implicit solvent GBMV-II calculations were used to estimate conformational energies of sampled structures, normalized per monomer for size-independent comparison. Cu^{2+} coordination altered thermodynamic stability trends across oligomer sizes, identifying trimer assemblies as particularly stable under certain Cu^{2+} concentrations.

5.4 Mechanistic Insights

The combined DFT–MD framework revealed that:

(1) Cu^{2+} stabilizes specific oligomer sizes *via* coordination-mediated inter-domain interactions.
(2) Surface confinement promotes CT-domain anchoring on Au(111), facilitating lateral protofibrillar growth.
(3) Cu^{2+} concentration modulates β-sheet retention and oligomer compactness.
(4) Trimers emerge as a minimal stable oligomeric nucleus under copper-rich conditions.

Importantly, simulated height distributions and morphology trends were consistent with AFM measurements, validating the mechanistic conclusions.

5.5 Broader Implications

This case study demonstrates how:

(1) DFT-derived coordination chemistry can be seamlessly integrated into atomistic MD.
(2) Surface models allow direct comparison with AFM experiments.
(3) Metal ion coordination modulates both thermodynamic stability and morphological evolution of amyloid assemblies.

Such multiscale workflows provide a mechanistic bridge between nanoscale experiments and atomic-level interpretation, particularly for metal-mediated aggregation phenomena in neurodegenerative disease contexts. Future extensions may include explicit redox cycling (Cu^{2+}/Cu^{+}), reactive oxygen species formation and hybrid QM/MM approaches for dynamically evolving coordination environments.

6 Case Study 2: Cu^{2+}-Modulated Adsorption of Aβ Peptides at a Polarized Liquid/Liquid Interface

To further illustrate the integration of electrochemical measurements with atomistic molecular dynamics (MD), we recently investigated Cu^{2+}-modulated adsorption of amyloid-β (Aβ) peptides at a polarized water/1,2-dichloroethane (DCE) interface. This study combined cyclic voltammetry (CV), electrochemical impedance spectroscopy

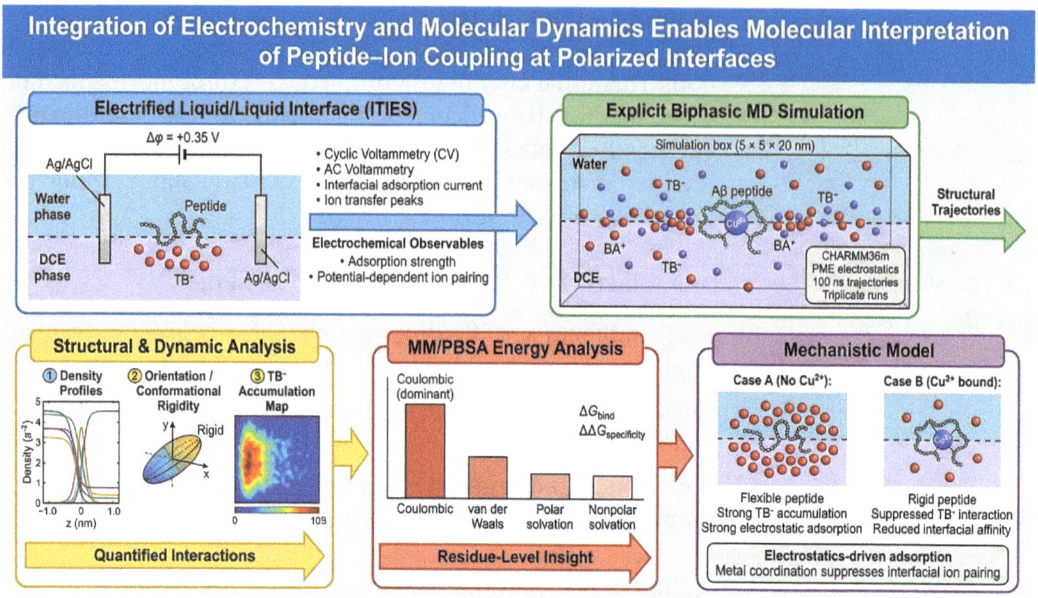

Fig. 6 Integrative electrochemistry–MD workflow for Cu(II)-modulated Aβ adsorption at polarized liquid/liquid interfaces. Conceptual multiscale workflow integrating electrochemical measurements at an electrified liquid/liquid interface (ITIES) with explicit biphasic molecular dynamics (MD) simulations to resolve Cu(II)-dependent modulation of amyloid-β (Aβ) peptide adsorption. At the experimental level (left), cyclic voltammetry (CV) and AC voltammetry quantify adsorption currents and potential-dependent ion pairing at $\Delta\varphi = +0.35\,V$ across the water/1,2-dichloroethane (DCE) interface. These observables are interpreted using atomistic biphasic MD simulations (top right) employing explicit water/DCE phases with CHARMM36m force field, PME electrostatics, and triplicate 100 ns trajectories. Structural trajectories enable analysis of (i) number-density profiles, (ii) peptide orientation and conformational rigidity, and (iii) accumulation of hydrophobic TB^- ions at the interface. MM/PBSA energy decomposition further quantifies Coulombic, van der Waals, and solvation contributions to binding free energies (ΔG_{bind}), revealing electrostatics as the dominant driving force. The resulting mechanistic model (right) demonstrates that in the absence of Cu^{2+}, Aβ peptides remain flexible and promote strong TB^- accumulation at the interface, leading to enhanced electrostatic adsorption. Upon Cu^{2+} coordination, peptide rigidity increases, TB^- association is suppressed, and overall interfacial affinity decreases. Together, this integrative framework provides a molecular-level interpretation of electrochemical signals in terms of peptide–ion coupling and metal-modulated adsorption at polarized soft biointerfaces

(EIS), and atomistic MD simulations to resolve how metal coordination alters peptide–ion interactions at an electrified soft interface [120]. Unlike the gold–water system in Case Study 1, this work interrogates peptide behavior at a liquid/liquid interface (ITIES) under applied Galvani potential, providing a model for polarized biological membranes (Fig. 6).

6.1 Experimental Motivation and Electrochemical Context

Protein adsorption at soft interfaces is governed by a combination of electrostatics, ion pairing, and conformational flexibility. At the ITIES, these processes can be tuned using an externally applied potential. Electrochemical experiments revealed that:

(1) $A\beta_{4-16}$ exhibits stronger adsorption at positive potential compared to $A\beta_{1-16}$.
(2) Interfacial accumulation of hydrophobic organic anions (TB^-) accompanies peptide adsorption.
(3) Cu^{2+} coordination significantly alters adsorption behavior.

This raised key mechanistic questions:

1. What molecular determinants govern peptide–ion coaccumulation at the interface?
2. How does Cu^{2+} coordination modify interfacial affinity?
3. Are electrostatic or hydrophobic interactions dominant under polarization?
4. Can MD simulations rationalize CV-derived adsorption signatures?

6.2 Multiscale System Construction: Electrochemistry–MD Integration

Electrochemical Framework

Experiments were conducted at a four-electrode ITIES system comprising:

(1) Aqueous phase: 10 mM LiCl (pH 5.5).
(2) Organic phase: DCE with tetrakis(pentafluorophenyl)borate (TB^-).
(3) Applied Galvani potential: +0.35 V.

Cyclic voltammetry and AC voltammetry resolved adsorption and ion-transfer processes.

Short N-terminal Aβ fragments ($A\beta_{1-16}$ and $A\beta_{4-16}$) were selected because:

(1) They contain the Cu^{2+} coordination site.
(2) They isolate metal-binding behavior from aggregation complexity.

Cu^{2+} complexes were formed in situ (1:0.9 Aβ:Cu ratio).

Atomistic MD Model Construction

To reproduce the electrified interface, a biphasic simulation box (5 × 5 × 20 nm) was constructed:

(1) Water phase (10 nm).
(2) DCE phase (10 nm).
(3) Explicit Li^+, Cl^- (aqueous).
(4) BA^+ and TB^- (organic).

The applied potential was mimicked by scaling interfacial ion concentrations to reproduce charge accumulation corresponding to +0.35 V.

Simulation details:

(1) CHARMM36m force field.
(2) 2 fs timestep.
(3) PME electrostatics.
(4) 100 ns production runs (triplicate).
(5) NPT ensemble.

Cu^{2+}-bound systems used RESP-derived partial charges from DFT calculations to preserve accurate coordination geometry.

6.3 Structural and Energetic Analyses

A multi-layer analysis pipeline was applied.

(i) **Interfacial Localization and Density Profiles**
Number density profiles along the z-axis revealed:

(1) $A\beta_{4-16}$ accumulates strongly at the interface.
(2) TB^- co-accumulates in the same interfacial region.
(3) BA^+ enrichment is secondary and electrostatically driven.
(4) $A\beta_{1-16}$ exhibited weaker and less stable interfacial localization.

This directly mirrors the stronger CV adsorption signals for $A\beta_{4-16}$.

(ii) **Binding Free Energy Decomposition (MM/PBSA)**
Binding free energies between peptide and TB^- were decomposed into:

(1) Coulombic electrostatics.
(2) van der Waals interactions.
(3) Polar solvation.
(4) Nonpolar solvation.

Key findings:

(1) Electrostatics dominate stabilization.
(2) $A\beta_{4-16}$ exhibits significantly more favorable ΔG_bind.
(3) MM/PBSA-derived $\Delta\Delta G_{specificity} \approx -1100\ kJ\ mol^{-1}$ confirms preferential TB^- binding.

(iii) **Residue-Level Contributions**
Per-residue free-energy decomposition showed that favorable residues include Phe4, Arg5, His6, His13, His14, and Lys16, while unfavorable residues include Asp7 and Glu11.

The absence of Glu3 in $A\beta_{4-16}$ removes an electrostatic penalty present in $A\beta_{1-16}$, explaining enhanced adsorption.

This provides a molecular explanation for differential CV behavior.

(iv) **Effect of Cu^{2+} Coordination**

Upon Cu^{2+} binding:

(1) Backbone flexibility decreases.

(2) TB^- accumulation at the interface is suppressed.

(3) Electrostatic stabilization is reduced.

(4) $\Delta\Delta G_{specificity} \approx -1250\,kJ\,mol^{-1}$ indicates a drastic reduction in affinity.

Mechanistically, Cu^{2+} chelation restricts His-mediated interactions and suppresses TB^- binding.

Electrochemical experiments independently corroborated reduced adsorption upon metal coordination.

(v) **Conformational Rigidity and Orientation**

Principal subspace analysis (block covariance overlap metric) demonstrated:

(1) $A\beta_{4-16}$ exhibits stable essential dynamics.

(2) Cu^{2+} coordination induces rigidification.

(3) Reduced reorientation at the interface corresponds to suppressed electrochemical response.

6.4 Mechanistic Insights

The integrated electrochemistry–MD workflow revealed that:

(1) Interfacial adsorption is electrostatically driven rather than hydrophobic.

(2) TB^- coaccumulation stabilizes peptide adsorption.

(3) $A\beta_{4-16}$ has intrinsically higher interfacial affinity due to the removal of acidic repulsion.

(4) Cu^{2+} coordination suppresses adsorption by rigidifying the N-terminal region and blocking favorable ion interactions.

(5) Electrochemical signals directly reflect molecular-scale peptide–ion coupling.

This establishes a direct mechanistic bridge between CV measurements and atomistic simulations.

6.5 Broader Implications

This case study demonstrates that:

(1) Electrified soft interfaces can model biologically relevant membrane polarization.

(2) CV and MD integration enables quantitative interpretation of adsorption energetics.

(3) Metal coordination modulates peptide behavior not only via aggregation kinetics but also through altered interfacial electrostatics.

(4) MM/PBSA decomposition provides molecular interpretation of electrochemical observables.

Together with Case Study 1, this work highlights how multiscale modeling frameworks can be extended from solid surfaces (Au) to soft, polarized interfaces (ITIES), offering complementary mechanistic perspectives on metal-mediated amyloid behavior. Future extensions could include:

(1) Explicit Cu^{2+}/Cu^{+} redox cycling.

(2) Constant-potential MD simulations.

(3) Reactive QM/MM models for dynamic coordination exchange.

(4) Inclusion of lipid bilayer mimetics to bridge ITIES and membrane systems.

7 Conclusions

Metal ion dysregulation is a hallmark of several neurodegenerative diseases, including Alzheimer's, Parkinson's, and tauopathies. Despite increasing structural and biophysical evidence of metal–amyloid interactions, the atomic-level mechanisms underlying aggregation, oligomerization, and redox-driven toxicity remain elusive. Computational modeling provides a uniquely powerful framework to bridge this knowledge gap [23, 55, 120]. Notably, a major limitation in the current structural biology of amyloid–metal systems is the absence of experimentally resolved metal-bound structures of full-length tau and α-synuclein in the Protein Data Bank (PDB). Unlike amyloid-β, which has several partial structures bound to Cu^{2+} and Zn^{2+}, tau and α-synuclein are significantly longer intrinsically disordered proteins (IDPs) with highly dynamic and flexible metal-binding modes that evade crystallographic or high-resolution cryo-EM capture. The available fibrillar forms are often metal-depleted due to sample preparation, and monomeric metal binding is transient and heterogeneous. This structural gap highlights the essential role of computational modeling in (a) Predicting plausible coordination geometries and (b) simulating conformational ensembles under metal-rich conditions.

In this chapter, I present a comprehensive overview of multiscale approaches for modeling amyloid–metal systems that leverage the strengths of both quantum mechanical (DFT) and classical (MD) techniques, and are reasonably transferable to metal-biomolecular systems, such as Cu-complex binding modes to different DNA morphologies [129]. I outline here procedures for constructing reliable coordination models, optimizing metal–ligand geometries, and performing long-timescale simulations to capture dynamic behavior under physiological conditions. Enhancements such as REST2 and surface simulations extend the applicability to disordered and surface-mediated aggregation.

The inclusion of two detailed case studies—one on Cu^{2+}-mediated α-synuclein aggregation at gold–water interfaces, and the other on Cu(II)-modulated amyloid-β adsorption at polarized liquid/liquid interfaces—demonstrates the power of multiscale computational workflows to bridge coordination chemistry, molecular dynamics, and experimental electrochemistry. In the first study, DFT-derived coordination geometries integrated into atomistic MD simulations revealed how copper stabilizes specific oligomeric states, modulates salt-bridge networks, and influences surface-confined protofibrillar morphology in quantitative agreement with AFM measurements [55]. In the second study, explicit biphasic MD simulations combined with cyclic voltammetry showed how Cu(II) coordination alters peptide conformational flexibility, suppresses electrostatically driven ion pairing at electrified interfaces, and reshapes interfacial binding thermodynamics as quantified by MM/PBSA analysis [120]. Together, these studies illustrate how quantum-informed parameterization, ensemble-based energetic evaluation, and direct comparison with nanoscale experiments enable mechanistic interpretation of metal-mediated amyloid behavior across both hard and soft biointerfaces. Such integrative strategies provide a robust framework for rationalizing experimental observations and uncovering previously inaccessible insights into the molecular determinants of metal-associated amyloid pathogenesis. As tools and force fields continue to improve, future studies will benefit from enhanced polarization models, explicit redox state transitions, hybrid QM/MM sampling, and integration with generative AI-driven conformational screening [130].

Looking forward, computational methods are poised to decipher more complex, multivalent metal-binding networks that underlie toxic amyloid polymorphs, particularly in heterogeneous environments such as membranes, organelles, and metal-rich plaque cores. This could aid in:

1. Predicting **novel coordination geometries** and their energetic stability.
2. Identifying **metal-dependent conformers** with distinct toxicity profiles.
3. Probing **transition states** involved in redox switching and metal exchange.

4. Developing **structure-based design rules** for chelators and modulators that selectively disrupt pathological coordination without affecting physiological metal balance.

Ultimately, the computational modeling of amyloid–metal systems not only complements experimental investigations but also guides hypothesis generation, toxicology predictions, and therapeutic design. I hope this chapter serves as a practical guide and source of inspiration for researchers exploring computational metallomics in neurodegeneration.

Acknowledgments

The author wishes to acknowledge the Irish Center for High-End Computing (ICHEC) for the provision of computational facilities and technical support that enabled several of the molecular dynamics and quantum chemical studies discussed in this chapter.

Conflicts of Interest The author has no conflicts of interest to declare that are relevant to the content of this chapter.

References

1. Kepp KP (2012) Bioinorganic chemistry of Alzheimer's disease. Chem Rev 112:5193–5239. https://doi.org/10.1021/cr300009x
2. Faller P (2009) Copper and zinc binding to amyloid-beta: coordination, dynamics, aggregation, reactivity and metal-ion transfer. Chembiochem 10:2837–2845. https://doi.org/10.1002/cbic.200900321
3. Barnham KJ, Masters CL, Bush AI (2004) Neurodegenerative diseases and oxidative stress. Nat Rev Drug Discov 3:205–214. https://doi.org/10.1038/nrd1330
4. Chia S et al (2024) A relationship between the structures and neurotoxic effects of abeta oligomers stabilized by different metal ions. ACS Chem Neurosci 15:1125–1134. https://doi.org/10.1021/acschemneuro.3c00718
5. Martinez Pomier K, Ahmed R, Huang J, Melacini G (2024) Inhibition of toxic metal-alpha synuclein interactions by human serum albumin. Chem Sci 15:3502–3515. https://doi.org/10.1039/d3sc06285f
6. Bergaglio T et al (2025) On levodopa interactions with brain disease amyloidogenic proteins at the nanoscale. ACS Omega 10:14487–14495
7. Nirmalraj PN, Bhattacharya S, Thompson D (2024) Accelerated Alzheimer's Abeta-42 secondary nucleation chronologically visualized on fibril surfaces. Sci Adv 10:eadp5059. https://doi.org/10.1126/sciadv.adp5059
8. Bartels T, Choi JG, Selkoe DJ (2011) Alpha-synuclein occurs physiologically as a helically folded tetramer that resists aggregation. Nature 477:107–110. https://doi.org/10.1038/nature10324
9. Gurry T et al (2013) The dynamic structure of alpha-synuclein multimers. J Am Chem Soc 135:3865–3872. https://doi.org/10.1021/ja310518p
10. Wang W et al (2011) A soluble alpha-synuclein construct forms a dynamic tetramer. Proc Natl Acad Sci USA 108:17797–17802. https://doi.org/10.1073/pnas.1113260108
11. Trexler AJ, Rhoades E (2012) N-Terminal acetylation is critical for forming alpha-helical oligomer of alpha-synuclein. Protein Sci 21:601–605. https://doi.org/10.1002/pro.2056
12. Bhattacharya S, Xu L, Arrue L, Bartels T, Thompson D (2024) Conformational selection of alpha-synuclein tetramers at biological interfaces. J Chem Inf Model 64:8010–8023. https://doi.org/10.1021/acs.jcim.4c01459
13. Xu L, Bhattacharya S, Thompson D (2022) Predictive modeling of neurotoxic alpha-synuclein polymorphs. Methods Mol Biol 2340:379–399. https://doi.org/10.1007/978-1-0716-1546-1_17

14. Xu L, Bhattacharya S, Thompson D (2019) On the ubiquity of helical alpha-synuclein tetramers. Phys Chem Chem Phys 21:12036–12043. https://doi.org/10.1039/c9cp02464f
15. Xu L, Bhattacharya S, Thompson D (2018) Re-designing the alpha-synuclein tetramer. Chem Commun 54:8080–8083. https://doi.org/10.1039/c8cc04054k
16. Bhattacharya S, Xu L, Thompson D (2020) Long-range regulation of partially folded amyloidogenic peptides. Sci Rep 10:7597. https://doi.org/10.1038/s41598-020-64303-x
17. Bhattacharya S, Xu L, Thompson D (2019) Molecular simulations reveal terminal group mediated stabilization of helical conformers in both amyloid-beta42 and alpha-synuclein. ACS Chem Neurosci 10:2830–2842. https://doi.org/10.1021/acschemneuro.9b00053
18. Bhattacharya S, Xu L, Thompson D (2022) Characterization of amyloidogenic peptide aggregability in helical subspace. Methods Mol Biol 2340:401–448. https://doi.org/10.1007/978-1-0716-1546-1_18
19. Banreti A et al (2022) Biological effects of the loss of homochirality in a multicellular organism. Nat Commun 13:7059. https://doi.org/10.1038/s41467-022-34516-x
20. Maraba O, Bhattacharya S, Conda-Sheridan M, Thompson D (2023) Modelling peptide self-assembly within a partially disordered tau filament. Nano Express 3:044004
21. Xu L, Bhattacharya S, Thompson D (2018) The fold preference and thermodynamic stability of alpha-synuclein fibrils is encoded in the non-amyloid-beta component region. Phys Chem Chem Phys 20:4502–4512. https://doi.org/10.1039/c7cp08321a
22. Bhattacharya S, Xu L, Thompson D (2018) Revisiting the earliest signatures of amyloidogenesis: roadmaps emerging from computational modeling and experiment. Wiley Interdiscip Rev Comput Mol Sci 8:e1359. https://doi.org/10.1002/wcms.1359
23. Bhattacharya S, Thompson D (2024) Recent advances in mapping protein self-assembly and aggregation for common proteinopathies. Acta Phys Polonica A 145:S37–S37
24. Nirmalraj PN et al (2020) Complete aggregation pathway of amyloid beta (1-40) and (1-42) resolved on an atomically clean interface. Sci Adv 6:eaaz6014. https://doi.org/10.1126/sciadv.aaz6014
25. Henry L, Bhattacharya S, Bergaglio T, Pinotsi D, Nirmalraj PN (2026) Differentiating Alzheimer's A beta isoforms coaggregated in cerebrospinal fluid via single-particle imaging. ACS Chem Neurosci 17:550–564. https://doi.org/10.1021/acschemneuro.5c00692
26. Atrian-Blasco E et al (2018) Cu and Zn coordination to amyloid peptides: from fascinating chemistry to debated pathological relevance. Coord Chem Rev 375:38–55. https://doi.org/10.1016/j.ccr.2018.04.007
27. Ali-Torres J, Marechal JD, Rodriguez-Santiago L, Sodupe M (2011) Three dimensional models of Cu(2+)-Abeta(1-16) complexes from computational approaches. J Am Chem Soc 133:15008–15014. https://doi.org/10.1021/ja203407v
28. Strodel B (2021) Amyloid aggregation simulations: challenges, advances and perspectives. Curr Opin Struct Biol 67:145–152. https://doi.org/10.1016/j.sbi.2020.10.019
29. Dong J et al (2003) Metal binding and oxidation of amyloid-beta within isolated senile plaque cores: Raman microscopic evidence. Biochemistry 42:2768–2773. https://doi.org/10.1021/bi0272151
30. Miller Y, Ma B, Nussinov R (2010) Zinc ions promote Alzheimer Abeta aggregation via population shift of polymorphic states. Proc Natl Acad Sci USA 107:9490–9495. https://doi.org/10.1073/pnas.0913114107
31. Miller Y, Ma B, Nussinov R (2012) Metal binding sites in amyloid oligomers: complexes and mechanisms. Coordination Chem Rev 256:2245–2252
32. Rezaei-Ghaleh N, Giller K, Becker S, Zweckstetter M (2011) Effect of zinc binding on beta-amyloid structure and dynamics: implications for Abeta aggregation. Biophys J 101:1202–1211. https://doi.org/10.1016/j.bpj.2011.06.062
33. Cheignon C, Faller P, Testemale D, Hureau C, Collin F (2016) Metal-catalyzed oxidation of Abeta and the resulting reorganization of Cu binding sites promote ROS production. Metallomics 8:1081–1089. https://doi.org/10.1039/c6mt00150e
34. Roy RG, Mandal PK, Maroon JC (2023) Oxidative stress occurs prior to amyloid abeta plaque formation and tau phosphorylation in Alzheimer's disease: role of glutathione and metal ions. ACS Chem Neurosci 14:2944–2954. https://doi.org/10.1021/acschemneuro.3c00486
35. Majernikova N et al (2024) The link between amyloid beta and ferroptosis pathway in Alzheimer's disease progression. Cell Death Dis 15: 782. https://doi.org/10.1038/s41419-024-07152-0
36. Marino T, Russo N, Toscano M, Pavelka M (2010) On the metal ion (Zn(2+), Cu(2+)) coordination with beta-amyloid peptide: DFT computational study. Interdiscip Sci 2:57–69. https://doi.org/10.1007/s12539-010-0086-x

37. Strodel B, Coskuner-Weber O (2019) Transition metal ion interactions with disordered amyloid-beta peptides in the pathogenesis of Alzheimer's disease: insights from computational chemistry studies. J Chem Inf Model 59:1782–1805. https://doi.org/10.1021/acs.jcim.8b00983
38. Bai J, Zhang Z, Liu M, Li C (2015) Alpha-synuclein-lanthanide metal ions interaction: binding sites, conformation and fibrillation. BMC Biophys 9:1. https://doi.org/10.1186/s13628-016-0026-1
39. Graves NJ, Jaafar AK, Gambin Y, Sierecki E (2025) Divalent and trivalent metallic ions differentially affect alpha-synuclein aggregation. ACS Chem Neurosci 16:3497–3512. https://doi.org/10.1021/acschemneuro.5c00160
40. Abrao-Nemeir I, Charles-Achille S, Cayrol B, Balme S (2025) Investigating the impact of Cu(2+) on alpha-synuclein aggregation: a single-molecule approach. Int J Biol Macromol 307:142161. https://doi.org/10.1016/j.ijbiomac.2025.142161
41. Synhaivska O, Bhattacharya S, Campioni S, Thompson D, Nirmalraj PN (2022) Single-particle resolution of copper-associated annular alpha-synuclein oligomers reveals potential therapeutic targets of neurodegeneration. ACS Chem Neurosci 13:1410–1421. https://doi.org/10.1021/acschemneuro.2c00021
42. Byrd EJ et al (2025) Zn(2+) Binding shifts the conformational ensemble of alpha-synuclein monomers toward accelerated amyloid formation. J Am Chem Soc 147:36464–36477. https://doi.org/10.1021/jacs.5c11056
43. Bolen ML, Menees KB, Dupreez AC, Tansey MG (2025) Iron mishandling in the brain and periphery in Parkinson's disease. NPJ Parkinsons Dis 11:246. https://doi.org/10.1038/s41531-025-01089-7
44. Ahmadi S et al (2019) Aggregation of microtubule binding repeats of tau protein is promoted by Cu(2). ACS Omega 4:5356–5366. https://doi.org/10.1021/acsomega.8b03595
45. Li X et al (2023) Zinc exacerbates tau-induced Alzheimer-like pathology in C57BL/6J mice. Int J Biol Macromol 242:124652. https://doi.org/10.1016/j.ijbiomac.2023.124652
46. Juarez-Romero GU et al (2026) Tau is a metal binding protein: structural insights into copper and zinc coordination sites and their impact on amyloid aggregation. Chem Sci 17(14):7277-96. https://doi.org/10.1039/d5sc08604c
47. Maina MB et al (2022) Dityrosine cross-links are present in Alzheimer's disease-derived tau oligomers and paired helical filaments (PHF) which promotes the stability of the PHF-core tau (297-391) in vitro. J Mol Biol 434:167785. https://doi.org/10.1016/j.jmb.2022.167785
48. Hatters DM (2021) Flipping the switch: how cysteine oxidation directs tau amyloid conformations. J Biol Chem 297:101309. https://doi.org/10.1016/j.jbc.2021.101309
49. Yamamoto A et al (2002) Iron (III) induces aggregation of hyperphosphorylated tau and its reduction to iron (II) reverses the aggregation: implications in the formation of neurofibrillary tangles of Alzheimer's disease. J Neurochem 82:1137–1147. https://doi.org/10.1046/j.1471-4159.2002.t01-1-01061.x
50. Roldan-Martin L, Sodupe M, Marechal JD (2024) Computational study of amyloid-beta(42) familial mutations and metal interaction: impact on monomers and aggregates dynamical behaviors. Inorg Chem 63:4725–4737. https://doi.org/10.1021/acs.inorgchem.3c04555
51. Rose F, Hodak M, Bernholc J (2011) Mechanism of copper(II)-induced misfolding of Parkinson's disease protein. Sci Rep 1:11. https://doi.org/10.1038/srep00011
52. Binolfi A et al (2010) Bioinorganic chemistry of Parkinson's disease: structural determinants for the copper-mediated amyloid formation of alpha-synuclein. Inorg Chem 49:10668–10679. https://doi.org/10.1021/ic1016752
53. Kállay C, Várnagy K, Micera G, Sanna D, Sóvágó I (2005) Copper (II) complexes of oligopeptides containing aspartyl and glutamyl residues. Potentiometric and spectroscopic studies. J Inorganic Biochem 99:1514–1525
54. Jing J et al (2021) Copper (Cu(2+)) ion-induced misfolding of tau protein R3 peptide revealed by enhanced molecular dynamics simulation. Phys Chem Chem Phys 23:11717–11726. https://doi.org/10.1039/d0cp05744d
55. Synhaivska O, Bhattacharya S, Campioni S, Thompson D, Nirmalraj PN (2022) Single-particle resolution of copper-associated annular α-synuclein oligomers reveals potential therapeutic targets of neurodegeneration. ACS Chem Neurosci 13:1410–1421
56. Bergaglio T, Bhattacharya S, Thompson D, Nirmalraj PN (2023) Label-free digital holotomography reveals ibuprofen-induced morphological changes to red blood cells. ACS Nanosci Au. https://doi.org/10.1021/acsnanoscienceau.3c00004
57. Jumper J et al (2021) Highly accurate protein structure prediction with AlphaFold. Nature 596:583–589. https://doi.org/10.1038/s41586-021-03819-2
58. Crescenzi O et al (2002) Solution structure of the Alzheimer amyloid beta-peptide (1-42)

in an apolar microenvironment. Similarity with a virus fusion domain. Eur J Biochem 269:5642–5648
59. Tomaselli S et al (2006) The alpha-to-beta conformational transition of Alzheimer's Abeta-(1-42) peptide in aqueous media is reversible: a step by step conformational analysis suggests the location of beta conformation seeding. Chembiochem 7:257–267. https://doi.org/10.1002/cbic.200500223
60. Ulmer TS, Bax A, Cole NB, Nussbaum RL (2005) Structure and dynamics of micelle-bound human alpha-synuclein. J Biol Chem 280:9595–9603. https://doi.org/10.1074/jbc.M411805200
61. Rao JN, Jao CC, Hegde BG, Langen R, Ulmer TS (2010) A combinatorial NMR and EPR approach for evaluating the structural ensemble of partially folded proteins. J Am Chem Soc 132:8657–8668. https://doi.org/10.1021/ja100646t
62. Arakhamia T et al (2021) Posttranslational modifications mediate the structural diversity of tauopathy strains. Cell 184:6207–6210. https://doi.org/10.1016/j.cell.2021.11.029
63. Zirah S et al (2006) Structural changes of region 1-16 of the Alzheimer disease amyloid beta-peptide upon zinc binding and in vitro aging. J Biol Chem 281:2151–2161. https://doi.org/10.1074/jbc.M504454200
64. La Penna G, Hureau C, Andreussi O, Faller P (2013) Identifying, by first-principles simulations, Cu[amyloid-beta] species making Fenton-type reactions in Alzheimer's disease. J Phys Chem B 117:16455–16467. https://doi.org/10.1021/jp410046w
65. Kohn W, Becke AD, Parr RG (1996) Density functional theory of electronic structure. J Phys Chem 100:12974–12980
66. Gaussian 16 (2016) Gaussian 16 Rev. C.01 v. C.01. Wallingford, CT. https://gaussian.com/
67. Neese F (2012) The ORCA program system. Wiley Interdiscip Rev Comput Mol Sci 2:73–78
68. Chung LW et al (2015) The ONIOM method and its applications. Chem Rev 115:5678–5796. https://doi.org/10.1021/cr5004419
69. Tirado-Rives J, Jorgensen WL (2008) Performance of B3LYP density functional methods for a large set of organic molecules. J Chem Theory Comput 4:297–306. https://doi.org/10.1021/ct700248k
70. Adamo C, Cossi M, Barone V (1999) An accurate density functional method for the study of magnetic properties: the PBE0 model. J Mol Struct Theochem 493:145–157
71. Zhao Y, Truhlar DG (2005) Design of density functionals that are broadly accurate for thermochemistry, thermochemical kinetics, and nonbonded interactions. J Phys Chem A 109:5656–5667. https://doi.org/10.1021/jp050536c
72. Check CE et al (2001) Addition of polarization and diffuse functions to the LANL2DZ basis set for p-block elements. J Phys Chem A 105:8111–8116
73. Tomasi J, Mennucci B, Cammi R (2005) Quantum mechanical continuum solvation models. Chem Rev 105:2999–3093. https://doi.org/10.1021/cr9904009
74. Klamt A, Schüürmann G (1993) COSMO: a new approach to dielectric screening in solvents with explicit expressions for the screening energy and its gradient. J Chem Soc Perkin Trans 2:799–805
75. Ali-Torres J, Mirats A, Marechal JD, Rodriguez-Santiago L, Sodupe M (2014) 3D structures and redox potentials of Cu2+ -Abeta(1-16) complexes at different pH: a computational study. J Phys Chem B 118:4840–4850. https://doi.org/10.1021/jp5019718
76. Lu Y et al (2019) Open-source, Python-based redevelopment of the ChemShell multiscale QM/MM environment. J Chem Theory Comput 15:1317–1328. https://doi.org/10.1021/acs.jctc.8b01036
77. Kuhne TD et al (2020) CP2K: an electronic structure and molecular dynamics software package-quickstep: efficient and accurate electronic structure calculations. J Chem Phys 152:194103. https://doi.org/10.1063/5.0007045
78. Jo S, Kim T, Iyer VG, Im W (2008) CHARMM-GUI: a web-based graphical user interface for CHARMM. J Comput Chem 29:1859–1865. https://doi.org/10.1002/jcc.20945
79. Case DA et al (2025) Recent developments in amber biomolecular simulations. J Chem Inf Model 65:7835–7843. https://doi.org/10.1021/acs.jcim.5c01063
80. Van Der Spoel D et al (2005) GROMACS: fast, flexible, and free. J Comput Chem 26:1701–1718. https://doi.org/10.1002/jcc.20291
81. Huang J et al (2017) CHARMM36m: an improved force field for folded and intrinsically disordered proteins. Nat Methods 14:71–73. https://doi.org/10.1038/nmeth.4067
82. Piana S, Lindorff-Larsen K, Shaw DE (2011) How robust are protein folding simulations with respect to force field parameterization. Biophys J 100:L47–49. https://doi.org/10.1016/j.bpj.2011.03.051
83. Best RB, Mittal J (2010) Protein simulations with an optimized water model: cooperative helix formation and temperature-induced unfolded state collapse. J Phys

Chem B 114:14916–14923. https://doi.org/10.1021/jp108618d
84. Maier JA et al (2015) ff14SB: improving the accuracy of protein side chain and backbone parameters from ff99SB. J Chem Theory Comput 11:3696–3713. https://doi.org/10.1021/acs.jctc.5b00255
85. Piana S, Donchev AG, Robustelli P, Shaw DE (2015) Water dispersion interactions strongly influence simulated structural properties of disordered protein states. J Phys Chem B 119:5113–5123. https://doi.org/10.1021/jp508971m
86. Abascal JL, Vega C (2005) A general purpose model for the condensed phases of water: TIP4P/2005. J Chem Phys 123:234505. https://doi.org/10.1063/1.2121687
87. Li P, Song LF, Merz KM, Jr (2015) Parameterization of highly charged metal ions using the 12-6-4 LJ-type nonbonded model in explicit water. J Phys Chem B 119:883–895. https://doi.org/10.1021/jp505875v
88. Li P, Merz Jr. KM (2016) MCPB.py: a Python based metal center parameter builder. J Chem Inf Model 56:599–604. https://doi.org/10.1021/acs.jcim.5b00674
89. Roldan-Martin L, Sodupe M, Marechal JD (2023) Computational assessment of the impact of Cu(II) and Al(III) on beta-amyloid(42) fibrils: binding sites, structural stability, and possible physiological implications. Front Neurosci 17:1110311. https://doi.org/10.3389/fnins.2023.1110311
90. Vanommeslaeghe K, Raman EP, MacKerell Jr. AD (2012) Automation of the CHARMM General Force Field (CGenFF) II: assignment of bonded parameters and partial atomic charges. J Chem Inf Model 52:3155–3168. https://doi.org/10.1021/ci3003649
91. Vanommeslaeghe K, MacKerell Jr. AD (2012) Automation of the CHARMM General Force Field (CGenFF) I: bond perception and atom typing. J Chem Inf Model 52:3144–3154. https://doi.org/10.1021/ci300363c
92. Michaud-Agrawal N, Denning EJ, Woolf TB, Beckstein O (2011) MDAnalysis: a toolkit for the analysis of molecular dynamics simulations. J Comput Chem 32:2319–2327. https://doi.org/10.1002/jcc.21787
93. McGibbon RT et al (2015) MDTraj: a modern open library for the analysis of molecular dynamics trajectories. Biophys J 109:1528–1532. https://doi.org/10.1016/j.bpj.2015.08.015
94. Humphrey W, Dalke A, Schulten K (1996) VMD: visual molecular dynamics. J Mol Graph 14:33–38
95. Schrodinger LLC (2015) The PyMOL molecular graphics system. Version 1:8. Schrodinger LLC
96. Roe DR, Cheatham TE 3rd (2013) PTRAJ and CPPTRAJ: software for processing and analysis of molecular dynamics trajectory data. J Chem Theory Comput 9:3084–3095. https://doi.org/10.1021/ct400341p
97. Leang SS, Rendell AP, Gordon MS (2014) Quantum chemical calculations using accelerators: migrating matrix operations to the NVIDIA Kepler GPU and the Intel Xeon Phi. J Chem Theory Comput 10:908–912. https://doi.org/10.1021/ct4010596
98. Showalter SA, Bruschweiler R (2007) Validation of molecular dynamics simulations of biomolecules using NMR spin relaxation as benchmarks: Application to the AMBER99SB force field. J Chem Theory Comput 3:961–975. https://doi.org/10.1021/ct7000045
99. Rahman MU, Rehman AU, Liu H, Chen HF (2020) Comparison and evaluation of force fields for intrinsically disordered proteins. J Chem Inf Model 60:4912–4923. https://doi.org/10.1021/acs.jcim.0c00762
100. Bayly CI, Cieplak P, Cornell W, Kollman PA (1993) A well-behaved electrostatic potential based method using charge restraints for deriving atomic charges: the RESP model. J Phys Chem 97:10269–10280
101. Lemkul JA, Huang J, Roux B, MacKerell AD, Jr (2016) An empirical polarizable force field based on the classical drude oscillator model: development history and recent applications. Chem Rev 116:4983–5013. https://doi.org/10.1021/acs.chemrev.5b00505
102. Ponder JW et al (2010) Current status of the AMOEBA polarizable force field. J Phys Chem B 114:2549–2564. https://doi.org/10.1021/jp910674d
103. Hanwell MD et al (2012) Avogadro: an advanced semantic chemical editor, visualization, and analysis platform. J Cheminform 4:17. https://doi.org/10.1186/1758-2946-4-17
104. Schrodinger LLC (2015) The AxPyMOL molecular graphics plugin for Microsoft PowerPoint. Version 1:8. Schrodinger LLC
105. Olsson MH, Sondergaard CR, Rostkowski M, Jensen JH (2011) PROPKA3: consistent treatment of internal and surface residues in empirical pKa predictions. J Chem Theory Comput 7:525–537. https://doi.org/10.1021/ct100578z
106. Gordon JC et al (2005) H++: a server for estimating pKas and adding missing hydrogens to macromolecules. Nucleic Acids Res 33:W368–371. https://doi.org/10.1093/nar/gki464
107. Breneman CM, Wiberg KB (1990) Determining atom-centered monopoles from molecular electrostatic potentials. The need for high sampling density in formamide conformational analysis. J Comput Chem 11:361–373

108. Hakkennes MLA, Buda F, Bonnet S (2023) MetalDock: an open access docking tool for easy and reproducible docking of metal complexes. J Chem Inf Model 63:7816–7825. https://doi.org/10.1021/acs.jcim.3c01582
109. Bussi G, Donadio D, Parrinello M (2007) Canonical sampling through velocity rescaling. J Chem Phys 126:014101. https://doi.org/10.1063/1.2408420
110. Corti DS (2001) Isothermal-isobaric ensemble for small systems. Phys Rev E Stat Nonlin Soft Matter Phys 64:016128. https://doi.org/10.1103/PhysRevE.64.016128
111. Berendsen HJ, Postma Jv, van Gunsteren WF, DiNola A, Haak JR (1984) Molecular dynamics with coupling to an external bath. J Chem Phys 81:3684–3690
112. Hess B, Bekker H, Berendsen HJ, Fraaije JG (1997) LINCS: a linear constraint solver for molecular simulations. J Comput Chem 18:1463–1472
113. Gaalswyk K, Haider A, Ghosh K (2023) Critical assessment of self-consistency checks in the all-atom molecular dynamics simulation of intrinsically disordered proteins. J Chem Theory Comput 19:2973–2984. https://doi.org/10.1021/acs.jctc.2c01140
114. Li L, Casalini T, Arosio P, Salvalaglio M (2022) Modeling the structure and interactions of intrinsically disordered peptides with multiple replica, metadynamics-based sampling methods and force-field combinations. J Chem Theory Comput 18:1915–1928. https://doi.org/10.1021/acs.jctc.1c00889
115. Zerze GH, Miller CM, Granata D, Mittal J (2015) Free energy surface of an intrinsically disordered protein: comparison between temperature replica exchange molecular dynamics and bias-exchange metadynamics. J Chem Theory Comput 11:2776–2782. https://doi.org/10.1021/acs.jctc.5b00047
116. Wang L, Friesner RA, Berne BJ (2011) Replica exchange with solute scaling: a more efficient version of replica exchange with solute tempering (REST2). J Phys Chem B 115:9431–9438. https://doi.org/10.1021/jp204407d
117. Choi YK et al (2022) CHARMM-GUI nanomaterial modeler for modeling and simulation of nanomaterial systems. J Chem Theory Comput 18:479–493. https://doi.org/10.1021/acs.jctc.1c00996
118. Gamero-Quijano A et al (2022) On the origin of chaotrope-modulated electrocatalytic activity of cytochrome c at electrified aqueous|organic interfaces. Chem Commun 58:3270–3273. https://doi.org/10.1039/d1cc05293d
119. Gamero-Quijano A et al (2021) Modulating the pro-apoptotic activity of cytochrome c at a biomimetic electrified interface. Sci Adv 7:eabg4119. https://doi.org/10.1126/sciadv.abg4119
120. Holzinger A, Bhattacharya S, Wiloch-Szyborska MZ, Jönsson-Niedziółka M (2026) Ion-mediated adsorption of amyloid-β peptides at the liquid/liquid interface: insights from electrochemistry and molecular dynamics simulations. ACS Electrochem ACS Electrochem. 2:286–296. https://doi.org/10.1021/acselectrochem.5c00358
121. Kabsch W, Sander C (1983) Dictionary of protein secondary structure: pattern recognition of hydrogen-bonded and geometrical features. Biopolymers 22:2577–2637. https://doi.org/10.1002/bip.360221211
122. Wang E et al (2019) Assessing the performance of the MM/PBSA and MM/GBSA methods. 10. Impacts of enhanced sampling and variable dielectric model on protein-protein interactions. Phys Chem Chem Phys 21:18958–18969. https://doi.org/10.1039/c9cp04096j
123. Xu L, Sun H, Li Y, Wang J, Hou T (2013) Assessing the performance of MM/PBSA and MM/GBSA methods. 3. The impact of force fields and ligand charge models. J Phys Chem B 117:8408–8421. https://doi.org/10.1021/jp404160y
124. Lee MS, Salsbury Jr FR, Brooks III CL (2002) Novel generalized Born methods. J Chem Phys 116:10606–10614
125. Ross GA et al (2023) The maximal and current accuracy of rigorous protein-ligand binding free energy calculations. Commun Chem 6:222. https://doi.org/10.1038/s42004-023-01019-9
126. Gaur P et al (2025) Molecularly engineered memristors for reconfigurable neuromorphic functionalities. Adv Mater e09143. https://doi.org/10.1002/adma.202509143
127. Malcomson T, Repiscak P, Erhardt S, Paterson MJ (2022) Protocols for understanding the redox behavior of copper-containing systems. ACS Omega 7:45057–45066. https://doi.org/10.1021/acsomega.2c05484
128. Jinich A et al (2018) Quantum chemistry reveals thermodynamic principles of redox biochemistry. PLoS Comput Biol 14:e1006471. https://doi.org/10.1371/journal.pcbi.1006471
129. Gibney A et al (2026) Expanding the DNA damaging potential of artificial metallo-nucleases with click chemistry. Nat Commun. https://doi.org/10.1038/s41467-026-68911-5
130. Zhu J et al (2025) Accurate generation of conformational ensembles for intrinsically disordered proteins with IDPFold. Adv Sci 12:e11636

Check for updates

Chapter 5

Fluorescent Probes for Metal Ion Detection

Andreas M. Grabrucker

Abstract

Fluorescent probes have become indispensable tools for detecting and quantifying metal ions in cells and tissues of the nervous system. Their high sensitivity, spatiotemporal resolution, and compatibility with diverse analytical platforms have driven extensive development over the past several decades, resulting in an evolving landscape of molecular sensors. This chapter provides a comprehensive review of fluorescent probes for metal ion detection, with an emphasis on commercially available indicators and widely adopted probe architectures. The chapter surveys fluorescent probes for alkali, alkaline earth, transition, and post-transition metal ions, detailing representative fluorophore scaffolds, binding affinities, spectral properties, and application domains. Practical considerations such as selectivity, reversibility, intracellular buffering, and quantitative limitations are critically examined. By integrating chemical design principles with application-driven performance metrics, this chapter aims to serve as a reference for researchers selecting fluorescent probes for metal ion sensing in the nervous system.

Keywords Fluorescent probes, Metal ion detection, Ratiometric sensors, CIEF and PET mechanisms, Zinc and calcium indicators, Transition metal sensing, Fluorescence imaging

1 Introduction

Fluorescence is a photophysical phenomenon in which a material, known as a fluorophore, absorbs light at a specific excitation wavelength and then re-emits light at a longer, lower-energy emission wavelength [1]. The fundamental working principle of a fluorescent probe-based metal detection method relies on a discernible change in the photophysical properties of the probe (fluorophore) before and after its specific binding to a metal. This alteration in the fluorescence signal forms the basis for detecting varying concentrations of specific metal ions [2].

Fluorescent probes are among the most versatile and widely used strategies for detecting metal ions in biological samples. Their appeal stems from high sensitivity, spatial and temporal resolution, compatibility with live systems, and adaptability to a wide range of detection platforms [3].

Andreas M. Grabrucker (ed.), *Brain Metallomics*, Neuromethods, vol. 237,
https://doi.org/10.1007/978-1-0716-5468-2_5,
© The Author(s), under exclusive license to Springer Science+Business Media, LLC, part of Springer Nature 2026

The field of fluorescent probe development is driven by several key demands on fluorescent metal sensors. First, there is a need for highly specialized sensors that can distinguish between different metals and not only visualize those but also accurately monitor their concentrations within specific physiological windows. This is possible because the intensity or lifetime of the emitted fluorescence is directly proportional to the concentration of a metal ion. Such probes require high sensitivity and exquisite selectivity to avoid misleading signals, particularly in cells and tissues where numerous metal ions coexist. In addition, they need to be stable to allow detection without significant photobleaching, which is the irreversible degradation of fluorophores upon prolonged light excitation. Furthermore, to measure metal dynamics, they need fast binding kinetics. Their affinity will determine whether they can visualize only "free" (aqueous) metal ions, or whether they can also scavenge metals bound more weakly to peptides or proteins within tissues. Finally, to bind metal ions within cells, they need to be cell membrane-permeable or expressed within cells or subcellular compartments [3–5] (Table 1).

If these criteria are met, fluorescent probes offer distinct advantages over traditional analytical methods such as atomic absorption spectrometry (AAS), inductively coupled plasma mass spectrometry (ICP-MS), X-ray fluorescence spectrometry, and ion chromatography [6]. Their benefits include high sensitivity, convenience, accuracy, cost-effectiveness, and reliability [6]. Indeed, fluorescent probes

Table 1
Criteria for effective fluorescent probes

High selectivity	Paramount for distinguishing the target ion from other coexisting metals, especially those with similar physico-chemical properties (e.g., Cd^{2+} vs Zn^{2+} K^{+} vs Na^{+} Mg^{2+} vs Ca^{2+}).
High sensitivity (low detection limits)	The ability to accurately detect metal ions at extremely low, trace levels.
Biocompatibility	For applications in biological systems and in vivo studies, probes must exhibit minimal toxicity to cells and tissues.
Solubility	Essential for practical applications in biological fluids, tissues, and cells.
Kinetics	Fast response times enable real-time monitoring of dynamic biological processes.
Photostability	Resistance to photobleaching.
pH stability	This is crucial for applications in biological systems where physiological pH can fluctuate, ensuring that the fluorescence signal is primarily due to metal ion binding rather than pH changes.
Tunable emission range	The ability to design probes with varying emission wavelengths, particularly in the red-to-near-infrared (NIR) region, is advantageous for multiplexing and for achieving deeper tissue penetration in biological imaging.

represent a significant paradigm shift in analytical methodology for metal ions. Their unique ability to provide real-time, in situ, and noninvasive detection with high spatial resolution [2] offers a level of information detail often inaccessible to traditional bulk analytical techniques. This enables dynamic studies of metal ion trafficking and homeostasis in living systems [7], which are crucial for advancing understanding of disease mechanisms and developing targeted therapeutic interventions, moving beyond simple quantitative measurements to profound mechanistic insights. Often, fluorescent sensors are integrated with high-resolution fluorescent microscopy techniques or even advanced microscopy techniques [e.g., stimulated emission depletion microscopy (STED), structured illumination microscopy (SIM), single-molecule localization microscopy (SMLM)] [8].

Fluorescent probes are categorized based on their fundamental fluorophore types, such as organic small molecules, various nanomaterials, and genetically encoded sensors. Generally, a functional fluorescent sensor for cations or anions comprises two essential components: a fluorescent carrier (the fluorophore itself) and an ionic carrier (the recognition unit). These two components can exist as independent chemical species or be covalently linked within a single molecular structure.

However, several modern fluorescent probes are designed with three key constituents: the recognition unit (or receptor), which selectively interacts with the target metal; the fluorescence unit (the fluorophore), responsible for the optical signal; and a connector or linker, which structurally bridges the recognition and fluorescence units [8]. The recognition unit's structure is paramount for ensuring selectivity towards the ion of interest. Once binding of a metal ion to a fluorescent probe occurs, changes in its fluorescence are induced in response through several possible mechanisms [7] (Fig. 1).

Photoinduced electron transfer (PET) is a prevalent mechanism in which an electron is transferred from the fluorophore to either the recognition unit or the bound metal ion, leading to fluorescence quenching. Upon binding of the target metal ion, this PET process is inhibited, resulting in a characteristic "turn-on" fluorescence signal [9]. Less frequently, fluorescence quenching itself is used as a detection method. Here, metal-binding leads to a decrease in fluorescence intensity due to electron or energy transfer by the bound metal ion [2], which is responsible for "turn-off" fluorescence responses.

Another mechanism to induce a fluorescent response is intramolecular charge transfer (ICT). Metal ion binding can alter the electronic distribution within the probe molecule, thereby affecting its intramolecular charge-transfer characteristics. This typically manifests as shifts in the emission or excitation intensities, which can be bathochromic (red-shift), hyperchromic (increased intensity), hypochromic (decreased intensity), or hypsochromic (blue-shift).

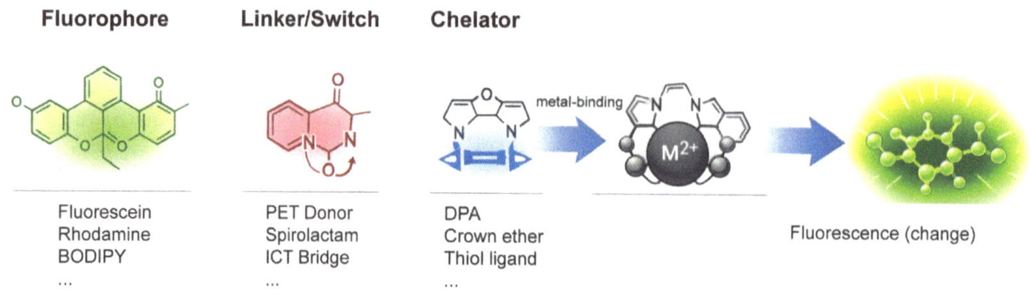

Fig. 1 General design of fluorescent metal ion probes: Modular architecture of fluorescent metal ion probes. A typical probe consists of a fluorophore core, a metal-binding unit, and a coupling element that translates metal coordination into an optical signal. Depending on probe design, metal-binding can alter fluorescence intensity, emission wavelength, or excitation profile. The modularity of this architecture enables systematic tuning of affinity, selectivity, and photophysical properties by independently modifying each component

Furthermore, in fluorescence resonance energy transfer (FRET)-based probes, energy is transferred from a donor fluorophore to an acceptor fluorophore. Metal ion binding can induce a conformational change that brings these two fluorophores into optimal proximity for FRET, or conversely, disrupts an existing FRET, leading to a ratiometric fluorescence signal.

Much less frequently found is aggregation-induced emission (AIE). Here, fluorophores exhibit minimal or no fluorescence when molecularly dispersed in solution but become highly emissive upon aggregation. Metal ion binding will induce this aggregation, thereby "turning on" the fluorescence. Besides, chelation-induced enhanced fluorescence (CIEF) is a mechanism in which the chelation of a metal ion by the probe can rigidify the fluorophore's structure, leading to an increase in fluorescence intensity.

Finally, in some designs, the metal ion acts as a catalyst for a chemical reaction involving the probe, leading to the formation of a fluorescent product, or it induces redox reactions that alter the oxidation state or electronic configuration of the fluorophore, thereby changing its fluorescent properties.

While, some of those mechanisms are reversible, some probes can act as chemodosimeters, undergoing an irreversible chemical reaction with the bound metal ion that results in a permanent change in the fluorescence signal. In addition, some probes can be used for ratiometric sensing. These probes exhibit two distinct emission or excitation wavelengths. Upon metal ion binding, the intensity ratio of these two wavelengths changes. A ratiometric readout provides an inherent self-calibration mechanism that effectively compensates for variations in probe concentration, excitation light fluctuations, and photobleaching, thereby enabling more robust, quantitative measurements [10–13].

Thus, for many metals, it is possible to select a probe tailored to achieve specific spectroscopic responses (e.g., turn-on, turn-off, and ratiometric) that are optimally suited to different analytical applications. For example, ratiometric probes are highly valued in biological systems because they inherently provide a self-calibrating signal (Fig. 2). Conversely, "turn-on" probes are often preferred when a low background signal is critical for high-contrast detection. This means that selecting a fluorescent probe is highly context-dependent. Researchers must consider the target metal ion's specific chemical properties, the complexity of the sample matrix, the required sensitivity, and selectivity, and the intended imaging modality (e.g., fluorescence microscopy vs flow cytometry) when choosing a probe.

2 Classes of Fluorescent Probes

Many traditional fluorescent probes are small organic molecules. However, several new nanomaterials and genetically encoded sensors have been developed in recent years. While organic probes offer chemical versatility, nanomaterials can provide enhanced photostability and a larger surface area for interaction, and genetically encoded sensors effectively overcome the challenges of cellular loading and precise targeting in live cells.

The continuous development of new fluorescent probe technologies is largely driven by the imperative to overcome fundamental limitations of earlier generations, such as photobleaching and nonspecific binding [14]. The introduction of nanomaterials and the engineering of genetically encoded sensors are direct responses to these practical issues. Nanomaterials, for instance, can offer improved

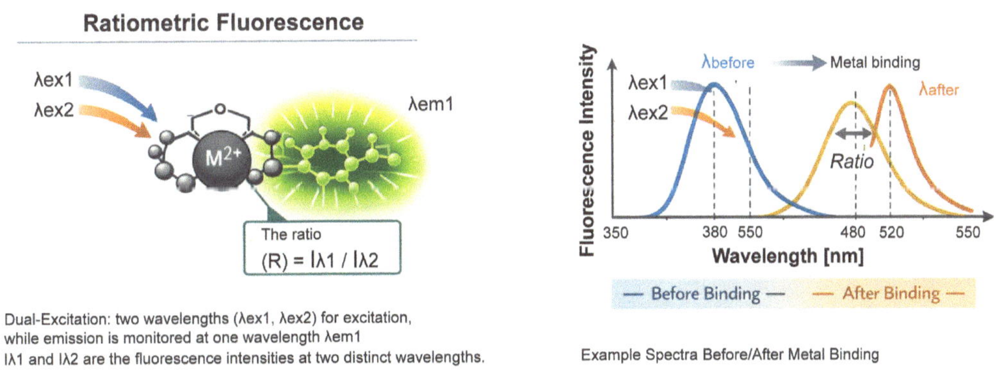

Fig. 2 Ratiometric sensing via excitation or emission shifts. Ratiometric fluorescence sensing of metal ions. Ratiometric probes exhibit wavelength shifts upon metal binding, enabling intensity-independent quantification. Benzofuran-based indicators such as Fura-2 and SBFI display metal-dependent excitation shifts, while other systems rely on emission ratiometry via ICT or FRET mechanisms. Ratiometric sensing is particularly valuable for quantitative intracellular measurements and long-term imaging

photostability [8]. Genetically, encoded sensors address the issue of nonspecific binding by being precisely expressed within specific cells or subcellular compartments, thereby localizing the signal [15].

2.1 Organic Small Molecule Fluorescent Probes

Organic small-molecule fluorescent probes constitute a widely developed class due to their inherent high fluorescence quantum yields, ease of chemical modification, good biocompatibility, high sensitivity, and suitability for fast, real-time detection [16]. Their general architecture typically involves a fluorophore covalently linked to an ion recognition carrier [2]. Common fluorophore examples include coumarin, rhodamine, fluorescein, BODIPY (difluoro-boron dipyrromethene), pyrene, quinoline, naphthalimide, anthracene, danyl, quinazoline, cyanine dyes, unquinoline, naphthalene, luciferin, and various phenol derivatives [2] (Table 2).

2.2 Nanofluorescent Probes

Nanomaterial-based probes, encompassing carbon quantum dots (CQDs), traditional quantum dots (QDs), gold nanoclusters (AuNCs), and metal-organic frameworks (MOFs), have gained significant prominence in recent years. This is attributed to their unique optical properties, biocompatibility, large specific surface area, high conductivity, inherent photostability, and generally low toxicity [17]. Carbon dots (CDs)/carbon quantum dots (CQDs) offer high sensitivity, selectivity, simplicity, and rapid monitoring capabilities for metal ion detection [17]. Specific examples include nitrogen-doped carbon dots (NCDs) for Fe(III) and Hg(II) [16]; B,N-CQDs for Cr^{3+}, Cu^{2+}, and Fe^{2+}; N-CQDs for Cr^{3+}, Cu^{2+}, Fe^{2+}, and Ni^{2+} [17]. Quantum dots (QDs) are frequently employed in ratiometric fluorescence nanosensors, such as cadmium telluride (CdTe) QDs combined with carbon nanodots for Cu^{2+}, and dual-emission QDs like CdSe@C QDs for enhanced ion recognition of Cu^{2+} [2]. Metal-organic frameworks (MOFs) are highly porous crystalline materials, with examples including UiO-66-$(OH)_2$ MOF for Fe^{3+} [2].

Table 2
Representative fluorophore cores in metal ion probes

Fluorophore core	Typical features
BODIPY	High quantum yield, narrow emission, amenable to derivatization
Rhodamine/RhodZin-like	Turn-on via spirolactam opening; visible emission
Fluorescein/Fluo-series	Bright green emission, widely commercialized
Cyanines	Tunable near-IR emission
Coumarins	Blue-green emitters with modular binding sites

2.3 Genetically Encoded Fluorescent Sensors (GEFS)

Genetically encoded fluorescent sensors represent powerful tools for elucidating metal dynamics within living systems. They enable the real-time visualization, quantification, and tracking of metal ions in various states: dissolved in solution, attached to cellular membranes, or within the cytoplasm of living cells [18]. A key advantage is that they do not require acute loading into cells; instead, their genes can be introduced into specific cell types or entire transgenic organisms [15]. GEFS can be single fluorescent proteins that embed a specific metal-binding site, thereby modulating their fluorescence upon metal ion binding. Examples include Camgaroos and G-CaMP for Ca^{2+}; GAF2-PEB and CreiLOVN41C for Zn^{2+}; MnLaMP1 and MnLaMP2 for Mn^{2+}; CreiLOV for Cu^{2+} [8]; and GINKO1 for K^+ [19].

Alternatively, GEFS can be FRET-based Systems. These multifluorescent protein probes leverage Förster resonance energy transfer (FRET). Two fluorescent proteins are linked by a metal binding site such that metal binding induces a conformational change, bringing them into optimal proximity for FRET, or conversely, disrupting an existing FRET, thereby generating a ratiometric signal [15]. Examples include Cameleons for Ca^{2+}; GINKO2 and GEPIIs for K^+ [19]; KIRIN1 and KIRIN1-GR for K^+; MerFS for Hg^{2+}; and SenALiB for As^{3+} [15].

Another class of GEFS is based on DNAzymes. These genetically encoded sensors utilize metal-specific RNA-cleaving DNAzymes to regulate the expression of fluorescent proteins, enabling ratiometric sensing. Examples include Mg^{2+}-specific 10–23 and Zn^{2+}-specific 8–17 DNAzymes [15].

3 Fluorescent Probes for Specific Metal Ions

From a practical perspective, commercially available probes (Table 3) fall into two broad categories. Ratiometric probes, exemplified by benzofuran-based indicators such as Fura-2 and SBFI, enable quantitative measurements that are independent of probe concentration and excitation intensity, albeit at the cost of UV excitation and reduced photon efficiency. Single-wavelength (turn-on or turn-off) probes, such as Fluo-4 or FluoZin-3, dominate high-throughput and imaging applications because of their high brightness, fast kinetics, and compatibility with standard optics, but require careful calibration and controls to ensure quantitative reliability.

3.1 Na^+ Detection

Na^+ and K^+ sensing remains challenging because of high intracellular background concentrations and the need for millimolar affinity. For detecting physiological concentrations of Na^+ in neuroscience applications, key probes include SBFI and Sodium Green. SBFI exhibits a dissociation constant (K_d) for Na^+ of 11.3 mM (in solutions with 135 mM combined Na^+ and K^+), demonstrating approximately

Table 3
Fluorophore summary table (quick reference)

Metal ion	Commercial probes/fluorophores
Na^+	CoroNa™ Green/Red, SBFI
K^+	PBFI, Asante Potassium Green
Ca^{2+}	Fura-2, Indo-1, Fluo-3/4, Calcium Green™, Cal-520™
Mg^{2+}	Mag-Fura-2, Magnesium Green™, Mag-Fluo-4
Zn^{2+}	Zinpyr-1/5, FluoZin™-3, Zinquin, TSQ
Cu^{2+}	CS1/CS3, Copper Green™, CuFL
Fe^{2+}/Fe^{3+}	Phen Green SK, FeRhoNox-1
Pb^{2+}	Leadfluor-1, rhodamine lead probes
Hg^{2+}	Rhodamine-Hg probes, dansyl/BODIPY chelators
Cd^{2+}	Zn-analogous tuned probes
Others	Lumogallion (Al^{3+}), specialized research probes

18-fold greater selectivity for Na^+ over K^+. This is important as both metals are positively charged and similar, and K^+ is often present at much higher concentrations, especially inside cells such as neurons. Without strong Na^+ selectivity, the fluorophore's signal would be dominated by K^+. In that respect, sodium green offers superior selectivity for Na^+ over K^+ (~41-fold vs. ~18-fold for SBFI) and a much higher fluorescence quantum yield (0.2 vs. 0.08 for SBFI). It can be excited at 488 nm, making it compatible with confocal laser-scanning microscopes. Thus, it is often preferred in applications. However, sodium green lacks the direct ratiometric readout capability of SBFI. Genetically encoded options for Na^+ include a crown-ether-containing ligand-based sensor [15] (Table 4).

3.2 K^+ Detection

PBFI is a synthetic dye for K^+. However, PBFI has low selectivity (only 1.5-fold more selective for K^+ than Na^+), low quantum yield (0.024), a small fluorescent response, and requires UV excitation (350 nm), which can potentially damage biological molecules [20]. Genetically encoded options include GINKO1, GINKO2, and GEPIIs. GINKO1, a single fluorescent protein-based sensor, is suitable for detecting K^+ at physiologically relevant concentrations and enables dual-color imaging [15]. GINKO2, an improved variant of GINKO1, offers higher sensitivity and specificity [19]. GEPIIs, which are FRET-based, enable quantitative real-time imaging [15] (Table 4).

3.3 Ca^{2+} Detection

Key organic small molecule fluorescent probes, primarily BAPTA-based, include Fura-2, Indo-1, Fluo-3, Fluo-4, and Calcium Green-1 [21]. These indicators typically respond to Ca^{2+} binding by an increase in

Table 4
Common commercial fluorescent probes for metal ion detection

Metal ion	Probe	Fluorophore core	λ_{ex}(nm)	λ_{em}(nm)	$K_{d\,(typical)}$	Detection mode
Na^+	SBFI	Benzofuran	340/380	505	3–25 mM	Ratiometric
K^+	PBFI	Benzofuran	340/380	505	6–15 mM	Ratiometric
Ca^{2+}	Fura-2	Benzofuran	340/380	510	140–250 nM	Ratiometric
Ca^{2+}	Fluo-4	Fluorescein	488	516	~345 nM	Turn-on
Mg^{2+}	Mag-Fura-2	Benzofuran	335/370	510	1–2 mM	Ratiometric
Zn^{2+}	FluoZin-3	Fluorescein	494	516	1–15 nM	CHEF
Zn^{2+}	Zinpyr-1	Fluorescein	515	530	~0.7 nM	PET suppression
Zn^{2+}	RhodZin-3	Rhodamine	550	575	~65 nM	Ring opening
Cu^{2+}	Phen Green FL	Fluorescein	490	525	~0.3–1 μM	Quenching
Fe^{2+}	FeRhoNox-1	Rhodamine	540	575	~20 μM	Redox-activated
Pb^{2+}	Leadmium Green	Fluorescein	490	525	1–10 nM	CHEF
Cd^{2+}	Leadmium Green	Fluorescein	490	525	10–50 nM	CHEF
Hg^{2+}	Rhodamine-Hg	Rhodamine	550	580	nM range	Thiophilic turn-on

quantum yield or a shift in emission/excitation wavelength. While BAPTA-based indicators are highly selective for Ca^{2+} over Mg^{2+}, many Ca^{2+} indicators that use chelating groups (like BAPTA or EGTA derivatives) are not perfectly selective. In particular, they might bind Zn^{2+} very effectively—sometimes even better than Ca^{2+}. Examples include Fura-2, Indo-1, Fluo-3, and Fluo-4. While designed for Ca^{2+}, they can bind Zn^{2+} with high(er) affinity. Thus, even trace Zn^{2+} levels can cause changes in fluorescence. Nevertheless, ratiometric indicators (Fura-2, Indo-1) remain the gold standard for quantitative Ca^{2+} measurements, whereas intensity-based probes (Fluo-4, Cal-520) dominate functional imaging and screening applications.

Genetically encoded calcium indicators (GECIs) have some advantages in terms of specificity, as they are generally more selective. Still, Zn^{2+} can interfere in some cellular contexts or at high Zn^{2+} levels. GECIs are broadly classified into single fluorescent protein systems (e.g., Camgaroos, G-CaMP, CatchER), paired fluorescent protein systems (e.g., Cameleons), Near-IR GECIs (e.g., RCaMPs, RGECOs), and special GECIs for active neurons (e.g., Eos, SynTagMA) [18] (Table 4).

3.4 Mg^{2+} Detection

Magnesium detection presents challenges because of its chemical similarity to Ca^{2+} and weaker ligand binding. Commercial Mg^{2+}

indicators, therefore, typically exhibit millimolar dissociation constants and reduced selectivity, limiting their use to relative rather than absolute quantification. Key organic small molecule fluorescent probes include the widely used Mag-Fura-2, Mag-Indo-1, Magnesium Green, and Mag-fluo-4. Mag-Fura-2 (K_d 1.9 mM), Mag-Indo-1 (K_d 2.7 mM), Magnesium Green (K_d ~1.0 mM), and Mag-fluo-4 (K_d 4.7 mM) are designed to respond to physiological Mg^{2+} concentrations [22]. Mag-fluo-4 exhibits a much more sensitive fluorescence response to Mg^{2+} binding than Magnesium Green, which is advantageous given the typically small physiological fluctuations of intracellular Mg^{2+} concentration [23]. Genetically encoded options include DNAzyme-based sensors [15] (Table 4).

3.5 Fe^{3+}/Fe^{2+} Detection

Distinguishing Fe^{2+} from Fe^{3+} is essential in neuroscience because the two oxidation states have fundamentally distinct biological roles. Fe^{2+} is redox-active and drives Fenton chemistry, oxidative stress, and ferroptotic cell death, whereas Fe^{3+} predominates in transport and storage (e.g., transferrin, ferritin). State-specific iron detection is therefore critical for interpreting iron accumulation in neurodegeneration, mitochondrial dysfunction, and activity-dependent redox signaling, where toxicity depends on iron speciation rather than total iron levels.

Many probes have been developed for Fe^{3+}, including various nanometer-scale fluorescent probes (such as semiconductor nanoparticles, quantum dots, and carbon dots (CDs)) and numerous organic small molecule fluorescent probes [2, 24]. Key organic small-molecule examples include the TB-BTB framework, EuL_3, Tricyanofuran hydrazone (TCFH), deoxycholic acid and rhodamine molecules, Trimeric phenolic coumarin (TPC), and vanilloid fluorescent probes [2]. Detection mechanisms for Fe^{3+} probes typically fall into turn-off, turn-on, ratiometric, and chemodosimeter categories. Many operate primarily via fluorescence quenching.

Probes for Fe^{2+} include N-oxides, nitroxyl radicals, endoperoxides, bionic ligands, imines, and various carbon dots (CDs) [2]. Specific examples encompass dual-correspondence luminescence probes with MOF, MIL-124 (a MOF-like layer), QM-Fe (an AIE-based probe), spiropyran-based fluorescent probes, and coumarin fluorescent probes. The commercially available indicator FeRhoNox-1 (a rhodamine-based Fe^{2+} sensor used for live-cell imaging) is widely used. Many constructed probes are of the "on-off" type due to the strong interaction between paramagnetic Fe^{2+} and the fluorophore, which can induce fluorescence quenching (Table 4).

3.6 Zn^{2+} Detection

A variety of fluorophore groups are employed for Zn^{2+} detection, including quinoline, rhodamine, naphthalene, coumarin, naphthalimide, pyrene, luciferin, and derivatives of phenol [2, 25, 26]. Examples of key cell-permeable fluorescent probes include FluoZin-3/

FluoZin-3 AM. FluoZin-3 AM is the acetoxymethyl ester (AM) derivative of FluoZin-3. The AM groups mask the negatively charged carboxylate groups of the parent molecule, making it neutral and membrane-permeable, allowing it to diffuse into living cells. Once inside, intracellular esterases cleave the AM groups, regenerating the charged FluoZin-3, which becomes trapped inside the cytosol and is now competent to bind Zn^{2+} and produce fluorescence. Similarly, Zinquin is available as Zinquin and Zinquin AM. Other intensity-based Zn^{2+} probes with high sensitivity include RhodZin-3, a Rhodamine platform that yields an orange-emitting Zn^{2+} indicator, and the DA ZP1 Series: derivatives of Zinpyr-1 (ZP1). Zinpyr-1 is a small-molecule fluorescent sensor for mobile Zn^{2+} based on a dichlorofluorescein fluorophore. In the DA-ZP1 compounds, the two phenolic hydroxyl groups on the ZP1 scaffold are diacetylated (DA) to mask fluorescence and modulate reactivity; Zn^{2+} binding promotes Zn^{2+}-triggered hydrolysis of the acetyl esters, restoring the fluorescein fluorescence and giving a fluorogenic turn-on response. This reaction-based mechanism gives DA-ZP1 high sensitivity (sub-nanomolar K_d for Zn^{2+}) and a low background signal, enabling imaging of mobile zinc in biological systems, including live cells and neural tissue slices at relatively low probe concentrations. In contrast, Newport Green DCF/PDX is a sensor with moderate affinity for Zn^{2+}. However, one advantage is that it is essentially insensitive to Ca^{2+} at physiological ranges. Genetically encoded options include GAF2-PEB and CreiLOVN41C, and DNAzyme-based sensors [15] (Table 4).

3.7 Cu^{2+} Detection

Copper and iron probes face the additional complexity of redox chemistry. Many commercially available Cu^{2+} and Fe^{2+} probes rely on fluorescence quenching or redox-activated mechanisms, which can complicate quantitative interpretation but provide oxidation-state specificity. Probes for Cu^{2+} include small polymer fluorescent probes (e.g., anthracene, danyl, pyrene, quinazoline, cyanine dyes, naphthalimide, quinoline, rhodamine, fluorescein, BODIPY), and those derived from natural sources (e.g., N,S co-doped CQD from banana juice). Key examples include Phen Green FL, a sensor responsive to Cu^{2+} and Cu^{+} that quenches baseline fluorescence, and Coppersensor-1 (CS1) or Coppersensor-3 (CS3). CS-1 is a small-molecule, membrane-permeable, turn-on fluorescent probe selective for Cu^{+} based on a BODIPY fluorophore coupled to a thioether-rich receptor. Fluorescence increases significantly upon Cu^{+} binding. CS-3 is a brighter, improved version of CS1 with a higher quantum yield and larger fluorescence turn-on amplitude. For live-cell or slice imaging, Phen Green FL (Table 4) and (CS1/CS3)-type probes dominate. Related variants are Mito-CS1, a mitochondria-targeted version of CS1 designed to visualize copper pools specifically in mitochondria, and RCS1 (ratiometric CS1), a ratiometric derivative of CS1 allowing internal referencing and improved quantification of Cu^{+} levels.

Further Cu sensors include the "turn-on" NIR probe DCM-Cu, used in neuroscience because it operates in the near-infrared (NIR) range, enabling deeper tissue penetration and lower "autofluorescence" (noise from natural brain tissue). It is used for real-time imaging of Cu^{2+} fluctuations in living neurons and brain slices. Ratiometric fluorescence nanosensors (CdTe QDs with CDs) are frequently used in cellular neuroscience. The ratiometric nature allows Cu levels to be measured accurately, regardless of the thickness of the brain cell or tissue being imaged. "Off-on" methods utilizing mesoporous silica (MS) spheres are also employed. These are often used as "nanocarriers." MS spheres can be loaded with fluorescent dyes and targeted to cross the blood–brain barrier (BBB) or enter specific neural cells to detect intracellular Cu. In addition, while often used in environmental water testing, specific derivatives of pyrimidine-based chemosensor (PyrCS) are used in cells because they are highly selective for Cu^{2+}. PyrCS sensors have a pyrimidine moiety that acts as a chelator for copper ions, often with additional nitrogen or sulfur donors to improve selectivity. Cu^{2+} or Cu^{+} binding induces a change in the electronic structure of the fluorophore. Most PyrCS probes are "turn-on" or ratiometric.

All fluorophores mentioned above detect essential metals in the human body. Probes for the three missing essential metals, Mn, Mo, and Co, are available but remain technically challenging for use in nervous tissue. Mn^{2+} indicators are the most developed and have seen limited neurobiological application, although paramagnetic quenching and poor selectivity over Ca^{2+}/Mg^{2+} constrain sensitivity and interpretation. Co^{2+} probes exist largely as fluorescence-quenching or displacement-based sensors with restricted cellular specificity, while Mo sensing is rare and typically targets molybdate via reaction-based mechanisms developed for non-neuronal contexts. Overall, the paramagnetism, redox activity, and complex speciation of these metals severely limit fluorophore performance in neural systems, restricting their use to specialized experimental questions rather than routine imaging.

Besides essential metals, the detection of Hg, Pb, and Cd is critical in neuroscience because these metals are highly neurotoxic even at trace levels, and their toxicity depends on chemical form and cellular localization. Hg, particularly methylmercury, crosses the blood–brain barrier, disrupts thiol-dependent proteins, and induces oxidative stress. Pb^{2+} interferes with calcium signaling, impairing synaptic transmission and plasticity, while Cd^{2+} perturbs metal homeostasis, generates reactive oxygen species, and triggers apoptosis [27]. Fluorescent probes enable sensitive, compartment-specific detection of these metals in neural tissue, providing essential insight into their mechanisms of neurotoxicity and roles in neurodevelopmental and neurodegenerative processes.

3.8 Hg^{2+} Detection

Rhodamine-Hg is a turn-on fluorescent chemosensor in which Hg^{2+} binding to a sulfur-containing receptor triggers opening of the rhodamine spirolactam ring, enabling selective detection of mercury ions in chemical and biological systems, including studies of neurotoxicity (Table 4). However, genetically encoded sensors are the gold standard for specific neuroscience applications: MerFS is a FRET-based sensor that uses a mercury-binding protein (MerR) fused to fluorescent proteins (such as CFP and YFP). When Hg^{2+} binds, it causes a conformational change that alters the FRET signal. NeMeHg and iNeMeHg are currently the cutting-edge "Next-Generation" sensors [15]. They are specifically designed to detect methylmercury (MeHg). iNeMeHg is particularly useful for in vivo imaging, allowing researchers to track mercury movement in living brain circuits. In addition, nanoparticles can be used. N-CQDs and N,S/C-dots are carbon-based dots often used in neuro-imaging because they have low toxicity and high photostability.

3.9 Pb^{2+} Detection

Leadmium Green AM is a key sensor for Pb^{2+}. However, it is responsive to Pb^{2+} and Cd^{2+} via thiol chelation (hence, the name Lead/(Cad)mium). Its intensity increases upon binding, but the selectivity between Pb^{2+} and Cd^{2+} remains an area of ongoing improvement (Table 4). Other probes for Pb^{2+} include de novo probe 5-[(anthracene-9-methylene) amino] quinolin-10-ol (ANQ) [28], used in neural cell cultures to study the kinetics of lead entry into cells, and BODIPY-based probes. BODIPY probes with polyamide receptors are used to track lead ions in complex biological environments, including neural tissues, due to their resistance to pH changes. The genetically encoded Met-lead and NIR-PbP are the most significant for direct brain imaging. Met-lead is a FRET-based biosensor. It consists of a Pb-binding protein motif (from the PbrR protein) fused between two fluorescent proteins (ECFP and cpVenus) and has been successfully used in Drosophila brains to monitor real-time lead accumulation and to study how the blood–brain barrier protects against heavy metal toxicity [29].

3.10 Cd^{2+} Detection

Leadmium Green exhibits excellent sensitivity but limited selectivity between chemically similar ions (i.e., Pb^{2+} and Cd^{2+}), a limitation that must be explicitly acknowledged in environmental and biological applications. Another significant challenge in developing selective Cd^{2+} probes stems from the very similar binding properties of Cd^{2+} and Zn^{2+}. Two-photon excited cadmium fluorescent probe (TPCd) and the Liu Cd-1 probe are the most relevant for modern neuroscience research. TPCd is a tool for brain imaging. TPCd allows researchers to visualize cadmium in deep brain slices or live tissue with high spatial resolution and minimal damage to neurons [30]. Liu Cd-1 is an intracellular fluorescent probe. It is highly selective for Cd^{2+} over other metals like Zn^{2+} [31]. Another probe for Cd^{2+}

Table 5
Application-driven probe selection (an example)

Application	Preferred probe type	Rationale
Live-cell Ca^{2+} imaging	Ratiometric (Fura-2)	Quantitative, motion-corrected
High-throughput screening	Turn-on (Fluo-4, Cal-520)	High signal, fast kinetics
Synaptic Zn^{2+}	High-affinity Zn probes	Detect nM Zn^{2+} transients
Redox biology	Rhodamine-based Fe^{2+}	Oxidation-state selectivity
In vivo imaging	NIR cyanine/BODIPY	Tissue penetration

is TPE-HC (TPE-His-Pro-Gly-Cys), which selectively detects Cd^{2+} by aggregation-induced emission effect [32].

Across all metals, probe performance is strongly context-dependent. Binding affinities, fluorescence response magnitudes, and selectivity profiles are influenced by pH, competing ligands, protein binding, and compartmentalization.

4 Conclusions

Fluorescent probes offer unparalleled sensitivity, selectivity, and the ability to provide real-time, non-invasive, high-resolution information. The field has matured into three primary classes of fluorophores: organic small molecules, nanofluorescent probes, and genetically encoded fluorescent sensors. Each class offers distinct strengths, from the chemical versatility and ease of modification of small molecules to the enhanced photostability and large surface area of nanomaterials, and the precise cellular targeting and in vivo applicability of genetically encoded systems. Rather than a singular "best" approach, the application will determine the specific fluorophore (Table 5), and future advances are likely to involve sophisticated hybrid designs that combine the benefits of multiple platforms.

A persistent challenge is achieving optimal selectivity and dynamic range, particularly for ions with similar chemical properties (e.g., Cd^{2+} and Zn^{2+}). The trade-off between high affinity for trace detection and a broad dynamic range for physiological monitoring remains a key consideration in probe design. Furthermore, issues such as photobleaching and non-specific binding, historically limiting factors, are being actively addressed through innovations in nanomaterial engineering and the inherent targeting capabilities of genetically encoded sensors. Continuous innovation in this domain promises to further refine neurometallomics and is indispensable for future breakthroughs in neuroscience.

Conflict of Interest The author has no conflict of interest to declare that are relevant to the content of this chapter.

References

1. De Acha N, Elosúa C, Corres JM, Arregui FJ (2019) Fluorescent sensors for the detection of heavy metal ions in aqueous media. Sensors 19(3):599. https://doi.org/10.3390/s19030599
2. Li L, Wang J, Xu S, Li C, Dong B (2022) Recent progress in fluorescent probes for metal ion detection. Front Chem 10:875241. https://doi.org/10.3389/fchem.2022.875241
3. Czarnik AW (1993) Fluorescent chemosensors for ion and molecule recognition, vol 538. ACS Symposium Series; American Chemical Society, Washington, DC
4. de Silva AP, Gunaratne HQ, Gunnlaugsson T, Huxley AJ, McCoy CP, Rademacher JT, Rice TE (1997) Signaling recognition events with fluorescent sensors and switches. Chem Rev 97(5):1515–1566. https://doi.org/10.1021/cr960386p
5. Valeur B, Leray I (2000) Design principles of fluorescent molecular sensors for cation recognition. Coord Chem Rev 205(1):3–40. https://doi.org/10.1016/S0010-8545(00)00246-0
6. Lai L, Yan F, Chen G, Huang Y, Huang L, Li D (2023) Recent progress on fluorescent probes in heavy metal determinations for food safety: a review. Molecules 28(15):5689. https://doi.org/10.3390/molecules28155689
7. Silswal A, Kavyashree P, Koner AL (2024) Review on lysosomal metal ion detection using fluorescent probes. ACS Omega 9(12):13494–13508. https://doi.org/10.1021/acsomega.3c08606
8. Sun X, Xiang T, Xie L, Ren Q, Chang J, Jiang W, Jin Z, Yang X, Ren W, Yu Y (2025) Recent advances in fluorescent nanomaterials designed for biomarker detection and imaging. Mater Today Bio 32:101763. https://doi.org/10.1016/j.mtbio.2025.101763
9. Shahbaz M, Dar B, Sharif S, Khurshid MA, Hussain S, Riaz B, Musaffa M, Khalid H, Ch AR, Mahboob A (2024) Recent advances in the fluorimetric and colorimetric detection of cobalt ions. RSC Adv 14(14):9819–9847. https://doi.org/10.1039/d4ra00445k
10. Tsien RY (1989) Fluorescent probes of cell signaling. Annu Rev Neurosci 12:227–253. https://doi.org/10.1146/annurev.ne.12.030189.001303
11. Grover K, Koblova A, Pezacki AT, Chang CJ, New EJ (2024) Small-molecule fluorescent probes for binding- and activity-based sensing of redox-active biological metals. Chem Rev 124(9):5846–5929. https://doi.org/10.1021/acs.chemrev.3c00819
12. Grynkiewicz G, Poenie M, Tsien RY (1985) A new generation of Ca2+ indicators with greatly improved fluorescence properties. J Biol Chem 260(6):3440–3450
13. Paredes RM, Etzler JC, Watts LT, Zheng W, Lechleiter JD (2008) Chemical calcium indicators. Methods 46(3):143–151. https://doi.org/10.1016/j.ymeth.2008.09.025
14. Fu Y, Zhang X, Wu L, Wu M, James TD, Zhang R (2025) Bioorthogonally activated probes for precise fluorescence imaging. Chem Soc Rev 54(1):201–265. https://doi.org/10.1039/d3cs00883e
15. Torres-Ocampo AP, Palmer AE (2023) Genetically encoded fluorescent sensors for metals in biology. Curr Opin Chem Biol 74:102284. https://doi.org/10.1016/j.cbpa.2023.102284
16. Xu M, Xing J, Yuan B, He L, Lu L, Chen N, Cai P, Wu A, Li J (2023) Organic small-molecule fluorescent probe-based detection for alkali and alkaline earth metal ions in biological systems. J Mater Chem B 11(15):3295–3306. https://doi.org/10.1039/d3tb00268c
17. Yan Z, Cai Y, Zhang J, Zhao Y (2021) Fluorescent sensor arrays for metal ions detection: A review. Measurement 187(5):110355. https://doi.org/10.1016/j.measurement.2021.110355
18. Jensen GC, Janis MK, Nguyen HN, David OW, Zastrow ML (2024) Fluorescent protein-based sensors for detecting essential metal ions across the tree of life. ACS Sens 9(4):1622–1643. https://doi.org/10.1021/acssensors.3c02695
19. Wu SY, Wen Y, Serre NBC, Laursen CCH, Dietz AG, Taylor BR, Drobizhev M, Molina RS, Aggarwal A, Rancic V, Becker M, Ballanyi K, Podgorski K, Hirase H, Nedergaard M, Fendrych M, Lemieux MJ, Eberl DF, Kay AR, Campbell RE, Shen Y (2022) A sensitive and specific genetically-encoded potassium ion biosensor for in vivo applications across the tree of life. PLoS Biol 20(9):e3001772. https://doi.org/10.1371/journal.pbio.3001772
20. Sambath K, Liu X, Wan Z, Hutnik L, Belfield KD, Zhang Y (2021). Potassium ion fluorescence probes: structures, properties and bioimaging. ChemPhotoChem 5(4):317-325. https://doi.org/10.1002/cptc.202000236
21. de Melo Reis RA, Freitas HR, de Mello FG (2020) Cell calcium imaging as a reliable method to study

neuron-glial circuits. Front Neurosci 14:569361. https://doi.org/10.3389/fnins.2020.569361
22. Günther T (2006) Concentration, compartmentation and metabolic function of intracellular free Mg^{2+}. Magnes Res 19(4):225–36
23. Lee S, Lee HG, Kang SH (2009) Real-time observations of intracellular Mg^{2+} signaling and waves in a single living ventricular myocyte cell. Anal Chem 81(2):538–42. https://doi.org/10.1021/ac8013324
24. Aron AT, Reeves AG, Chang CJ (2018) Activity-based sensing fluorescent probes for iron in biological systems. Curr Opin Chem Biol 43:113–118. https://doi.org/10.1016/j.cbpa.2017.12.010
25. Fahrni CJ, O'Halloran TV (1999) Aqueous coordination chemistry of quinoline-based fluorescence probes for the biological chemistry of zinc. J. Amer Chem Soc 121(49):11448–11458
26. Nolan EM, Lippard SJ (2009) Small-molecule fluorescent sensors for investigating zinc metalloneurochemistry. Acc Chem Res 42(1):193–203. https://doi.org/10.1021/ar8001409
27. Błażewicz A, Grabrucker AM (2022) Metal profiles in autism spectrum disorders: a crosstalk between toxic and essential metals. Int J Mol Sci 24(1):308. https://doi.org/10.3390/ijms24010308
28. Anand T, Sivaraman G, Mahesh A, Chellappa D (2015) Aminoquinoline based highly sensitive fluorescent sensor for lead(II) and aluminum(III) and its application in live cell imaging. Anal Chim Acta 853:596–601. https://doi.org/10.1016/j.aca.2014.11.011
29. Yang DM, Manurung RV, Lin YS, Chiu TY, Lai WQ, Chang YF, Fu TF (2020) Monitoring the heavy metal lead inside living Drosophila with a FRET-based biosensor. Sensors 20(6):1712. https://doi.org/10.3390/s20061712
30. Juvekar V, Park SJ, Yoon J, Kim HM (2021) Recent progress in the two-photon fluorescent probes for metal ions. Coord Chem Rev 427:213574. https://doi.org/10.1016/j.ccr.2020.213574
31. Ding X, Zhang F, Bai Y, Zhao J, Chen X, Ge M, Sun W (2017) Quinoline-based highly selective and sensitive fluorescent probe specific for Cd^{2+} detection in mixed aqueous media. Tetrahedron Lett 58(40):3868–3874. https://doi.org/10.1016/j.tetlet.2017.08.068
32. Zhou J, Wang X, Wang D, Yuan L, Hao J, Hou J, Wu J, Lin P (2024) Design and application of Cd2+polypeptide fluorescent probes based on aggregation induced emission (AIE). J Mol Histol 55(4):581–588. https://doi.org/10.1007/s10735-024-10201-9

Chapter 6

Quantification of Metals in Brain Cells and Tissues: Atomic Absorption Spectroscopy (AAS)

Marta Marina-Latorre, Rosario Pereiro and Lara Lobo

Abstract

Two main techniques for metals analysis are based on atomic absorption spectroscopy (AAS): Flame AAS (FAAS) and Electrothermal AAS (ETAAS). ETAAS, with detection limits in the low μg L^{-1} range or even below, offers much higher analytical sensitivity than FAAS. This chapter presents an outline of the basic principles and instrumentation required for AAS analyses. Next, sampling procedures for brain tissues and cultured cells are described, along with sample mineralization protocols to mitigate potential chemical interferences arising from the samples' organic matrices. Given the lower limits of detection of ETAAS, an ETAAS method is detailed for the analysis of metals of biomedical relevance in specific regions of brain tissue. Throughout the chapter, specific details are provided to achieve high-quality analysis.

Keywords Atomic absorption spectrometry, Flame atomization, Electrothermal atomization, Elemental analysis, Sample mineralization, Quality assurance

1 Introduction

Atomic absorption spectroscopy (AAS) is based on the absorption of monochromatic radiation by free gaseous atoms in the ground state. Therefore, a main requirement for AAS measurements is that the sample containing the metallic elements to be analyzed must be heated to convert them to atomic form. The atoms of the different metals will absorb light at characteristic wavelengths corresponding to their electronic transitions. Since each element exhibits unique electronic transitions, the use of monochromatic radiation of the proper wavelength of the analyte will allow for highly selective detection [1].

The absorbance (Abs) measures the capacity of a substance to absorb light of a specified wavelength (see Eq. 1):

$$\text{Abs} = \log I_0/I \tag{1}$$

Andreas M. Grabrucker (ed.), *Brain Metallomics*, Neuromethods, vol. 237,
https://doi.org/10.1007/978-1-0716-5468-2_6,
© The Author(s), under exclusive license to Springer Science+Business Media, LLC, part of Springer Nature 2026

being I_0 the intensity of the incoming radiation and I the intensity emerging from the atomizer.

Under constant instrument operation conditions, the absorbance is proportional to the concentration of the absorbing analyte (Beer–Lambert law) in the absence of interferences. This proportionality can be easily quantified through a previous instrumental calibration of the analyte, measuring standards of concentrations within the linear dynamic range.

The basic AAS instrument design is illustrated in a simplified scheme in Fig. 1. AAS instruments are rather simple devices not requiring a vacuum for operation as compared to mass spectrometers. The indispensable components to perform AAS measurements are: (i) a light source, (ii) a high temperature atomizer, (iii) a monochromator that allows for minimizing spectral interferences, (iv) a detector (a photomultiplier tube or a solid-state device) and a readout system. In addition, many AAS instruments have a built-in background corrector. As can be seen in Fig. 1, the light source, the atomizer, and the monochromator are aligned.

Typically, line sources such as hollow cathode or electrodeless discharge lamps that emit monochromatic radiation at the appropriate wavelength for the element to be analyzed are employed, allowing single-element detection (i.e., analysis of other elements requires substituting the lamp). More recently, AAS instruments with a continuum radiation source, a high-resolution monochromator, and an array detector have been developed, allowing for the sequential and simultaneous determination of chemical elements from the same sample aliquot [2].

Two main AAS analytical techniques have been developed depending on the type of atomizer employed: flame AAS (FAAS) and electrothermal AAS (ETAAS), which is also called graphite furnace AAS (GFAAS) because the atomizer commonly consists of a graphite-coated furnace.

In FAAS, acetylene is used as the fuel, and air as the oxidant, to analyze most metals. When compounds are insufficiently dissociated in such an atmosphere, a higher-temperature flame (nitrous oxide–acetylene) is used to provide sufficient thermal energy to achieve complete atomization. FAAS instruments are widely employed because they have a comparatively low cost, analytical methods are well established, and they allow high-throughput analysis of single elements. However, its main limitation is the relatively high detection

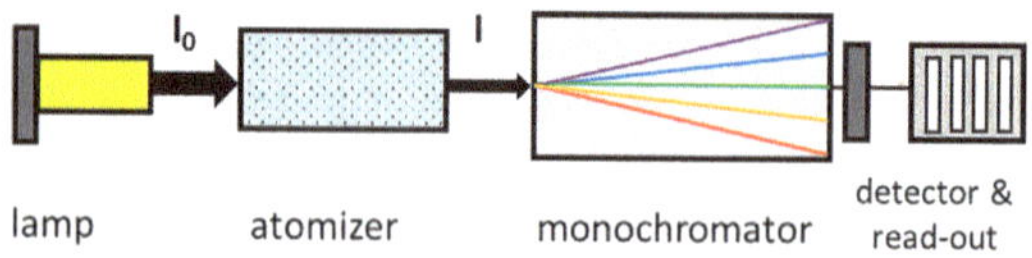

Fig. 1 Schematics of the configuration of an AAS instrument with the basic components

limits, being typically applied when concentrations exceed 1 mg kg^{-1}. Examples of FAAS applications in the analysis of brain tissues include the determination of cadmium [3], iron [4], and lead [5], among others.

ETAAS offers higher sensitivity than FAAS, enabling analysis of metal concentrations at or below 0.01 mg kg^{-1} at a modest cost per analysis, and it is compatible with very small sample volumes (in the order of a few microliters). In ETAAS, the furnace consists of an open tube (made of pyrolytic carbon-coated graphite), a few cm long and a few mm in diameter, with an open orifice in the central part for sample introduction. The tube has an inner platform, typically called "L'Vov platform", which contributes to providing a constant temperature environment [1]. Depending on the ETAAS instrument, the graphite tube can be heated longitudinally (electrical current flows along the length of the tube) or transversally (electrical current flows perpendicular to the optical path, simultaneously along the entire tube length). Transversal heating maintains an isothermal atmosphere along the entire tube length, minimizing chemical interferences caused by condensation of matrix components, which are more pronounced at the tube ends than at its center. An ETAAS analysis begins by introducing a fixed volume of sample into the furnace at room temperature. The tube is then subjected to a well-defined, multistep temperature program by increasing the applied electric current to the furnace. The heating steps include: (i) sample drying, (ii) pyrolysis to remove as many matrix components as possible, and (iii) analyte atomization; in this last step, the light absorbed is measured. Parameters to be considered at each step include final temperature, ramp time, hold time, and argon flow through the tube (applied in all steps except atomization). The analysis ends with a short, high-temperature clean-out step to avoid cross-contamination between successive analyses. An ETAAS analysis requires about 3–4 min.

An optimization of the temperature program is needed for accurate elemental quantification. Moreover, the analysis of volatile and moderately volatile elements by ETAAS requires the use of chemical modifiers, either to thermally stabilize the analyte, allowing for higher pyrolysis temperatures, or to remove sample matrix components. For such purposes, Mg, Pd, Rh, or Ru is mixed into the sample solution, or permanent modifiers such as Pd, W–Rh, or Zr–Pd are coated on pyrolytic graphite tubes [6].

Finally, it is important to note that ETAAS instruments require a background corrector (also recommended in FAAS) to remove residual spectral interferences. The two most widely employed are the Deuterium and the Zeeman background correctors [1, 2]. In the first case, the light from a deuterium lamp (emitting a continuum in the spectral region of interest) and the light from the analytical source cross sequentially the atomizer and the transmitted light is detected and compared. This is a simple and relatively cheap accessory; however, it does not allow correction of the spectral background of the analytical line itself; instead, the spectral background is corrected by averaging the absorption over its surroundings. On the contrary, the background corrector based on the Zeeman effect, which requires the use of a magnetic field, offers improved analytical performance, enabling background correction at the exact wavelength of the analytical line.

Representative applications related to the quantification of different metals in human brain sections by ETAAS [7–19] are collected in Table 1. In addition, as ETAAS offers analytical sensitivity enough, examples of applications can also be found in the literature for the analysis of cells, such as endothelial cells that line the cerebral vasculature [20] and cell cultures [21].

Table 1
Selected applications related to metal determinations in sections of human brain tissues by ETAAS

Elements	Brain regions	Context	Ref.
Al	Temporal, frontal, parietal, and occipital lobes	Familial Alzheimer's disease	[7]
Al	Cortex	Familial Alzheimer's disease	[8]
Al	Temporal, frontal, parietal, and occipital lobes	Multiple sclerosis	[9]
Al	Temporal, frontal, parietal, and occipital lobes	Autism	[10]
Al, Fe, Cu	Temporal, frontal, parietal, and occipital lobes	Aged human brains	[11]
Cd, Pb, Hg[a]	Tumor tissue	Brain tumors	[12]
Fe, Cu, Zn	Gray matter and white matter	Cerebral amyloid angiopathy in Alzheimer's disease	[13]
Fe, Cu	Frontal cortex and hippocampus	Alzheimer's disease and related dementias	[14]
Fe	Frontal cortex, superior and middle temporal, caudate nucleus, putamen, globus pallidus, cingulate gyrus, hippocampus, inferior parietal lobule, visual cortex of the occipital lobe, midbrain, pons, medulla, and cerebellum	Evaluation of anatomical region differences and age-related changes	[15]
Mn	Frontal cortex, encephalic trunk, vermis, nucleus caudatus, cerebellum, hypothalamus	Healthy individuals who suffered sudden, accidental death	[16]
Ni	Cerebral cortex	Non-occupationally exposed subjects	[17]
Sr	Cortex frontalis parasaggitalis, cortex frontalis basalis, lobus occipitalis, lobus parietal, putamen	Alzheimer's disease	[18]
Zn	Hippocampal brain tissue	Alzheimer's disease	[19]

[a]Hg was determined by vapor generation—AAS

In this chapter, the procedure for sampling brain tissues and cell cultures is first described, along with the sample pretreatment steps, which usually include mineralization to avoid potential interferences from the sample's organic components and further conversion to a liquid form. Then, a detailed method for calibration and elemental determination of brain digests by ETAAS, measuring aluminum, iron, and copper as model elements. Specific details for achieving high-quality determinations are provided throughout the chapter.

2 Materials

2.1 Equipment

Electrothermal atomic absorption spectrometer with transversely heated furnace and Zeeman background corrector, laboratory grade microwave oven, drying oven or freeze dryer, hot plate with temperature control, analytical balance, pipettes, PFA or PTFE containers and beakers.

2.2 Reagents

High purity reagents, including 70% (w/v) HNO_3, 35% (w/v) HCl, and 30% (w/v) H_2O_2, are commonly used for mineralization of the samples (as an alternative to H_2O_2, some procedures also collect the use of $HClO_4$, but precautions with this acid must be taken as it might be explosive in contact with organic matter), distilled and deionized ultrapure water, and standard solutions containing 1000 mg L^{-1} of the metals to be analyzed.

3 Methods

3.1 Sampling

Brain Tissues

Analytical information about specific brain regions is required in most studies; therefore, as a first step, it is necessary to isolate those regions to be analyzed. However, the elements of interest for measurement may be present in the tool used for brain dissection, which could act as a contamination source; for example, metals such as Fe and Cr are present in stainless-steel knives. To avoid this risk, ceramic knives are recommended. Once tissues are obtained from the donor, the samples must be frozen and stored at −80°C until analysis. It is not convenient to analyze paraffined/deparaffined samples because metal contamination with chemicals employed, or leaching, can occur during these processes.

Most analytes are at trace or ultratrace levels, so there is also a risk of sample contamination from the storage container. Therefore, they must preferably be made of materials such as PFA or PTFE. In addition, a previous cleaning step of the containers is recommended, using a 5% (w/v) HNO_3 solution, or alternatively 5% (w/v) HCl, for 24 h and then rinsing with ultrapure water.

Cell Cultures

When aiming to provide a given element concentration in a cell culture, the number of cells to be digested must be considered. Traditionally, cultured cell suspensions and adherent cells (after trypsinization) can be accurately counted using a hemocytometer. Basically, an aliquot of a homogeneous cell suspension is mixed with Trypan blue, deposited into the chamber, and counted on the grid. The number of live cells mL^{-1} in the original cell suspension is thus obtained by considering the average number of cells counted in the measured grid set, its volume, and the cell suspension dilution (1:5) with Trypan blue. As an alternative to manual counting, automated cell counters can be used as well. In all cases, issues dealing with contamination related to the reagents used during sample preparation should be considered by preparing the corresponding procedural blank (*see* **Note 1**).

3.2 Sample Mineralization

For brain tissue elemental analysis, before introducing the sample into the instrument for analytical measurements, sample preparation is recommended to eliminate organic matrix interference. Different strategies can be employed, such as conventional acid mineralization on a hot plate or microwave-assisted digestion. Regardless of the procedure selected, it is imperative to ensure that the brain samples are previously weighed, with accuracy and precision, and water must be removed either employing a drying oven [15] or lyophilized with a freeze dryer [14] until constant weight. This step allows us to express the results in terms of dry tissue, especially important because the brain tissues have a high water content and, if not eliminated, would induce important errors in the accuracy of the analytical results.

In the literature, most of the authors report rather similar mineralization methods using microwave-assisted digestion ovens, as it is collected in Table 2. As a sample-preparation example, wet brain tissue samples are placed in a drying oven at 37°C until constant weight [11]. Then, the dried tissues are weighed (see Table 2) and introduced into the microwave oven with a mixture of 1 mL 70% (w/v) HNO_3 and 1 mL 30% (w/v) H_2O_2.

The sample digestion can be performed with the following "power (W)/time (min)" program: 250/1, 0/2, 250/5, 400/5, and 600/5 [15]. The procedure is complete when a colorless or light-yellow, transparent solution free of residues is obtained. Upon cooling, each digest is diluted to a fixed volume with ultrapure water. For quality assurance purposes, in studies that require analysis of a large number of brain tissue samples, each batch of digests should include brain tissue samples, procedural blanks, standard reference materials, and reagent spikes for each element.

Laboratories without a microwave-assisted digestion equipment available can employ a conventional acid digestion procedure on a

Table 2
Selected microwave-assisted digestion protocols were employed for the mineralization of human brain tissues

Dry to constant weight in an incubator at 37°C	0.3 g wet weight	1 mL 70% w/v HNO_3 + 1 mL 30% w/v H_2O_2	7, 8, 9, 10, 11
Dry to constant weight in a dry oven at 110°C	0.1–0.5 g dry weight	2.5 mL 70% w/v HNO_3 + 1 mL 30% w/v H_2O_2	15
Homogenization of the tissue with distilled, deionized water	2 mL brain homogenate (50/60% w/v)	5 mL 70% w/v HNO_3	16
Tissue homogenization	1.5 g wet organ homogenate	Mixture of HNO_3 + H_2O_2 (v/v 5:1)	17
Dry to constant weight	0.025 g dry weight	2 mL 70% w/v HNO_3 + 0.5 mL 30% w/v H_2O_2	18
Dry to constant weight in a dry oven at 80°C	Not specified	0.5 mL 30% w/v H_2O_2 + 0.1 mL 70% w/v HNO_3	19

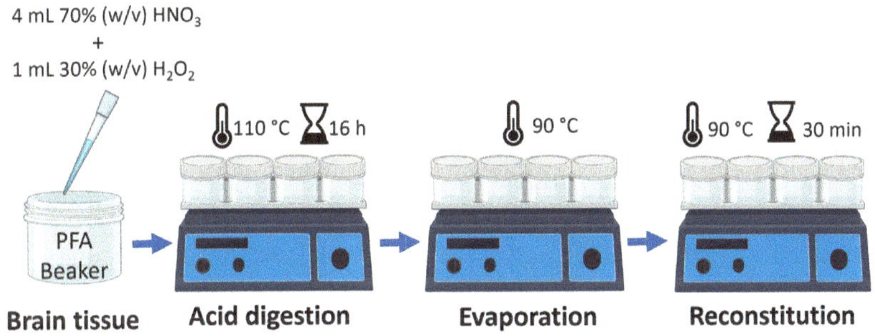

Fig. 2 Representation of the conventional acid digestion protocol on a hot plate employed to digest brain tissues

hot plate, as schematized in Fig. 2. In a typical protocol, an accurately weighed sample amount in the order of 10–100 mg of the sample is typically added to a Savillex® beaker with 4 mL 70% (w/v) HNO_3 and 1 mL 30% (w/v) H_2O_2 [22]. Then, the beakers are closed and placed on a hot plate at 110°C for 16 h to ensure complete digestion. Once this is accomplished and the samples have cooled, the beakers can be opened and placed on a hot plate at 90°C until dry. Then, the residue can be redissolved in the desired solvent, typically 1% (w/v) HNO_3 in ultrapure water solution. As it is collected in **Note 2**, special cautions must be taken to avoid sample contamination (due to reagents and the recipients employed for sample dissolution), as well as the risk of overpressures produced during the mineralization process using acids at high temperatures.

In addition, as described in **Note 3**, ETAAS allows for the direct analysis of solids, typically using the slurry sampling technique [23] as an alternative to the mineralization process. In such cases, no special equipment is needed for sample digestion, and greener sample processes are feasible as compared to mineralization, since fewer reagents are employed. However, it is important to note that, for many applications, ETAAS procedures based on slurry sampling are not as well validated as those involving mineralization processes.

3.3 Calibration and Analyte Determination

In this section, the determination of aluminum, iron, and copper in the temporal, frontal, occipital, and parietal lobes of brain digests prepared by microwave digestion of the samples as described above [11] will be detailed. The selection of these three analytes for ETAAS analyses is representative, since they require different pyrolysis and atomization temperatures: when comparing the three elements, copper is the one requiring the lowest temperatures, while aluminum requires the highest.

ETAAS instruments with longitudinally heated graphite tubes or a deuterium background-corrector implemented could be employed. Nevertheless, in this application, the use of an ETAAS instrument with transversal heating of the graphite tube (allowing for a homogeneous furnace heating) and with a Zeeman background corrector is proposed, offering higher capabilities to subtract background signals right at the analytical emission wavelength [11].

Table 3 lists the temperature program used to measure iron. For copper, the temperature program is the same except for the pyrolysis and the atomization temperatures (1100°C and 1900°C, respectively). On the other hand, for aluminum, pyrolysis occurs in two 10-s steps (500°C and 1200°C), and the atomization step is at 2300°C. For the analysis of the three elements, an argon flow of 250 mL min^{-1} is employed in all steps except during atomization. Sample volumes introduced for each analysis are 20 μL for iron and copper, and 30 μL for aluminum. Each experimental extract

Table 3
Temperature program for the measurement of iron by ETAAS with a transversally heated tube

Step	T°C	Ramp (s)	Hold (s)	Ar flow (mL min^{-1})
Drying	110	10	40	250
Drying	130	15	40	250
Pyrolysis	1300	10	20	250
Atomization	2100	0	5	0
Clean-out	2500	1	5	250

datum must be the arithmetic mean of three injections with a relative standard deviation <10%. In some other research [15], 15 μL of the sample is injected followed by 5 μL of magnesium nitrate solution (employed as a chemical modifier) for the determination of iron in 14 areas of the brain.

Once the procedural blank values are subtracted from the total Abs measured for each sample, the analyte concentration can be obtained using a calibration method. To construct calibration plots, known concentrations of the metal to be analyzed, diluted in the same medium as the sample, are measured (see blue plot termed "external standards" of Fig. 3). Typically, the ETAAS instrument can be calibrated by a serial dilution of 40, 60, and 100 mg L^{-1} solutions of iron, aluminum, and copper with 1% (w/v) HNO_3 in ultrapure water. In this respect, care must be taken to check the linearity of the calibration graph (*see* **Note 4**).

In addition, it must be noted that the remaining sample matrix may sometimes modify the slope of this relationship. This can be checked by using the so-called "standard additions" method (*see* **Note 5**). In that event, matrix-matched standards must be used for calibration [8–10]. However, exact matrix-matching is not always feasible, as the exact matrix composition may not be known; in such cases, it is necessary to resort to the standard addition method for each sample to be analyzed.

3.4 Certified Reference Materials

Because brain tissue is not commercially available as a certified reference material (CRM) for elemental determinations, CRM from tissues of different animal origins is used as an alternative. Some

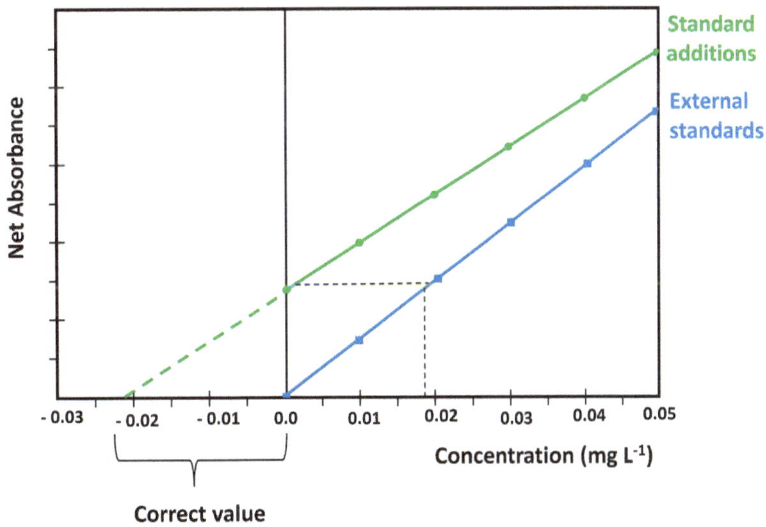

Fig. 3 Main calibration modes in AAS. The blue calibration corresponds to conventional calibration with external standards. The green calibration refers to the standard addition method

examples employed in literature for elemental analysis of human brain tissue are DOLT-4 dogfish liver and DORM-3 fish protein (National Research Council, Canada) for Fe [15]; SRM 1566B oyster tissue from National Institute of Standards and Technology (NIST) and IAEA-407 fish homogenate (International Atomic Energy Agency) for Al, Fe, and Cu [11]; SRM 8414 bovine muscle powder (NIST) for Cd, Hg, and Pb [12]; and SRM 1577a bovine liver (NIST) for Mn [16], Sr [18], and Zn [19]. In addition to CRMs, spiked solutions with known analyte concentrations can also be used for quality control experiments. Both CRMs and spiked samples should undergo the same pretreatment and be used at the same concentrations as the samples.

4 Notes

1. In the case of adherent cells, it should not be overlooked that the trypsinization step required to obtain a cell suspension prior to cell counting might introduce contamination with certain trace elements. In this case, cell counting can be done in a well of the plate, using the remaining wells for digestion (assuming the same number of cells in the digested wells as in the well used for counting). As an alternative to this procedure, cells could be directly removed from the well by adding a small amount of the acid used for mineralization, thereby expressing the analyte concentration with respect to the mass of digest.
2. High purity trace grade analysis reagents (often specifically labeled as ultrapure) are recommended, mainly when dealing with the determinations of elements at ultratrace level. High-purity distilled and deionized water must be employed throughout the process. Moreover, special care must be taken to ensure the beakers used for sample mineralization are cleaned thoroughly, especially when reused in sequential analysis.

For this purpose, a four-step cleaning protocol is typically applied: clean the beakers rigorously with ultrapure water three times, then fill the beakers with a 50% (w/v) HNO_3 solution in ultrapure water, close them and heat them at 110°C for 24 h on a hot plate. These steps should be repeated once more. Then, after rinsing with ultrapure water, the beakers are filled with 50% (w/v) HCl in ultrapure water, closed and heated to 110°C for 24 h on a hot plate. Again, these steps should be repeated once. As an additional tip, if the samples are very fatty, the fat residue can stick to the beaker's edges. In such cases, a preliminary cleaning step should be applied, cleaning the beakers with a soap solution.

In addition, it must be taken into account that the material employed during the sample preparation (flasks, beakers, pipettes, etc.) could contain trace levels of the elements of analytical interest, so a previous cleaning must be carried out, commonly using 5% (w/v) HNO_3 or 5% (w/v) HCl diluted in ultrapure water for 24 h and then rising with ultrapure water. Also, it is highly advisable that the sample preparation be performed in a clean-room laboratory with HEPA/ULPA filters and a laminar-flow hood with positive air pressure. Given the high risk of contamination, it is very important to prepare multiple procedural blanks, which will be measured interspersed throughout the full set of tissue digests.

In addition, it must be considered that overpressures could occur during the mineralization protocol, so it is crucial to avoid them. In the conventional digestion procedure, this could be done by adding the H_2O_2 solution gradually and allowing the reaction to proceed before closing the beakers.

3. Quantification of metals in biological tissues by ETAAS can also be carried out after a slurry sampling preparation procedure. The slurry can be prepared with a protocol like the following [16, 24]: (a) An appropriate aliquot of homogenate is taken, depending on the desired concentration of the slurry (typically 2% w/v), and is transferred to a volumetric flask. (b) Deionized water is added to the sample in the flask to fill approximately half of its capacity, and the flask is slightly agitated for a few seconds to form a homogeneous slurry. (c) In the case of acid addition, the desired amount of nitric acid is added to the flask. (d) The calibrated flask with the slurry is finally filled to the labeled volume with ultrapure water. Then, the sample slurries in the flask are treated for 10 min in an ultrasonic bath at 25°C.

The authors [16, 24] remark that this procedure must be followed in strict sequence to get stable and homogeneous slurries; when the steps are altered, it is common to observe the instantaneous denaturalization of the homogenate and the formation of particles of undesirable size.

4. AAS techniques often suffer from short linear ranges, typically limited by matrix effects, atomizer efficiency, detector saturation, and self-absorption. Therefore, as a first step, it is mandatory to carry out a calibration graph with increasing concentrations of standards to check the dynamic linear range of the method. Appropriate dilutions of the sample with 1% (w/v) HNO_3 are required to bring the analyte concentration within the linear range, or nonlinear calibration approaches must be used for high concentrations [25].
5. In the standard additions method, the sample is spiked with a few additions of known amounts of the element to be analyzed. Therefore, the sample matrix can be considered constant across all solutions to be measured. Any slope differences, comparing the slope observed for pure standards (blue plot) with that for analyte spikes added to the sample (green plot), are proof of matrix interferences. As shown in the green plot of Fig. 3, by extrapolation

back to the negative side of the abscissa, the correct value of the analyte in the sample may be calculated by the standard additions method: 0.021 mg L^{-1} rather than the 0.019 mg L^{-1} obtained using the blue plot.

Acknowledgments

Financial support by project PID2022-137319OB (-C21 and -C22) funded by MCIN/AEI/10.13039/501100011033/FEDER, European Union (EU), and project IDE/2024/000694 of the Government of the Principality of Asturias and FEDER Program of the EU through SEKUENS Agency is appreciated. M. Marina-Latorre acknowledges the "Severo Ochoa" Grant with Ref. PA-23-BP22-183 (Principality of Asturias, Spain).

Conflict of Interest The authors have no conflicts of interest to declare that are relevant to the content of this chapter.

References

1. Sanz-Medel A, Pereiro R (2014) Atomic absorption spectrometry: an introduction, 2nd edn. Momentum Press, LLC, New York
2. Ferreira SLC, Bezerra MA, Santos AS, dos Santos WN, Novaes CG, de Oliveira OMC, Oliveira ML, Garcia RL (2018) Atomic absorption spectrometry–a multi element technique. Trends in Anal Chem 100:1–6. https://doi.org/10.1016/j.trac.2017.12.012
3. Lech T, Sadlik JK (2017) Cadmium concentration in human autopsy tissues. Biol Trace Elem Res 179:172–177. https://doi.org/10.1007/s12011-017-0959-5
4. Litwin T, Gromadzka G, Szpak GM, Jabłonka-Salach K, Bulska E, Członkowska A (2013) Brain metal accumulation in Wilson's disease. J Neurol Sci 329:55–58. https://doi.org/10.1016/j.jns.2013.03.021
5. Haraguchi T, Ishizu H, Takehisa Y, Kawai K, Yokota O, Terada S, Tsuchiya K, Ikeda K, Morita K, Horike T, Kira S, Kuroda S (2001) Lead content of brain tissue in diffuse neurobrillary tangles with calcification (DNTC): the possibility of lead neurotoxicity. Neuroreport 12:3887–3890. https://doi.org/10.1097/00001756-200112210-00006
6. Acar O (2024) The use of chemical modifiers in electrothermal atomic absorption spectrometry. Appl Spectrosc Rev 59:340–354. https://doi.org/10.1080/05704928.2022.2147537
7. Mirza A, King A, Troakes C, Exley C (2017) Aluminium in brain tissue in familial Alzheimer's disease. J Trace Elem Med Biol 40:30–36. https://doi.org/10.1016/j.jtemb.2016.12.001
8. Mold M, Linhart C, Gómez-Ramírez J, Villegas-Lanau A, Exley C (2020) Aluminum and amyloid-β in familial Alzheimer's disease. J Alzheimers Dis 73:1627–1635. https://doi.org/10.3233/JAD-191140
9. Mold M, Chmielecka A, Rodriguez MRR, Thom F, Linhart C, King A, Exley C (2018) Aluminium in brain tissue in multiple sclerosis. Int J Environ Res Public Health 15:1777. https://doi.org/10.3390/ijerph15081777
10. Mold M, Umar D, King A, Exley C (2018) Aluminium in brain tissue in autism. J Trace Elem Med Biol 46:76–82. https://doi.org/10.1016/j.jtemb.2017.11.012
11. House E, Esiri M, Forster G, Ince PG, Exley C (2012) Aluminium, iron and copper in human brain tissues donated to the medical research council's cognitive function and ageing study. Metallomics 4:56–65. https://doi.org/10.1039/c1mt00139f
12. Al-Saleh I, Shinwari N (2001) Levels of cadmium, lead, and mercury in human brain tumors. Biol Trace Elem Res 79:197–203. https://doi.org/10.1385/BTER:79:3:197
13. Schrag M, Crofton A, Zabel M, Jiffry A, Kirsch D, Dickson A, Mao XW, Vinters HV, Domaille

DW, Chang CJ, Kirsch W (2011) Effect of cerebral amyloid angiopathy on brain iron, copper, and zinc in Alzheimer's disease. J Alzheimers Dis 24:137–149. https://doi.org/10.3233/JAD-2010-101503

14. Magaki S, Raghavan R, Mueller C, Oberg KC, Vinters HV, Kirsch WM (2007) Iron, copper, and iron regulatory protein 2 in Alzheimer's disease and related dementias. Neurosci Lett 418:72–76. https://doi.org/10.1016/j.neulet.2007.02.077
15. Ramos P, Santos A, Pinto NR, Mendes R, Magalhães T, Almeida A (2014) Iron levels in the human brain: a post-mortem study of anatomical region differences and age-related change. J Trace Elem Med Biol 28:13–17. https://doi.org/10.1016/j.jtemb.2013.08.001
16. Marcó LM, Hernández Caraballo EA, Pascusso C, Alvarado J (2003) Determination of manganese in brain samples by slurry sampling graphite furnace atomic absorption spectrometry. Talanta 59:897–904. https://doi.org/10.1016/S0039-9140(02)00652-5 16
17. Dudek-Adamska D, Lech T, Konopka T, Kościelniak P (2021) Nickel content in human internal organs. Biol Trace Elem Res 199:2138–2144. https://doi.org/10.1007/s12011-02002347-w
18. Peltz-Császma I, Andrási E, Lásztity A, Kosel S (2005) Determination of strontium and its relation to other alkaline earth elements in human brain samples. Microchem J 79:375–381. https://doi.org/10.1016/j.microc.2004.06.006
19. Bjorklund NL, Sadagoparamanujam V-M, Taglialatela G (2012) Selective, quantitative measurement of releasable synaptic zinc in human autopsy hippocampal brain tissue from Alzheimer's disease patients. J Neurosci Methods 203:146–151. https://doi.org/10.1016/j.jneumeth.2011.09.008
20. Bhattacharjee S, Zhao Y, Hill JM, Culicchia F, Kruck TPA, Percy ME, Pogue AI, Walton JR, Lukiw WJ (2013) Selective accumulation of aluminum in cerebral arteries in Alzheimer's disease (AD). J Inorg Biochem 126:35–37. https://doi.org/10.1016/j.jinorgbio.2013.05.007
21. Song H, Zheng G, Shen X-F, Zhao Z-H, Liu Y, Liu Y, Liu Y-Y, Kang J-J, Chen JY, Luo W-J (2020) An efficient autometallography approach to localize lead at ultrastructural levels of cultured cells. Biophys Rep 6:211–221. https://doi.org/10.1007/s41048-020-00116-9
22. Hobin K, Costas-Rodriguez M, Van Wonterghem E, Vandenbroucke RE, Vanhaecke F (2023) Alzheimer's disease and age-related changes in the Cu isotopic composition of blood plasma and brain tissues of the APPNL-G-F murine model revealed by multi-collector ICP-mass spectrometry. Biology 12:857. https://doi.org/10.3390/biology12060857.
23. Campillo N, Muñoz-Delgado E, López-García I, Baeza-Albarracín Y, HernándezCórdoba M (2010) Suspensions of biological tissues in alkaline medium for the determination of copper, manganese and cobalt by electrothermal atomic absorption spectrometry. Microchim Acta 171:71–79. https://doi.org/10.1007/s00604-010-0411-4
24. Marcó LM, Greaves ED, Alvarado J (1999) Analysis of human blood serum and human brain samples by total reflection X-ray fluorescence spectrometry applying Compton peak standardization. Spectrochim Acta B 54:1469–1480. https://doi.org/10.1016/S0584-8547(99)00085-3
25. Andersen JET (2008) Investigation of the performance of the instrument by nonlinear calibration. Microchim Acta 160:89–96. https://doi.org/10.1007/s00604-007-0804-1

Chapter 7

Inductively Coupled Plasma Mass Spectrometry in Brain Metallomics

Anna Błażewicz

Abstract

Inductively coupled plasma mass spectrometry (ICP-MS) has become the method of choice for determining the elemental composition of biological samples, offering precise isotopic analysis and the ability to distinguish chemical species of elements. ICP-MS, coupled with separation techniques, is a powerful approach for detecting and quantifying metalloproteins and metal-binding proteins, offering high sensitivity and specificity in biomolecular detection and quantification. Laser ablation ICP-MS enables the analysis of small sample volumes with minimal destruction, preserving the integrity of valuable or limited samples. Single-cell (SC) ICP-MS provides insights into cellular heterogeneity, allowing scientists to study variations in metal uptake, distribution, and metabolism at the single-cell level, whereas single-particle (SP) ICP-MS delivers essential information on the number concentration and the size distribution of nanoparticles, which is crucial for understanding the properties and behavior of nanomaterials, including metallotherapeutics. This chapter presents the basic principles of ICP-MS and procedures for preparing brain tissue and cell samples. Typical interferences that can affect measurements and ways to reduce them are discussed. Particular emphasis is placed on speciation analysis, elemental mapping, and recent applications of ICP-MS-based techniques for analyzing brain tissue and cerebrospinal fluid. New research trends are presented, including the integration of ICP-MS with other analytical and omics methodologies in brain metallomics.

Keywords ICP-MS, LA-ICP-MS, SC-ICP-MS, SP-ICP-MS, Bioimaging, Speciation

Abbreviations

2D	Two-dimensional
3D	Three-dimensional
AAS	Atomic absorption spectrometry
AC	Alternating current
AD	Alzheimer's disease
AES	Atomic emission spectrometry
BBB	Blood–brain barrier
CE	Capillary electrophoresis

Andreas M. Grabrucker (ed.), *Brain Metallomics*, Neuromethods, vol. 237,
https://doi.org/10.1007/978-1-0716-5468-2_7,
© The Author(s), under exclusive license to Springer Science+Business Media, LLC, part of Springer Nature 2026

CNS	Central nervous system
CRC	Collision/reaction cell
CRMs	Certified reference materials
CSF	Cerebrospinal fluid
Cys	Cysteine
DC	Direct current
DRC	Dynamic reaction cell
ETAAS	Electrothermal atomic absorption spectrometry
FAAS	Flame atomic absorption spectrometry
GBCA	Gadolinium-based contrast agent
GC	Gas chromatography
GE	Gel electrophoresis
GPC	Gel permeation chromatography
GPX	Glutathione-peroxidase
HAS	Human serum albumin
HPLC	High-performance liquid chromatography
HR	High-resolution
IC	Ion chromatography
ICP-MS	Inductively coupled plasma mass spectrometry
IEC	Ion exchange chromatography
ISO	International Standards Organization
ISs	Internal standards
KED	Kinetic energy discrimination
LA	Laser ablation
LC	Liquid chromatography
LOD	Limit of detection
m/z	Mass-to-charge ratio
MALDI	Matrix-assisted laser desorption/ionization
MC	Multi-collector
Met	Methionine
MS/MS	Tandem mass spectrometry
NanoSIMS	Nanoscale secondary ion mass spectrometry
NTS	Nontarget screening
OES	Optical emission spectrometry
PD	Parkinson's disease
PIXE	Proton-induced X-ray emission
ppb	Parts-per-billion
ppm	Parts-per-million
ppq	Parts-per-quadrillion
ppt	Parts-per-trillion
QQQ	Triple-quadrupole
RF	Radio-frequency
RMs	Reference materials
RP	Reversed phase
SC	Single cell
SEC	Size-exclusion chromatography
SelenoP	Selenoprotein P
SF	Sector field

SIS	Sample injection systems
SP	Single particle
TOF	Time-of-flight
TQ	Triple-quadrupole
TXNRD	Thioredoxin reductase
UV	Ultraviolet
XFM	X-ray fluorescence microscopy

1 Introduction

Inductively coupled plasma mass spectrometry (ICP-MS), developed since the 1980s [1], is currently one of the leading multi-element techniques for the determination of elements in research laboratories [2]. The most popular medical research areas where ICP-MS is commonly used include clinical, toxicological, and pharmaceutical analyses. Although ICP-MS was primarily utilized to study metals and semimetals, which ionize very efficiently due to their low ionization potential, nowadays, ICP-MS can determine around 70% of the elements in the periodic table. Not only naturally occurring elements may be determined, but many radiogenic isotopes, such as technetium, neptunium, plutonium, and americium. Elements that ICP-MS cannot measure are hydrogen and helium (are below the mass range of the mass spectrometer), argon, nitrogen, and oxygen (are present in plasma and air), and neon (has a higher ionization energy than argon). Even fluorine, which poses a challenge for ICP-MS due to its high first ionization potential (higher than that of argon), can be measured by ICP tandem mass (M/M) spectrometry (ICP-MS/MS) using an indirect method [3]. ICP-MS provides quantitative information (mass concentration) as well as qualitative information, including the identification of chemical species (i.e., the specific form of an element defined by its isotopic composition, electronic or oxidation state, and/or complex or molecular structure [4]).

ICP-MS has several advantages over other elemental analysis techniques, such as atomic absorption spectrometry (AAS), ICP optical emission spectrometry (ICP-OES), and ICP atomic emission spectrometry (ICP-AES). Simultaneous, rapid data acquisition (approx. 2-min acquisition time for each sample) for multiple elements is a significant advantage over other techniques, such as single-element flame AAS (FAAS) or electrothermal AAS (ETAAS). Other advantages of ICP-MS are a wide linear dynamic range (over several orders of magnitude), straightforward interpretation of mass spectra, and high precision measurement of isotope ratios [5]. Compared with ETAAS and ICP-OES, the relative sensitivity of ICP-MS is better. ICP-OES cannot measure down to the same sub-ppb concentration levels as ICP-MS. High-resolution (HR) ICP-MS can determine even mid and low parts-per-quadrillion (ppq) concentrations.

The excellent detection capability and increased sample throughput make ICP-MS an indispensable instrument for the routine analysis in environmental, pharmaceutical, and biomonitoring applications [6]. Research laboratories use ICP-MS capabilities to combine elemental analysis with other techniques, such as liquid chromatography (LC) or gas chromatography (GC) for chromatographic separations or laser ablation (LA) for direct sampling of solids. Speciation analysis, single-cell (SC) and single-particle (SP) ICP-MS, and bioimaging of metals are currently the main research directions utilizing ICP-MS. Many studies are integrating central nervous system (CNS) metallomics with omics methods using ICP-MS-based techniques.

2 The Principles of ICP-MS

ICP-MS is based on the ionization of the sample by an inductively coupled plasma, followed by the separation and detection of the resulting ions according to their mass-to-charge ratio (*m/z*). The high argon plasma temperature (~6000–8000 K) is sufficient to atomize and ionize almost all elements. Plasma is formed at the outlet of the ICP torch by the interaction of an intense magnetic field produced by radiofrequency (RF) passing through a copper coil on a tangential flow of argon gas. There are three concentric quartz tubes through which argon flows. The central tube delivers the sample as an aerosol, while the middle tube (the injector) delivers the larger amount of gas needed to create the plasma. The outer tube flows argon, which acts as a cooling agent and prevents the quartz from melting [7]. Samples that are introduced into a high-energy RF plasma undergo desolvation, atomization, and ionization.

Liquid samples are introduced into the plasma by spraying. Solids can be introduced directly into the plasma after laser or electrothermal evaporation or after they have been carried into solution, whereas gaseous substances can be introduced directly [8].

The atomized sample in the nebulizer passes through the mist chamber to the burner and through the interface (connector). It comprises two cones that lower the pressure, which mechanical and turbomolecular pumps maintain. The low pressure is necessary to produce a stream of gaseous ions, which runs on specific tracks through a system of electric and magnetic fields. By the use of electrostatic ion optics elements, the generated ions are transferred to a high-vacuum system, then, before their detection, are separated by *m/z* using a quadrupole, magnetic sector, or time-of-flight (TOF) mass analyzer [9]. The sample introduction system and plasma are operated at atmospheric pressure, whereas the ion optics, mass discrimination, and ion detection areas are operated under high vacuum (10^{-7}-10^{-10} Pa).

Nowadays, most instruments have a collision/reaction cell (CRC) placed between the ion optics elements and the mass spectrometer to reduce the problem of mass interference. The generated ion beam passes through the CRC, which reacts preferentially with interfering ions with the same target *m/z* or with the analyte, which can then be distinguished from the primary interferences. The advantages of the different interference removal approaches are described in **Sect. 2.1**.

The separation of ions from the plasma is often the biggest problem in the ICP-MS technique, and the interface between the torch and the mass analyzer is the critical point in determining how much of the resulting ions will be transferred to the analyzer.

The quadrupole mass analyzer (with CRC or interfaces) is the most common type of analyzer used in clinical applications [5, 6]. It consists of four parallel hyperbolic or cylindrical metal rods to which alternating current (AC) and direct current (DC) RF potentials are applied, creating a time-varying electric field at the center through which the ions pass. For ions with a certain *m/z*, only specific combinations of AC and DC potentials result in a stable trajectory for the ions to fly through the quadrupole. In contrast, ions with unstable trajectories collide with the rods and are not carried through the quadrupole. With the ability to rapidly increase the voltage, scanning the entire mass range is done in a few milliseconds [7]. The resolution at which quadrupole analyzers operate (typically around 0.75 amu) is adequate for most clinical applications, although one should be aware of spectroscopic interferences that arise, such as when non-analyte ions have the same *m/z* as the analyte.

The ICP-MS triple-quadrupole (QQQ) spectrometer includes all the features of single quadrupole analyzers and an additional quadrupole mass filter. An additional mass analyzer filters the ion beam before the ions reach the CRC, thus providing full control over what ions reach the cell. In the ICP-MS tandem configuration (ICP-MS/MS), the reaction gases used in the collision cell can reliably remove interferences by reacting selectively with ions of a specific element [10]. This eliminates the uncertainty and variability seen with reactive cell gases on single-quadrupole ICP-MS systems. In ICP-MS/MS, two operational modes—on-mass and mass-shift—are utilized to enhance analytical specificity and sensitivity by addressing spectral interferences. On-mass mode maintains the analyte at its original *m/z* while removing interferences, whereas mass-shift mode alters the analyte's *m/z* through chemical reactions, facilitating interference-free detection [9, 10].

Regarding ion separation in the TOF-MS analyzer, it uses the principle of different times for given ions to travel the same distance in an area without the influence of any field. Ions of different masses but with the same kinetic energy move at different speeds. The time to reach the detector is inversely proportional to the mass of the ions. The voltage that accelerates the ions is applied periodically so that the ion flux has a pulsating form, allowing separation by velocity. The average transit time through the analyzer is in the range of 10–100 μs.

In order to minimize the number of collisions of gas molecules with ions, a low pressure (10^{-5} Pa) is set in the analyzer [11].

An electron multiplier is the most commonly used detector, which allows signal amplification on the order of 10^6–10^8. Since it operates in sequential mode, it can only collect ions with a specific m/z value at a specific time. Dual detectors (operating in both digital and analog mode) can automatically switch from pulse to analog mode when the signal intensity exceeds a certain threshold. This allows the linear dynamic range of the detector to be extended to about 8–12 orders of magnitude [5, 6].

Quantification, imaging, and isotopic analysis of solid samples can be done with LA-ICP-MS. It allows direct analysis without intensive preparation of such samples [12]. It is a microanalysis technique using high-irradiance ultraviolet lasers to analyze very small spots of solid samples. It enables the user to obtain micron-resolution spatial information in both a lateral and depth direction. Laser operating parameters, such as wavelength, beam power, pulse duration, or beam width, significantly influence the ablation process. The frequency of laser strokes and the energy of the radiation determine the depth and profile of the crater formed after ablation, thus determining the analytical signal intensity of the measured isotopes of selected elements. The laser spot size can be adjusted according to the required spatial resolution. Modern LA systems offer spot sizes down to 1 μm.

A laser beam removes tiny materials from the surface of a solid sample, and the resulting aerosols are directly transmitted to the ICP source for ionization and analysis [13]. The strengths of the LA-ICP-MS method are ultra-traceable sensitivity, large dynamic range, and high spatial resolution (from 100 micrometers to sub-micrometers) [14]. Repeating the ablation analysis process in known x, y, and z coordinates on the sample surface (two-dimensional, 2D) or in the sample volume (three-dimensional, 3D), elemental maps may be generated.

The integrity of tissue samples is essential during the ablation process to meet the requirement for good resolution and reliable elemental analysis. The low temperature must be controlled so that no additional thermal effects disrupt the structure and composition of the sample. A cryogenic LA cell with a small internal volume has been designed and successfully tested for LA-ICP-MS imaging studies of biological samples by Konz et al. [15]. Although there are many different types of cryogenic LA cells, the small size of the cooling electronics and its circular arrangement enabled much better visualization of structurally complex tissue samples. Figure 1 provides a schematic representation of the LA-ICP-MS [16].

More details of the technical design and construction of the various components of modern ICP-MS instruments can be found on instrument manufacturers' websites or elsewhere.

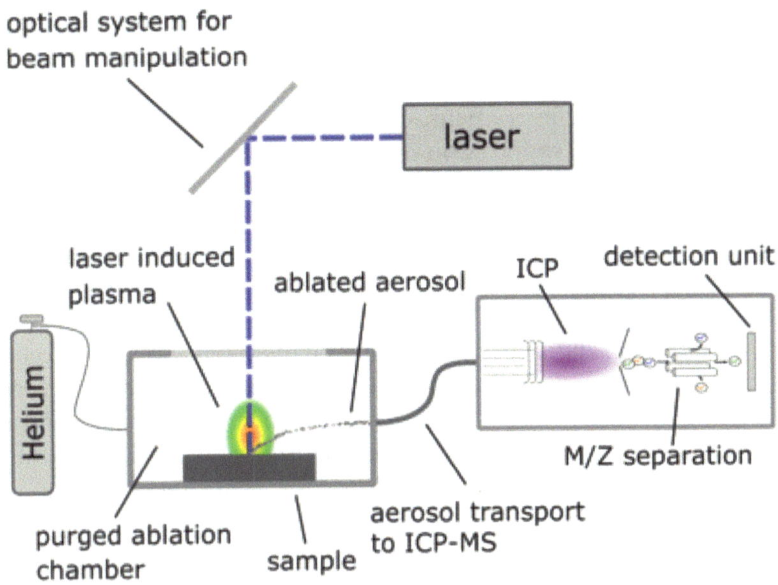

Fig. 1 Schematic representation of the LA-ICP-MS (reproduced by permission of the author Lukas Brunnbauer, TU Wien, Austria; available at [16])

Table 1
Typical limits of detection for the chemical elements are determined by using ICP-MS

Detection limit[a]		
µg/L	**µg/g**	**Elements**
0.1	0.05	Li, Be, B, Mg, Al, Sc, Ti, V, Cr, Mn, Co, Ni, Cu, Zn, Ga, Ge, As, Se, Rb, Sr, Y, Zr, Nb, Mo, Ru, Rh, Pd, Ag, Cd, In, Sn, Sb, Te, Cs, Ba, La, Ce, Pr, Nd, Sm, Eu, Gd, Tb, Dy, Ho, Er, Tm, Yb, Lu, Hf, Ta, W, Re, Os, Ir, Pt, Au, Hg, Tl, Pb, Bi, Th, U
1	0.5	Na, Si, P, K, Ca, Fe, Br, I
10	5	Cl

[a]These detection limits are theoretical best-case scenarios assuming no spectral interferences (based on [19]).

With the push toward lower detection capabilities, especially in the field of clinical toxicology, the main challenge of ICP-MS is providing high-sensitivity analyses with extremely low background noise. Limits of detection (LODs) are in the low parts-per-trillion (ppt) or even ppq range for many elements [17]. Although the terms "limit of detection" and "sensitivity" are not synonymous, an analytical method is often said to be sensitive when the LOD is low [18]. Typical LODs for the chemical elements determined by using ICP-MS are included in Table 1.

The LODs depend on several factors, including the element being measured (elements with low ionization potentials or poor ionization efficiencies may present higher LODs), the sensitivity of the instrument, the sample matrix (high matrix loads can cause signal suppression or enhancement), and the specific measurement conditions (e.g., dilutions used in sample preparation, design of the sample introduction system, or temperature of the plasma). To achieve low LODs, it is essential to control the sensitivity of the ion source carefully, the performance of the detector, mass spectrometer resolution (HR-ICP-MS offers better sensitivity), and ion transmission efficiency. Depending on the application, advanced systems with HR mass spectrometers or CRC technology may be used to improve LODs [20]. ICP-QQQ has much higher sensitivity and a much lower background than single-quadrupole ICP-MS, providing about ten times lower detection limits for most elements [21].

3 Main Analytical Challenges

3.1 Interferences

Conventional ICP-MS measurements are prone to various interferences. Biological tissue samples have complex matrices, and interfering elements are often present at concentrations significantly different from the expected concentrations of analytes. Therefore, prior knowledge of the interferences associated with a specific set of samples is necessary. It allows for selecting the appropriate sample preparation procedure and analytical methodology for determining analytes. As shown in Fig. 2, interferences in ICP-MS are generally classified into three main categories: spectral, matrix, and physical. They may be reduced by appropriate instrumentation and software solutions.

Isobaric interferences occur when ions of different elements have the same m/z, leading to potential misidentification or inaccurate quantification of analytes. These may result from overlapping isotopes with identical m/z values (e.g., isotopes Ba^+, Ce^+, and La^+ occur at m/z 138, causing potential overlap) and from overlapping of doubly charged ions with single-charge ions of other elements (e.g., Pb^{2+} at m/z 103 may interfere with Rh^+ measurements). Because most elements have several isotopes, one isotope without isobaric interferences may be selected to avoid inter-elemental isobaric interferences. When the source of interference is an isotope with approximately the same or lower peak intensity, it is possible to make a correction by measuring the intensity of another isotope of the interfering element. Then, the appropriate correction factor must be subtracted from the intensity of the interfering isotope [22]. However, for some elements, such as ^{87}Rb on ^{87}Sr and ^{204}Hg on ^{204}Pb, isobaric interferences are challenging to remove, even using HR instruments. The removal of

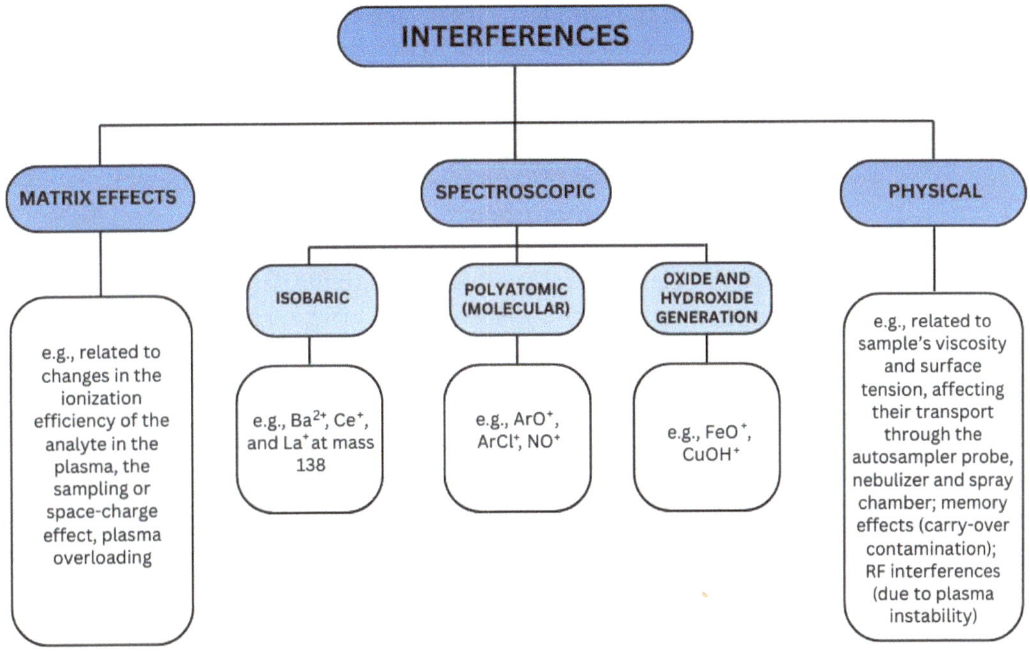

Fig. 2 Main types of interferences in ICP-MS

isobaric interferences in this case may be achieved by using reactive gases (O_2 and NH_3) inside the CRC of the QQQ ICP-MS system, as in the studies of Kutscher and Nelms [23], where the control of undesirable side reactions was provided by an additional quadrupole placed upstream of the CRC.

Another type of spectral interference results from the formation of polyatomic ions between the chemical species present in the Ar plasma and in the sample matrix. Polyatomic interferences are particularly important for m/z values below 82 (e.g., $^{40}ArH^+$, $^{40}Ar^{2+}$, $^{16}OH^+$, $^{16}O^{2+}$, etc.). The most abundant isotope of Ca, ^{40}Ca, has a spectrum identical to that of Ar. Therefore, to avoid interference from ^{40}Ar during ^{40}Ca measurements, ozone was used as the reagent gas in the ICP-MS/MS analysis [24]. Isotopes of heavier elements (e.g., lanthanides) are susceptible to molecular oxide formation. Oxide and hydroxide interferences, caused by both analyte and matrix components, are particularly troublesome in ICP-MS.

Many molecular interferences can be reduced or even eliminated by using cool plasma, CRC, desolvation, and combining ICP-MS with separation techniques. The plasma's oxygen content is reduced by lowering the Ar flow rate, reducing the formation of metal oxides and hydroxides. Furthermore, finding the optimal RF power to achieve efficient ionization without excessive oxidation is crucial.

Modern ICP-MS instruments usually come with an integrated cell that uses either non-reactive collision gas (e.g., helium) or reaction gases (e.g., hydrogen, ammonia, or methane), which help break down interfering species like metal oxides and hydroxides [25]. Due to their inherently larger size, multi-atom interferents are separated from the analyte of interest by kinetic energy discrimination (KED) [26].

Elements from groups 4–6 of the periodic table (e.g., As, Se, Sb, Bi) may react easily with hydrogen in the plasma, so formed hydrides (e.g., AsH^+, SeH^+) can interfere with the detection, especially when the m/z is the same as for another analyte. Hydride-related interferences can be reduced by implementing adequate sample preparation strategies or employing hydride generation techniques to convert the analytes into gaseous hydrides introduced into the plasma. Therefore, to enhance sample analysis and provide complementary information on elements that readily form volatile hydrides, hydride generation (HG)-ICP-MS may be applied.

HR ICP-MS instruments (e.g., sector-field ICP-MS) allow for better separation of ion species. For instance, Fe forms FeO, but an HR instrument can distinguish between the Fe^+ signal and the FeO^+ signal, helping to eliminate interferences. By carefully controlling the ICP-MS operating conditions, the formation of oxide and hydroxide interferences may be effectively reduced. When HR ICP-MS does not provide sufficient precision, a multi-collector (MC) ICP-MS may offer enhanced capabilities. MC-ICP-MS integrates a unique pre-cell mass filter with a dedicated CRC to effectively separate isobaric interferences and matrix elements before isotope ratio analysis. This technique enables the determination of isotopic compositions of the sample with a measurement precision reaching 0.001% [27]. Such high precision is paramount in clinical studies. The results of research studies suggest that stable isotope fractionation may serve as a valuable biomarker for diagnosing and understanding the progression of neurodegenerative diseases such as Alzheimer's disease (AD) [28].

To ensure high-quality analysis, optimizing the sample introduction system is as important as resolution and precision. This optimization ensures that the sample matrix is clean and free from excessive amounts of water, salts, or other substances that may cause the formation of hydroxides or oxides. Using internal standards (ISs) that are free from interference can help correct for any minor variations in the plasma conditions that may cause increased oxide or hydroxide formation. Common ISs used in ICP-MS include elements like Rh, In, or Te, which help normalize signal variations caused by matrix effects.

Some matrix components, like alkali metals, can enhance the ionization of certain analytes, leading to falsely elevated signals. One of the most effective ways to control matrix effects is to use matrix-matched calibration standards that closely match the composition of the sample matrix. This helps account for ionization differences caused by the matrix components. If the matrix contains high concentrations of certain elements that may cause ionization suppression, then dilution of the sample may help reduce matrix effects. Clinical samples usually have complex matrices that may cause ionization suppression or enhancement. Therefore, matrix-matching standards and appropriate dilution techniques are essential to mitigate these effects.

Low analyte levels in a sample can result in signals near the instrument's detection limits, increasing the potential for background noise interference. Handling small sample amounts, in turn, increases the risk of contamination from external sources, including laboratory equipment and the environment. On the other hand, introducing small sample amounts can help mitigate matrix effects. Therefore, more and more micro-flow sample injection systems (SISs) are commercially available. Such systems coupled to ICP-MS/MS were successfully applied to analyze micro samples of biological fluids with a volume of up to 0.4 μL [29].

By selecting appropriate calibration strategies tailored to the sample matrix and potential interferences, analysts can enhance the accuracy and precision of ICP-MS measurements, ensuring that matrix effects are effectively compensated for during analysis. Several calibration strategies have been adopted, including internal standardization, external calibration with certified reference materials (CRMs), internal standards or online addition calibration, and isotope dilution quantification strategies (involving enrichment of the isotopic ratio of the calibration elements with known amounts of a single isotope, so that the resulting measured ratio indicates the concentration of the element that is already present in the sample) [30].

3.2 The Paucity of Certified Reference Materials in Brain Metallomics

CRMs are necessary to evaluate the accuracy of measurements, calibrate instruments, and check the reliability of analytical methods. Unfortunately, so far, no reference material has been produced and certified by national metrological laboratories in accordance with guidelines established by *the* International Standards Organization (ISO) specifically for trace element content in human brain tissue. Due to that limitation, researchers often utilize CRMs from other biological matrices to ensure the accuracy and reliability of their analyses. The absence of a CRM tailored for human brain tissue presents challenges in establishing standardized reference values for trace elements in the brain. This limitation underscores the need for developing such materials to enhance the precision and comparability of trace element analyses in brain metallomics. A typical alternative in the absence of CRMs is to compare results obtained by another method and use reference materials (RMs) other than brain tissues (usually bovine liver or other mammalian tissues) with known analyte concentrations. Commercially available matrices for trace element measurements in clinical samples include frozen blood, serum, urine, lyophilized human blood, or lyophilized bovine blood.

Unfortunately, the paucity of CRMs is one of the reasons for some discrepancies between published data on elemental content in the human brain determined by the use of ICP-MS. Ramos et al. [31] provided a systematic review summarizing the available qualitative and quantitative data on trace element contents in different human brain regions of individuals without any pathologies. Of the 55 eligible studies published over 60 years included in the systematic review, many were conducted using ICP-MS and different RMs. Notably, the authors considered many confounding factors that could undermine the reliability of the results. For instance, studies of samples subjected to any fixation method that could affect the level of the determined elements were excluded. Robust reference values for the elements in healthy brain tissues were reported [31]. Aging and neurodegenerative processes can have different effects on different brain regions and, thus, on measuring the content and distribution of individual elements [32]. Moreover, chemical elements are usually absorbed to varying degrees, which depend on various factors (e.g., physiological conditions, diet, gastrointestinal health, elemental interactions, or blood–brain barrier (BBB) permeability). The aforementioned factors make analyzing brain tissue samples quite a complex task.

4 Sample Preparation

The detection limit can be significantly influenced by the quality of the sample preparation process, which also plays a crucial role in ensuring the accuracy, precision, and reliability of the analytical results obtained via ICP-MS. Inefficient sample preparation can introduce contamination or remove analytes from the sample. After collection using "metal-free" devices, brain tissue samples (stored as frozen or fixed in 4% formaldehyde) are usually subjected to microwave digestion. While the boiling point limits the temperature in an open-vessel digestion system, the temperature that can be reached in a closed-vessel digestion system can be much higher. As a result, faster and more efficient digestion of samples is possible. Modern microwave digestion systems provide overpressure release mechanisms. However, typically, the mass of the tissue sample is not large (e.g., 30 mg brain sample [33]), so overpressure should not be a problem. Usually, such systems are designed to withstand high pressures and temperatures (up to 200 bar and 300°C) [34].

Usually, microwave digestion of clinical samples in a closed system is carried out in Teflon vessels with the addition of a solution of strong oxidizing acid, such as concentrated nitric acid (HNO_3). The carbon-hydrogen bonds are converted to CO_2 and H_2O, and NO_x is released during the digestion of samples. To minimize the volume of concentrated HNO_3, which can corrode the cones at higher concentrations and increase the digestion mixture's oxidizing power, a

30% solution of hydrogen peroxide (H_2O_2) is usually added. Since the H_2O_2 is reduced to water, it does not introduce any additional interference in the preparation procedure and subsequent analysis [35]. Sulfuric acid (H_2SO_4) is not a recommended reagent because it decomposes in the plasma to sulfur-oxygen species, interfering with copper at mass 65 and zinc at mass 64–70, and may hamper the determination of elements forming low-solubility sulfates. Blank samples should be digested in parallel under the same conditions as the samples. An appropriate dilution step may be applied whenever necessary.

Microwave digestion systems are faster than conventional non-microwave-assisted methods and virtually contamination-free (minimizing the risk of cross-contamination between samples by using sealed digestion vessels and automated sample handling systems, ensuring the integrity of analytical results) [36].

All the sample preparation protocols vary depending on the sample introduction system used. As mentioned earlier, LA-ICP-MS of solid samples does not require digestion or dissolution of samples. This poses a significant advantage, as both the introduction of contaminants and the loss of volatiles are minimized.

Particular attention should be paid to soft tissue samples undergoing analysis. Due to microbiological instability and high water content, soft tissue samples must be treated appropriately before and during the ablation procedure. Water removal can change the composition and distribution of elements in biological material, as can the introduction of fixatives. It was assumed that ablation in a cryo-chamber is the standard procedure, but the high equipment cost has led to a preference for ablation at room temperature after prior water removal from the sample [37]. Loss of analytes during freezing and thawing of the sample can be reduced by fixing tissues in formalin (an aqueous formaldehyde solution) and embedding in paraffin [38]. Unfortunately, the problem remains with alkali and alkaline earth metals, which easily dissolve in aqueous formaldehyde solution, and their loss during the material preparation for testing can reach over 80% [39]. Omitting the steps of additional rinsing, fixation, and minimal processing better reflects actual in vivo conditions. Thus, tissue samples are often frozen in liquid nitrogen, and thin sections of a few micrometers are prepared in a cryotome without a fixation step. Storage of such samples requires deep freezing (min −80°C) to avoid proteolytic reactions [38]. Cryo-preparation is currently considered a reference method [37].

In order to visualize the distribution of elements by the use of LA-ICP-MS as closely as possible to represent their actual content in the solid sample, both the preparation and analysis of standards for the preparation of the calibration curve must be carried out carefully, taking into account the nature, structure, and composition of the sample [40].

For single cell (SC) ICP-MS (SC-ICP-MS)—the method described in the next section—the complex preparation procedure (e.g., creating the cell stock solution after removing cell aggregation, washing the cells multiple times from the original exposure medium, and resuspending the cells in a medium free from large amounts of metal analyte) must be performed with extreme caution because cells can be easily damaged at any stage of preparation. The immobilization of cells is crucial to achieving stable ablation. Adherent cultured cells can be analyzed immediately after removal of the medium and rinsing, for example, with saline. Various microfluidic devices are used to align the cells [41].

As for quantitative elemental analysis of nanoparticles and/or cell suspensions that have been previously digested, this is possible using ICP-MS, but the ability to obtain spatial information on the distribution of nanoparticles in a single cell is irreversibly lost.

5 Single-Cell and Single-Particle ICP-MS

ICP-MS has become the method of choice when information about the elemental content of a biological sample is needed, as well as isotopic information and quantitative differentiation of the chemical species of elements. As technology advances, more analytical techniques and models will likely reveal more profound insights into the fundamental processes occurring within a single cell. However, analyzing the content of a given element/metal at the single-cell level remains a major challenge. Assessing the full metallome of an SC—meaning the full complement of metal ions and metal-containing biomolecules—is an emerging area of research that combines advanced analytical techniques with sophisticated bioinformatics tools. LA-ICP-MS analysis of single cells, while more time-consuming than tissue sample analysis, offers the advantage of selectively targeting and measuring cells based on specific size and morphological characteristics.

The study by Haraguchi et al. [42] is considered pioneering in the field of SC-ICP-MS. They conducted a comprehensive metallomic analysis of salmon egg cells, employing both ICP-AES and ICP-MS after microwave-assisted acid digestion to determine the elemental composition of individual cells. This approach allowed for the simultaneous detection of multiple elements within a single biological cell.

Technological advances in SC-ICP-MS have recently accelerated [43–47]. However, preventing analyte loss during cell preparation—such as washing, fixation, permeabilization, drug incubation, and antibody labeling—remains crucial [43, 44]. Due to the small size of the cells, they must be filtered through specialized chambers that allow droplets to enter the ICP. Since most droplets are smaller than

1 μm, a special introduction system was developed to allow larger, micron-sized units to enter the ICP [45]. This system was achieved by modifying the spray chamber design and incorporating new flow patterns inside the chamber, as detailed in a white paper issued by a company specializing in analytical equipment production [45].

Rapid LA-ICP-TOF-MS analysis of hundreds of blood cells at the SC level without additional nonstandard sample preparation was described by Theiner et al. [46]. However, similar studies on single brain cells are rare. Such analyses would aid in understanding the mechanisms of metal uptake and elimination by individual brain cells, determining the distribution of elements within specific cell populations, and elucidating the interactions between metal-containing molecules (like metallotherapeutics) and particular cell types. SC analysis in brain research using ICP-MS may provide insights into how specific neurons or glial cells accumulate these metals, contributing to the development of targeted treatments [47]. Recently, Tanaka and Ogra [48] provided an overview of elemental analysis of single cells using ICP-MS, highlighting advances in instrumentation.

The development of new metallotherapeutics, formulated in nanoparticle form to enhance their efficacy and targeting capabilities, has led to increased interest in single-particle (SP) ICP-MS (SP-ICP-MS) [49]. Incorporating metal-based drugs into nanoparticles offers several advantages, including improved solubility, controlled release, and targeted delivery to specific cells or tissues. SP-ICP-MS enables detection, particle size characterization, quantification of their number and mass concentrations, and dissolved forms of the target element(s). Such an analytical approach utilizes the excellent detection capabilities of ICP-MS. Therefore, studies on the accumulation and neurotoxicity of single nanoparticles may be conducted. This is particularly important when studying the brain's ability to absorb metallotherapeutics and respond to treatment or delivering therapeutic agents across the BBB. Ongoing research aims to optimize nanoparticle design for safe and effective brain-targeted therapies. Recent studies have highlighted the potential of theranostic nanoparticles (integrating therapeutic and diagnostic functions to facilitate simultaneous detection, targeted drug delivery, and therapeutic efficacy monitoring) in preclinical central nervous system models, combining precise imaging with site-specific treatment [50].

Some analytical challenges in measuring nanoparticles and microparticles with SP-ICP-MS include transport efficiency into the ICP. For particles with diameters ≤100 nm, transport efficiency is typically size-independent. However, larger particles exhibit different transport dynamics, affecting measurement accuracy [51]. This may make it difficult to obtain consistent results across different particle sizes.

In recent years, the development of SC-ICP-MS and SP-ICP-MS has accelerated in technical aspects (e.g., modifications to nebulizers, spray chamber design, and the incorporation of new flow patterns within the spray chamber) and in the development of RMs [52]. Although both SC-ICP-MS and SP-ICP-MS are analytical approaches that share the core ICP-MS technology, they differ in their applications, methodologies, and data interpretation [41–53]. Table 2 presents a comparative overview of SC-ICP-MS and SP-ICP-MS.

Table 2
A comparative overview of SC-ICP-MS and SP-ICP-MS

	SC-ICP-MS	SP-ICP-MS
Primary application	Analysis of elemental content within a single biological cell, providing insights into cellular functions and heterogeneity	Characterization of metal-based nanoparticles, including determination of particle size, concentration, and composition
Sample type	Individual cells from biological samples	Metal-containing nanoparticles in suspension
Measurement focus	Quantification of intracellular elements	Measurement of nanoparticle size distribution and concentration
Data acquisition mode	Time-resolved analysis with rapid data acquisition to capture transient signals from single cells	Detection of discrete bursts of signal produced by a particle, enabling size and concentration measurements (typically, for particles with a diameter ≤100 mm, the signal intensity is proportional to the mass of an element)
Sample introduction	For example, flow cytometry or microfluidic devices	Nanoparticle suspensions are introduced directly into ICP-MS
Data interpretation	Analysis of signal intensity and distribution to determine elemental content variations among individual cells, facilitating studies on cellular heterogeneity and function	Evaluation of signal intensity and frequency to calculate nanoparticle size and concentration is essential for nanoparticle characterization and quality assessment
Advantages	High sensitivity for low-abundance elements, minimal sample preparation, preserving cell integrity and the capability to analyze multiple elements simultaneously	Rapid analysis of nanoparticle populations, the ability to detect nanoparticles in complex matrices, provides both qualitative and quantitative data on nanoparticle characteristics
Limitations	Requires optimization of sample introduction to handle single cells effectively, potential challenges with complex biological matrices	Lack of cellular reference materials necessitates careful calibration to distinguish between ionic and particulate signals and may require sample dilution to prevent multiple particles from being detected simultaneously

6 Clinical Utility of ICP-MS

The elemental analysis technique, across various analytical configurations, enables the investigation of the relationships between the content and distribution of chemical elements in tissues, cells, and body fluids and their associations with various diseases, including those affecting the nervous system. Many studies employing ICP-MS have examined the distribution of essential metals across various brain regions, enhancing our understanding of their roles in neurological functions, as well as the presence of toxic metals and their involvement in CNS disorders [54–58].

The versatility of the described technique is expanding, as exemplified by its application for the determination of fluoride, considered an environmental contaminant [59]. Fluoride compounds easily cross the BBB and affect numerous cellular pathways. They play a role in microglial activation and oxidative stress [60]. Over the years, increasing attention has been paid to the effects of fluoride on brain metabolism and cognitive functions, especially in relation to environmental exposure to fluoride via drinking water. Several studies have focused on how fluoride affects the developing brain of children and its possible neurotoxic effects in adults [61, 62]. From an analytical perspective, fluoride measurement poses challenges for ICP-MS. Due to the properties of the plasma, metal fluoride cations (most intense BaF^+) can be formed, allowing indirect detection of fluoride content in brain tissue, with a reported detection limit of about 0.05 mg/L [63].

In the field of trace element analysis, ICP-MS studies are usually performed using mice or rat brain tissues. Animal models of diseases play a crucial role in experimental studies that cannot be conducted on humans due to ethical concerns. DeBenedictis et al. [63] utilized ICP-MS to analyze Fe, Zn, and Cu in fresh brain tissues from diverse model species, including Chordata (vertebrates like mammals and fish), Annelida, Arthropoda (insects), and Mollusca (snails). Exploration of the trace metal distribution in non-mammalian brain tissues emphasized the significant role of trace metal biology across the animal kingdom, not just within mammalian nervous systems.

The subsequent sections focus on developing research areas, including bioimaging and speciation analysis, as well as the integration of ICP-MS with omics methodologies, which highlights the clinical relevance of ICP-MS in the field of brain metallomics.

6.1 Mapping Metal Ion Distribution in the Brain

The development of imaging techniques that provide spatial information about the metal content of individual cells allows for the mapping of specific brain regions and neuronal networks. Understanding how the distribution of metals correlates with neuronal connectivity can lead to new insights into brain function and disorders. Over the years, various techniques have been developed to determine elements both in cell culture samples and under conditions as close as possible to those occurring in vivo [64]. Worth highlighting here is the use of LA-ICP-MS for precise anatomical localization and quantification of trace elements in human and laboratory animal brains ex vivo [65–68]. LA-ICP-MS is useful for investigating the regional and cellular distribution of metals and related metal-containing biomolecules [68, 69]. Moreover, it can shed light on other features, such as the cross-sectional distribution and the presence and distribution of discrete functional units, such as nanoparticles or single cells. The non-destructive nature of LA-ICP-MS facilitates the study of living cell interactions with nanoparticles by preserving the native state of the cells, thus enabling spatially resolved molecular analysis [70].

Of particular importance is the ability of LA-ICP-MS to discriminate stable isotopes, which drives the development of studies of brain physiology after injury or during neurodegeneration, metal binding dysregulation, and the distribution of metal-containing drugs [67, 69]. Using LA-MC-ICP-MS, which combines the advantages of superior ionization of an ICP source and the precise measurements of a magnetic sector MC-MS, it was possible to follow the fate of zinc (by imaging the ratio of tracer isotopes of Zn, i.e., ^{67}Zn and ^{70}Zn to ^{64}Zn) in the rat brain [71]. It turned out that different brain regions differed in the kinetics of zinc metabolism. The results of Solovyev et al. [72], who also applied MC-ICP-MS for measuring the isotope ratios, indicated significant pathology-specific changes in brain Fe and Zn homeostasis in mouse models of AD. It was suggested that the associated changes in isotopic composition may serve as a marker of the proteinopathies underlying AD and other types of dementia.

LA-ICP-MS imaging has been applied to a wide range of sample types and matrices in various disease states, including animal models of neurological disorders [69, 73]. By combining high analysis sensitivity, high scanning speed, and short integration time, LA-ICP-MS helps simultaneously generate maps of many elements from single brain sections. Hare et al. [73] described a detailed protocol for bio-imaging iron, copper, manganese, and phosphorus in a 6-hydroxydopamine lesioned mouse brain. Heterogeneous distribution of these elements throughout the brain and the abnormalities that occurred after injury or intoxication were revealed. The 3D maps showing the variability of trace element concentrations in the mouse brain were generated by LA of consecutive, serial slices quantified against tissue standards from the same animal [74]. The demonstration of quantitative differences and spatial distribution of elements could then be used to track the expression and functionality of metal-binding proteins.

LA-ICP-MS is particularly useful in detecting light ions (low-energy X-ray emission), in which it has advantages over proton-induced X-ray emission (PIXE) and X-ray fluorescence microscopy (XFM) [75]. Figure 3 compares common ex vivo direct elemental mapping

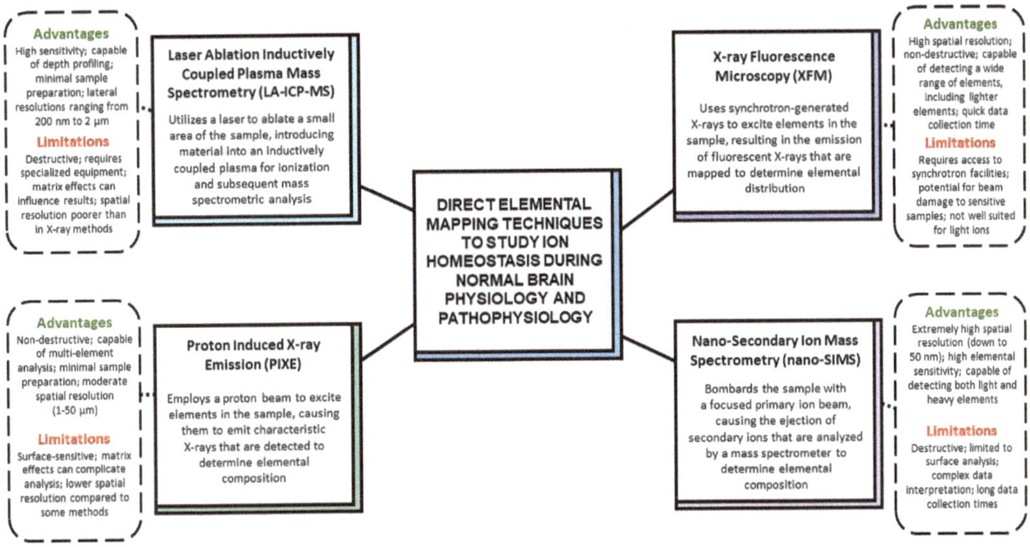

Fig. 3 Comparison of direct elemental mapping techniques to study ion homeostasis during normal brain physiology and pathophysiology

techniques, including nanoscale secondary ion mass spectrometry (NanoSIMS), used to study ion homeostasis in normal brain physiology and pathophysiology.

6.2 Speciation Analysis, Hyphenated Techniques, and Integration of ICP-MS with Omics Methodologies

It was not until the 20th century that scientists began to recognize that the biological effects of chemical elements could vary depending on their chemical forms (species). The development of high-performance liquid chromatography (HPLC) coupled with ICP-MS and other analytical techniques allowed the study of the distribution and effects of different species of elements in biological samples such as biological fluids or tissues [75, 76]. This has contributed to the start of clinical research into the role of various metal species in diseases such as cancer, cardiovascular disease, poisoning, and neurological disorders. Decades of research on chemical speciation have proven that the concept of speciation is fundamental to understanding metal neurotoxicity [76–90].

Contemporary metallomics extends beyond measuring the total content of biologically relevant metals or studying their localization at the tissue or cellular level. Researchers engaged in metallomics and speciation analysis focus on chemical compounds and elements characterized by their isotopic composition, oxidation state, and structural configuration. The redox state of elements influences their ability to traverse biological barriers, including the BBB and the blood-cerebrospinal fluid (CSF) barrier. Compromise of these barriers can lead to chemical-induced neurotoxicity and contribute to various neurological disorders [80, 81]. Identifying the specific speciation forms of elements in the tissue or cell provides knowledge that is useful for understanding the metabolism of metals and metalloids. Moreover, speciation analysis brings further possibilities for understanding the complex interactions of the metallome with the genome, proteome, or lipidome. The term "speciomics" is an umbrella concept encompassing all omics approaches dedicated to speciation analysis [91].

Because plasma conditions can destroy molecular information in a sample, combining techniques is often necessary in speciation studies. A prime example is chromatography in its various forms, which separates compounds in a sample for subsequent elemental analysis using ICP-MS. Furthermore, techniques coupled with ICP-MS enable the detection of analytes at lower concentration thresholds. As mentioned earlier, ICP-MS can identify trace species in a sample down to subparts per billion levels, while LC-ICP-MS typically offers lower detection limits than LC-MS alone.

Regarding miniaturized separation techniques, when coupling nano- or micro-HPLC with ICP-MS, efficiently introducing the flow effluent into the plasma is crucial. This requires specialized interfaces, such as dedicated micro-nebulizer/spray chamber combinations and adding sheath flows, to ensure effective nebulization and minimize sample loss. A group of Chinese researchers has developed a "zero-interfacing" system to couple micro-HPLC with ICP-MS, aiming to minimize dead volume and molecular diffusion [92]. This system features a custom-designed direct injection nebulizer with a tapered nozzle, accommodating a capillary chromatographic column as the sample capillary. Argon gas flow is directed through the nozzle to efficiently transport analytes from the micro-HPLC directly into the ICP-MS plasma for analysis. A critical factor is the distance of the nebulizer nozzle from the plasma base, as too close a distance to the hot plasma can cause the nebulizer nozzle to melt [92].

It should be noted that speciation studies of metals, metalloids, and non-metals in brain tissues or cells are less frequent than in other tissues due to several important factors. Firstly, the brain's complex structure and the dynamic nature of the metallome make it difficult to accurately identify and quantify the specific chemical forms of elements. Secondly, techniques such as LC-ICP-MS are powerful but require meticulous sample preparation to prevent changes in metal speciation during analysis [80]. Preparing brain tissue samples (e.g., extraction, clean-up, pre-concentration, derivatization) can alter the speciation forms of metals. Therefore, handling and analyzing samples carefully is crucial to prevent such alterations. Furthermore, many analytical methods lack the sensitivity and specificity needed to detect trace amounts of metal species in complex biological matrices like brain tissue. Additionally, the non-covalent nature of many metal species and their low concentrations in the brain make them challenging to study [86].

Strzeminska et al. [87] investigated the presence of gadolinium (Gd) in different chemical forms in animal brains, taking advantage of the fact that the solubility of the macrocyclic and linear Gd complexes drastically differs in their in-body stability. This study was a preliminary attempt to evaluate some potential properties of Gd species with prolonged residence time in the brain. ICP-MS was applied to determine total Gd concentrations, whereas size-exclusion chromatography (SEC) was coupled to ICP-MS to analyze soluble Gd species. The insoluble Gd species were analyzed by SP-ICP-MS.

Frenzel et al. [88] conducted research to understand the mechanisms of Gd deposition in the brain following the administration of various Gd-based contrast agents (GBCAs). To assess the chemical forms of residual Gd after repeated GBCA administrations in magnetic resonance imaging (MRI) studies, they employed centrifugation to separate soluble and insoluble fractions of Gd compounds. Subsequently, soluble fractions were separated using gel permeation chromatography (GPC). The Gd concentration in all tissue fractions and the GPC eluate was then measured using ICP-MS.

Speciation studies are often performed on CSF samples because CSF is a readily accessible biological material that provides valuable insights into neurological conditions. CSF samples are most often collected by lumbar puncture for well-defined medical indications. Like any invasive procedure, it carries the risk of medical complications. However, CSF seems to be a practical alternative for obtaining molecular information without surgery. CSF analysis can help track toxic metal exposure and its potential role in the development of neurodegenerative diseases or cognitive impairments [83–85]. The metal profiles in CSF can serve as biomarkers for diagnosing or monitoring disease progression. Moreover, CSF can provide real-time data on how metals are transported and regulated in the brain. In addition, the dynamics of metal homeostasis, which is essential for understanding how metals influence processes such as neurotransmission, enzyme activity, and oxidative stress, may also be captured.

Vinceti et al. [89] employed ion exchange chromatography (IEC) hyphenated with ICP-MS to analyze selenium species in human CSF samples from patients with mild cognitive impairment. Their findings suggest the relation between concentrations of various Se species in CSF and the risk of conversion to Alzheimer's dementia or other dementia. Research has demonstrated that among various selenium species (including selenite (Se(IV)), selenate (Se(VI)), selenomethionine-bound selenium (Se–Met), selenocysteine-bound selenium (Se–Cys), thioredoxin reductase-bound selenium (Se–TXNRD), glutathione peroxidase-bound selenium (Se–GPX), selenoprotein P-bound selenium (Se–SelenoP), and human serum albumin-bound selenium (Se–HSA)) the inorganic hexavalent selenate (Se(VI)), which lacks a direct physiological role, was associated with AD.

Nischwitz et al. [83] provided data on total concentrations, as well as speciation information of elements such as Al, As, Ca, Cd, Cu, Fe, Mg, Mn, Hg, Pb, Se, and Zn in CSF. Total concentrations of all these elements were determined by the use of ICP-MS, HR-ICP-MS, or ICP-DRC-MS. The authors highlighted the potential of hyphenated techniques as diagnostic tools for analyzing CSF to determine both covalent and non-covalent metal species. ICP-MS coupled with an HR separation technique (e.g., reversed-phase (RP) RP-HPLC-ICP-MS or IEC-ICP-MS) was the most useful in most speciation studies reported. However, the variations in concentrations obtained from different studies are noticeable. The authors pointed out numerous interferences in their identification using ICP-MS in standard and HR mode (an example is vanadium and $(^{35}Cl^{16}O)^+$ interferences at $^{51}V^+$ caused by high NaCl concentration in the CSF).

A combination of SEC with ICP-DRC-MS and capillary electrophoresis (CE) coupled with ICP-DRC-MS in a 2D approach has been employed to study Mn speciation in paired serum/CSF samples [90]. Such an analytical approach enabled the separation and identification of Mn species based on their size and charge, providing insights into their distribution and potential roles in neurotoxicity. Research indicates that Mn-induced neurotoxicity is influenced by chemical speciation and by factors such as brain cell types, their origins, and developmental stages. Certain Mn species may be more readily taken up by specific neuronal populations, leading to differential susceptibility to Mn-induced damage. The findings of Michalke et al. [93] suggested that elevated serum Mn-citrate levels may be a valuable marker of increased risk for neurological disorders such as manganism.

Targeted analysis of Mn speciation may have diagnostic and therapeutic benefits for the broad human population, as Mn dyshomeostasis has been linked to the increasing incidence of Parkinson's disease (PD) and other neurodegenerative disorders such as AD disease and Huntington's disease [93, 94]. Proven analytical tools such as ICP-MS and well-designed sample analysis protocols are essential. Developing efficient extraction methods that maintain species stability during extraction is crucial [95]. Regarding the future development of speciation assays for clinical samples such as brain tissues and single cells, the need to produce certified and commercially available standards is highlighted [96].

Employing ICP-MS as a detector in LC or GC enhances compound identification confidence [97–99]. Moreover, combining it as an element-specific detector with other techniques, such as 2D-GE or ion chromatography (IC), greatly expands its analytical capabilities and practical applications in a metallomic context [97]. Commonly used in environmental applications, IC is becoming increasingly valuable for clinical applications [97–99]. Coupling an IC with ICP-MS provides both selective ion separation and precise identification and quantification via mass spectrometry. Moreover, improved precision and repeatability of the determinations are anticipated. Employing IC-ICP-MS, different valence states of Cr, Se, or As in the form of organic and inorganic species can be sensitively and unambiguously identified in one single run [97].

Recently, Iwase et al. [100] have developed an integrated LA/LC-ICP-MS approach, enabling the direct ablation of rat brain tissue sections followed by separation and identification of trace element species through LC-ICP-MS. Both imaging and speciation studies were conducted simultaneously.

A deeper insight into species structure is possible by combining ICP-MS with MS/MS. The main advantages of such a combination compared to conventional single-quadrupole ICP-MS include, first and foremost, much better separation of adjacent masses due to two mass-filtering steps. Moreover, better detection limits can be achieved due to higher sensitivity and lower background. A recent publication by Niehaus et al. [101] presented multimodal analytical tools for molecular and elemental characterization of lesions in brain tissue from patients with multiple sclerosis. The research methods used included ICP-MS/MS and LA-ICP-TOF-MS, which enabled highly sensitive quantitative analysis and spatial microanalysis of endogenous and xenobiotic elements within and around the lesions.

Advances in metal determination strategies have expanded ICP-MS applications beyond direct profiling of the metallome to include analysis of biomolecules such as proteins and nucleic acids [102]. In 2013, Hu et al. [103] introduced a method combining continuous-flow GE with ICP-MS to identify metal-associated proteins within cells. The applied strategy enabled the simultaneous identification of intracellular proteins via matrix-assisted laser desorption/ionization (MALDI)-TOF-MS analysis and the quantification of protein-bound metals by ICP-MS. The coupled techniques were successfully applied to identify protein targets of a bismuth-based drug in *Helicobacter pylori* and to analyze the metal-binding properties of the protein SlyD in *Escherichia coli*. Since then, numerous studies have been conducted to identify the protein targets of metallodrugs in human cells, contributing to the rapid development of metalloproteomics [102–105].

The SEC-ICP-MS method was developed to profile metal-binding proteins in primary cultures of neurons and astrocytes [106]. This technique enabled the differentiation of metalloprotein profiles, highlighting both common and unique metalloprotein species [106]. SEC is a widely used technique for separating proteins based on their molecular size, preserving their native state and weak metal–protein interactions. However, its lower resolution and peak capacity can be limiting compared to other chromatographic methods. To maintain protein integrity, minimizing sample handling stresses is essential, avoiding metal chelators like ethylene diamine tetra-acetic acid (EDTA) and limiting freeze-thaw cycles [107].

ICP-MS has become increasingly valuable in identifying proteins that bind heteroatom-containing drugs, such as platinum, iodine, or copper-based therapeutics. By detecting the elemental signatures of these drugs, ICP-MS facilitates the identification of their protein targets, thereby elucidating their mechanisms of action and potential side effects [108]. In addition, some metabolomics studies incorporate metallomics to investigate susceptibility and resistance to metal-based anticancer drugs [105].

ICP-MS enhances compound identification and quantification in nontarget screening (NTS) by providing element-specific responses and precise compositional data. Feldmann et al. [109] discussed the challenges and future directions of ICP-MS development in NTS, which gained popularity in environmental monitoring and environmental neurochemistry based on the proven influence of environmental factors on brain health [110].

The future of the ICP-MS technique also lies in the simplification of biomolecular analyses, which may offer efficient and accurate methods for studying biomolecular interactions (e.g., antibody-antigen binding, nucleic acid base pairing, enzyme-substrate interactions, and aptamer-affinity binding) without the complications of labeling [111]. Both the development of label-free assays and the use of metal nanoparticle labels continue to enhance the applicability of ICP-MS in biomolecular interaction studies [104].

The widespread implementation of LC-ICP-MS in routine clinical practice faces several challenges, including high equipment and maintenance costs, which may pose barriers for many healthcare facilities. Additionally, obtaining regulatory approval for new diagnostic methods can be a lengthy and complex process. Despite these challenges, the growing recognition of the importance of speciation analysis and mass spectrometry-coupled techniques in clinical science is encouraging [112]. Advances in microfluidics, improved sample preparation methods, the development of modern sample introduction systems, and more sensitive ICP-MS instruments are expected to enhance the resolution and sensitivity of single-cell analysis. Future studies are anticipated to combine data from ICP-MS analyses of SC or SP with other single-cell omics technologies, such as transcriptomics, proteomics, or metabolomics. This integrated approach should provide a more comprehensive understanding of the effects of essential and toxic elements on brain function at the molecular level, both in health and various disease states. Data obtained through various ICP-MS methodologies, enhanced by continually evolving bioinformatics tools, are expected to broaden our understanding of brain function and associated diseases. Integrating metallomics data with other omics datasets will illuminate previously unknown cellular functions and regulatory mechanisms involving metal ions.

Conflict of Interest

The author has no conflicts of interest to declare that are relevant to the content of this chapter.

References

1. Houk RS, Fassel VA, Flesch GD et al (1980) Inductively coupled argon plasma as an ion source for mass spectrometric determination of trace elements. Anal Chem 52(14):2283–2289. https://doi.org/10.1021/ac50064a012
2. Yang HS, LaFrance DR, Hao Y (2021) Elemental testing using inductively coupled plasma mass spectrometry in clinical laboratories: an ACLPS critical review. Am J Clin Pathol 156(2):167–175. https://doi.org/10.1093/ajcp/aqab013
3. Jamari NLA, Dohmann JF, Raab A et al (2017) Novel non-target analysis of fluorine compounds using ICPMS/MS and HPLC-ICPMS/MS. J Anal Atom Spectrometry 32:942–950. https://doi.org/10.1039/C7JA00051K
4. IUPAC Compendium of Chemical Terminology (2025) International Union of Pure and Applied Chemistry Online Version 5.0.0. 5th edn. https://doi.org/10.1351/goldbook.CT06859
5. Wilschefski SC, Baxter MR (2019) Inductively coupled plasma mass spectrometry: introduction to analytical aspects. Clin Biochem Rev 40(3):115–133. https://doi.org/10.33176/AACB-19-00024
6. Thomas R (2013) Practical guide to ICP-MS: a tutorial for beginners. 3rd edn. CRC Press, Taylor & Francis CRC Press Group, ISBN-13:978-1-4665-5544-0 (eBook-PDF)
7. Jurkowski K, Piekoszewski W (2015) Spektrome-tria mas sprzężona z plazmą wzbudzaną indukcyjnie z mikropróbkowaniem za pomocą odparowania laserowego (LA ICP-MS). Wydawnictwo Scientiae et Didactics, Kraków. ISBN: 978-83-941637-0-9
8. Todol JL, Mermet JM (2008) Liquid sample introduction in ICP spectrometry. Elsevier. https://doi.org/10.1016/B978-0-444-53142-1.X0001-5
9. Cygański A (2017) Metody spektroskopowe w chemii analitycznej. Wydawnictwo WNT, Warszawa
10. Clases D, Gonzalez de Vega R (2022) Facets of ICP-MS and their potential in the medical sciences—part 1: fundamentals, stand-alone and hyphenated techniques. Anal Bioanal Chem 414:7337–7361. https://doi.org/10.1007/s00216-022-04259-1
11. Mamyrin BA (2001) Time-of-flight mass spectrometry concepts, achievements, and prospects. Int J Mass Spectrom 206(3):251–266. https://doi.org/10.1016/S1387-3806(00)00392-4
12. Pozebon D, Scheffler GL, Dressler VL et al (2014) Review of the applications of laser ablation inductively coupled plasma mass spectrometry (LA-ICP-MS) to the analysis of biological samples. J Anal Atom Spectrometry 29:2204. https://doi.org/10.1039/c4ja00250d
13. Russo RE, Mao X, Gonzalez JJ, et al. (2013) Laser ablation in analytical chemistry. Anal Chem 85(13):6162–77. https://doi.org/10.1021/ac4005327
14. van Malderen SJM, Managh AJ, Sharp BL et al (2016) Recent developments in the design of rapid response cells for laser ablation-inductively coupled plasma-mass spectrometry and their impact on bioimaging applications. J Anal Atom Spectrometry 31:423–439. https://doi.org/10.1039/C5JA00430F
15. Konz I, Fernández B, Fernández ML et al (2014) Design and evaluation of a new Peltier-cooled laser ablation cell with on-sample temperature control. Analyt Chim Acta 809:88–96. https://doi.org/10.1016/j.aca.2013.11.040
16. TU Wien. (2025). Laser Ablation–Inductively Coupled Plasma–Mass Spectrometry (LA-ICP-MS). https://www.tuwien.at/en/tch/lalibs/techniques/la-icp-ms#c507609 (accessed 1 Mar 2025)
17. Al-Hakkani MF (2019) Guideline of inductively coupled plasma mass spectrometry "ICP–MS": fundamentals, practices, determination of the limits, quality control, and method validation parameters. SN Appl Sci 1:791. https://doi.org/10.1007/s42452-019-0825-5
18. Massart DL, Vandeginste BGM, Deming SN, Michotte Y, Kaufman L (2003) Chapter 7: Sensitivity and Limit of Detection. In: Chemometrics: A Textbook. Data Handling in Science and Technology, vol. 2. Elsevier, pp. 107–114.
19. EAG Laboratories (2017) ICP-OES and ICP-MS detection limit guidance. Sunnyvale, CA: EAG Laboratories. https://www.eag.com/wp-content/uploads/2016/09/application-note-icp-oes-and-icp-ms-detection-

limit-guidance-M-000516.pdf(Accessed: 1 March 2025).
20. Jackson B, Liba A, Nelson J (2014) Advantages of reaction cell ICP-MS on doubly charged interferences for arsenic and selenium analysis in foods. J Anal Atom Spectrometry. https://doi.org/10.1039/C4JA00310A.
21. Koch J, Günther D (2017) Laser ablation inductively coupled plasma mass spectrometry. In: Lindon JC, Tranter GE, Koppenaal DW (eds) Encyclopedia of spectroscopy and spectrometry (3rd edn). Academic Press, pp. 526–532. https://doi.org/10.1016/B978-0-12-803224-4.00024-8
22. Taylor HE (2001) Inductively coupled plasma mass-spectrometry, practices and techniques. Academic Press, New York
23. Kutscher D, Nelms S (2017) Effective removal of isobaric interferences on strontium and lead using triple-quadrupole ICP-MS. Spectroscopy 32(10):38–41
24. Asakawa D, Zhu Y (2024) Determination of calcium isotopes using inductive coupled plasma-tandem mass spectrometry with ozone-induced oxidation. Anal Chem 15,96(41):16315–16321. https://doi.org/10.1021/acs.analchem.4c03391
25. Koppenaal DW, Eiden GC, Barinaga CJ (2004) Collision and reaction cells in atomic mass spectrometry: development, status, and applications. J Anal Atom Spectrometry 19:561–570. https://doi.org/10.1039/B403510K
26. Yamada N (2015) Kinetic energy discrimination in collision/reaction cell ICP-MS: theoretical review of principles and limitations. Spectrochim Acta Part B Atomic Spectrosc 110:31–44. https://doi.org/10.1016/j.sab.2015.05.008
27. Sullivan K, Moore RET, Vanhaecke F (2023) The influence of physiological and lifestyle factors on essential mineral element isotopic compositions in the human body: Implications for the design of isotope metallomics research. Metallomics 15(3). https://doi.org/10.1093/mtomcs/mfad012
28. Moynier F, Borgne ML, Lahoud E et al (2020) Copper and zinc isotopic excursions in the human brain affected by Alzheimer's disease. Alzheimer's Dementia 12(1):e12112. https://doi.org/10.1002/dad2.12112
29. Liu T, Bolea-Fernandez E, Vanhaecke F (2022) Analysis of microsamples of biological fluids via inductively coupled plasma-tandem mass spectrometry using a micro-flow injection system coupled to a high-efficiency sample introduction setup. Spectrochim Acta Part B Atom Spectrosc 198:106546. https://doi.org/10.1016/j.sab.2022.106546
30. Pan H, Feng L, Lu Y et al (2022) Calibration strategies for laser ablation ICP-MS in biological studies: a review. Trends Analyt Chem 156:116710. https://doi.org/10.1016/j.trac.2022.116710
31. Ramos P, Santos A, Pinto NR et al (2014) Anatomical region differences and age-related changes in copper, zinc, and manganese levels in the human brain. Biol Trace Elem Res 161:190–201. https://doi.org/10.1007/s12011-014-0093-6
32. Ashraf A, Michaelides C, Walker TA et al (2019) Regional distributions of iron, copper and zinc and their relationships with glia in a normal aging mouse model. Front Aging Neurosci 11. https://doi.org/10.3389/fnagi.2019.00351
33. Bellouard M, de la GrandMaison GL, Cappy J et al (2022) Trace elements repartition in body fluids, hair and organs in an autopsied population evaluated by ICP-MS high resolution. Environ Toxicol Pharmacol 95:103978. https://doi.org/10.1016/j.etap.2022.103978
34. Stadler A, Michaelis M (2021) A chemist's guide to sample preparation, 3rd edn, vol. 21. Anton Paar Publishing. Graz. https://www.anton-paar.com/corp-en/a-chemists-guide-to-sample-preparation/
35. Krachler M, Radner H, Irgolic KJ (1996) Microwave digestion methods for the determination of trace elements in brain and liver samples by inductively coupled plasma mass spectrometry. Fresenius' J Analyt Chem 355(2):120–128. https://doi.org/10.1007/s0021663550120
36. Godshaw J, Hopfer H, Nelson J et al (2017) Comparison of dilution, filtration, and microwave digestion sample pretreatments in elemental profiling of wine by ICP-MS. Molecules 22(10):1609. https://doi.org/10.3390/molecules22101609
37. Jagielska A, Ruszczynska A, Wagner B et al (2020) Mikropróbkowanie laserowe w układzie LA-ICP_MS w wielopierwiastkowym obrazowaniu probek klinicznych. In: Buszewski B, Staneczko-Baranowska I (eds.). Bioanalityka, PWN. 553–563
38. Berg D, Malinowsky K, Reischauer B et al (2011) Use of formalin-fixed and paraffin-embedded tissues for diagnosis and therapy in routine clinical settings. Methods Mol Biol 785:109–22. https://doi.org/10.1007/978-1-61779-286-1_8
39. Bonta M, Török S, Hegedus B et al (2017) A comparison of sample preparation strategies for biological tissues and subsequent trace element analysis using LA-ICP-MS. Anal Bioanal Chem 409:1805–1814. https://doi.org/10.1007/s00216-016-0124-6
40. Sajnóg AJ, Hanć A, Barałkiewicz D (2020) Bioobrazowanie pierwiastków w tkankach klinicznych techniką LA-ICPMS. In:

Baranowska I, Buszewski B (eds) Bioanalityka w nauce i życiu. 1, Nowe wyzwania w bioanalizie klinicznej i ocenie naturalnych surowców leczniczych, 1, Warszawa, Wydawnictwo Naukowe PWN, 445, ISBN 9788301212834
41. Zheng LN, Sang YB, Luo RP et al (2019) Determination of silver nanoparticles in single cells by microwell trapping and laser ablation ICP-MS determination. J Anal Atom Spectrometry 34:915–921. https://doi.org/10.1039/C8JA00438B
42. Haraguchi H, Ishii A, Hasegawa T et al (2008) Metallomics study on all-elements analysis of salmon egg cells and fractionation analysis of metals in cell cytoplasm. Pure Appl Chem 80(12):2595–2608
43. Meyer S, López-Serrano A, Mitze H et al (2018) Single-cell analysis by ICP-MS/MS as a fast tool for cellular bioavailability studies of arsenite. Metallomics 10(1):73–76. https://doi.org/10.1039/c7mt00285h
44. Schoeberl A, Theiner S, Koellensperger G (2020) Single-cell analysis by ICP-MS—current status and future trends. Spectroscopy 7:17–20
45. www.perkinelmer.com (2017) Single cell ICP-MS analysis: quantification of metal content at the cellular analysis: quantification of metal content at the cellular level. www.perkinelmer.com; https://pepolska.pl/wp-content/uploads/2019/11/Broszura-Analiza-SC-ICP-MS-EN.pdf
46. Theiner S, Schweikert A, Van Malderen SJM et al (2019) Single-cell trace element imaging by laser ablation mass spectrometry and inductively coupled plasma in clinical practice. Analyt Chem 91(13):8207–8212. https://doi.org/10.1021/acs.analchem.9b00698
47. Ellison G, Hollings AL, Hackett MJ (2022) A review of the "metallome" within neurons and glia, as revealed by elemental mapping of brain tissue. BBA Advances 2:100038. https://doi.org/10.1016/j.bbadva.2021.100038
48. Tanaka Y, Ogra Y (2025) Single-cell elemental analysis using inductively coupled plasma mass spectrometry. Metallom Res 5(1):AWrev01-AWrev10. https://doi.org/10.11299/metallomicsresearch.MR202415
49. Laborda F, Abad-Alvaro I, Jimenez MS et al (2023) Catching particles by atomic spectrometry: Benefits and limitations of single particle-inductively coupled plasma mass spectrometry. Spectrochim Acta Part B 199:106570. https://doi.org/10.1016/j.sab.2022.106570
50. Tan Q, Zhao S, Xu T, Wang Q, Zhang M, Yan L, Chen X, Lan M (2023) Inorganic nano-drug delivery systems for crossing the blood–brain barrier: advances and challenges. Coord Chem Rev 494:215344. https://doi.org/10.1016/j.ccr.2023.215344
51. Lomax-Vogt M, Carter LM, Wielinski J et al (2025) Challenges in measuring nanoparticles and microparticles by single particle ICP-QMS and ICP-TOFMS: size-dependent transport efficiency and limited linear dynamic range. J Analyt Atom Spectrometry 40(8):848
52. Davison C, Beste D, Bailey M et al (2023) Expanding the boundaries of atomic spectroscopy at the single-cell level: critical review of SP-ICP-MS, LIBS and LA-ICP-MS advances for the elemental analysis of tissues and single cells. Anal Bioanal Chem 415:6931–6950. https://doi.org/10.1007/s00216-023-04721-8
53. Chronakis MI, Meermann B, von der Au M (2025) The evolution of data treatment tools in single-particle and single-cell ICP-MS analytics. Anal Bioanal Chem 417:7–13. https://doi.org/10.1007/s00216-024-05513-4
54. Junceda S, Cruz-Alonso M, Fernandez B et al (2024) Iron dysregulation in Alzheimer's disease: LA-ICP-MS bioimaging of the distribution of iron and ferroportin in the CA1 region of the human hippocampus. Biomolecules 14(3):295. https://doi.org/10.3390/biom14030295
55. Agarwal P, Ayton S, Agrawal S et al (2022) Brain copper may protect from cognitive decline and Alzheimer's disease pathology: a community-based study. Mol Psychiatry 27(10):4307–4313. https://doi.org/10.1038/s41380-022-01802-5
56. Hutchinson RW, Cox AG, McLeod CW et al (2005) Imaging and spatial distribution of beta-amyloid peptide and metal ions in Alzheimer's plaques by laser ablation-inductively coupled plasma-mass spectrometry. Anal Biochem 346(2):225–33. https://doi.org/10.1016/j.ab.2005.08.024
57. Ha Y, Tsay OG, Churchill DG (2012) ICP-MS for the neurodegenerative and brain sciences. In: Linert W, Kozlowski H (eds.). Metal ions in neurological systems. Springer, Vienna
58. Pamphlett R, Buckland ME, Bishop DP (2023) Potentially toxic elements in the brains of people with multiple sclerosis. Sci Rep 13: 655. https://doi.org/10.1038/s41598-022-27169-9
59. Jha S, Mishra V, Sharma D et al (2011) Fluoride in the environment and its metabolism in humans. Rev Environ Contamin Toxicol 211:121–42. https://doi.org/10.1007/978-1-4419-8011-3_4
60. Żwierełło W, Maruszewska A, Skórka-Majewicz M et al (2023) Fluoride in the central nervous system and its potential influence on the development and invasiveness of brain tumours-a research hypothesis. Int J Mol Sci 24(2):1558. https://doi.org/10.3390/ijms24021558

61. Bashash M, Thomas D, Hu H et al (2017) Prenatal fluoride exposure and cognitive outcomes in children at 4 and 6–12 years of age in Mexico. Environ Health Perspect 125(9):097017. https://doi.org/10.1289/EHP655
62. Grandjean P (2019) Developmental fluoride neurotoxicity: an updated review. Environ Health 18(1):110. https://doi.org/10.1186/s12940-019-0551-x
63. DeBenedictis CA, Raab A, Ducie E et al (2020) Concentrations of essential trace metals in the brain of animal species—a comparative study. Brain Sci 10(7):460. https://doi.org/10.3390/brainsci10070460
64. Hare DJ, Kysenius K, Paul B et al (2017) Imaging metals in brain tissue by laser ablation-inductively coupled plasma-mass spectrometry (LA-ICP-MS). J Vis Exp 22(119):55042. https://doi.org/10.3791/55042
65. Hare DJ, Raven EP, Roberts BR et al (2016) Laser ablation-inductively coupled plasma-mass spectrometry imaging of white and gray matter iron distribution in Alzheimer's disease frontal cortex. NeuroImage 137:124–131. https://doi.org/10.1016/j.neuroimage.2016.05.057
66. Feng L, Wang J, Li H et al (2017) A novel absolute quantitative imaging strategy of iron, copper and zinc in brain tissues by isotope dilution laser ablation ICP-MS. Analyt Chim Acta 984:66–75. https://doi.org/10.1016/j.aca.2017.07.003
67. Becker JS (2013) Imaging of metals in biological tissue by laser ablation inductively coupled plasma mass spectrometry (LA-ICP-MS): state of the art and future developments. J Mass Spectrometry 48(2):255–68. https://doi.org/10.1002/jms.3168
68. Portbury SD, Hare DJ, Sgambelloni C et al (2016) A time-course analysis of changes in cerebral metal levels following a controlled cortical impact. Metallomics 8(2):193–200. https://doi.org/10.1039/c5mt00234f
69. Becker JS, Matusch A, Palm C et al (2010) Bioimaging of metals in brain tissue by laser ablation inductively coupled plasma mass spectrometry (LA-ICP-MS) and metallomics. Metallomics 2:104–111. https://doi.org/10.1039/b916722f
70. Drescher D, Giesen C, Traub H, Panne U, Kneipp J, Jakubowski N (2012) Quantitative imaging of gold and silver nanoparticles in single eukaryotic cells by laser ablation ICP-MS. Anal Chem 84(22):9684–9688. https://doi.org/10.1021/ac302639c
71. Urgast DS, Hill S, Kwun I-S, Beattie JH, Goenaga-Infante H, Feldmann J (2012) Zinc isotope ratio imaging of rat brain thin sections from stable isotope tracer studies by LA-MC-ICP-MS. Metallomics 4(10):1057–1063. https://doi.org/10.1039/c2mt20119d
72. Solovyev N, El-Khatib AH, Costas-Rodríguez M et al (2021) Cu, Fe, and Zn isotope ratios in murine Alzheimer's disease models suggest specific signatures of amyloidogenesis and tauopathy. J Biol Chem 296:100292. https://doi.org/10.1016/j.jbc.2021.100292
73. Hare DJ, Lei P, Ayton S et al (2014) An iron-dopamine index predicts risk of Parkinsonian neurodegeneration in the substantia nigra pars compacta. Chem Sci 5(6):2160–2169. https://doi.org/10.1039/C3SC53461H
74. Hare DJ, George JL, Grimm R et al (2010) Three-dimensional elemental bio-imaging of Fe, Zn, Cu, Mn and P in a 6-hydroxydopamine lesioned mouse brain. Metallomics 11:745–753. https://doi.org/10.1039/c0mt00039f
75. Hartnell D, Andrews W, Smith N et al (2020) A review of ex vivo elemental mapping methods to directly image changes in the homeostasis of diffusible ions (Na+, K+, Mg2 +, Ca2 +, Cl−) within brain tissue. Front Neurosci 13:1415. https://doi.org/10.3389/fnins.2019.01415
76. Barałkiewicz D, Bulska E (2009) Specjacja chemiczna, problemy i możliwo ci, Wydawnictwo MALAMUT, Warszawa
77. Templeton DM, Ariese F, Cornelis R et al (2000) Guidelines for terms related to chemical speciation and fractionation of elements: definitions, structural aspects, and methodological approaches. Pure Appl Chem 72:1453–1470. https://doi.org/10.1351/pac200072081453
78. Zoorob GK, McKiernan JW, Caruso JA (1998) ICP-MS for elemental speciation studies. Mikrochim Acta 128:145–168. https://doi.org/10.1007/BF01243044
79. Templeton DM (2015) Speciation in metal toxicity and metal-based therapeutics. Toxics 3(2):170–186. https://doi.org/10.3390/toxics3020170
80. Templeton DM (2012) Elemental speciation in clinical sciences. In: Caroli S, Záray G (eds) Analytical techniques for clinical chemistry: methods and applications. John Wiley and Sons, Hoboken, NJ, USA, pp. 159–177
81. Templeton DM (1999) Biomedical aspects of trace element speciation. Fresenius J Anal Chem 363:505–511. https://doi.org/10.1007/s002160051234
82. Zheng W, Aschner M, Ghersi-Egea JF (2003) Brain barrier systems: a new frontier in metal neurotoxicological research. Toxicol Appl Pharmacol 192(1):1–11. https://doi.org/10.1016/s0041-008x(03)00251-5
83. Nischwitz V, Berthele A, Michalke B (2008) Speciation analysis of selected metals and determination of their total contents in paired

serum and cerebrospinal fluid samples: an approach to investigate the permeability of the human blood-cerebrospinal fluid-barrier. Analyt Chim Acta 627(2):258–269. https://doi.org/10.1016/j.aca.2008.08.018

84. Pamphlett R, Bishop DP (2023) The toxic metal hypothesis for neurological disorders. Front Neurol 14:1173779. https://doi.org/10.3389/fneur.2023.1173779
85. Hollings AL, Lam V, Takechi R et al (2020) Revealing differences in the chemical form of zinc in brain tissue using K-edge X-ray absorption near-edge structure spectroscopy. Metallomics 12:2134–2144. https://doi.org/10.1039/d0mt00198h
86. Michalke B, Nischwitz V (2010) Review on metal speciation analysis in cerebrospinal fluid—current methods and results: a review. Analyt Chim Acta 682(1–2):23–36. https://doi.org/10.1016/j.aca.2010.09.054
87. Strzeminska I, Factor C, Jimenez-Lamana J et al (2022) Comprehensive speciation analysis of residual gadolinium in deep cerebellar nuclei in rats repeatedly administered with gadoterate meglumine or gadodiamide. Invest Radiol 57(5):283–292. https://doi.org/10.1097/RLI.0000000000000846
88. Frenzel T, Apte C, Jost G et al (2017) Quantification and assessment of the chemical form of residual gadolinium in the brain after repeated administration of gadolinium-based contrast agents: comparative study in rats. Invest Radiol 52(7):396–404. https://doi.org/10.1097/RLI.0000000000000352
89. Vinceti M, Chiari A, Eichmüller M et al (2017) A selenium species in cerebrospinal fluid predicts conversion to Alzheimer's dementia in persons with mild cognitive impairment. Alz Res Therapy 9:100. https://doi.org/10.1186/s13195-017-0323-1
90. Michalke B, Lucio M, Berthele A et al (2013) Manganese speciation in paired serum and CSF samples using SEC-DRC-ICP-MS and CE-ICP-DRC-MS. Anal Bioanal Chem 405:2301–2309. https://doi.org/10.1007/s00216-012-6662-7
91. Arruda MAZ, de Jesus JR, Blindauer CA, Stewart AJ (2022) Speciomics as a concept involving chemical speciation and omics. J Proteomics 263:104615. https://doi.org/10.1016/j.jprot.2022.104615
92. Zhou Y, Song X, Yan X, Yang L, Chen S, Wang Q (2022) Zero-interfacing μHPLC to ICPMS. Anal Chem 94(49):16975–16979. https://doi.org/10.1021/acs.analchem.2c03951
93. Michalke B (2012) Manganese speciation related to neurotoxicity in humans. In: Li Y, Zhang J (eds.). Metal ion in stroke. Springer Series in Translational Stroke Research, Springer, New York, NY. https://doi.org/10.1007/978-1-4419-9663-3_28
94. Budinger D, Barral S, Soo AKS et al (2021) The role of manganese dysregulation in neurological disease: emerging evidence. Lancet Neurol 20(11):956–968. https://doi.org/10.1016/S1474-4422(21)00238-6
95. Diederich J, Michalke B (2011) Enhanced extract preparation of native manganese and iron species from brain and liver tissue. Anal Bioanal Chem 399:1799–1806. https://doi.org/10.1007/s00216-010-4162-1
96. Feldmann J, Hansen HR, Karlsson TM et al. (2024) Challenges and future developments for elemental analysis by ICP-MS in non-target screening. Environ Sci Technol 58(4):2345–2356. https://doi.org/10.1021/acs.est.4c00504
97. Michalski R, Błażewicz A, Kończyk J et al (2019) Ion chromatography for small molecule determination in clinical and pharmaceutical studies. LCGC Europe December, https://cdn.sanity.io/files/0vv8moc6/chroma/1c3e-b5fba2539f3c52bfd1b5ccb9cff71ee04b05.pdf/LCGCEurope_December2019.pdf
98. Sanz-Medel A, Montes-Bayón M, Fernández Sánchez LM (2003) Trace element speciation by ICP-MS in large biomolecules and its potential for proteomics. Anal Bioanal Chem 377:236–247. https://doi.org/10.1007/s00216-003-2082-z
99. Michalski R (2014) Application of ion chromatography in clinical studies and pharmaceutical industry. Mini Rev Med Chem 14(10):862–72.
100. Iwase M, Tanaka Y-K, Fukumoto Y et al (2025) Laser ablation coupled with LC-ICP-MS for local speciation of trace elements in tissues. Analyt Chem 97(9):5306–5312. https://doi.org/10.1021/acs.analchem.5c00153
101. Niehaus P, Gonzalez de Vega R, Haindl MT et al (2024) Multimodal analytical tools for the molecular and elemental characterisation of lesions in brain tissue of multiple sclerosis patients. Talanta 270:125518. https://doi.org/10.1016/j.talanta.2023.125518
102. Coverdale JPC, Polepalli S, Arruda MAZ et al (2024) Recent advances in metalloproteomics. Biomolecules 14(1):104. https://doi.org/10.3390/biom14010104
103. Hu L, Cheng T, He B et al (2013) Identification of metal-associated proteins in cells by using continuous-flow gel electrophoresis and inductively coupled plasma mass spectrometry. Angew Chem Int Edn 52(18):4916–4920. https://doi.org/10.1002/anie.201300794
104. Xu X, Wang H, Li H et al. (2020) Metalloproteomic approaches for matching metals to proteins: the power of inductively coupled plasma mass spectrometry (ICP-MS). Chem Lett 49:697–704. https://doi.org/10.1246/cl.200155

105. Galvez L, Rusz M, Schwaiger-Haber M et al (2019) Preclinical studies on metal based anticancer drugs as enabled by integrated metallomics and metabolomics. Metallomics 11(10):1716–1728. https://doi.org/10.1039/c9mt00141g
106. Hare DJ, Grubman A, Ryan TM et al (2013) Profiling the iron, copper and zinc content in primary neuron and astrocyte cultures by rapid online quantitative size exclusion chromatography-inductively coupled plasma-mass spectrometry. Metallomics. 5(12):1656–1662. https://doi.org/10.1039/c3mt00227f
107. Lothian A, Roberts BR (2016) Standards for quantitative metalloproteomic analysis using size exclusion ICP-MS. J Vis Exp 110:e53737. https://doi.org/10.3791/53737
108. Pavelicova K, Do T, Vejvodova M et al (2024) Joint forces of mass spectrometric techniques (ICP-MS and MALDI-TOF-MS) and fluorescence spectrometry in the study of platinum-based cytostatic drugs interactions with metallothionein MT2 and MT3. Talanta 274:125920. https://doi.org/10.1016/j.talanta.2024.125920
109. Feldmann J, Hansen HR, Karlsson TM et al (2024) ICP-MS as a contributing tool to nontarget screening (NTS) analysis for environmental monitoring. Environ Sci Technol 58(29):12755–12762. https://doi.org/10.1021/acs.est.4c00504
110. Polemiti E, Hese S, Schepanski K et al (2024) How does the macroenvironment influence brain and behaviour—a review of current status and future perspectives. Mol Psychiatry 29:3268–3286. https://doi.org/10.1038/s41380-024-02557-x
111. Hu J, Yan X, Le XC (2024) Label-free detection of biomolecules using inductively coupled plasma mass spectrometry (ICP-MS. Analyt Bioanalyt Chem 416(11):2625–2640. https://doi.org/10.1007/s00216-023-05106-7
112. Tsikas D (2024) Perspectives of quantitative GC-MS, LC-MS, and ICP-MS in the clinical medicine science—the role of analytical chemistry. J Clin Med 13(23):7276. https://doi.org/10.3390/jcm13237276

Chapter 8

Detailed Workflow for Cryo-Correlative Light and Synchrotron X-Ray Fluorescence Imaging of Proteins and Metals in Primary Rat Hippocampal Neurons

Ines Kelkoul, Noémie Pied, Mónica Fernández-Monreal, Richard Ortega and Asuncion Carmona

Abstract

Some metals, such as iron, copper, zinc, and manganese, play essential roles in biological processes, and their dysregulation has been implicated in several neurological disorders, including Alzheimer's and Parkinson's diseases, as well as neurodevelopmental conditions. Therefore, understanding the precise distribution and interactions of metals within neurons is essential to elucidate their physiological functions and dysfunctions in neuropathologies. We present a detailed step-by-step protocol for the correlative nano-imaging of metals and proteins under native cryogenic conditions, combining cryo-fluorescence light microscopy (cryo-FLM) and cryo-synchrotron X-ray fluorescence (cryo-SXRF) imaging. Rapid plunge-freezing of samples enables the preservation of native biological structures and chemical states. In addition, cryogenic mode during FLM observations reduces photobleaching and increases fluorescence signal stability. Cryo-SXRF provides highly sensitive, quantitative elemental mapping at high spatial resolution, enabling precise localization of metals within biological samples. For example, by integrating these two techniques, we investigated metal–protein interactions in cultured primary rat hippocampal neurons. This methodological framework provides a powerful tool for studying, under native conditions, the neurobiological functions of metals and metal dyshomeostasis in neurodegenerative and neurodevelopmental disorders.

Keywords Metal dyshomeostasis, Neurodegenerative disorders, Cryogenic nano-imaging, Cryo-fluorescence light microscopy (cryo-FLM), Synchrotron X-ray fluorescence (SXRF), Metal–protein interactions

1 Introduction

Essential trace metals such as iron, copper, zinc, and manganese are fundamental for numerous biological and neurobiological processes, including enzymatic activity, redox homeostasis, and neurotransmission [1, 2]. However, the dyshomeostasis of these metals in the brain is associated with neurological impairments. Previous studies have identified dysregulation of metal homeostasis as a hallmark of neurodegenerative diseases, with alterations in iron, copper, and zinc

Andreas M. Grabrucker (ed.), *Brain Metallomics*, Neuromethods, vol. 237,
https://doi.org/10.1007/978-1-0716-5468-2_8,
© The Author(s), under exclusive license to Springer Science+Business Media, LLC, part of Springer Nature 2026

metabolism contributing to protein misfolding, oxidative stress, and neuronal dysfunction [3]. For instance, manganese accumulation disrupts dopaminergic signaling, contributing to Parkinson-like symptoms [4, 5]. Similarly, elevated levels of iron and copper have been observed in the brains of Alzheimer's disease patients, promoting oxidative stress and protein aggregation [6–8]. Despite these associations, the precise mechanisms underlying metal–protein interactions in neuronal cells remain poorly understood, necessitating advanced imaging techniques for detailed investigation.

To address these challenges, correlative nano-imaging under cryogenic conditions, combining cryo-fluorescence light microscopy (cryo-FLM) with synchrotron X-ray fluorescence (SXRF) imaging, can be employed. Cryo-FLM enables high-resolution visualization of intraneuronal structures while preserving their native state, minimizing photobleaching, and reducing molecular diffusion artifacts [9]. SXRF, on the other hand, provides highly sensitive, quantitative elemental mapping, enabling precise localization and quantification of metals within biological specimens [10]. By integrating these techniques, correlative nanoimaging offers a powerful approach to studying metal–protein interactions in neurons, elucidating their role in neurodegenerative and neurodevelopmental disorders [11].

Understanding the interplay between neurotoxic metals and cellular components is crucial for developing therapeutic strategies to mitigate metal-induced toxicity. By leveraging advanced cryogenic nano-imaging techniques, we aim to provide novel insights into the role of metal–protein interactions in neuronal health and disease, paving the way for targeted interventions in neurotoxicity-related disorders.

Studying metal–protein interactions requires a specific preparation protocol to preserve native chemical element distributions and cellular structures. This chapter describes the experimental methods for the cryocorrelative imaging of primary rat hippocampal neurons as a model system. Their well-characterized synaptic architecture provides an ideal framework for studying metal distribution at the subcellular and synaptic levels. A detailed, step-by-step description of the culture protocol is provided in the "Methods" section, adapted from our previous work [12, 13]. Figure 1 illustrates the workflow for primary rat hippocampal neurons.

While the present protocol is optimized for the cryo-correlative imaging of primary rat hippocampal neurons using a dedicated synchrotron-based setup, its methodological framework is transferable to a variety of experimental contexts. With appropriate adjustments—such as modifying the cell culture substrate, adapting cryo-fixation procedures, or optimizing imaging parameters—this approach can be applied to other cell types (e.g., induced pluripotent stem cell-derived neurons or heterologous expression systems), and to other imaging platforms. These modifications, if they preserve the methodology's

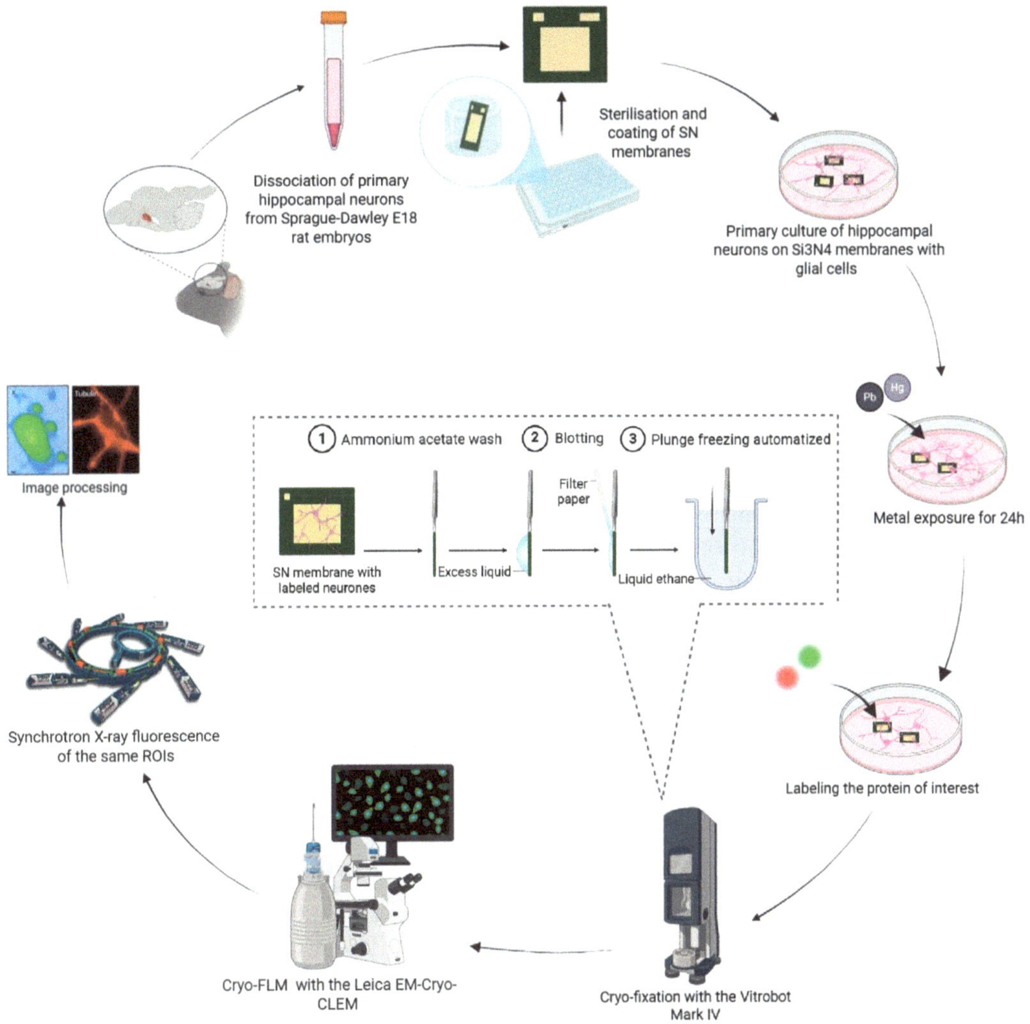

Fig. 1 Schematic overview of the workflow for cryo-correlative light and synchrotron X-ray fluorescence (SXRF) imaging of proteins and metals in primary rat hippocampal neurons, including the cryofixation protocol

fundamental principles, will enable broader applications for the study of metal–protein interactions and trace metal homeostasis across various biological models and experimental environments.

2 Materials for Cryo-Correlative Light and Synchrotron X-ray Fluorescence Imaging of Proteins and Metals in Primary Rat Hippocampal Neurons

2.1 Culture of Primary Neurons on Silicon Nitride Membranes

The culture of primary neurons is a cornerstone of studying neuronal development, synaptic function, and cellular interactions. Primary neurons are cocultured with glial cells to promote optimal survival and maturation, providing essential trophic support and growth factors. The protocol is adapted from the method described by Kaech and Banker [14], with modifications to accommodate the specific requirements of synchrotron analysis, particularly the use of silicon nitride (SN) membranes as culture substrates. These membranes provide an ultra-thin, flat, X-ray-transparent substrate that minimizes background signal and enables high-resolution elemental imaging.

For our studies, SN membranes are manufactured by Silson (UK) (Fig. 2A) and feature a silicon frame surrounding the inner SN membrane. This silicon frame enhances handling and stability and can be customized in size and thickness, along with the dimensions of the inner membrane. The membrane specifications can be tailored to individual synchrotron beamline requirements. For our experiments, we typically use membranes with a 5 × 5 mm silicon frame, 200 μm thick, and a central square SN membrane measuring 2 × 2 mm with a thickness of 500 nm. This configuration offers an optimal balance between stability, ease of handling, and a large surface area for neuronal culture, as neurons are cultured within the central square. The membranes have two distinct surfaces: flat and recessed, resulting from the thickness difference between the silicon frame and the central SN membrane. Neurons are cultured on the flat side of the central square. To ensure correct orientation during imaging, we designed membranes with two orientation markers in the silicon frame: a rectangle in the upper-left corner and a square in the upper-right corner (Fig. 2A).

To enable cell culture and promote neuronal adhesion, SN membranes are sterilized and coated. After the coating process, neurons are seeded onto the membranes and subsequently transferred to a Petri dish containing glial cells. The use of SN membranes marks a significant difference from conventional coverslip-based culture methods. In standard protocols, coverslips are inverted to position neurons close to glial cells, enabling direct face-to-face interactions [14]. However, to preserve the mechanical integrity of SN membranes, they must remain upright, with the flat side—where neurons are cultured—facing upwards. The thickness of the silicon membrane is lower than that of glass coverslips, and the interaction between glial cells and neurons is maintained. The coculture is then maintained until the desired developmental stage, after which metal exposure

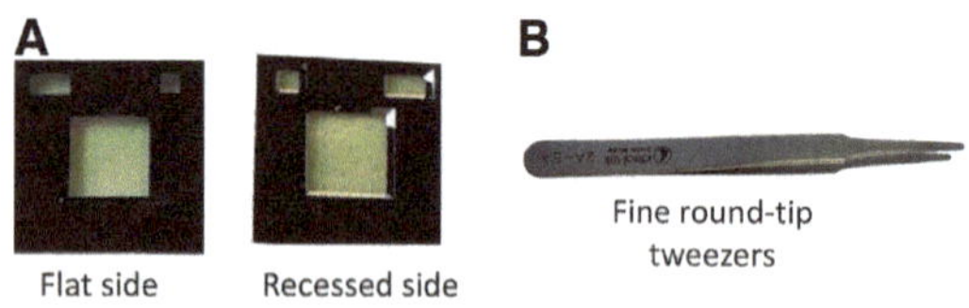

Fig. 2 (**A**) Silicon nitride membranes were used for primary neuron culture and adapted for SXRF analysis. (**B**) Fine, round-tip tweezers are used for handling the membranes

and protein labeling are performed based on the specific objectives of the study. For example, neurons cultured for 3 days are ideal for examining developing growth cones [15], whereas neurons cultured for 21 days develop fully mature synapses, suitable for studies of synaptic function and network formation [16, 17].

2.2 Cryo-Fluorescence Light Microscopy (Cryo-FLM)

Cryogenic fluorescence microscopy is a powerful imaging technique that allows the visualization of fluorescently labeled biological structures at low temperatures, minimizing structural alterations and photobleaching while preserving sample integrity. Traditional fluorescence microscopy often relies on chemical fixation methods, which can introduce artifacts by cross-linking biomolecules and altering their native conformation and by permeabilizing the plasma membrane, resulting in the release of intracellular elements [18]. In contrast, cryo-microscopy prevents ice crystal formation and cell damage after rapid plunge-freezing of samples and preserves ultrastructural details with minimal perturbation [19]. Furthermore, cryogenic conditions significantly reduce photobleaching and enhance fluorescence signal stability, both of which are crucial for high-resolution imaging of subcellular structures.

Cryo-FLM is performed on a modified commercial wide-field microscopy system, the Leica EM-Cryo-CLEM (Fig. 3A). This system comprises a cryogenic chamber under an upright microscope (Fig. 3B), which ensures cryogenic preservation of the sample throughout the acquisition process. The microscope is equipped with a dedicated 50× cryo-objective (HC PL APO 50×/0.90 CRYO CLEM) featuring a numerical aperture of 0.9 and a working distance of 0.28 mm, and a tube lens of 1.2× magnification, allowing for high-resolution imaging in liquid nitrogen vapor. To maintain the required cryogenic environment, an automatic pump system is connected to a liquid nitrogen dewar, regulating temperatures in the microscope cryogenic chamber to 77.15 K.

Sample handling is facilitated by a dedicated Leica cryo-transfer shuttle, which ensures that the sample remains at cryogenic temperature and is protected from ice contamination during the transfer of the SN membranes (Fig. 3C). Leica has developed a dedicated cryo-transfer cartridge for these membranes, enabling precise sample positioning while avoiding exposure to ambient conditions. To minimize the risk of contamination from ambient humidity, a custom-designed plexiglass chamber was developed to fit the microscope tightly. This chamber maintains a controlled humidity level (~30%), significantly reducing the risk of ice crystal formation. Such ice formation can degrade image quality, obscure fine structural details, and compromise the integrity of cryo-preserved samples. In addition, ice can hinder the visualization of cellular structures in bright field imaging and impair the accurate localization of regions of interest—an essential requirement for correlative workflows involving multiple

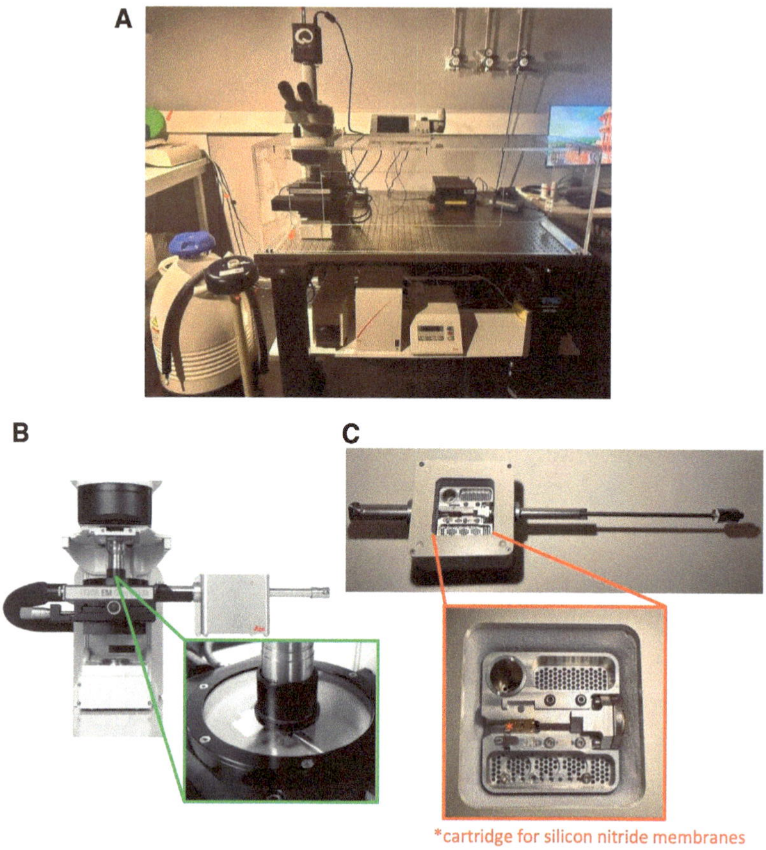

Fig. 3 Leica EM-Cryo-CLEM cryomicroscope and cryo-transfer components. (**A**) Overview of the Leica EM-Cryo-CLEM system, featuring the automatic filling dewar with an integrated pump and the custom-designed plexiglass chamber to reduce ambient humidity and prevent ice formation. (**B**) Close-up of the cryo-chamber, designed for maintaining cryogenic temperatures and minimizing exposure to ambient conditions during sample imaging. (**C**) The cryo-transfer shuttle has a dedicated cartridge for handling SN membranes (red star), ensuring precise and contamination-free sample transfer under cryogenic conditions

imaging modalities. To further limit humidity, silica desiccant beads can be placed inside the cryo-chamber (Fig. 3A).

The microscope is controlled by the MetaMorph (Molecular Devices, v7.10.5) software, allowing for automated image acquisition across multiple wavelengths using dedicated fluorescence Leica filter cubes (A4, Ex: BP360/40, Em: BP470/40; GFP, Ex: BP470/40, Em:525/50; N3, Ex: BP546/12 Em:600/40; Y5, Ex: 620/660 Em:700/75). The system features a motorized XY stage, enabling large-scale mosaic imaging through ND-scan software (Gataca systems). One corner of the SN membrane is designated as the origin coordinate (x0, y0). It can be identified, respectively, with the orientation frames, allowing precise registration of regions of interest for subsequent analysis at synchrotron facilities.

For image acquisition, the microscope is equipped with a high-speed Hamamatsu Orca-FusionBT sCMOS camera (89.1 frames/s at 16 bits, 2304 × 2304 pixels), providing a resolution of 350 nm and matching the system's expected performance limits. The camera's high sensitivity and low noise properties make it well-suited for cryogenic fluorescence imaging, ensuring optimal signal detection with minimal photodamage.

The Leica EM-Cryo-CLEM microscope is specifically designed for stable long-term imaging under cryogenic conditions, preserving mechanical and thermal stability throughout data acquisition. Its integration with SN membrane supports enables seamless correlation between fluorescence microscopy and synchrotron-based X-ray fluorescence (SXRF) analyses. Overall, this system provides a robust platform for high-resolution cryo-FLM imaging, minimizing artifacts while preserving ultrastructural details.

2.3 Cryo-Synchrotron X-ray Fluorescence Imaging (Cryo-SXRF)

SXRF is a powerful analytical technique that uses synchrotron radiation for X-ray fluorescence —an extraordinarily bright X-ray source, billions of times brighter than conventional X-ray sources. This brightness allows for highly efficient excitation of atoms and the detection of extremely low concentrations of elements. Therefore, SXRF imaging enables mapping chemical elements at low trace concentrations, such as trace metals within cells and tissues.

The methodology presented here is primarily based on procedures conducted at the ID16A beamline [20] of the European Synchrotron Radiation Facility (ESRF) in Grenoble, France. While specific details may vary between synchrotron facilities and beamlines, the general workflow remains consistent across most facilities.

The ID16A beamline is dedicated to nano-imaging, providing a high-brilliance X-ray beam focused on nanometric dimensions [21]. This beamline is optimized for hard X-ray imaging and X-ray fluorescence microscopy, with an extreme focusing capability down to 13 nm and a very high photon flux (up to 10^{12} photons/s) with a spatial resolution of up to 15 nm. It operates at two energy levels, 17.1 keV and 33.6 keV, and is tailored for applications in biomedicine, materials science, and nanotechnology. The X-ray beam energy is selected according to the element(s) of interest. In our studies, a 17.1 keV beam is suitable to reach the excitation energy of K-shells of most physiological metals such as Mn, Fe, Cu, and Zn. This beamline is also suitable for analyzing samples under cryogenic conditions. For our studies, the X-ray beam was focused to a size of 40 nm, providing a flux of $2.2 \cdot 10^{11}$ photons/s. The 40nm spatial resolution allows for a detailed study of element distributions within neurons, including in small compartments such as synaptic compartments in the micrometer range. At ID16A, the analysis chamber operates under a vacuum of ~10^{-8} mbar and is cooled to a temperature of 108 K by an unforced flow of mixed liquid/gaseous nitrogen.

Special care must be taken to ensure that samples remain frozen during storage, transport, and measurement. The membranes are transported to the synchrotron facility using a dry-shipper system (Fig. 4A). This system incorporates a porous material that absorbs liquid nitrogen, preserving the samples by releasing nitrogen vapors, while preventing liquid spillage during transit. The ice layer covering the sample must be carefully controlled when working with cryo-fixed, frozen, hydrated biological samples. It should be thin enough to allow emitted X-rays to pass through and reach the detector, yet thick enough to prevent the beam from melting the ice and damaging the sample. The ice layer thickness can be measured at the beamline before analysis, with a thickness of less than 10 μm being ideal. A preliminary visual inspection also provides a good indication: if the cells are not visible under the light microscope, it suggests the presence of excessive ice.

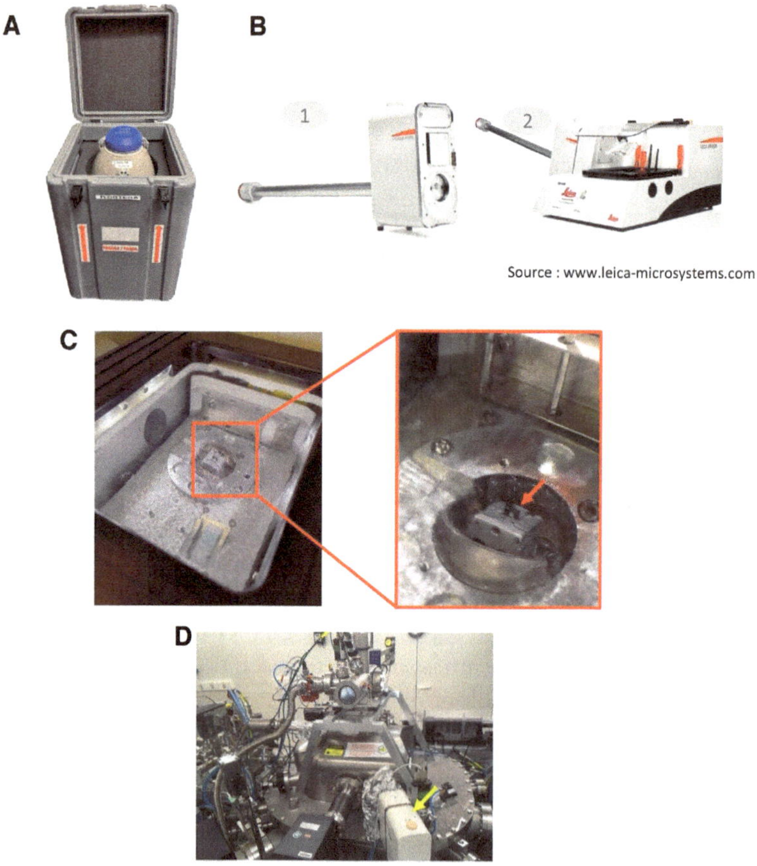

Fig. 4 (**A**) Dry shipper dewar used to maintain samples in a hydrated and vitrified state during shipment. (**B**) Leica EM VCM (2) and EM VCT500 (1) cryo-transfer systems. (**C**) Loading chamber of the EM VCM with a silicon nitride membrane in place (red arrow), (**D**) ready for transfer via the EM VCT500 (yellow arrow) system to the sample chamber of the ID16A beamline

To maintain cryogenic conditions, the samples are mounted in the ID16A sample chamber using Leica's EM VCM and EM VCT500 systems (Fig. 4B). The Leica EM VCM (Vacuum Cryo Manipulation System) is a modular platform designed for the precise handling and transfer of samples under vacuum or controlled atmospheric conditions, capable of maintaining cryogenic temperatures. It comprises a vacuum cryo-loading station (Fig. 4C) that facilitates contamination-free sample handling by enabling sample loading under vacuum. The system has illumination, magnification tools, a tool dryer, and automated liquid nitrogen refilling. The Leica EM VCT500 (Vacuum Cryo Transfer System 500) is engineered to transfer samples between different preparation and analysis instruments while maintaining stringent environmental conditions, including high vacuum, controlled atmospheres, and cryogenic temperatures. It features a hermetically sealed transfer chamber with an innovative valve concept to prevent sample contamination during handling. The integrated active sample-cooling system provides rapid cooling and maintains thermal stability throughout the transfer process. Both the EM VCM and EM VCT500 systems have been adapted by the manufacturer to include SN membranes.

At ID16A, the SXRF signal is detected using two large solid-angle multielement silicon drift diode detectors positioned at approximately 90° on either side of the incident beam (Fig. 4D). This configuration necessitates that neurons grow on the flat side of the silicon membranes. If neurons are on the wrong side, the membrane substrate frame may obstruct part of the X-ray fluorescence signal. During analysis, the area of interest is first subjected to a coarse scan at a lower X-ray flux and lower spatial resolution than the fine scan. For these coarse scans, typical parameters include an integration time of 100 ms per pixel and a step size of 500 nm. A smaller region of interest is selected for high-resolution fine scanning based on the resulting images. For fine scans, samples are raster-scanned in continuous mode with an integration time of 100 ms per pixel and a step size of 15–80 nm. This approach optimizes beam time, as high-resolution scans are time-consuming. For example, a 10 × 10 μ m scan with a 50 nm step size takes approximately 1 h and 15 min. Doubling the scan dimensions to 20 × 20 μm quadruples the time to 5 h. Similarly,halving the step size (e.g., to 25 nm) for a 10 × 10 μm area also increases scan time fourfold, resulting in a duration of 5 h.

3 Methods for Cryo-Correlative Light and Synchrotron X-ray Fluorescence Imaging of Proteins and Metals in Primary Rat Hippocampal Neurons

3.1 Primary Rat Hippocampal Culture on Silicon Nitride Membranes

This section provides a step-by-step protocol for preparing primary cultures of hippocampal neurons on SN membranes. As previously mentioned, those membranes feature a dark frame measuring 5 × 5 mm with a thickness of 200 μm, surrounding a central square of 2 × 2 mm with a thickness of 500 nm (Fig. 2A). Neurons are cultured on the flat side of this central square. Glial cells are harvested from the hippocampus and cultured in monolayer in 60 cm^2 Petri dishes for 15 days prior to their interaction with neurons. Primary neurons are isolated from the hippocampus of 18-day-old embryonic rats, dissociated, and maintained in a neurobasal medium supplemented with 2 mM L-glutamine and the neuronal supplement B27.

The process described involves three main steps: membrane sterilization, application of a poly-lysine coating to promote cell adhesion, and seeding of neurons onto the membranes, followed by coculturing with glial cells. To prevent contamination, all procedures must be performed under sterile conditions in a biosafety cabinet, and all materials and solutions must be sterilized prior to use.

Sterilization of Membranes

1. Using fine round-tip tweezers (Fig. 2B) (Ideal TEK, 2A-SA), hold the membranes by their frame without touching the inner square and place them vertically in a well of a 96-well plate. Place one membrane per well.
2. Add 200 μL of absolute ethanol (Sigma-Aldrich, 493511) to each well to cover the entire membrane.
3. Close the lid of the 96-well plate to avoid ethanol evaporation and leave the plate under the biosafety cabinet for 3 h.
4. Remove the ethanol using a pipette with a sterile filter tip.
5. Leave the lid slightly open and wait 1 h for the membranes to dry and evaporate the remaining rest of the ethanol.

The membranes are now sterile. Be careful to maintain sterility throughout the following steps. Close the lid and store the dish at room temperature. The next step, coating, should ideally be performed immediately, but can also be done within the next few days.

Coating of Membranes

1. Prepare a 1 mg/mL poly-L-lysine solution in 0.1 M borate buffer (pH 8.3).
2. Sterilize the poly-L-lysine solution using a 0.22 μm filter (Merck, SLGS033). The solution can then be stored in the refrigerator for future use.

3. In the 96-well plate, where the membranes have been sterilized and placed vertically, add 200 µL of the poly-L-lysine solution per well containing a membrane, ensuring the solution fully covers the membrane.
4. Incubate overnight at 37°C at 5% CO_2 atmosphere.
5. Remove the poly-L-lysine solution using a pipette with a sterile filter tip.
6. Perform two washes with Neurobasal medium (Gibco, 21103049).
7. Add 200 µL of medium to each well to prevent the membranes from drying. Store the SN membranes until seeding. Dry membranes tend to float during seeding and require manipulation to submerge, increasing the risk of damage or breakage. Keeping the coated membranes immersed in the medium under sterile conditions prevents this risk.

Seeding of Neurons

Neurons are seeded at a density of 11,000 cells/cm^2. This ensures optimal axonal and dendritic development until DIV21 (Days In Vitro), without network overlap, to enable imaging of isolated neurites. If culture time is reduced, neuronal density can be slightly increased in other developmental stages. Membranes are placed in 12-well plates, 5–6 membranes per well. Each well has a surface area of 3.6 cm^2. To achieve the desired density, 40,000 cells per well are required.

1. Add 2 mL of equilibrated Neurobasal medium supplemented with B-27™ Plus (NB B27+) (Gibco, A3582801) and 1% penicillin–streptomycin (Merck, P4333) to a well of a 12-well plate.
2. Using thin, round-tip tweezers, place the coated membranes horizontally in the well containing the medium, with the flat side facing up. Ideally, four membranes should be placed per well to avoid overlapping and facilitate handling, but up to 6 can be placed.
3. Add 0.5 mL of equilibrated NB B27+ to the Eppendorf tube containing the neurons.
4. Gently homogenize using a pipette with a sterile filter tip and distribute the neurons into the well containing the membranes. Slightly shake the 12-well plate.
5. Incubate for 2–4 h at 37°C with 5% CO_2 to allow the neurons to adhere to the membranes.
6. Using thin round-tip tweezers, transfer the membranes into a Petri dish containing the glial cells, ensuring the flat side remains facing up. One Petri dish can accommodate more than ten membranes.

7. Incubate the glia dish with the membranes for 3 days at 37°C + 5% CO_2.
8. At DIV3, add 2 µM cytosine arabinoside (Ara C, C1768, Sigma) to stop glial cell growth.
9. Return the glia dish with the membranes to the incubator at 37°C with 5% CO_2 until use at the desired development stage (e.g., DIV21).
10. Once a week, add 1 mL of equilibrated NB B27+ to compensate for culture medium evaporation.

3.2 Metal Exposure and Neuron Labeling with Fluorescent Probes

This section outlines the procedure for exposing neurons to neurotoxic metals when required to study metal overexposure and labeling the proteins or organelles of interest. For metal weighing, use disposable plastic spatulas and plastic weighing dishes. Work under a fume hood while wearing gloves, a lab coat, and an FFP2 mask to ensure the operator's safety. Neuron exposure to metals should then be performed under sterile conditions in the biosafety cabinet, using pipettes dedicated exclusively to metal handling (to avoid metal contamination coming from the degradation of the mechanical system of pipetting if the pipette is used for other chemicals), and sterile, filter-equipped cones. A stock solution of the metal of interest is prepared in BrainPhys medium (StemCell, ref # 05791), which has been pretested to ensure optimal dissolution of the metals. If only physiological elements are of interest, neurons do not need to be exposed to any metals, and the proteins or organelles of interest can be labeled directly.

Metal Exposure

1. Prepare a 1 mM stock solution of the desired metal compound (e.g., $MnCl_2$).
2. Weigh the required amount of metal under a fume hood and transfer the powder into a 30 mL Corning tube.
3. Add the appropriate volume of prewarmed and equilibrated BrainPhys medium (i.e., incubated at 37°C in 5% CO_2 for at least 4 h to allow gas exchange and pH stabilization) to prepare the stock solution.
4. Vortex for 2 min to ensure complete dissolution of the metal.
5. Use a 0.22 µm filter syringe and filter the stock solution into a sterile plastic tube.
6. Perform exposure in a Petri dish containing membranes and glial cells, while keeping 5 mL of the culture medium. Use the same medium to minimize disruption to the culture and maintain optimal survival conditions.
7. Calculate the volume of the stock solution needed to achieve the desired metal concentration in 5 mL of medium.

8. Adjust the culture medium volume in the Petri dish containing neurons on SN membranes and the glial cells feeder layer to 5 mL minus the volume of the stock solution to be added.
9. Add the stock solution to the Petri dish using a pipette exclusively dedicated to metal manipulation, equipped with a filter tip.
10. Incubate the dish at 37°C with 5% CO_2 for the desired time.

Fluorescent Labeling of Neuron Structures

This step depends on the type of fluorochrome used to label the structure of interest. Exposure time and reagent concentration must be adjusted according to the manufacturer's recommendations. This step must be carried out under a biosafety cabinet to maintain sterile conditions. The fluorescent markers must be compatible with live-cell microscopy and should not contain permeabilizing agents to prevent loss of intracellular components. Examples of suitable exogenous markers are those designed for live-cell imaging and include the ones we have used as SiR and SiR700 actin or tubulin dyes (Spirochrome), LIVE-Actin dye (Aberrior), MitoTracker™ (Invitrogen), LysoTracker (Invitrogen), CellLight® reagents, such as for Golgi labeling (Molecular Probes) or Hoechst-33342 (Sigma Aldrich).

Another option is to perform a transfection or electroporation prior to cell seeding to induce the expression of fluorescently tagged proteins. These expression vectors allow the synthesis of fusion proteins, enabling visualization of specific targets within live cells via their fluorescence. Transfection can be achieved using chemical, lipid-based, or physical methods to deliver the desired material into cells, whereas electroporation uses electrical pulses to create temporary pores in the cell membrane, allowing for material uptake. For example, the XpH20 construct can be used to label postsynaptic density protein-95 (PSD95), as previously reported [13, 22]. However, these methods can reduce cell viability, which must be considered when planning experiments. Neurons should therefore be seeded at a higher density.

3.3 Plunge-Freezing

In this section, we will describe the cryofixation steps using the Vitrobot Mark IV system from FEI (Fig. 5A), but it can be applied to any other automated plunge-freezing system. This system includes various accessories, such as an ethane tray, a holder that surrounds the central part (designed to support cryotubes to store the samples), a brass ethane cup placed in the central part, a cooling spindle positioned above the ethane cup, and an anti-contaminator ring that maintains a cold atmosphere by controlling nitrogen vapor (Fig. 5B). Cryofixation involves vitrifying biological samples by plunging them into liquid ethane cooled with liquid nitrogen. This step is critical to ensuring the native state of the samples and minimizing ice formation. To achieve this, it is crucial to carefully dry the samples

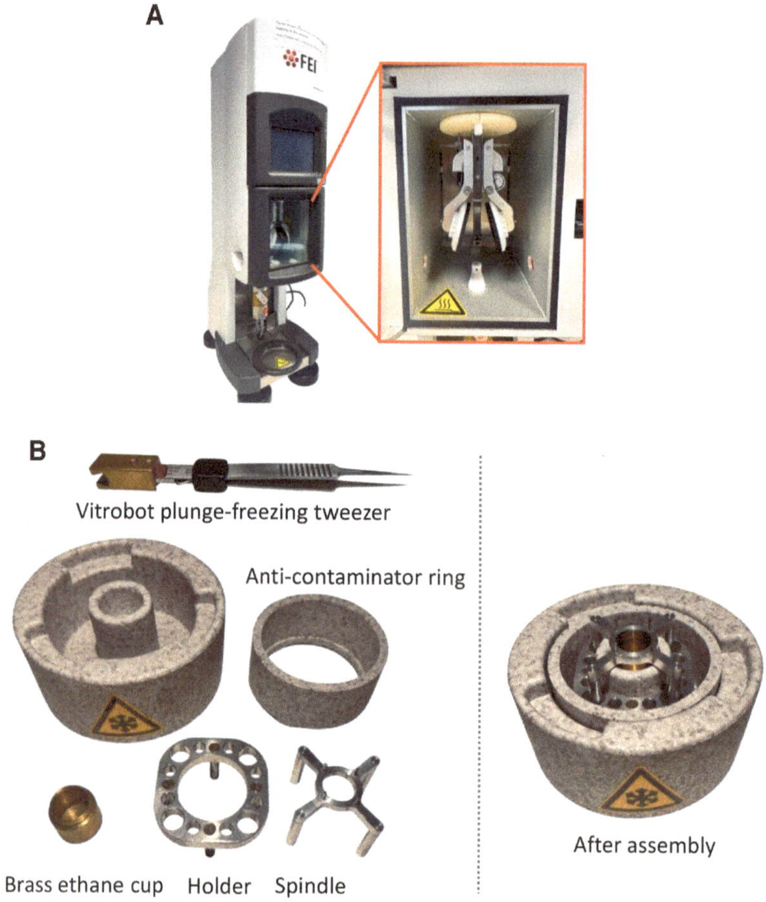

Fig. 5 (**A**) The Vitrobot Mark IV cryofixation system. (**B**) All associated tools before and after assembly

and prevent cell damage. Cryofixation must be performed at low ambient humidity to avoid ice formation, and the operator should wear a mask to avoid breathing humidity.

Turning on the Vitrobot and Material Preparation

1. Prepare 160 mM solution of ammonium acetate (Sigma-Aldrich, 431311) in ultra-trace elemental analyses grade water (Fisher Scientific, 10728098) (300 mOsm, pH 7.4). This solution will be used to rinse the neurons prior to cryofixation to remove the extracellular medium.
2. Vortex and let the solution rest at room temperature overnight. Then, measure the pH and osmolarity of the solution and adjust them to match the parameters of the culture medium used (in our case, 300 mOsm, pH 7.4; these parameters are medium-dependent and should be verified by the user).

3. Filter and store the solution at +4°C for further utilization. For long-term storage, check the osmolarity and pH before use.
4. The day before cryofixation, equilibrate some milliliters of the ammonium acetate solution at 37°C with 5% CO_2 overnight, or at least 2 h.
5. Fill a nitrogen dewar with liquid nitrogen, filter it through a clean tissue (Nonwoven wipes, Ecotex, 08EC971000TA), and transfer some nitrogen into a smaller dewar for easy pouring into the ethane tray.
6. Assemble the ethane container according to the manufacturer's instructions.
7. Pour liquid nitrogen into the ethane tray and add a small amount to the brass ethane cup. Ensure that the tray remains filled and that the anti-contaminator ring is elevated.
8. Wait for the liquid nitrogen in the brass ethane cup to fully evaporate before proceeding to liquefy ethane in the brass ethane cup under a fume hood.
9. Use large forceps to remove the spindle; otherwise, the ethane will solidify.
10. Purge the ethane bottle. For safety reasons, ethane should not remain in the pipes.
11. Keep the ethane tray covered with a polystyrene box, when it is not in use to prevent ice formation and protect it from humidity. Add liquid nitrogen every 10 min, to the tray to maintain a proper temperature.
12. Turn on the Vitrobot and add 40 mL of sterilized distilled water to the humidifier container. Place the two circular filter papers 55/20 mm (Ted Pella Inc., 47000-100) in the paper holder in the Vitrobot chamber (Fig. 5A).
13. Set the following parameters:
    - In the "CONSOLE" tab, set the temperature to 27°C and the humidity to 80%.
    - In the "OPTION" tab, set the following parameters: 4 s for blotting time, 0 for blot force, 2 for the total number of blotting, and 0 for wait and drain time.

Performing Cryofixation

14. Add 1 mL of the preheated ammonium acetate solution to a well of a 24-well plate. This will facilitate the washing step.
15. Prepare a cryogenic-compatible storage container, such as a Nunc™ cryotube (Thermo Scientific™, 10674511), to store the SN membrane. Place the cryotube holder into the ethane

tray, ensuring contact with liquid nitrogen, to allow proper cooling. The holder is designed to accommodate cryotubes in a vertical position.

16. On the Vitrobot screen, select "place new grid" to prepare the Vitrobot arm to receive the tweezers and membrane for plunge-freezing.
17. Using thin, round-tip tweezers, handle the membrane by its black edge, avoiding the central area where hippocampal neurons from rat embryos are cultured and labeled with fluorescent tags. Immerse it in the ammonium acetate solution for a few seconds to rinse.
18. Use the specific tweezer designed for the Vitrobot to grip the membrane by its black edge, again avoiding contact with the central area.
19. Lower the clamp mechanism of the tweezers to lock the membrane in position.
20. Place a piece of filter paper (Whatman, 1002-055) on one edge of the membrane to blot excess liquid by capillary action.
21. Attach the tweezers holding the membrane to the Vitrobot arm. On the screen, press "start process."
22. Place the ethane tray in its designated position on the Vitrobot (make sure the outer ring is still full of liquid nitrogen and that the cooling spindle has been removed). Allow the process to run automatically.
23. Once the process is complete, add liquid nitrogen to the tray to ensure the container is full, and carefully detach the tweezers to ensure the sample remains submerged in the ethane.
24. Slowly lift the tweezers to ensure ethane evaporation while remaining in liquid nitrogen vapors.
25. Transfer the membrane to the cryo-storage box by carefully releasing the clamp mechanism of the tweezers, ensuring the sample remains within the anti-contaminant ring and is continuously exposed to nitrogen vapor to maintain cryogenic conditions.
26. Store the cryo-storage box in a long-term storage system adapted to cryogenic conditions.
27. Repeat for as many samples as required.

Turn off the Vitrobot

28. On the Vitrobot screen, press "exit."
29. Once the screen is black, turn off the Vitrobot by pressing the button at the back of the device.
30. Remove the filter papers.

31. Empty the humidifier, then place it vertically on absorbent paper with the opening facing downward to allow it to dry.
32. Remove the brass ethane cup using tweezers and let the ethane evaporate in the fume hood.
33. Disassemble the device, empty the liquid nitrogen from the ethane tray, and let all components dry on absorbent paper.
34. Once all components are dry, store them in a dust-free environment.

The samples are now cryofixed and must be stored either in liquid nitrogen vapor or direct contact with liquid nitrogen. As long as the samples remain under cryogenic conditions, there is no specific time limit for storage.

3.4 Imaging Fluorescent Markers by cryo-FLM

This section will explain step-by-step how to use a Leica EM-Cryo-CLEM microscope. Specialized cartridges compatible with SN membranes are used for sample observation.

Microscope Switch-on

1. Fill the microscope dewar with liquid nitrogen filtered through a clean tissue.
2. Gently and slowly lower the pump into the liquid nitrogen dewar, ensuring gradual movement to avoid sudden temperature changes that could damage the fragile pump or cause projections. The pump system must be properly positioned before the cooling process begins, as it solidifies upon activation, making further adjustments impossible. Verify that the insulation nozzle is securely attached.
3. Ensure the objective is lowered to at least 2 cm above the sample to properly seal the aperture between the objective and the chamber, preventing nitrogen leakage and ice formation.
4. Turn on the pump controller using the rear power button, then press the blue "Cool" button to initiate cooling.
5. Allow a few jets of vapor to escape to eliminate moisture in the pump tubing.
6. Attach the pump to the cryo-chamber of the microscope and lock it in place by tightening the screw. Once attached, the pump will freeze and must not be moved to avoid damage.
7. Allow the system to cool down for approximately 10 min. The pump controller should indicate a temperature of 77, 15 K(−196°C). To ensure the system is ready, the pump flux has to be lower than 50%; otherwise, wait until the system is stabilized.
8. Turn on the microscope, camera, and computer.

9. Once the camera indicator light turns green, open the MetaMorph software on the computer (or the corresponding software).

Transfer Shuttle Preparation

1. Pour liquid nitrogen into a small dewar that has been filtered with a clean cloth beforehand.
2. Carefully position the Leica SN membrane cartridge into the designated slot within the transfer shuttle (Fig. 3C) and secure the position by locking the small latch.
3. Gradually pour liquid nitrogen into the transfer shuttle to avoid splashing or thermal stress.
4. Place the transparent lid on top to seal it and prevent humidity contamination.
5. Allow the transfer shuttle to cool for 10 min.
6. Submerge the rounded-tip forceps in liquid nitrogen within the transfer shuttle and let them cool to cryogenic temperature before use.

Membrane Placement

1. Place the transfer shuttle inside the plexiglass chamber in a controlled humidity atmosphere.
2. Place the cryo-storage box with the membrane of interest inside the transfer shuttle.
3. Using precooled round-tipped tweezers, carefully remove the membrane from the cryo-storage box and place it in the Leica Si3N4 membrane cartridge. Ensure that the flat side, which contains the cells, faces upward. To ensure proper orientation and facilitate subsequent correlative imaging, note the membrane's orientation and, ideally, always position the membrane in the same way, with the two markers at the top.
4. Make sure the objective is at least 2 cm above the sample.
5. Let the transfer shuttle rod be in the closed position.
6. Attach the transfer shuttle securely to the cryomicroscope using the loading port.
7. Open the lockers on both the cryo-chamber and the transfer shuttle
8. Perform the transfer by sliding the transfer shuttle rod through the loading port. Make sure that the membrane is correctly positioned.
9. Move the rod to the open position and carefully withdraw it from the cryo-chamber, which releases the cartridge.
10. Close the lockers on both the cryo-chamber and the transfer shuttle

11. Remove the transfer shuttle from the loading port.
12. For short-time acquisition, every 10 min, add liquid nitrogen to the transfer shuttle to ensure sufficient liquid nitrogen remains and no humidity forms. The transfer shuttle can be emptied, thoroughly dried, and kept at room temperature for long-term acquisition. It can then be refilled with liquid nitrogen when sample removal is required. This approach minimizes ice contamination caused by prolonged exposure of the transfer shuttle to ambient humidity.

Coordinate Definition

To facilitate correlative imaging and subsequent analysis, defining a precise coordinate system for the sample is essential. The microscope is first manually focused on the upper edge of the SN membrane, ensuring that the selected edge is square or rectangular to account for potential image inversion induced by the optical system. This point is then set as the origin (0,0) in the imaging software to provide a consistent reference for further acquisitions. Once the origin is defined, the software enables automated imaging using a mosaic acquisition approach. The region of interest (ROI) can be selected based on the sample's spatial distribution, with a typical acquisition area covering approximately 3 mm in length and 2.4 mm in width. This allows for a complete high-resolution scan of the membrane while ensuring precise spatial correlation across the entire dataset. The mosaic acquisition function enables seamless stitching of individual images, creating a comprehensive sample map.

Additionally, a focal stack (Z-series acquisition) can be performed to enhance image accuracy, capturing multiple focal planes within a defined range. This is particularly useful when working with samples where organelles may overlap, ensuring that all structures of interest remain in focus even if there is potential for overlapping features within the volume. By following this approach, the exact x and y coordinates of each acquired image are recorded, allowing precise localization of specific ROIs. This spatial reference system is crucial for correlative analyses, as it enables the retrieval of the same ROIs during synchrotron X-ray fluorescence (SXRF) mapping, ensuring accurate overlay and interpretation of multi-modal datasets.

Remove the Membrane and Switch off the Microscope

1. Position the objective upwards to ensure it does not touch the membrane during retrieval.
2. Attach the transfer shuttle to the cryochamber's loading port. Open the lockers on both the cryochamber and the transfer shuttle.
3. Set the transfer shuttle to the open position and slide the rod through the loading port.

4. Retrieve the membrane, place the rod in the closed position, and remove the membrane from the cryo-chamber by sliding the rod through the loading port.
5. Close the lockers on both the cryo-chamber and the transfer shuttle and remove the shuttle from the loading port.
6. Secure the position by locking the small latch.
7. Place the membrane back into the cryo-storage box using the pre-cooled round-tipped tweezers. Once the samples have been observed under cryo-microscopy, they can be stored in liquid nitrogen vapor until the same areas are observed by SXRF.
8. Empty the liquid nitrogen from the transfer shuttle and let it dry.
9. On the pump controller, press the red "heat" button.
10. Wait 30 min to 1 h for the system to return to room temperature.
11. Turn off the microscope and camera.
12. Once the system has reached room temperature, slowly remove the pump from the liquid nitrogen dewar and turn off the pump controller using the button at the back.

3.5 Imaging Metals by Cryo-synchrotron X-ray Fluorescence

This section focuses on the ID16A beamline at the ESRF in Grenoble, France, a state-of-the-art synchrotron station dedicated to nano-imaging. Notably, it is optimized for analyzing samples under cryogenic conditions. The cryogenic transfer system consists of two main components: the EM VCM and the EM VCT500 (Fig. 4B). The EM VCM is the stationary unit that ensures proper positioning of the SN membranes within the sample port while controlling the shuttle's cooling and vacuum conditions. The EM VCT500, on the other hand, serves as the mobile unit that physically transfers the sample holder into the analysis chamber of the beamline. First, we will describe, step by step, how to perform the sample transfer under cryogenic and vacuum conditions. Then, we will provide a more general overview of the SXRF analysis. Samples are shipped to the beamline using a dry shipper dewar (Fig. 4A), preserving them in a frozen, hydrated state during transport.

Turning on the EM VCM and Positioning the EM VCT500

1. In the loading station, position the sample carrier using the two screwdrivers provided with the EM VCM so that it is perfectly aligned with the hole corresponding to the entry point of the EM VCT500 rod
2. Gradually fill the liquid nitrogen reservoir in small increments, then securely close its lid after filling.

3. Turn on the Leica EM VCM and allow the startup process to complete.
4. Press the "Cooling" button and wait for the loading station to fill with liquid nitrogen and cool down.
5. Regularly check the liquid nitrogen reservoir and refill it if necessary.
6. Fill the mobile part (EM VCT500) with liquid nitrogen until the temperature displayed on the EM VCM control screen reaches −196°C.

Sample's Positioning

1. Place round-tipped tweezers into the tool holder to allow it to cool, then use them to carefully remove the membrane from its storage system and transfer it onto the sample carrier. The smooth side should face the manipulator. Note the orientation in which the membrane was placed, particularly where the square and rectangular markers are located.
2. Secure the membrane by tightening the screws on the sample carrier.
3. Load the sample into the EM VCT500, the mobile component of the system, using the rod.
4. In the EM VCM control screen, initiate the vacuum process for the EM VCT500 (check the pressure on the screen).
5. Once the system is under vacuum, detach the EM VCT500 by pressing "detach" on the control screen.
6. Detach and bring it to the ID16A beamline and transfer the sample carrier to the analysis chamber. The membranes remain frozen as the entire chamber is maintained at 108 K.

Spotting ROIs, Running Coarse and Fine Scans

During Cryo-FLM, an overview image of the entire membrane was acquired. At the beamline, the coordinates of the four corners of the inner square of the membrane (where the neurons are cultured) are visually identified and matched with their respective positions in the overview image acquired by cryo-FLM. This process establishes a spatial correlation between the features visible on the beamline interface and their counterparts in the fluorescence overview image. Using the fluorescence overview image, the coordinates of the regions of interest (ROIs) are calculated relative to one of the membrane's corners. At the beamline, the stage is aligned with the designated corner and then moved by the calculated distances (x, y) to reach the region of interest.

Once the area of interest is identified, a coarse scan is performed first under low X-ray flux conditions and at a lower resolution than the fine scans. After this initial step, a smaller, more precise fine scan is selected on the region of interest based on the coarse scan images.

This approach optimizes beam time, as high-resolution scans are time-intensive. Given the value of beam time, it is crucial to carefully optimize the scan positions and areas before conducting fine scans.

For neuronal or cell cultures, fine scans of the samples are performed using a raster scanning approach in continuous mode, with an integration time of 100 ms/pixel and a step size of 50 nm/pixel. For quantitative analyses, calibration is essential and should be conducted using standards similar to the sample's characteristics. For thin biological samples, the AXO standard from the Dresden-based company is commonly employed.

Sample Removal

The process follows the reverse order of the loading procedure. First, the Leica systems EM VTC500 and EM VCM are cooled using liquid nitrogen. The vacuum is established in the mobile EM VTC500 connected to the beamline. The sample carrier is carefully removed from the analysis chamber and placed into the VTC500. Next, the mobile unit is transferred to the Leica EM VCM system, where the vacuum is released. The sample carrier is positioned in the precooled loading station, allowing the membrane to be safely extracted, stored in a cryogenic box, and returned to the liquid nitrogen dewar for long-term preservation.

Data Treatment

The X-ray fluorescence spectra are processed and calibrated using Python scripts built on the PyMCA library [23]. Following the fitting of the AXO standard, the sample data is fitted, and the resulting maps, expressed in areal mass density, are saved as 32-bit TIFF images. The image fitting process is independent of data acquisition and can be performed post-measurement (https://www.silx.org/doc/PyMca/latest/xrf/material-definition/index.html).

3.6 Processing of Cryo-Correlative Imaging

Once the results are obtained as TIFF images for both techniques, the ICY software is used to correlate the two images. ICY is an open-source software platform developed in 2011 by the Institut Pasteur [24] (version 2.4.3, available at http://icy.bioimageanalysis.org), dedicated to the analysis and processing of biomedical images in 2D, 3D, and 4D. It provides an intuitive interface for visualizing, processing, and analyzing images, supports numerous formats, and allows the integration of custom plugins. Among its tools, the eC-CLEM (Correlative Light and Electron Microscopy) plugin [25] facilitates the alignment and overlay of data from light and electron microscopy by enabling the import of multimodal images, correction of scales, rotations, and distortions, as well as the precise superimposition of fluorescent signals and ultrastructural features. The ICY software and the ec-CLEM plugin must be installed for the following protocol. Here, we will describe how to use this software to overlay two different images.

Running ICY and eC-CLEM

Before starting, open all the cryo-FLM images acquired simultaneously and merge them into a single image. Similarly, load the cryo-SXRF images and merge them into a single image, including elements with high neuronal concentrations, such as potassium. These merged images will provide the general shape of the cell and serve as reference points for correlation. The scaling of each image should be updated during this merging process.

1. Open the software.
2. Load the merged images generated from both techniques (cryo-fluorescence and cryo-SXRF).
3. In the ICY search bar, type "eC-CLEM" to launch the plugin. A window similar to Fig. 6 will appear on the screen.
4. In tab number 1, select "2D (X, Y[T])". Meaning that no deformation should be introduced in the images.
5. In tab number 2, select the cryo-FLM merge images under "Image that will be transformed and resized," and the cryo-SXRF merge image under "Image that will not be modified."
6. In tab number 3, select "Show difference in positions" and "Show predicted error in positions."
7. Select Start in tab number 5.
8. Using the SXRF image as a reference, place initial points on recognizable neuronal structures based on their shape (Fig. 7A).
9. After placing three points, the image will automatically resize. Add additional points and adjust their positions as needed to improve alignment accuracy.

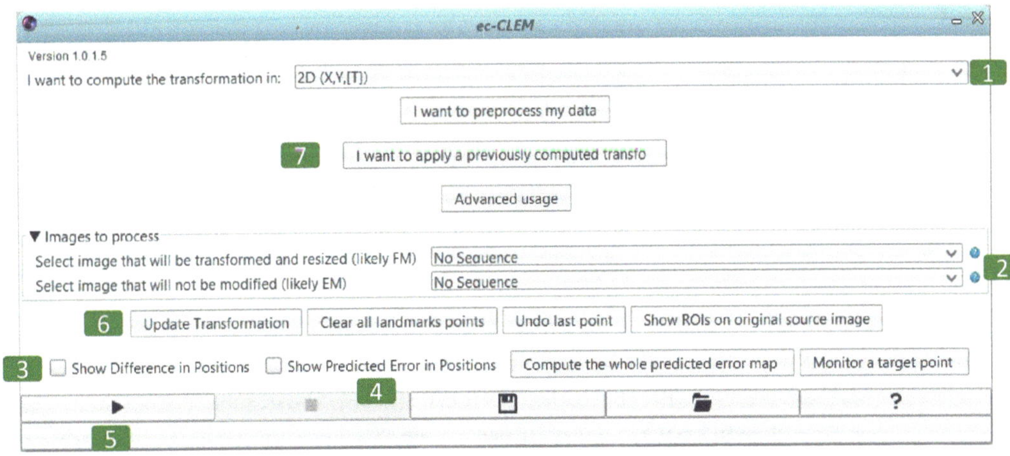

Fig. 6 Graphical user interface of eC-CLEM in ICY software for image superimposition

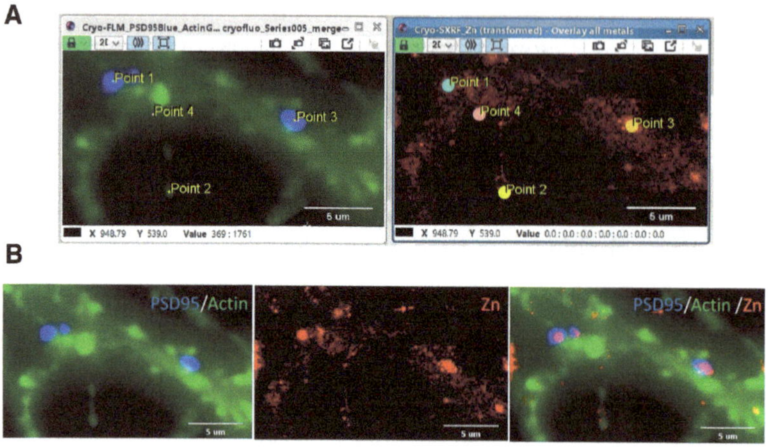

Fig. 7 Example of Cryo-FLM and SXRF image overlay using the ec-CLEM plugin. Images were acquired from primary rat hippocampal neurons cultured on silicon nitride membranes. Actin (green) and PSD95 (blue) were visualized by Cryo-FLM, while zinc (red) distribution was mapped by cryo-SXRF at beamline ID16A (ESRF). The datasets were aligned using the ICY software with the ec-CLEM plugin. (**A**) Point selection for image superimposition with the eC-CLEM plugin. (**B**) Cryo-FLM and cryo-SXRF images before and after overlay

10. Use the "Show predicted error in positions" and "Show difference in positions" options to refine point placement based on the visual feedback provided by the software.
11. Once the points are correctly placed, click "STOP" (tab number 4, Fig. 6). A new image will appear with both channels, displaying the superimposed SXRF and cryo-fluorescence images. If there is a misalignment, the procedure can be restarted by clicking "START." Previously placed points will remain visible and can be adjusted as needed.

The final image can be saved in TIFF format with all channels accurately superimposed. These correlated images enable a wide range of analyses by combining elemental distributions obtained from cryo-SXRF with molecular and structural information from cryo-FLM. Specifically, this approach quantifies metal concentrations within neuronal structures, providing insights into their potential roles in physiological or pathological conditions. The spatial colocalization of metals with specific organelles or proteins can be assessed, enabling the study of protein–metal interactions.

4 Conclusions

Cryo-correlative imaging combining fluorescence microscopy and synchrotron X-ray fluorescence (SXRF) represents a major advancement in studying metal–protein interactions within neurons. The methodology detailed in this chapter enables high-resolution imaging of essential metals such as iron, copper, and zinc while preserving the structural integrity of cells under cryogenic conditions. The application of this technique to primary hippocampal neuron cultures has demonstrated its effectiveness in mapping essential metals at the subcellular level and correlating them with key neuronal structures. For instance, cryo-correlative imaging combining STED and SXRF revealed a co-segregation of zinc and copper with actin in synaptic compartments, suggesting their involvement in synaptic transmission and cytoskeletal remodeling [16]. Similarly, cryo-SXRF studies have shown that iron accumulates in growth cones, highlighting a potential role in axonal development [15]. In dendritic protrusions and axonal varicosities, copper was also found to colocalize with actin and tubulin, further supporting its implication in cytoskeletal dynamics and neuronal signaling [12]. Altogether, these findings emphasize the power of this correlative approach to explore the spatial organization of metals and their functional relationship with neuronal architecture.

Beyond its application to primary neurons, this methodological framework is transferable to other biological systems. With appropriate adaptations—including adjustments of cell culture substrates, cryo-fixation protocols, or imaging parameters—this approach can be implemented in various experimental setups. Such flexibility expands its relevance to both basic and translational research. Importantly, this correlative approach holds great potential for studying neurological disorders related to metal dyshomeostasis. Analyzing the abnormal distribution of these elements could provide valuable insights into pathologies such as Alzheimer's disease, where metal accumulation in specific brain regions is suspected to exacerbate oxidative stress and neurodegeneration, especially in synaptic compartments [12]. Likewise, using SXRF in experimental neurotoxicity models could help assess the impact of toxic metal exposure on neuronal architecture. In conclusion, combining cryo-fluorescence microscopy and SXRF constitutes an innovative, versatile and adaptable approach for exploring metal–protein interactions in neuronal cells. Its future development, particularly through the integration of additional spectroscopic techniques or X-ray tomography, will open new perspectives for biomedical research and neuroscience.

Acknowledgments

The authors acknowledge the support of the French Agence Nationale de la Recherche (ANR) under grant ANR-21-CE34-0011 (project SuperResMetalToxSyn). The experiments were performed on beamline ID16A at the European Synchrotron Radiation Facility (ESRF), Grenoble, France. We are grateful to Dr. Peter Cloetens and Dr. Murielle Salomé at the ESRF for assisting in using beamline ID16A. Light microscopy was performed at the Bordeaux Imaging Center, a member of the National Infrastructure FranceBioImaging, supported by the French National Research Agency (ANR-10-INBS-04). We thank the IINS Cell Biology Facility, used in this study and supported by the GPR BRAIN_2030, especially Natacha Retailleau and Adrien Caralp, for the production of cellular and/or molecular tools.

Conflict of Interest The authors have no conflicts of interest to declare that are relevant to the content of this chapter.

References

1. Maret W (2016) The metals in the biological periodic system of the elements: concepts and conjectures. Int J Mol Sci 17(1):66. https://doi.org/10.3390/ijms17010066
2. White AR (éd) (2017) Metals in the brain, vol 124. Springer, New York
3. Barnham KJ, Bush AI (2014) Biological metals and metal-targeting compounds in major neurodegenerative diseases. Chem Soc Rev 43(19):6727–6749. https://doi.org/10.1039/C4CS00138A
4. Aschner M, Guilarte TR, Schneider JS, Zheng W (2007) Manganese: recent advances in understanding its transport and neurotoxicity. Toxicol Appl Pharmacol 221(2):131–147. https://doi.org/10.1016/j.taap.2007.03.001
5. Gonzalez-Cuyar LF, Nelson G, Criswell SR, Ho P, Lonzanida JA, Checkoway H, Seixas N, Gelman BB, Evanoff BA, Murray J, Zhang J, Racette BA (2014) Quantitative neuropathology associated with chronic manganese exposure in South African mine workers. NeuroToxicology 45:260–266. https://doi.org/10.1016/j.neuro.2013.12.008
6. Brewer GJ (2012) Copper excess, zinc deficiency, and cognition loss in Alzheimer's disease. BioFactors 38(2):107–113. https://doi.org/10.1002/biof.1005
7. Li DD, Wang Z-Y, Wei Z, Zhao P (2017) Serum copper, zinc, and iron levels in patients with Alzheimer's disease: a meta-analysis of case-control studies. Front Aging Neurosci 9(300). https://doi.org/10.3389/fnagi.2017.00300
8. Ward RJ, Zucca FA, Duyn JH, Crichton RR, Zecca L (2014) The role of iron in brain ageing and neurodegenerative disorders. Lancet Neurol 13(10):1045–1060. https://doi.org/10.1016/S1474-4422(14)70117-6
9. Kaufmann R, Hagen C, Grünewald K (2014) Fluorescence cryo-microscopy: current challenges and prospects. Curr Opin Chem Biol 20:86–91. https://doi.org/10.1016/j.cbpa.2014.05.007
10. Pushie MJ, Sylvain NJ, Hou H, Hackett MJ, Kelly ME, Webb SM (2022) X-ray fluorescence microscopy methods for biological tissues. Metallomics 14(6):mfac032. https://doi.org/10.1093/mtomcs/mfac032
11. Kelkoul I, Muñoz VP, Ortega R, Carmona A (2025) Synchrotron-based correlative imaging of metals and proteins in neuronal cells: state of the art and future challenges in neurometallomics. Metallomics 17(2):mfaf003. https://doi.org/10.1093/mtomcs/mfaf003
12. Ortega R, Fernández-Monreal M, Pied N, Roudeau S, Cloetens P, Carmona A (2024) Native cryo-correlative light and synchrotron X-ray fluorescence imaging of proteins and essential metals in subcellular neuronal compartments. Chem Biomed Imaging 2(11):744–754. https://doi.org/10.1021/cbmi.4c00038
13. Ortega R, Roudeau S, Carmona A (2022) Correlative nano-imaging of metals and proteins in primary neurons by synchrotron X-ray fluorescence and STED super resolution microscopy: experimental validation. J Neurosci Method 381:109702. https://doi.org/10.1016/j.jneumeth.2022.109702
14. Kaech S, Banker G (2006) Culturing hippocampal neurons. Nat Protocol 1(5):2406–2415. https://doi.org/10.1038/nprot.2006.356
15. Carmona A, Chen S, Domart F, Choquet D, Ortega R (2022) Imaging the structural organization of chemical elements in growth cones of developing hippocampal neurons. Metallomics 14(1):mfab073. https://doi.org/10.1093/mtomcs/mfab073
16. Domart F, Cloetens P, Roudeau S, Carmona A, Verdier E, Choquet D, Ortega R (2020) Correlating STED and synchrotron XRF

nano-imaging unveils cosegregation of metals and cytoskeleton proteins in dendrites. eLife 9:e62334. https://doi.org/10.7554/eLife.62334

17. Grabrucker A, Vaida B, Bockmann J, Boeckers TM (2009) Synaptogenesis of hippocampal neurons in primary cell culture. Cell Tissue Res 338(3):333–341. https://doi.org/10.1007/s00441-009-0881-z
18. Perrin L, Carmona A, Roudeau S, Ortega R (2015) Evaluation of sample preparation methods for single cell quantitative elemental imaging using proton or synchrotron radiation focused beams. J Analyt Atom Spectromet 30(12):2525–2532. https://doi.org/10.1039/C5JA00303B
19. Schwartz CL, Sarbash VI, Ataullakhanov FI, Mcintosh JR, Nicastro D (2007) Cryo-fluorescence microscopy facilitates correlations between light and cryo-electron microscopy and reduces the rate of photobleaching. J Microsc 227(2):98–109. https://doi.org/10.1111/j.1365-2818.2007.01794.x
20. Da Silva JC, Pacureanu A, Yang Y, Fus F, Hubert M, Bloch L, Salome M, Bohic S, Cloetens P (2017) High-energy cryo X-ray nano-imaging at the ID16A beamline of ESRF. In: Lai B, Somogyi A (éds) X-ray nanoimaging: instruments and methods III. SPIE, p 14
21. Cesar Da Silva J, Pacureanu A, Yang Y, Bohic S, Morawe C, Barrett R, Cloetens P (2017) Efficient concentration of high-energy X-rays for diffraction-limited imaging resolution. Optica 4(5):492. https://doi.org/10.1364/OPTICA.4.000492
22. Rimbault C, Breillat C, Compans B, Toulmé E, Vicente FN, Fernandez-Monreal M, Mascalchi P, Genuer C, Puente-Muñoz V, Gauthereau I, Hosy E, Giannone G, Chamma I, Mackereth CD, Poujol C, Choquet D, Sainlos M (2021) Engineering paralog-specific PSD-95 synthetic binders as potent and minimally invasive imaging probes. Neuroscience. https://doi.org/10.1101/2021.04.07.438431
23. Solé VA, Papillon E, Cotte M, Walter PH, Susini J (2007) A multiplatform code for the analysis of energy-dispersive X-ray fluorescence spectra. Spectrochim Acta Part B Atom Spectrosc 62(1):63–68. https://doi.org/10.1016/j.sab.2006.12.002
24. De Chaumont F, Dallongeville S, Chenouard N, Hervé N, Pop S, Provoost T, Meas-Yedid V, Pankajakshan P, Lecomte T, Le Montagner Y, Lagache T, Dufour A, Olivo-Marin J-C (2012) Icy: an open bioimage informatics platform for extended reproducible research. Nat Method 9(7):690–696. https://doi.org/10.1038/nmeth.2075
25. Paul-Gilloteaux P, Heiligenstein X, Belle M, Domart M-C, Larijani B, Collinson L, Raposo G, Salamero J (2017) eC-CLEM: flexible multidimensional registration software for correlative microscopies. Nat Method 14(2):102–103. https://doi.org/10.1038/nmeth.4170

Chapter 9

Crosstalks Between Neurometals and Amyloidogenic Proteins at Synapses and Their Involvement in the Pathogenesis of Neurodegenerative Diseases

Masahiro Kawahara, Ken-ichiro Tanaka and Midori Kato-Negishi

Abstract

Trace elements such as iron, zinc, copper, and manganese, termed "neurometals" play central roles in normal brain functions. Their deficiency causes various adverse effects, such as learning disabilities, whereas their excess often produces reactive oxygen species and damages the brain. Increasing evidence suggests that the dyshomeostasis of these trace elements, termed "neurometals," is involved in the pathogenesis of various neurodegenerative diseases. In conformational diseases such as Alzheimer's disease, prion diseases, and Lewy body diseases, disease-related amyloidogenic proteins can regulate metal homeostasis at the synapses; thus, the loss of the protective functions of normal proteins causes neurodegeneration. Moreover, metal-induced conformational changes in the amyloidogenic proteins increase their neurotoxicity. Zn and Cu are secreted to synaptic clefts and contribute to neural information processing and memory formation. However, excess Zn and Cu after transient global ischemia play central roles in the pathogenesis of vascular-type senile dementia (VD). We present here an overview of the roles of these neurometals in the synapse and their connections with the pathogenesis of neurodegenerative diseases based on metal–protein and metal–metal interactions.

Keywords Amyloid, Neurotoxicity, Conformation, Reactive oxygen species (ROS), Gastrointestinal absorption, Ca^{2+} homeostasis, Pore, Synapse

Abbreviations

AD	Alzheimer's disease
AβP	Alzheimer's β-amyloid protein
ALS	amyotrophic lateral sclerosis
AMPA	α-amino-3-hydroxy-5-methylisoxazole-4-propionic acid
BBB	blood–brain barrier
CSF	cerebrospinal fluid
CJD	Creutzfeldt-Jakob disease
CTR1	copper transporter 1
DFO	deferoxamine
DLB	dementia with Lewy bodies

Andreas M. Grabrucker (ed.), *Brain Metallomics*, Neuromethods, vol. 237,
https://doi.org/10.1007/978-1-0716-5468-2_9,
© The Author(s), under exclusive license to Springer Science+Business Media, LLC, part of Springer Nature 2026

DMT1	divalent metal transporter 1
ER	endoplasmic reticulum
FPN	ferroportin
GABA	γ-aminobutyric acid
IRE	iron-responsive element
IRP	iron regulatory protein
LBD	Lewy body disease
LTP	long-term potentiation
MT	metallothionein
NMDA	N-methyl-d-aspartate
NAC	nonamyloid component
NBTI	non-transferrin-bound iron
PD	Parkinson's disease
PrP	prion protein
ROS	reactive oxygen species
TF	transferrin
TSE	transmissible spongiform encephalopathy
VD	vascular-type senile dementia

1 Introduction

Several trace elements, such as iron (Fe), zinc (Zn), copper (Cu), and manganese (Mn), are essential for life beings and play inevitable roles in various organs. Among them, the brain is a highly peculiar organ. It is an energy-demanding organ that uses approximately 20% of the body's total oxygen despite its small volume, which is only 2% of the human body weight [1]. Thus, a considerable amount of Fe and Cu exist in the brain since both are related to enzymes for mitochondrial energy production processes. Moreover, the brain requires these trace elements, termed "neurometals," for its intrinsic functions, such as myelination and synthesis of various neurotransmitters. Additionally, Zn and Cu play essential roles in neural information processing and the formation of new memories because they are released from synaptic vesicles activity-dependent and regulate neuronal excitability [2, 3].

These neurometals are obtained mainly from food or drinking water and can be absorbed via the gastrointestinal tract. However, their absorption rates are generally low and influenced by their chemical speciation and the substances coexisting ineurotoxicity of PrP106–126 and the formation ofn food. Thus, these neurometals tend to become deficient, and their deficiency causes various adverse effects on normal brain functions, particularly in the developing brain.

By contrast, an excess of neurometals is toxic. Because of its disproportionate oxygen and blood flow, the brain is susceptible to oxidative stress and reactive oxygen species (ROS) caused by Fe, Cu, and Mn [4]. Furthermore, neurons are susceptible to the accumulation of toxic metals, including mercury, aluminum, lead, and cadmium, because neurons do not generally regenerate but are maintained throughout an individual's entire life span. Although the blood–brain barrier (BBB), which is composed of tight junctions between endothelial cells, restricts the entry of toxic elements into the brain, this barrier system is imperfect.

Increasing evidence suggests that dyshomeostasis of these neurometals, deficiency and excess, is involved in the pathogenesis of various neurodegenerative diseases such as Alzheimer's disease (AD), prion diseases, and Lewy body diseases (LDBs) [including Parkinson's disease (PD) and dementia with Lewy bodies (DLB)], vascular-type senile dementia (VD), Wilson's disease, Menkes disease, and amyotrophic lateral sclerosis (ALS) [5–9]. These diseases are mainly divided into two categories. AD, prion diseases, and LDBs are included in the category of conformational diseases [10], which commonly involve the abnormal deposition of disease-related proteins, termed "amyloidogenic proteins" [AβP in AD, prion protein (PrP) in prion disease, and α-synuclein in LDBs (PD and DLB)]. Meanwhile, VD, Menkes disease, and Wilson disease are caused by an excess or deficiency of neurometals.

In these conformational diseases, there are at least two possible pathogenic pathways; the first supports "the loss of the normal, protective functions," and the second supports "the gain of toxic functions" [11]. The amyloidogenic proteins commonly tend to form oligomers with β-pleated sheet structures and to form amyloid fibrils. Metal ions cause the oligomerization of amyloidogenic proteins and induce their conformational changes. Moreover, these amyloidogenic proteins all colocalize at the synapse and regulate metal homeostasis. Therefore, it is possible that the loss of functions of these proteins causes the dyshomeostasis of neurometals, produces reactive oxygen species (ROS), and triggers neurodegenerative processes. Moreover, these amyloidogenic proteins or their fragment peptides commonly form Ca^{2+}-permeable channels (amyloid pores) on membranes and exhibit synaptotoxicity and neurotoxicity [12]. The gain of toxic functions of amyloidogenic proteins will cause Ca dyshomeostasis and trigger various neurodegenerative pathways. Metals reportedly influence these processes. Meanwhile, excess Zn or Cu are secreted to the synaptic clefts after transient global ischemia and play critical roles in the neurodegeneration and the pathogenesis of VD [13].

From these perspectives, we review the contribution of neurometals to the pathogenesis of several neurodegenerative diseases, focusing on metal–metal interactions and metal–protein interactions in the synapse based on our recent findings and those of others.

2 Characteristics of Neurometals and Their Roles in Brain Functions

2.1 Iron

Iron is the most abundant trace element in the brain and in the whole body [14]. The human body contains approximately 3–5 g of Fe, and approximately 70% of total Fe is contained in heme proteins such as hemoglobin and myoglobin. Fe is essential for numerous biological functions in most organs, such as oxygen transport, oxidative phosphorylation, DNA synthesis, mitochondrial respiration, and metabolism, by acting as a cofactor or a component of enzymes or functional proteins, including cytochrome C oxidase, catalase, cytochrome P-450, tyrosine hydroxylase, and succinate dehydrogenase. Fe is essential for brain functions such as myelination and neurotransmitter synthesis. Fe generally forms ferric iron (Fe^{3+}) and ferrous iron (Fe^{2+}). The chemical speciation of Fe is regulated by ferroxidases, which oxidize Fe^{2+} ions to Fe^{3+}, and ferrireductase (ferric iron reductase), which reduce Fe^{3+} ions to Fe^{2+}.

Among orally administered Fe, heme-Fe, mainly from meats, is absorbed via transporters such as the heme carrier protein 1 (HCP-1) with a relatively high absorption rate (10%–20%) [15]. Moreover, nonheme Fe, mainly as Fe^{3+}, is reduced to Fe^{2+} by ferrireductases (such as duodenal cytochrome b) or gastric acid and is subsequently absorbed through the divalent metal transporter 1 (DMT1) [16]. The absorption rate of non-heme Fe is low (2%–5%) and is influenced by numerous dietary components, such as phytate, oxalate, ascorbic acid, and polyphenols. Absorbed Fe^{2+} is exported to the extracellular space via ferroportin (FPN), a Fe^{2+} exporter protein [17]. Hepcidin, a peptide hormone produced in the liver, plays a key role in Fe metabolism by regulating FPN. Excess Fe is transported to the liver, oxidized, and stored in ferritin, a Fe storage protein. Fe^{2+} ions are oxidized to Fe^{3+} in the bloodstream by ferroxidases (such as ferritin or ceruloplasmin), and then Fe^{3+} is transported by binding to transferrin (TF), an iron-binding protein that binds two Fe^{3+} ions. TF-bound Fe crosses the BBB via its receptors and enters neurons or glial cells. Non-transferrin-bound iron (NBTI), mainly Fe^{3+} ions that bind to small molecules such as citrates, are reduced to Fe^{2+} ions by ferrireductases and enter cells via DMT1. Intracellular Fe^{2+} is transferred to enzymes in the mitochondria, endoplasmic reticulum (ER), or Golgi apparatus, which require bioactive Fe^{2+} as a cofactor. In the brain, Fe mainly accumulates in the globus pallidus, putamen, and substantia nigra. Fe deficiency causes anemia, loss of the ability to exercise, and decreased intellectual ability. Fe supplementation reportedly improves the working memory of adolescent girls [18].

Fe regulates the expression of various Fe-related proteins [19]. The mRNA of these proteins possesses an iron-responsive element (IRE) in the 3′-UTR or 5′-UTR. For example, the mRNA of ferritin has an IRE motif in its 5′-UTR. Under Fe-deficient conditions, iron-responsive proteins (IRPs; IRP1 and IRP2) bind to the IRE motif and inhibit the expression of ferritin, which eventually promotes the circulation of Fe in the blood. Under conditions with an excess of Fe, Fe binds to IRPs, promotes their dissociation from ferritin mRNA, and enhances ferritin translation, ultimately decreasing the concentration of free Fe. By contrast, the IRE motif at the 3′ -UTR of transferrin mRNA binds to IRPs under Fe-deficient conditions, protects mRNAs from endonuclease cleavage, and promotes the translation of transferrin and the circulation of Fe. This IRE-IRP signaling pathway controls the level of Fe in the brain and other organs.

Despite the importance of Fe in the brain, excess Fe is toxic because it is a redox-active metal, and Fe^{2+} forms ROS through Fenton reactions. Thus, FPN exports excess Fe^{2+} to the extracellular space in neurons. Not only the total amount of Fe but also its chemical speciation, namely, the ratio of Fe^{2+}/Fe^{3+}, is strictly regulated. Recent studies have suggested that ferroptosis, a nonapoptotic form of programmed cell death caused by excess Fe, is associated with the pathogenesis of various diseases [20].

Figure 1 shows the absorption and transportation of Fe and Zn in the body.

2.2 Zinc

Zinc is the second most abundant element in the body [21]. Zn plays essential roles in various physiological functions such as mitotic cell division, protein synthesis, immunological responses, and DNA and RNA synthesis as a cofactor of more than 300 enzymes or metalloproteins. Zn is also responsible for the DNA-binding ability of many

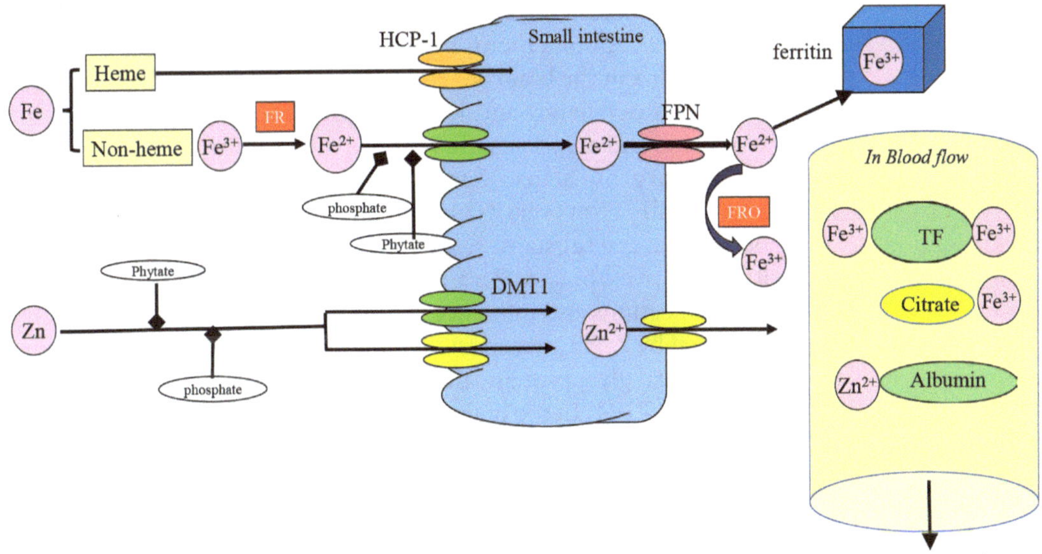

Fig. 1 Absorption and transportation of neurometals in the body. Orally administered metals can be absorbed mainly via the gastrointestinal pathway. Among dietary Fe, heme-Fe is absorbed via heme carrier protein (HCP-1) in the small intestine. Nonheme Fe is primarily in the form of Fe^{3+} and is reduced to Fe^{2+} by ferrireductase (FR) or ascorbic acid in food. Thereafter, Fe^{2+} is absorbed via a divalent metal transporter (DMT1). However, its absorption rate is influenced by phosphate, phytate, and other divalent cations in food. Some of the absorbed Fe^{2+} is transported to the liver and stored in ferritin as Fe^{3+}. Other Fe^{2+} is exported via ferroportin (FPN), and then, the Fe^{2+} is oxidized to Fe^{3+} by ferroxidase (FRO), binds to transferrin (TF) or citrate in the blood, and is transported to the brain. Zn is absorbed via DMT1 or ZIP4 in the small intestine. Its absorption is influenced by phytate and phosphate. The absorbed Zn^{2+} is excreted by ZnT1 and transported by binding with albumin and α 1-acid glycoprotein in the blood

transcription factors, and approximately 10% of the human genome possesses the Zn-binding motif. Zn^{2+} also acts as a signaling molecule to facilitate the transcription of the signaling cascades, such as Ca^{2+} [22]. Therefore, Zn is essential for normal growth and development. Its deficiency causes dwarfism with retardation of physical and sexual development in humans [23].

The maintenance of Zn homeostasis is regulated by metallothioneins (MTs) and two-types of Zn transporters: ZnT transporters and Zrt- and Irt-like protein (ZIP) transporters. Metallothioneins (MTs) are ubiquitous proteins with 20 cysteine residues that can bind seven atoms of Zn, Cu, Cd, and other metals [24]. There are three types of metallothioneins: MT1, MT2, and MT3. MT1 and MT2 are ubiquitously expressed throughout the body, whereas MT3 is primarily localized in the central nervous system.

ZnT transporters facilitate the efflux of Zn^{2+} when it is in excess, whereas ZIP transporters facilitate its influx when it is deficient [25]. Among the ten types of ZnT transporters in mammals, ZnT1 plays a pivotal role in Zn^{2+} efflux and protection from excess Zn^{2+} in the brain. ZnT3 is localized in the membranes of presynaptic vesicles, transports Zn^{2+} into synaptic vesicles, and maintains high Zn^{2+} concentrations in these vesicles. There are 14 ZIP transporters distributed in membranes of subcellular organelles (the Golgi apparatus or ER) or cells. Among them, ZIP4 predominantly regulates intracellular Zn influx in the brain. Genetic dysregulation of these transporters causes Zn deficiency and leads to diseases such as Ehlers–Danlos syndrome and acrodermatitis enteropathica in humans.

Dietary Zn is absorbed in the gastrointestinal tract via ZIP4 and DMT1. However, its absorption rate is generally as low as 30% because several substances, such as phytate in plants or phosphates in food additives, bind to Zn and inhibit its absorption. Cu^{2+} and other divalent cations also influence the absorption of Zn. Absorbed Zn is excluded by ZnT1, transported in the blood by binding with albumin and α1-acid glycoprotein, and thereafter transferred into neurons via DMT1 or ZIP4. In the brain, high concentrations of Zn are observed in the hippocampus, amygdala, cerebral cortex, thalamus, and olfactory cortex. A substantial fraction (approximately 10% or more) of brain Zn is stored in the presynaptic vesicles of excitatory glutamatergic neurons [26]. Synaptic Zn is secreted into synaptic clefts with glutamate during neuronal excitation as free Zn ions (Zn^{2+}) or is in a loosely bound form. The secreted Zn^{2+} regulates overall brain excitability by binding with receptors such as *N*-methyl-D-aspartate (NMDA)-type glutamate receptors, α-amino-3-hydroxy-5-methylisoxazole-4-propionic acid (AMPA)-type glutamate receptors, γ-aminobutyric acid (GABA) receptors, P2-type purinergic receptors, and glycine receptors [27–28]. Synaptically released Zn^{2+} also modulates dendritic functions [29] and contributes to the induction of long-term potentiation (LTP) [30], a form of synaptic information storage well known as a paradigm of the mechanism by which memory is formed. Ueno et al. demonstrated that Zn released from mossy fibers in the hippocampus modulates the activity of neuronal circuits [31]. Therefore, synaptic Zn^{2+} acts as an endogenous neuromodulator that provides spatiotemporal information regarding neuronal plasticity and is essential for memory formation. Consistent with these findings, Zn deficiency in childhood causes delayed mental development and learning disabilities [32]. Zn deficiency causes odor and smell, taste, and fear-conditioning disorders in adults [33]. Zn deficiency is also associated with seizures, epilepsy, and convulsions [34, 35].

Owing to the abundance of Zn in the synapses, ZnT1 and ZIP4 reportedly localize to synapses [36, 37]. Elevated levels of Zn induce the expression of ZnT1 and the translocation of ZnT-1 on membrane surfaces [38]. In addition to Zn regulation, ZnT1 regulates NMDA-R and Ca^{2+} channels in postsynaptic membranes [36, 39].

2.3 Copper

Copper is the third most abundant element in the body [40]. Dietary Cu is absorbed via DMT1 in the small intestine, mainly in the duodenum, transported to the liver, and distributed by ceruloplasmin in the blood. The required amount of Cu is not high, and the daily intake of Cu is as low as 0.7 mg; therefore, Cu deficiency under physiological conditions is rare. Meanwhile, excess Cu from drinking water is associated with non-Indian cirrhosis in children [41]. Cu homeostasis is regulated by copper-transporting ATPases (ATP7A and ATP7B). ATP7A is a Cu efflux protein that delivers Cu to Cu-containing enzymes in the trans-Golgi network. ATP7B has functional similarity to ATP7A and exports Cu into bile for excretion. A deficiency or excess of Cu caused by genetic disorders associated with these transporters can cause neurodegenerative diseases such as Menkes disease and Wilson's disease [42]. Excess intracellular Cu can bind to both MTs and Zn. The Cu transporter 1 (Ctr1), which transports Cu^{+} and DMT1, also contributes to Cu homeostasis.

In the brain, Cu is localized in the thalamus, substantia nigra, striatum, and hippocampus. Cu is involved in normal brain functions, such as neurotransmitter synthesis and myelination, by acting as a cofactor for numerous enzymes, including cytochrome C, superoxide dismutase, lysyl oxidase, and tyrosinase. Cu exerts neuroprotective effects as a component of Cu/Zn superoxide dismutase (Cu/Zn-SOD), an endogenous antioxidant. Cu is also involved in Fe homeostasis as a component of ceruloplasmin and other ferroxidases, termed "multicopper ferroxidases," which convert Fe^{2+} to Fe^{3+} [43]. A considerable amount of intracellular Cu accumulates in presynaptic domains and is released into synaptic clefts during neuronal excitation, similar to Zn^{2+} [44]. The released Cu^{2+} modulates neuronal excitability by binding to NMDA-type glutamate receptors, AMPA-type glutamate receptors, and GABA receptors [3, 44, 45]. However, excess free Cu is toxic, because it exists as oxidized Cu^{2+} and reduced Cu^{+}, producing ROS. Ctr1 also reportedly exists in synapses [46]. ATP7A is localized mainly in the Golgi apparatus and reportedly plays an essential role in axonal targeting and synaptogenesis [47, 48].

The levels of Cu and Zn in synapses are of interest; however, they are controversial. The concentrations of both elements in the cerebrospinal fluid (CSF) of healthy individuals are less than 1 μM [49]. The free intracellular free Zn^{2+} level is estimated to be 1–10 nM [50]. However, the synaptic cleft is a small compartment, which can be conceptualized as a cylinder with a 120-nm radius and height of 20 nm [51]. The total volume of synaptic clefts is estimated to account for approximately 1% of the brain's extracellular space, which is similar to the total CSF volume. Therefore, Zn and Cu may be concentrated in the synaptic clefts. Indeed, Vogt et al. estimated that the Zn concentration in synapses is 1–100 μM [52]. Using atomic absorption spectroscopy, Kardos et al. found that approximately 100 μM Cu is released into the synaptic cleft [53]. A study using a Cu-sensitive fluorescent probe demonstrated that approximately 3 μ M Cu is released into the synaptic cleft [54].

2.4 Manganese

Manganese is essential for the function of various enzymes, including hydrolases, lyases, and ligases, is involved in metabolic processes, such as protein glycosylation and lipid synthesis, and influences neurotransmitter synthesis as a component of glutamine synthetase [55]. Mn also functions as a cofactor of superoxide dismutase (Mn-SOD), which facilitates the detoxification of free radicals.

Orally administered Mn becomes Mn^{2+} in the presence of gastric acid, is oxidized to Mn^{3+}, and is transported by binding to TF or citrate because the chemical characteristics of Mn^{3+} and Fe^{3+} are similar. TF-bound Mn can pass through the BBB and enter the brain via TF receptors, citrate transporters, and other transporters such as DMT1 [56]. Some Zn transporters, such as ZnT10, ZIP8, and ZIP14, can also allow Mn to enter the brain [57].

Although Mn deficiency is rare under physiological conditions, excess Mn from occupational exposure or environmental contamination causes neurotoxicity called "manganism," which involves symptoms similar to PD [58]. Mn also causes oxidative stress and inflammatory responses. Additionally, Mn influences Fe homeostasis and the expression of Fe-related proteins by affecting the IRE-IRP pathways [59].

2.5 Interactions between Neurometals and Other Metals

The interactions between one metal ion and another are particularly interesting. The interaction between metal-related proteins and metal ions is generally not strictly regulated. The binding site of such proteins often shares one metal ion and other similar metal ions, even with less binding affinity. For example, Zn and Cu can share the binding sites of many proteins, such as metallothioneins. Thus, ingesting excess Zn can cause Cu deficiency and vice versa. Some

Zn transporters, such as ZIP8, ZIP14, and ZnT10, can transport Fe and Mn [57, 60, 61]. DMT1 can transport ten metal ions, except for Fe^{2+} [62]. Al^{3+} can bind to Fe-binding proteins, such as TF or Fe chelators [63], because Al^{3+} has chemical characteristics similar to Fe^{3+}. Therefore, non-essential, toxic metal elements, such as Al, Pb, or Cd, can enter the brain and cause adverse effects by affecting important biological pathways.

Among such toxic elements, Al is particularly interesting because it has been associated with various neurodegenerative disorders, including AD, ALS in the Kii Peninsula and Guam, Gulf War syndrome, and autism spectrum disorder [64].

Al is the third most abundant element in the Earth's crust and is widely distributed throughout the environment. Compounds containing Al (e.g., clay, glass, and alum) have been used as materials for centuries. However, Al was first isolated in 1827 as a silvery metal, and its industrial use began only in 1886. In this context, Al is an old and new metal. Despite being widespread in the environment, Al is not essential for life, and no known biological reaction requires Al.

Nevertheless, Al is known to inhibit more than 300 biologically significant reactions owing to its peculiar chemical characteristics. Al is amphoteric, and its solubility at physiologically neutral pH is low. The ligand-exchange rate of Al^{3+} is 10^5 times slower than that of Mg^{2+} and 10^8 times slower than that of Ca^{2+}. Therefore, Al^{3+} cannot participate in Ca^{2+}- or Mg^{2+}-related enzymatic reactions but inhibits numerous Ca^{2+}- or Mg^{2+}-related enzymes. Al tightly binds to phosphate and inhibits kinases or phosphatases. Although Al^{3+} is not redox-active, Al is known to enhance Fe-induced lipid peroxidation.

Al is naturally present in several types of food, such as seaweed and tea leaves. Al can also enter the human body through food additives (such as alum and baking powder), pharmacological agents (such as antacids, antiperspirants, and vaccines), contamination from cooking utensils, and drinking water, which is purified with Al coagulants. Although the daily intake of Al is estimated to be about 20–40 mg, the absorption rate from the gastrointestinal tract is generally low and can be altered by its chemical speciation and coexisting substances. However, some orally administered Al can be absorbed from the gastrointestinal pathway, bind to TF in the blood, pass through the BBB, and enter the brain. A study using ^{26}Al (a non-radioactive isotope) and accelerated mass spectrometry estimated that as little as 1% of the Al in diets is absorbed via the gastrointestinal tract, and approximately 0.0036% of the Al is transported into the brain [65]. However, the biological half-life of Al is relatively long, and the level of Al in the brain does not change after 1 month of administration [66]. Owing to these adverse characteristics, Al is a well-established neurotoxin. Al causes dialysis encephalopathy in hemodialysis patients as a component of hemodialysis solutions or pharmacological compounds. The interactions between Al and other elements in neurodegenerative diseases will be discussed in Sect. 4.

3 Roles of Neurometals in the Pathogenesis of Neurodegenerative Diseases

3.1 Alzheimer's Disease

Among the neurodegenerative diseases that link neurometals, AD, VD, and DLB are included in senile-type dementia, which is characterized by profound memory loss and an inability to form new memories in older adults. The prevalence of senile dementia increases with age, and the number of dementia patients in Japan was estimated to be more than 7 million in 2020.

AD is responsible for approximately half of all cases of senile-type dementia and is characterized by the loss of neurons and synapses in the hippocampus or cerebral cortex, the extracellular accumulation of senile plaques, and the intraneuronal accumulation of neurofibrillary tangles (NFTs) [67]. The principal constituent of senile plaques is β-amyloid protein (AβP), while that of NFTs is phosphorylated tau protein. AβP is a small peptide with 39–43 amino acid residues that is derived from a large precursor protein [amyloid precursor protein (APP)] by the enzymatic cleavage at its N-terminus by β-secretase and by γ-secretase at its C-terminus. AβP(1–40), comprising the first 40 amino acids of AβP, is reportedly neurotoxic both in vitro and in vivo. AβP tends to form insoluble aggregates (oligomers) in solution. Freshly prepared AβP(1–40), which can form an β-helix or random-coiled structure, is nontoxic. However, oligomerized AβP(1–40) with β-pleated sheet structures is toxic, and the β-sheet content ratio correlates with its neurotoxicity. Thus, numerous biochemical and genetic studies have supported the amyloid cascade hypothesis, which proposes that the conformational changes in oligomerized AβP and subsequent neurodegeneration are crucial for the pathogenesis of AD [68].

Considering that AβP also exists in the CSF of young individuals [69], factors that accelerate or inhibit the oligomerization of AβP may play critical roles in the pathogenesis of AD [70]. Rifampicin, transthyretin, curcumin, aspirin, β-sheet breaker peptide, and carnosine inhibited the oligomerization. Several factors, such as the concentration of peptides, pH, the composition of solvents, the temperature, and the oxidation, mutation, and racemization of Aβ P, accelerate its oligomerization. Among these acceleratory factors, metal ions are particularly interesting because they can bind to several amino acid residues, such as histidine, arginine, and tyrosine, or phosphorylated amino acids and cause conformational changes by crosslinking with them. It is known that the deposition of senile plaques rarely occurs in rodent brains compared with primate brains, although the primary sequences of primate AβP and rodent AβP differ in only three amino acids (Arg^5, Tyr^{10}, and His^{13}) [71]. Since all three of these amino acids can bind metal ions, metals may contribute significantly to the oligomerization of AβP and the deposition of senile plaques. Studies by us and numerous others have demonstrated that metal ions, including Al^{3+}, Fe^{3+}, Zn^{2+}, Cu^{2+}, and Mn^{2+}, bind to Aβ P and enhance its oligomerization [72–77]. Indeed, the deposition of these elements was observed in senile plaques [78, 79]. However, the contribution of metals to AβP-induced neurotoxicity is still controversial. Morphological analysis using atomic force microscopy (AFM) revealed that the characteristics (size and shape) of the AβP oligomers formed in the presence of Al^{3+}, Zn^{2+}, Cu^{2+}, and Fe^{3+} are unique [80]. Sharma et al. revealed that Zn^{2+}-aggregated AβPs are less toxic than Cu^{2+}-aggregated AβPs [81].

APP, the precursor of AβP, is also a metal-binding protein. Zn and Cu bind to the N-terminus of APP and cause its dimerization [82, 83]. Zn and/or Cu influence the expression of APP, its trafficking and processing, and the production of AβP [84]. APP can reportedly convert oxidized Cu^{2+} to reduced Cu^{+} [85] and regulate Fe^{2+} efflux from neurons by stabilizing FPN [86]. Moreover, Fe regulates the expression of APP because the mRNA of APP contains an IRE domain in its 5′-UTR similar to that of ferritin [87]. Therefore, excess Fe promotes the translation of APP, and thereafter, the export of Fe^{2+} by FPN can be upregulated to maintain intracellular Fe levels. High-Fe diets reportedly cause the cleavage of APP and the accumulation of AβP in the brain [88]. Other studies have suggested the involvement of Fe homeostasis in AD. The genotype of transferrin C2 is a risk factor for AD [89], and Fe supplementation reportedly has beneficial effects on cognitive functions in AD patients [90].

Presenilins (presenilin-1 and -2), risk factors for early-onset familial AD, are also metal-binding proteins. Presenilins are involved in Ca homeostasis in the ER [91] and in Zn and Cu cellular uptake [92]. Moreover, Zn and Cu affect several functions of presenilins, including trafficking pathways and its γ-secretase activity [84]. We summarize the links between neurometals and AβP/APP in Fig. 2.

The molecular mechanism of AβP-induced neurotoxicity is of particular interest. AβP causes various adverse changes, such as ROS production, the induction of cytokines and ER stress, and the disruption of Ca^{2+} homeostasis [93–95]. Among these, the disruption of Ca^{2+} homeostasis, especially the abnormal increase in intracellular Ca^{2+} levels ($[Ca^{2+}]_i$), is considered a key pathogenetic event because Ca^{2+} ions play critical roles in normal brain functions. Ca^{2+} dyshomeostasis ultimately leads to membrane disruption, lipid peroxidation, ROS production, and apoptosis.

In 1993, Arispe et al. reported that AβP(1–40) was incorporated into artificial planar lipid membranes and formed giant multilevel pores (approximately 5 nS) with cation (including Ca^{2+}) selectivity [96]. Other neurotoxic peptide fragments of AβP, AβP(25–35), and AβP(1–42) also form pores in membranes [97]. Numerous studies using 3-D computer simulations, nuclear magnetic resonance spectroscopy, and AFM have demonstrated that AβPs form pore-like structures on membranes composed of 5–8-mer oligomers, termed

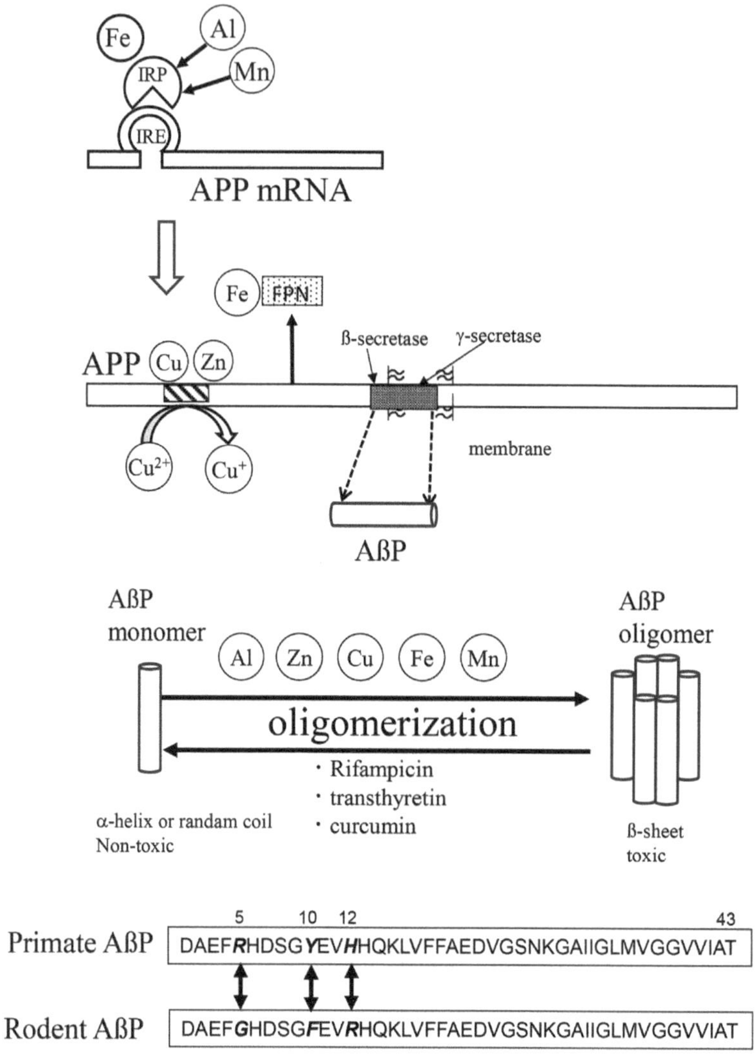

Fig. 2 Structural and metal-binding properties of amyloid precursor protein and β-amyloid protein. The level of Fe regulates the expression of amyloid precursor protein (APP) through the iron-responsive element-iron-responsive protein (IRE-IRP) pathway. Al^{3+} and/or Mn^{3+} can influence the IRE-IRP pathway. APP possesses Cu- and/or Zn-binding domains and reduces Cu^{2+} to Cu^{+}. APP also regulates Fe^{2+} efflux by stabilizing ferroportin (FPN). β-Amyloid protein (AβP) is derived from APP by the enzymatic cleavage of β-secretase and γ-secretase. Monomeric AβP forms α-helix or random-coil structures and is nontoxic, whereas the AβP oligomer exhibits β -pleated sheet structures and is toxic. Several factors, including metals (such as Al, Fe, Zn, Cu, and Mn), influence the oligomerization process. The primary sequences of primate AβP and rodent AβP differ by only three metal-binding amino acids (Arg^5, Tyr^{10}, and His^{13}; shown in italics and bold). All of these amino acids can bind to metals

"amyloid pores" [98–100]. We have demonstrated that AβPs form pores on neuronal cell membranes and liposomes using membrane patches from immortalized hypothalamic neurons (GT1–7 cells)

[101]. We also demonstrated that exposure to AβP(1–40) and other amyloidogenic peptides caused an increase in $[Ca^{2+}]_i$ in GT1-7 cells using multisite fluorometry with fura-2, a Ca^{2+}-sensitive fluorescent dye [102–105]. As shown in Figs. 3A, B(a), AβP(1–40) exhibits the elevations in $[Ca^{2+}]_i$ in many cells [12]. However, AβP(40–1), a reverse-sequence peptide with no neurotoxicity, caused no change in $[Ca^{2+}]_i$ (Fig. 3B(b)). Moreover, D-AβP (1–40), composed of D-amino acid residues only, also caused an increase in $[Ca^{2+}]_i$, similar to L-AβP (1–40) (Fig. 3B (c)). Therefore, it is suggested that AβP-induced $[Ca^{2+}]_i$ changes are not mediated by endogenous receptor-mediated pathways but rather by unregulated amyloid channels (pores). Based on this evidence, studies by us and others support the "amyloid channel hypothesis" that pore formation by AβPs on the neural membrane and the resulting change in $[Ca^{2+}]_i$ may be a primary pathogenetic event in AβP neurotoxicity [106–108]. Considering the structure of AβPs on membranes, the His^{12}–His^{13} residues, which can bind Zn^{2+}, are exposed to the inner surfaces of amyloid pores [98]. The channel activity was inhibited by adding Zn^{2+} and recovered by the Zn^{2+} chelator *o*-phenanthroline [97, 101, 108]. Therefore, Zn^{2+} may play a protective role in the neurotoxicity of AβP.

3.2 Prion Diseases

Prion diseases are fatal neurodegenerative diseases and include scrapie in sheep, bovine spongiform encephalopathy (BSE) in cattle, chronic wasting disease in elk, Creutzfeldt-Jakob disease (CJD),

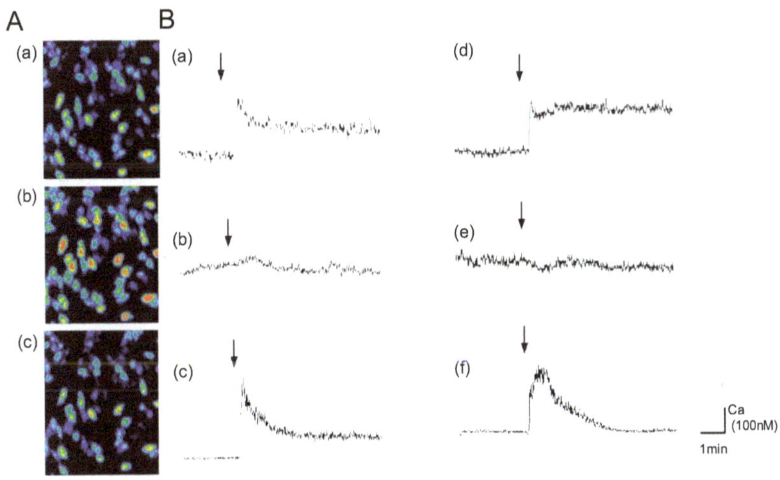

Fig. 3 Temporal changes in $[Ca^{2+}]_i$ after exposure to amyloidogenic proteins. (**A**) Temporal changes in $[Ca^{2+}]_i$ shown as pseudocolor images. AβP(1–40) (10 μM) was applied to GT1-7 cells, and temporal changes in fluorescence intensities corresponding to changes in $[Ca^{2+}]_i$ were analyzed by multisite fluorometry using fura-2. (a): 1 min before exposure to AβP(1–40); (b): 20 s after exposure; (c): 5 min after exposure. (**B**) Typical time course of $[Ca^{2+}]_i$ changes before and after exposure to peptides (each 10 μM). (a) AβP(1–40); (b) AβP(40–1); (c) D-AβP(1–40); (d) PrP106–126; (e) scramble PrP106–126; (f) NAC. The arrow indicates the time of peptide addition. (Modified from [12])

Gerstmann-Sträussler-Scheinker syndrome (GSS), and kuru in humans [109]. The pathological hallmarks of prion diseases are spongiform degeneration of glial cells and neurons, synaptic degeneration, and the accumulation of abnormal scrapie-like prion protein (PrP^{Sc}) in the brain. Normal prion protein (PrP^{C}) consists of a 30–35 kDa glycoprotein anchored at the plasma membrane with a glycosylphosphatidylinositol (GPI) domain (Fig. 4). It is widely distributed throughout the body, including the liver, heart, and brain. PrP^{C} and PrP^{Sc} share the same chemical characteristics with the same primary sequence, except that PrP^{Sc} is protease resistant, insoluble, and has a high β-sheet content. Thus, the conformational conversion of PrP^{C} to pathogenic PrP^{Sc} is central to the pathogenesis of prion diseases. When misfolded PrP^{Sc} enters the body via ingestion of contaminated food or iatrogenic contamination, the protease-resistant PrP^{Sc} invades the brain and induces neighboring PrP^{C} molecules to misfold and aggregate. Thus, prion diseases are categorized as conformational diseases called transmissible spongiform encephalopathies (TSEs).

PrP^{C} possesses an octarepeat domain composed of multiple tandem copies of an eight-residue sequence (PHGGGWGQ) (residues 66–90) and can bind metal ions such as Cu^{2+}, Zn^{2+}, Mn^{2+}, and Ni^{2+} (Fig. 4). Jackson et al. reported that four Cu atoms bind to the His residues in the octarepeat domain, and two other Cu atoms to other His residues (His^{96} and His^{111}) [110]. Brown et al. demonstrated that PrP-knockout mice exhibited decreased Cu levels and reduced activity of Cu-dependent enzymes in the brain. PrP^{C} regulates Cu influx into neurons, and Cu-bound PrP^{C} exerts neuroprotective effects by exhibiting Cu/Zn SOD-like activity [111]. PrP^{C} binds to AMPA-type and/or NMDA-type glutamate receptors in postsynaptic membranes and controls their functions in a Cu-dependent manner [112]. We have shown that peptide fragments of PrP^{C} (PrP73–88) attenuated the Cu-induced neuronal death of cultured hippocampal neurons [113].

Meanwhile, Cu reportedly affects the gene expression and cellular trafficking of PrP^{C} [114].

PrP^{C} also regulates Zn homeostasis. Bioinformatic studies revealed that genes for PrP^{C} have evolutionary similarities to ZIP-type Zn transporters such as ZIP5, ZIP6, and ZIP10 [115]. PrP^{C} acts as a Zn^{2+} sensor in synapses with the AMPA-type glutamate receptors and controls the cellular uptake of Zn^{2+} into postsynaptic neurons [116]. Increasing evidence suggests that Cu may be involved in the conformational conversion of PrP^{C} and the transmission of prion diseases [117]. PrP^{C} comprises a flexible N-terminal domain and the α-helix-rich C-terminal domain that changes to the β-sheet in the conformational changes of PrP^{Sc}. When Cu binds to the octarepeat region in the flexible N-terminal domain, Cu crosslinks with His residues in the C-terminal domain and changes PrP^{C} to a β-sheet-rich

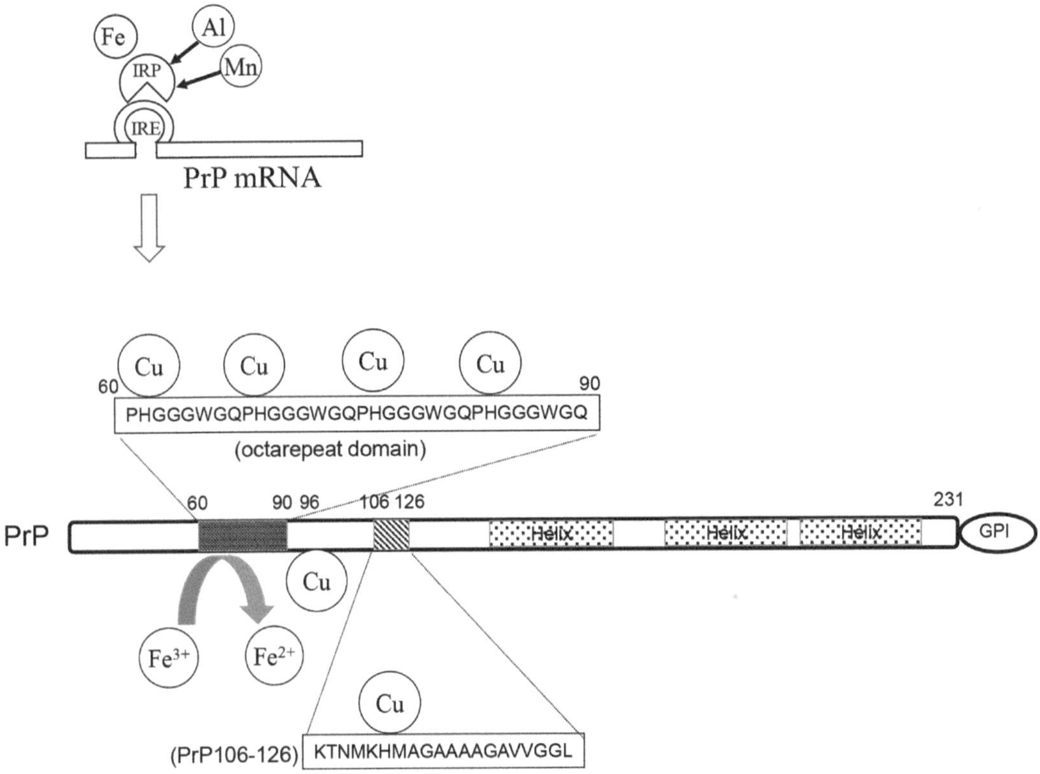

Fig. 4 Structural and the metal-binding properties of prion proteins. Normal prion protein (PrP^C) comprises a 30–35 kDa glycoprotein anchored at the plasma membrane with a glycosylphosphatidylinositol (GPI) domain. The level of Fe regulates the expression of PrP^C through the iron-responsive element-iron-responsive protein (IRE-IRP) pathway. PrP^C is composed of a flexible N-terminal domain and an α-helix-rich C-terminal domain. At the N-terminus, PrP^C possesses a highly conserved octarepeat domain composed of multiple tandem copies of an eight-residue sequence (PHGGGWGQ) (residues 66–90), among which His residues are critical for Cu and other metals, such as Zn, Mn, and Ni. Four Cu atoms can bind to the octarepeat domain in PrP^C, and two Cu atoms can bind to other His residues (His^{96} and His^{111}). PrP106–126, a neurotoxic fragment peptide, can bind to Cu and other metal ions. PrP^C possesses the ferrireductase activity, which converts Fe^{3+} to Fe^{2+}. The C-terminal domain is changed to a β-sheet and is involved in the conformational changes of PrP^{Sc}

PrP^{Sc}. Indeed, mutation of ATP7A delays the onset of prion disease by disrupting Cu homeostasis [118]. The binding of Zn^{2+} to the octarepeat domain reportedly affects the conformational changes of the C-terminal domain and also contributes to the conversion of PrP^C to PrP^{Sc} [119].

PrP^C is involved in Fe homeostasis. Notably, PrP^C acts as a ferrireductase that converts Fe^{3+} to Fe^{2+} [120]. The reduced Fe^{2+} is subsequently transported to the intracellular space via the complex of DMT1 and ZIP14 and is provided to enzymes that require bioactive Fe^{2+} [121]. Moreover, the Fe level controls the expression of PrP^C because its mRNA possesses an IRE motif, as does the APP gene [122].

Although PrP^C can bind Mn, Mn has been suggested to facilitate the pathogenesis of prion diseases. Mn enhanced the survival of PrP in model soils and increased its infectivity [123]. The risk of chronic wasting disease in deer was shown to be associated with increased Mn with Mg deficiency [124]. An epidemiological study in Slovakia suggested a relationship between the pathogenesis of CJD and an imbalance in the Mn/Cu ratio in food [125].

Based on these findings, the loss of protective functions of normal PrP^C may cause dyshomeostasis of neurometals and trigger neuronal death. Indeed, knockout mice lacking PrP^C presented altered Fe metabolism and several neurological dysfunctions, including the death of Purkinje neurons in the cerebellum, synaptic dysfunction, demyelination, depressive-like behavior, and memory loss [126]. By contrast, metals can contribute to the toxic functions of PrP^{Sc}. A fragment peptide of PrP (PrP106–126) has been used as a model peptide of PrP^{Sc} because it is neurotoxic to cultured hippocampal neurons and can form fibrils with β-sheet structures [127]. PrP106–126 also possesses the ability to bind Cu and Zn.

Similarly to AβP, PrP106–126 reportedly forms cation-permeable pores in artificial lipid bilayers [128]. Kourie et al. demonstrated that the channels formed by PrP106–126 are Cu-sensitive cation channels [129]. PrP^{Sc} also reportedly forms amyloid channels (pores) [130]. We demonstrated that PrP106–126 caused increases in $[Ca^{2+}]_i$ in GT1-7 cells (Fig. 3B(d)), although a scrambled PrP106–126 (a peptide with a random sequence) did not cause notable changes (Fig. 3B(e)) [12]. Moreover, we demonstrated that Zn and Cu attenuated the neurotoxicity of PrP106–126 and the formation of β-sheet structures [113]

3.3 Lewy Body Diseases

LBDs include PD, DLB, and multiple system atrophy. The common pathological hallmark of LBD is the abnormal intracellular aggregates termed "Lewy bodies," composed mainly of α-synuclein [131]. α-Synuclein is a 140–amino acid protein that is abundant in presynaptic terminals, particularly in the membranes of synaptic vesicles (Fig. 5). It plays critical roles in synaptic functions such as endocytosis, exocytosis, and the maintenance of synaptic plasticity. The fragment peptide of α-synuclein termed the "non-amyloid component (NAC)," coaccumulates with AβP in the senile plaques of patients with AD [132]. The oligomerization of α-synuclein or NAC contributes to neurotoxicity and the pathogenesis of LBD.

The link between α-synuclein and neurometals has been extensively studied because PD patients exhibit damage in Fe-rich regions (such as the substantia nigra). α-Synuclein reportedly binds Cu^{2+}, Zn^{2+}, and Mn^{2+} via N-terminal and C-terminal domains [133]. Uversky et al. revealed that Al and other metal ions enhance the oligomerization of α-synuclein [134]. Cu affects the binding of α -synuclein to lipid membranes and regulates their functions [135].

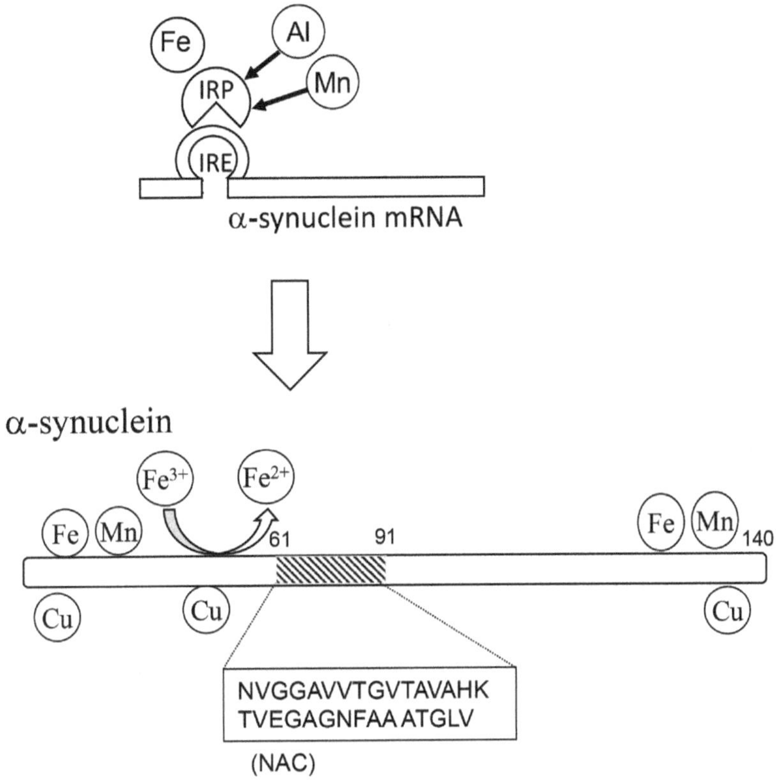

Fig. 5 Structural and metal-binding properties of α-synuclein. α-Synuclein is a 140-amino-acid protein. Its expression is regulated by The IRE-IRP pathway and APP and PrP^{C}. α-Synuclein binds Cu^{2+}, Zn^{2+}, Mn^{2+}, and other metal ions in its N-terminal and C-terminal domains. Like PrP^{C}, α-synuclein possesses the ferrireductase activity, which converts Fe^{3+} to Fe^{2+}

The His^{50} residue is essential for Cu binding to α-synuclein. Considering that a point mutation in His^{50} is observed in familial-type PD patients, the binding of Cu to α-synuclein is thought to be critical for the pathogenesis [136].

Interestingly, α-synuclein is also involved in Fe homeostasis. α-Synuclein possesses ferrireductase activity that converts Fe^{3+} to Fe^{2+}, as well as PrP^{C} [137]. This activity indicates that α-synuclein controls energy production and neurotransmitter synthesis in the presynaptic domain by providing bioavailable Fe^{2+} to enzymes requiring it in the mitochondria or ER. This ferrireductase activity was reported to be reduced in PD patients [138]. Moreover, the expression of α -synuclein is controlled by Fe levels because its mRNA possesses an IRE motif in the 5′-UTR similar to that of APP and PrP^{C} [139]. Mn reportedly induces the overexpression of α-synuclein by affecting the IRE-IRP pathway [140]. This finding is interesting because the neurotoxic characteristics of Mn resemble those of PD.

α-Synuclein also forms pores on lipid membranes, as well as AβP and PrP106–126 [141]. α-Synuclein reportedly forms pores on mitochondrial membranes, causing membrane disruption and degeneration [142]. We have demonstrated that NAC also increased $[Ca^{2+}]_i$ in GT1-7 cells (Fig. 3B(f)).

As described previously, amyloidogenic proteins share several common characteristics, including the formation of β-pleated sheet structures, neurotoxicity, and the ability to bind metals. Furthermore, they all colocalize at synapses, where metals are abundantly present. This is interesting since synaptic degeneration (synaptotoxicity) occurs in the early stages of these neurodegenerative diseases. Table 1 presents the sequences of these amyloidogenic proteins (fragment peptides) and their characteristics.

3.4 Vascular-Type Senile Dementia

VD is a cerebrovascular disease that accounts for approximately one-third of senile dementia cases. The risk factors for VD include older age, male sex, diabetes, and high blood pressure. The most common type of VD is caused by a series of small strokes or ischemia [143]. Following transient global ischemia or stroke, the interruption of blood flow and the resulting oxygen-glucose deprivation induces long-lasting membrane depolarization and excessive release of glutamate into synaptic clefts. Thereafter, excess glutamate causes overstimulation of glutamate receptors, and the resulting Ca^{2+} influx triggers delayed neuronal death in the hippocampus or cerebral cortex. Finally, the development of an infarct and subsequent cognitive dysfunction mark the pathogenesis of VD in older adults. Approximately 30% of stroke patients show symptoms of dementia within 3 months of the initial stroke.

Under ischemic conditions, excess Zn and glutamate are secreted into synaptic clefts. Zn translocation, namely, the increase in intracellular Zn^{2+} levels caused by the movement of chelatable Zn from presynaptic terminals to postsynaptic cell bodies, has been reported to occur in degenerated hippocampal neurons after ischemia [144]. The administration of Ca ethylenediaminetetraacetic acid (Ca-EDTA), a membrane-impermeable chelator of Zn^{2+}, has been shown to block the translocation of Zn, protect hippocampal neurons after transient global ischemia, and reduce the infarct volume [145]. These results strongly implicate Zn as a key player in delayed neuronal death after transient global ischemia and, ultimately, in the pathogenesis of VD [13, 146]. We investigated the molecular mechanism of Zn-induced neurotoxicity and found that sublethal concentrations of Cu^{2+} significantly exacerbated Zn-induced neurotoxicity [147]. Our results suggested that there are at least five pathways involved in Cu-enhanced Zn-induced neurotoxicity: the ER stress pathway, the stress-activated protein kinase/c-Jun amino-terminal kinase (SAPK/JNK) signaling pathway, the loss of energy production in mitochondria, the production of ROS, and the disruption of Ca^{2+} homeostasis [13]. Considering that Cu produces ROS [148] and ROS induces various apoptotic pathways such as the ER stress pathway and the

Table 1
Characteristics of amyloidogenic proteins

Disease	Amyloidonic protein or its related protein (fragment peptide)	β-sheet formation	cytotoxicity	Pore-formation or Ca^{2+} influx	Metal binding
Alzheimer's disease	AβP(1-42): ***DAEFRHDSGYEVHHQKLVFFAEDVGSN KGAIIGLMVGGVVIA***	+	+	+	Al, Zn, Cu, Fe, Mn
Prion diseases	PrP^{C}: MANLGCWMLVLFVATWSDLGLCKKRPKPGG WNTGGSRYPGQGSPGGNRYPPQGGGGWGQPHG GGWGQPHGGGWGQPHGGGWGQPHGGGWGQGG GTHSQWNKPSKP***KTNMKHMAGAAAAGAVVGGL***GGYVL GSAMSRPIIHFGSDYEDRYYRENMHRYPNQVYYRPMD EYSNQNNFVHDCVNITIKQHTVT PrP106-126: ***KTNMKHMAGAAAAGAVVGGL***	+	+	+	Cu, Zn, Fe, Mn
Lewy body diseases	α-synuclein : MDVFMKGLSKAKEGVVAAAEKTKQGVAEAA GKTKEGVLYVGSKTKEGVVHGVTTVAEKTK***EQVSNVGGA vvtgvtavahktvegagnfaaatglv***KKDQKNESGFGPE GTMENSENMPVNPNNETYEMPPEEEYQDYDPEA NAC: ***NVGGAVVTGVTAVAHKTVEGAGNFAA ATGLV***	+	+	+	Cu, Fe, Mn, Al

* The sequence of fragment peptide of each amyloidogenic protein (PrP106-126, NAC) is indicated in italic and bold form.

Disease	The primary sequence of amyloidogenic protein or its fragment peptide
	APP
Prion diseases	TTTKGENFTE TDVKMMERVVEQMCITQYERESQAYYQRGS SMVLFSSPPVILLISFLIFL IVG PrP106-126
LBD	NAC

		Binding Metal	ß-sheet formation	Deamidation	Functions of amyloidogenic proteins or their precursors
Alzheimer's disease	Al Zn Cu	Fe +		+ Neuronal proliferation and development	
Prion disease	Prion protein: *PrP106-126*	Zn Cu Fe Mn	+	+	
Lewy body disease	α-synuclein; *NAC* (a *fragment of α-synuclein*)	Cu Fe Al	+	+	
Type 2 daibetics	Islet amyloid peptide (amylin; human)	Al Cu Zn	+	+	unknown
Dialysis amyloidosis	beta2-miucroglobulin	Al	+	+	
ALS	Cu,Zn-SOD1	Cu Zn	+	+	·SOD activity ·cellular defence

SAPK/KNK pathway, it is probable that the coexistence of Zn and Cu in the same synapse under ischemic conditions triggers neurodegenerative processes and finally leads to VD pathogenesis.

4 Crosstalks between Neurometals and Amyloidogenic Proteins under Physiological Conditions

Increasing evidence supports the idea that neurometals play critical roles in neuronal functions through the cross-talk between amyloidogenic proteins. Notably, amyloidogenic and metal-related proteins are colocalized in synapses where metals are abundant. APP is located mainly in presynaptic membranes, PrP^C is located in postsynaptic membranes, and α-synuclein is located in the presynaptic cytosol or membranes of synaptic vesicles. Moreover, metal-related proteins such as ZnT1, ZIP4, DMT1, and Ctr1 are also localized in synapses.

Figure 6 presents an overview of the entry of trace elements into neurons and their functions in synapses. Considering that the synapse is an energy-demanding area in the brain [1], a considerable amount of Fe is required. TF-bound Fe can enter neurons or glial cells via TF receptors. Thereafter, intracellular Fe^{3+} is reduced to bioactive Fe^{2+} by ferrireductase and then transferred to enzymes that require Fe^{2+} as a cofactor in the mitochondria or the ER. In the presynaptic domain, α-synuclein acts as a ferrireductase. PrP^C acts as a ferrireductase in the postsynaptic domain to transport Fe^{2+} via DMT1/ZIP14 into postsynaptic neurons. Excess Fe^{2+} is exported to extracellular space via FPN and then oxidized to Fe^{3+} by ferroxidase to enter circulation. APP stabilizes FPN and plays a key role in Fe export. Meanwhile, Fe controls the expression of APP, PrP^C, and β-synuclein, given that their mRNA possesses the IRE motif in the 5'-UTR. Moreover, DMT1 is involved in the processing of APP and the production of AßP [149].

Zn is transported into neurons or glial cells via DMT1 or ZIP4 and accumulates in presynaptic vesicles by ZnT3. Cu is transported by binding with ceruloplasmin and then enters cells. Synaptic Zn and Cu are secreted into synaptic clefts during neuronal excitation alongside glutamate, bind to NMDA-type glutamate receptors (NMDA-Rs), AMPA-kainate-type glutamate receptors (A/K-Rs,) and other receptors, and regulate their functions. Both excitatory glutamate and inhibitory Zn (or Cu) are simultaneously secreted into synaptic clefts and diffusely spill over to neighboring synapses. This spillover of Zn and Cu can then inhibit or activate neuronal firing in a biphasic manner, depending on their concentrations, and contribute to synaptic plasticity. Secreted Zn is translocated to postsynaptic neurons by PrP^C with A/K-Rs as a ZIP analog. The elevated level of Zn causes the translocation of ZnT1 from the cytosol to synaptic membranes. The ZnT1 binds to NMDA-R and voltage-gated Ca^{2+} channels (VGCC) and regulates their functions. Cu is also

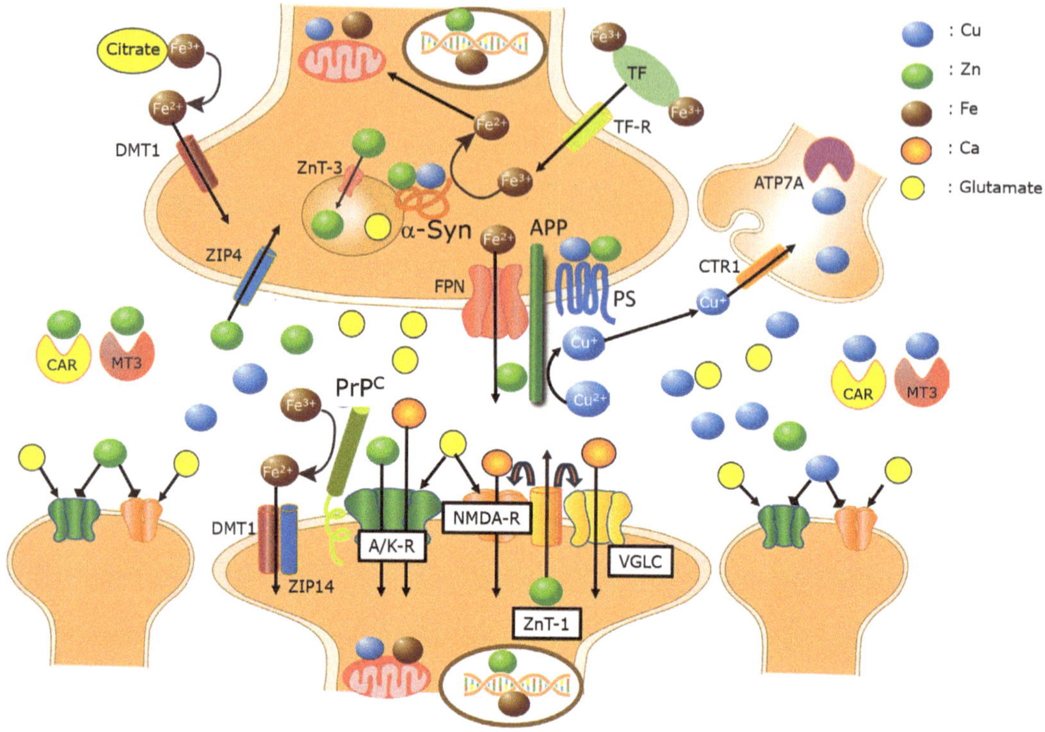

Fig. 6 Hypothetical scheme of the roles of trace elements in synapses. APP is localized primarily to the presynaptic membrane. PrP^C is localized to the postsynaptic membrane. α-Synuclein is localized mainly to the presynaptic domain and membranes of synaptic vesicles. Orally administered Fe is absorbed in the gastrointestinal tract and transported in the blood as Fe^{3+} by binding to transferrin (TF) or citrate. TF-bound Fe^{3+} can enter neurons or glial cells via its receptor (TF-R). Thereafter, Fe^{3+} is converted to Fe^{2+} by ferrireductases (FRs). α-Synuclein acts as a ferrireductase in the presynaptic domain and provides bioavailable Fe^{2+} to enzymes, requiring it in mitochondria and other organelles. PrP^C acts as a ferrireductase in the postsynaptic domain, and the converted Fe^{2+} is transported into cells by the complex of ZIP14 and divalent metal transporter 1 (DMT1). Non-transferrin bound Fe (Fe^{3+} that binds with citrate) is also reduced to Fe^{2+} by the FR and enters cells via DMT1. Excess intracellular Fe^{2+} is exported via ferroportin (FPN). APP stabilizes FPN and regulates Fe^{2+} efflux from cells. α-Synuclein and FPN coexist in presynaptic vesicles and control Fe homeostasis in the vesicles. Moreover, Fe controls the expression of various proteins. Orally administered Zn and Cu enter the neurons or glial cells, accumulate in synaptic vesicles, and are released in an activity-dependent manner into the synaptic clefts. Secreted Zn and Cu bind to NMDA-type glutamate receptors (NMDA-Rs), AMPA-kainate type glutamate receptors (A/K-Rs) or other receptors, and control neuronal excitability. Zn increases the expression of the zinc transporter ZnT-1 or the translocation of ZnT-1 to postsynaptic membranes. ZnT-1 subsequently binds to NMDA-Rs and L-type voltage-gated Ca^{2+} channels (L-VGCCs) to regulate overall brain excitability and maintain Ca homeostasis. PrP^C acts as a Zn sensor and regulates Zn^{2+} influx by binding with A/K-Rs. Secreted Cu regulates NMDA-Rs and A/K-Rs by binding with the normal cellular prion protein (PrP^C). The and transports Zn to postsynaptic neurons. Secreted Zn and Cu spill over to neighboring synapses and inhibit or potentiate neuronal excitability in a biphasic dose-dependent manner. Finally, Zn and Cu contribute to information processing and memory formation. ZIP4 can reuptake secreted Zn, and CTR1 can reuptake Cu once it has been reduced to Cu+ by APP. The subtle balance of Zn, Cu, and Ca maintains synaptic functions; Zn transporters, such as ZnT1 and ZIP4, contribute to regulating Zn levels in the synaptic clefts. *NMDA-R* NMDA-type glutamate receptor, *A/K-R* AMPA-kainate type glutamate receptor, *FPN* ferroportin, *TF* transferrin, *VDCC* voltage-dependent calcium channel, colored circles represent Zn, Cu, Fe, Mn, Ca, and glutamate

transported via the PrP^C into the postsynaptic domain and acts to synthesize neurotransmitters and protect neurons as Cu/Zn SOD. Moreover, Cu binds to PrP^C and differentially regulates NMDA-Rs and A/K-Rs. Therefore, Zn and Cu may collaborate in the synapse to regulate overall excitability and maintain Ca^{2+} levels. Secreted Zn can be rapidly removed by ZIP4 and then transferred to presynaptic vesicles via ZnT3. Similarly, CTR1 reuptakes secreted Cu once it has been reduced to Cu^+ by APP.

Other factors, such as MT3 and carnosine (β-alanyl histidine), also control neurometals in the synapses. MT3, a brain-specific metallothionein, is synthesized in neurons or glial cells [150]. Interestingly, MT3 is decreased in the brains of patients with AD. Carnosine is an endogenous dipeptide with various beneficial characteristics, such as anti-glycation, anti-stress, and antioxidant properties; hydroxyl radical scavenging; maintenance of pH balance; and chelation of metals, including Zn^{2+} and Cu^{2+} ions [151]. Carnosine is reportedly synthesized in oligodendrocytes and secreted into the synaptic cleft [152].

5 Hypothetical Scheme of the Pathogenesis of Neurodegenerative Diseases

These concerted metal–protein and metal–metal interactions at synapses are highly complex. Therefore, the disorder of these subtle interactions may cause dyshomeostasis of neurometals and the production of ROS and finally trigger neurodegenerative processes related to "the loss of function pathway." These interactions are vulnerable to perturbation by various genetic and environmental factors. As such perturbation factors, the mutations in APP, reportedly influence the stability of FPN and the export of Fe^{2+}. A decreased level of FPN, which is found in the brains of people with AD [153], causes the accumulation of Fe^{2+} and the production of ROS. Other environmental factors, such as high-Fe diets, reportedly cause the cleavage of APP and the accumulation of AβP in the brain [88]. In addition to Fe, Al can enter neurons via TF receptors and influence various Fe-related pathways, such as the IRE-IRP pathway, thus influencing the expression of APP and the production of AβP [154, 155]. Eventually, the accumulation of AβPs triggers "the gain of toxic functions" pathway (Fig. 7). Al, Mn, and other metal ions enhance the oligomerization of AβPs—the oligomerized AβPs form Ca^{2+}-permeable amyloid pores (AβP pores) on membranes. Unregulated Ca^{2+} flowing through amyloid pores can trigger various neurodegenerative processes, including ROS formation, disturbances in capacitive Ca^{2+} in the ER, and tau phosphorylation.

Additionally, the increase in $[Ca^{2+}]_i$ induces the overproduction of APP and increased production of AβP [156], ultimately inducing a vicious cycle of unregulated Ca^{2+} influx and finally leading to

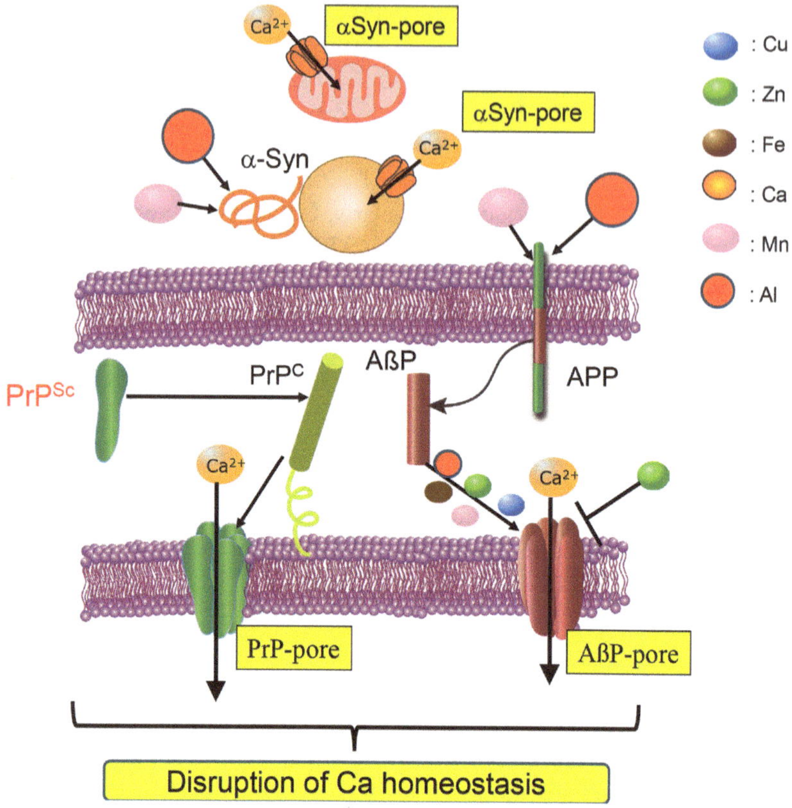

Fig. 7 Hypothesis: gain of toxic function of amyloid oligomers at the synapse. Perturbation factors such as the mutation of APP or the overexpression of APP cause the increased secretion of AβP, and oligomerized AβPs can be directly incorporated into membranes and form amyloid channels (AβP pores). Metals such as Al, Zn, Cu, and Mn enhance the oligomerization of AβP and facilitate the process. Thereafter, disruption of Ca^{2+} homeostasis by these AβP pores triggers various apoptotic pathways and causes synaptotoxicity and neurotoxicity, and finally contribute to the pathogenesis of AD. Contaminated PrP^{Sc} can invade into the brain, convert normal PrP^{C} in the brain to PrP^{Sc}. The loss of functions of normal PrP^{C} causes the metal dyshomeostasis and the production of ROS, and finally triggers neurodegenerative processes. Increased PrP^{Sc} forms pores on membranes (PrP pores), triggers Ca^{2+} dyshomeostasis and finally leads to the pathogenesis of prion diseases. Mutations inα -synuclein or the exposure to excess Mn causes the formation of Ca^{2+}-permeable pores (Syn pores) and trigger the pathogenesis of LBDs

the pathogenesis of AD. In this context, Zn has paradoxical roles in the pathogenesis of AD: Zn enhances the oligomerization of Aβ P, whereas Zn inhibits the channel activity of AβP pores. Indeed, Zn attenuated AβP-induced neurotoxicity in vitro [157]. Thus, Zn may have both protective and toxic functions, like Janus, the ancient Roman god with two faces [146].

Contaminating PrP^{Sc} from food or an iatrogenic device, which is resistant to digestion by proteases, passes into the brain, causes the conversion of normal PrP^{C} to PrP^{Sc}, leads to the loss of neuroprotective functions of normal PrP^{C}, which induces altered metal homeostasis and production of ROS. Indeed, PrPC-knock-out mice exhibit synaptic and neurological dysfunctions. The increased PrP^{Sc} also forms pores on membranes (PrP pores) and triggers the disruption of Ca^{2+} homeostasis. Therefore, the combination of "the loss of function of normal PrP^{C} pathway" and "the gain of toxic function by PrP^{Sc} pathway" leads to the pathogenesis of prion diseases.

Mutations in α-synuclein may influence its binding to Cu, affecting the membrane binding and ferrireductase activity and causing changes in the Fe^{2+}/Fe^{3+} ratio, ultimately producing ROS. Meanwhile, excess Mn from environmental exposure influences the IRE-IRP pathway, causes the accumulation of Fe and the overexpression of α-synuclein [140], and induces neurodegeneration. Moreover, mutated α-synuclein forms pores (□Syn pores) on membranes of synaptic vesicles or mitochondria, triggers Ca^{2+} dyshomeostasis and neurodegeneration and finally leads to the pathogenesis of LBD.

Considering the height of the synaptic clefts (the distance from presynaptic membranes to postsynaptic membranes) is as narrow as 20 nm, these amyloidogenic proteins may interact at the synapse. Indeed, PrP^{C} reportedly acts as a toxic receptor of AβP and contributes to its toxicity [157]. α-Synuclein increases AβP secretion by promoting APP processing by γ-secretase [158]. Presenilins also regulate the secretion of APP by binding to Zn and/or Cu [84].

As to the pathogenesis of VD, we proposed the hypothetical scheme in Fig. 8. Under normal physiological conditions, as previously described, Zn and Cu are secreted from presynaptic vesicles to the synaptic clefts during neuronal firings. Secreted Zn and Cu regulate brain excitability, contributing to neural information processing and memory formation. In general, secreted Zn and Cu are rapidly reuptaken by ZIP-4 and CTR1. However, under transient global ischemia conditions, widespread, prolonged excitation occurs in the brain, and both excess Zn and Cu can be secreted to synaptic clefts with glutamate. Secreted excess Zn causes the acute reduction of the GluR2 subunit of A/K-R, and Ca^{2+}-permeable A/K-R (Ca-A/K-R) appears. Both Ca^{2+} and Zn^{2+} are intracellularly translocated via Ca-A/K-R, NMDA-R, and L-VGCC. Increased $[Ca^{2+}]i$ and intracellular Zn^{2+} levels ($[Zn^{2+}]i$) then cause various neurodegenerative pathways, including the ER stress pathway, the SAPK/JNK signaling pathway, the loss of energy production in mitochondria, the production of ROS, and the disruption of Ca^{2+} homeostasis [13]. In ischemic conditions, a considerable amount of Zn and Cu are colocalized at the same synapse. Cu exacerbates Zn neurotoxicity by producing ROS, ultimately triggering the pathogenesis of VD. Indeed, epidemiological studies have suggested that Cu is a risk factor for stroke [159, 160].

Based on these working hypotheses, a metal-based therapeutic pathway may become a good candidate for the treatment/prevention of these neurodegenerative diseases. Clioquinol, a chelator of Zn and

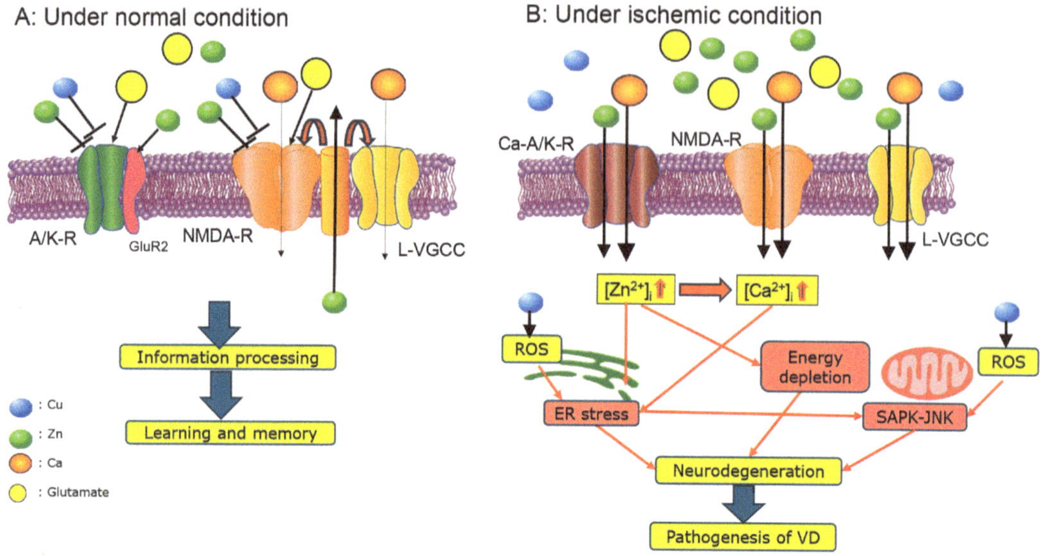

Fig. 8 Zn and Cu in the synapse under physiological and pathophysiological conditions. (**A**) Roles of Zn and Cu under normal physiological conditions. Zn and Cu are secreted from presynaptic vesicles to the synaptic clefts with glutamate during neuronal firing. Secreted Zn binds to NMDA-type glutamate receptors (NMDA-R) and AMPA/kainite type glutamate receptors (A/K-R), and inhibits their functions. Zn enhances the expression of the zinc transporter ZnT-1 or the translocation of ZnT-1 to synaptic membranes. Thereafter, ZnT-1 binds to NMDA-R and L-type voltage-gated Ca^{2+} channels (L-VGCC) to regulate overall brain excitability and maintain Ca homeostasis. Secreted Cu also regulates NMDA-R and A/K-R. Finally, Zn and Cu contribute to information processing and memory formation. The subtle balance of Zn, Cu, and Ca maintains synaptic functions. (B) Roles of Zn and Cu under ischemic, pathophysiological conditions. In transient global ischemia conditions, widespread, prolonged excitation occurs in the brain, and excess Zn and Cu can be secreted to synaptic clefts with glutamate. Secreted excess Zn causes the acute reduction of the GluR2 subunit of A/K-R, and Ca^{2+}-permeable A/K-R (Ca-A/K-R) appears. Both Ca^{2+} and Zn^{2+} are intracellularly translocated via Ca-A/K-R, NMDA-R, and L-VGCC. Increased intracellular Ca^{2+} levels ($[Ca^{2+}]_i$) and intracellular Zn^{2+} levels ($[Zn^{2+}]_i$) then cause various neurodegenerative pathways, including the ER stress pathway, the SAPK/JNK pathway, the loss of energy production, the production of ROS, ultimately triggering the pathogenesis of VD. Colored circles represent Zn, Cu, Ca, and glutamate

Cu, and its analog PBT2 have been examined for possible treatment of AD [161], although the results remain controversial [162]. Exley et al. demonstrated the efficacy of silicate, which chelates Al, as a possible treatment for AD [163]. Deferoxamine (DFO), a chelator of Fe and Al, has beneficial effects on maintaining the daily living skills of AD patients [164]. Recent approaches demonstrated that the intranasal DFO treatment attenuated synapse loss in the brain of a mouse model of AD and inhibited Fe-induced APP cleavage [165]. DFO and other Fe chelators are also considered possible treatments for PD [166, 167]. Moreover, clioquinol reportedly rescues the memory impairment associated with scrapie [168], and a Cu^{2+}-specific chelator (D-(−)-penicillamine) reportedly inhibits the pathogenesis of prion diseases in vivo [169]. We focused on carnosine as a possible treatment for VD based on its neuroprotective effect against Zn-induced neurotoxicity [170]. Carnosine also reportedly inhibits the oligomerization of AβP, attenuates neurodegeneration in a mouse model of AD [171], and protects against ischemia-induced neuronal death in vivo [172]. We have also demonstrated that carnosine attenuates PrP106–126-induced neurotoxicity [113]. Considering these beneficial characteristics of carnosine, we have published a patent for carnosine as a possible drug for treating VD [173].

6 Conclusions

In this chapter, we reviewed the enigmatic roles of trace elements in the brain and the pathogenesis of various neurodegenerative diseases based on common characteristics of amyloidogenic proteins and the cross-talk between metal ions at synapses. Our working hypothesis concerning the common aspects of neurodegenerative processes may shed light on their pathogenesis. The hypothesis regarding the link between trace elements and the pathogenesis of neurodegenerative diseases may lead to the development of drugs. However, this hypothesis has not yet been proven. In particular, Zn, Cu, and Fe levels in synapses are controversial and under investigation. The functions of amyloidogenic proteins and the localizations of metal transporters remain unsolved. Since we previously published the prototype of this hypothesis in 2017 [174], substantial progress has been made in this research field, particularly concerning the functions of Fe and Zn and the roles of metal transporters. We hope future studies on metals in synapses will support this hypothesis and aid in treating these neurodegenerative diseases. [175]; [176]

Acknowledgments

We thank Mikako Shimoda at Musashino University for the technical assistance. We also thank Robin James Storer, PhD, from Edanz (https://jp.edanz.com/ac) for editing a draft of this manuscript.

Conflict of Interest The authors have no conflicts of interest to declare that are relevant to the content of this chapter.

References

1. Faria-Pereira F, Morais VA (2022) Synapses: the brain's energy-demanding sites. Int J Mol Sci 23:3627
2. Bizup B, Tzounopoulos T (2024) On the genesis and unique functions of zinc neuromodulation. J Neurophysiol 132:1241–1254

3. Kardos J, Héja L, Simon A, Jablonkai I, Kovács R, Jemnitz K (2018) Copper signalling: causes and consequences. Cell Commun Signal 16: 71
4. Cobley JN, Fiorello ML, Bailey DM (2018) 13 reasons why the brain is susceptible to oxidative stress. Redox Biol 15:490–503
5. Adlard PA, Bush AI (2018) Metals and Alzheimer's disease: how far have we come in the clinic?. J Alzheimers Dis 62:1369–1379
6. Brown DR (2010) Metalloproteins and neuronal death. Metallomics 2:186–194
7. Jiang H, Song N, Jiao Q, Shi L, Du X (2019) Iron pathophysiology in Parkinson diseases. Adv Exp Med Biol 1173:45–66
8. Weiss JH, Sensi SL, Koh JY (2000) Zn^{2+}: a novel ionic mediator of neural injury in brain disease. Trends Pharmacol Sci 21:395–401
9. Kawahara M, Kato-Negishi M, Tanaka K-I (2023) Dietary trace elements and the pathogenesis of neurodegenerative diseases. Nutrients 15:2067
10. Carrell RW, Lomas DA (1997) Conformational disease. The Lancet 350:134–138
11. Winklhofer KF, Tatzelt J, Haass C (2008) The two faces of protein misfolding: gain- and loss-of-function in neurodegenerative diseases. EMBO J 27:336–49
12. Kawahara M, Kato-Negishi M, Tanaka K-I (2020) Amyloids: regulators of metal homeostasis in the synapse. Molecules 25:E1441
13. Kawahara M, Tanaka K-I, Kato-Negishi M (2024) Zinc, Copper, and Calcium: a triangle in the synapse for the pathogenesis of vascular-type senile dementia. Biomolecules 4:773
14. Wang Y, Wu Y, Li T, Wang X, Zhu C (2019) Iron metabolism and brain development in premature infants. Front Physiol 10:463
15. Collings R, Harvey LJ, Hooper L et al. (2013) The absorption of iron from whole diets: a systematic review. American J Clin Nutr 98:65–81
16. Skjørringe T, Burkhart A, Johnsen KB, Moos T (2015) Divalent metal transporter 1 (DMT1) in the brain: implications for a role in iron transport at the blood-brain barrier, and neuronal and glial pathology. Front Mol Neurosci 8:19
17. Sandnes M, Reikvam H (2024) Hepcidin as a therapeutic target in iron overload. Expert Opin Ther Targets 28:1039–1046
18. Silitonga HTS, Salim LA, Nurmala I, Wartiningsih M (2023) Compliance of iron supplementation and determinants among adolescent girls: a systematic review. Iran J Public Health 52:37–48
19. Zhou ZD, Tan EK (2017) Iron regulatory protein (IRP)-iron responsive element (IRE) signaling pathway in human neurodegenerative diseases. Mol Neurodegener 12:75
20. Wang C, Liu H, Xu S, Deng Y, Xu B, Yang T, Liu W (2023) Ferroptosis and neurodegenerative diseases: Insights into the regulatory roles of SLC7A11. Cell Mol Neurobiol 43:2627–2642
21. Hambidge M (2000) Human zinc deficiency. J Nutr 130:1344S–1349S
22. Hojyo S, Fukada T (2016) Roles of Zinc signaling in the immune system. J Immunol Res 2016:6762343
23. Prasad AS (2009) Impact of the discovery of human zinc deficiency on health. J Am Coll Nutr 28:257–265
24. Hamer DH (1986) Metallothionein. Annu Rev Biochem 55:913–951
25. Hara T, Takeda T, Takagishi T, Fukue K, Kambe T, Fukada T (2017) Physiological roles of zinc transporters: molecular and genetic importance in zinc homeostasis. J Physiol Sci 67:283–301
26. Frederickson CJ, Suh SW, Silva D, Frederickson CJ, Thompson RB (2000) Importance of zinc in the central nervous system: the zinc-containing neuron. J Nutr 130:1471S–1483S
27. Pochwat B, Nowak G, Szewczyk B (2015) Relationship between Zinc (Zn^{2+}) and glutamate receptors in the processes underlying neurodegeneration. Neural Plast 591563
28. Richards-Williams C, Contreras JL, Berecek KH, Schwiebert EM (2008) Extracellular ATP and zinc are co-secreted with insulin and activate multiple P2X purinergic receptor channels expressed by islet beta-cells to potentiate insulin secretion. Purinergic Signal 4:393–405
29. Morabito A, Zerlaut Y, Serraz B, Sala R, Paoletti P, Rebola N (2022) Activity-dependent modulation of NMDA receptors by endogenous zinc shapes dendritic function in cortical neurons. Cell Rep 38:110415
30. Sullivan JA, Zhang XL, Sullivan AP, Vose LR, Moghadam AA, Fried VA, Stanton PK (2018) Zinc enhances hippocampal long-term potentiation at CA1 synapses through NR2B containing NMDA receptors. PLoS One 13:e0205907
31. Ueno S, Tsukamoto M, Hirano T, Kikuchi K, K-Yamada M, Nishiyama N, Nagano N, Matsuki N, Ikegaya Y (2002) Mossy fiber Zn^{2+} spillover modulates heterosynaptic N-methyl-D-aspartate receptor activity in hippocampal CA3 circuits. J Cell Biol 158:215–220
32. Black MM (1998) Zinc deficiency and child development. Am J Clin Nutr 68:464S–469S
33. Takeda A, Tamano H (2017) The impact of synaptic Zn^{2+} dynamics on cognition and its decline. Int J Mol Sci 18:E2411

34. Chuang SH, Reddy DS (2019) Zinc reduces antiseizure activity of neurosteroids by selective blockade of extrasynaptic GABA-A receptor-mediated tonic inhibition in the hippocampus. Neuropharmacology 148:244–256
35. Heydarian F, Nakhaei AA, Majd HM, Bakhtiari E (2020) Zinc deficiency and febrile seizure: a systematic review and meta-analysis. Turk J Pediatr 62:347–358
36. Mellone M, Pelucchi S, Alberti L, Genazzani AA, Luca MD, Gardoni F (2015) Zinc transporter-1: a novel NMDA receptor-binding protein at the postsynaptic density. J Neurochem 132:159–168
37. De Benedictis CA, Haffke C, Hagmeyer S, Sauer AK, Grabrucker AM (2021) Expression analysis of zinc transporters in nervous tissue cells reveals neuronal and synaptic localization of ZIP4. Int J Mol Sci 22:4511
38. Nishito Y, Kambe T (2019) Zinc transporter 1 (ZNT1) expression on the cell surface is elaborately controlled by cellular zinc levels. J Biol Chem 294:15686–15697
39. Mor M, Beharier O, Cook DI, Campbell CR, Gheber LA, Katz A, Moran A, Etzion Y (2023) ZnT1 induces a crosstalk between T-type and L-type calcium channels through interactions with Raf-1 kinase and the calcium channel β2 subunit. Metallomics 15:031
40. Scheiber IF, Mercer JFB, Dringen R (2014) Metabolism and functions of copper in brain. Prog Neurobiol 116:33–57
41. Zietz BP, de Vergara JD, Dunkelberg H (2003) Copper concentrations in tap water and possible effects on infant's health-results of a study in Lower Saxony, Germany. Environ Res 92:129–138
42. Kodama H, Fujisawa C, Bhadhprasit W (2012) Inherited copper transport disorders: biochemical mechanisms, diagnosis, and treatment. Curr Drug Metab 13:237–250
43. Helman SL, Zhou J, Fuqua BK, Lu Y, Collins JF, Chen H, Vulpe CD, Anderson GJ, Frazer DM (2022) The biology of mammalian multi-copper ferroxidases. Biometals https://doi.org/10.1007/s10534-022-00370-z
44. D'Ambrosi N, Rossi L (2015) Copper at synapse: release, binding and modulation of neurotransmission. Neurochem Int 90:36–45
45. Aedo F, Delgado R, Wolff D, Vergara C (2007) Copper and zinc as modulators of neuronal excitability in a physiologically significant concentration range. Neurochem Int 50:591–600
46. Fransson AE, Kisiel M, Pirttilä K, Pettersson C, Pierre PV, Laurell GFE (2017) Hydrogen inhalation protects against ototoxicity induced by intravenous cisplatin in the Guinea pig. Front Cell Neurosci 11:280
47. Hartwig C, Méndez GM, Bhattacharjee S, Vrailas-Mortimer AD, Zlatic SA et al. (2021) Golgi-dependent copper homeostasis sustains synaptic development and mitochondrial content. J Neurosci 41:215–233
48. Meskini RE, Kelli J, Crabtree L, Cline LB, Mains RE, Eipper A, Ronnett GV (2007) ATP7A (Menkes protein) functions in axonal targeting and synaptogenesis. Mol Cell Neurosci 34:409–421
49. Roos PM, Vesterberg O, Syversen T, Flaten TP, Nordberg M (2013) Metal concentrations in cerebrospinal fluid and blood plasma from patients with amyotrophic lateral sclerosis. Biol Trace Elem Res 151:159–170
50. Frederickson CJ, Koh JY, Bush AI (2005) The neurobiology of zinc in health and disease. Nat Rev Neurosci 6:449–462
51. Schikorski T, Stevens CF (1997) Quantitative ultrastructural analysis of hippocampal excitatory synapses. J Neurosci 17:5858–5867
52. Vogt K, Mellor J, Tong G, Nicoll R (2000) The actions of synaptically released zinc at hippocampal mossy fiber synapses. Neuron 26:187–196
53. Kardos J, Kovács I, Hajós F, Kálmán N, Simonyi M (1989) Nerve endings from rat brain tissue release copper upon depolarization. A possible role in regulating neuronal excitability. Neurosci Lett 103:139–144
54. Hopt A, Korte S, Fink H, Panne U, Niessner R, Jahn R, Herms J (2003) Methods for studying synaptosomal copper release. J Neurosci Methods 128:159–172
55. Chen P, Chakraborty S, Mukhopadhyay S et al. (2015) Manganese homeostasis in the nervous system. J Neurochem 134:601–610
56. Nyarko-Danquah I, Pajarillo E, Digman A, Soliman KFA, Aschner M, Lee E (2020) Manganese accumulation in the brain via various transporters and its neurotoxicity mechanisms. Molecules 25:5880
57. Fujishiro H, Kambe T (2022) Manganese transport in mammals by zinc transporter family proteins, ZNT and ZIP. J Pharmacol Sci 148:125–133
58. Dirandeh E, Palizgir A, Kassiri N (2022) An overview of the relationship between occupational manganese exposure and Parkinsonism. Cureus 14:e32161
59. Khan MA, Walden WE, Theil EC, Goss DJ (2017) Thermodynamic and kinetic analyses of iron response element (IRE)-mRNA binding to iron regulatory protein, IRP1. Sci Rep 7: 8532
60. Wang CY, Jenkitkasemwong S, Duarte S, Sparkman BK, Shawki A, Mackenzie B, Knutson MD (2012) ZIP8 is an iron and zinc transporter whose cell-surface expression is

up-regulated by cellular iron loading. J Biol Chem 287:34032–34043

61. Pinilla-Tenas JJ, Sparkman BK, Shawki A, Illing AC, Mitchell CJ, Zhao N, Liuzzi JP, Cousins RJ, Knutson MD, Mackenzie B (2011) Zip14 is a complex broad-scope metal-ion transporter whose functional properties support roles in the cellular uptake of zinc and nontransferrin-bound iron. Am J Physiol Cell Physiol 301:C862–C871
62. Garrick MD, Dolan KG, Horbinski C, Ghio AJ, Higgins D, Porubcin M, Moore EG, Hainsworth LN, Umbreit JN, Conrad ME, Feng L, Lis A, Roth JA, Singleton S, Garrick LM (2003) DMT1: a mammalian transporter for multiple metals. Biometals 16:41–54
63. Moshtaghie AA, Ani M, Bazrafshan MR (1992) Comparative binding study of aluminum and chromium to human transferrin. Biol Trace Elem Res 32:39–46
64. Kawahara M, Tanaka K-I, Kato-Negishi M (2021) Neurotoxicity of Aluminum and its link with neurodegenerative diseases. Metallomics Res MR202104
65. Jouhanneau P, Raisbeck GM, Yiou F, Lacour B, Banide H, Drueke TB (1997) Gastrointestinal absorption, tissue retention, and urinary excretion of dietary aluminum in rats determined by using ^{26}Al. Clin Chem 43:1023–1028
66. Kobayashi K, Yumoto S, Nagai H, Hosoyama Y, Imamura M, Masuzawa S, Koizumi Y, Yamashita H (1990) ^{26}Al tracer experiment by accelerator mass spectrometry and its application to the studies for amyotrophic lateral sclerosis and Alzheimer's disease I. Proc Japan Acad Ser B 66:189–192
67. Selkoe DJ (1991) The molecular pathology of Alzheimer's disease. Neuron 6:487–498
68. Selkoe DJ, Hardy J (2016) The amyloid hypothesis of Alzheimer's disease at 25 years. EMBO Mol Med 8:595–608
69. Fukuyama R, Mizuno T, Mori S, Nakajima K, Fushiki S, Yanagisawa K (2000) Age-dependent change in the levels of Aß40 and Aß42 in cerebrospinal fluid from control subjects, and a decrease in the ratio of Aß42 to Aß40 level in cerebrospinal fluid from Alzheimer's disease patients. Eur Neurol 43:155–160
70. Kawahara M, Negishi-Kato M, Sadakane Y (2009) Calcium dyshomeostasis and neurotoxicity of Alzheimer's beta-amyloid protein. Expert Rev Neurother 9:681–693
71. Dyrks T, Dyrks E, Masters CL, Beyreuther K (1993) Amyloidogenicity of rodent and human beta A4 sequences. FEBS Lett 324:231–236
72. Exley C, Price NC, Kelly SM, Birchal JD (1993) An interaction of beta-amyloid with aluminium in vitro. FEBS Lett 324:293–295
73. Kawahara M, Muramoto K, Kobayashi K, Mori H, Kuroda Y (1994) Aluminum promotes the aggregation of Alzheimer's amyroid ß-protein in vitro. Biochem Biophys Res Commun 198:531–535
74. Mantyh PW, Ghilardi JR, Rogers S, DeMaster E, Allen CJ, Stimson ER, Maggio JE (1993) Aluminum, iron, and zinc ions promote aggregation of physiological concentrations of beta-amyloid peptide. J Neurochem 61:1171–1174
75. Bush AI, Pettingell WH, Multhaup G, d Paradis M, Vonsattel JP, Gusella JF, Beyreuther K, Masters CL, Tanzi RE (1994) Rapid induction of Alzheimer Ab amyloid formation by zinc. Science 265:1464–1467
76. Atwood CS, Moir RD, Huang X, Scarpa RC, Bacarra NM, Romano DM, Hartshorn MA, Tanzi RE, Bush AI (1998) Dramatic aggregation of Alzheimer aß by Cu(II) is induced by conditions representing physiological acidosis. J Biol Chem 273:12817–12826
77. Wallin C, Kulkarni YS, Abelein A, Jarvet J, Liao Q, Strodel B, Olsson L, Luo J, Abrahams JP, Sholts SB, Roos PM, Kamerlin SC, Gräslund A, Wärmländer SK (2016) Characterization of Mn(II) ion binding to the amyloid-β peptide in Alzhei-mer's disease. J Trace Elem Med Biol 38:183–193
78. Mold M, Linhart C, Gómez-Ramírez J, Villegas-Lanau A, Exley C (2020) Aluminum and Amyloid-β in Familial Alzheimer's Disease. J Alzheimers Dis 73:1627–1635
79. James SA, Churches QI, de Jonge MD, Birchall IE, Streltsov V, McColl G, Adlard PA, Hare DJ (2017) Iron, copper, and zinc concentration in Aβ plaques in the APP/PS1 mouse model of Alzheimer's disease correlates with metal levels in the sur-rounding neuropil. ACS Chem Neurosci 8:629–637
80. Chen WT, Liao YH, Yu HM, Cheng IH, Chen YR (2011) Distinct effects of Zn^{2+}, Cu^{2+}, Fe^{3+}, and Al^{3+} on amyloid-beta stability, oligomerization, and aggregation: amyloid-beta destabilization promotes annular protofibril formation. J Biol Chem 286:9646–9656
81. Sharma AK, Pavlova ST, Kim J, Kim J, Mirica LM (2013) The effect of Cu(2+) and Zn(2+) on the Aβ42 peptide aggregation and cellular toxicity. Metallomics 5:1529–1536
82. White AR, Multhaup G, Maher F, Bellingham S, Camakaris J, Zheng H, Bush AI, Beyreuther K, Masters CL, Cap-pai R (1999) The Alzheimer's disease amyloid precursor protein modulates copper-induced toxicity and

oxidative stress in primary neuronal cultures. J Neurosci 19:9170–9179
83. Baumkötter F, Schmidt N, Vargas C, Schilling S, Weber R, Wagner K, Fiedler S, Klug W, Radzimanowski J, Nickolaus S, Keller S, Eggert S, Wild K, Kins S (2014) Amyloid precursor protein dimerization and synaptogenic function depend on copper binding to the growth factor-like domain. J Neurosci 34:11159–11172
84. Gerber H, Wu F, Dimitrov M, Garcia Osuna GM, Fraering PC (2017) Zinc and copper differentially modulate amyloid precursor protein processing by γ-secretase and amyloid-ß peptide production. J Biol Chem 292:3751–3767
85. Multhaup G, Schlicksupp A, Hesse L, Beher D, Ruppert T, Masters CK, Beyreuther K (1996) The amyloid precursor protein of Alzheimer's disease in the reduction of copper(II) to copper(I). Science 271:1406–1409
86. Wong BX, Tsatsanis A, Lim LQ, Adlard PA, Bush AI, Duce JA (2014) β-Amyloid precursor protein does not possess ferroxidase activity but does stabilize the cell surface ferrous iron exporter ferroportin. PLoS One 9:e114174
87. Rogers JT, Randall JD, Cahill CM, Eder PS, Huang X, Gunshin H, Leiter L, McPhee J, Sarang SS, Utsuki T, Greig NH, Lahiri DK, Tanzi RE, Bush AI, Giordano T, Gullans SR (2002) An iron-responsive element type II in the 5'-untranslated region of the Alzheimer's amyloid precursor protein transcript. J Biol Chem 277:45518–45528
88. Chen M, Zheng J, Liu G et al. (2019) High dietary iron disrupts iron homeostasis and induces amyloid-β and phospho-tau expression in the hippocampus of adult wild-type and APP/PS1 transgenic mice. J Nutrition 149:2247–2254
89. Namekata K, Imagawa M, Terashi A, Ohta S, Oyama F, Ihara Y (1997) Association of transferrin C2 allele with late-onset Alzheimer's disease. Hum Genet 101:126–129
90. Imagawa M, Naruse S, Tsuji S, Fujioka A, Yamaguchi H (1992) Coenzyme Q10, iron, and vitamin B6 in genetically-confirmed Alzheimer's disease. Lancet 340:671
91. Pizzo P, Basso E, Filadi R, Greotti E, Leparulo A, Pendin D, Redolfi N, Rossini M, Vajente N, Pozzan T, Fasolato C (2020) Presenilin-2 and calcium handling: molecules, organelles, cells and brain networks. Cells 9:2166
92. Greenough MA, Volitakis I, Li QX, Laughton K, Evin G, Ho M, Dalziel AH, Camakaris J, Bush AI (2011) Presenilins promote the cellular uptake of copper and zinc and maintain copper chaperone of SOD1-dependent copper/zinc superoxide dismutase activity. J Biol Chem 286:9776–9786
93. Madhu P, Mukhopadhya S (2021) Distinct types of amyloid-β oligomers displaying diverse neurotoxicity mechanisms in Alzheimer's disease. J Cell Biochem 122:1594–1608
94. Small DH, Mok SS, Bornstein JC (2001) Alzheimer's disease and Aß toxicity: from top to bottom. Nat Rev Neurosci 2:595–598
95. Sivanesan S, Tan A, Rajadas J (2013) Pathogenesis of Abeta oligomers in synaptic failure. Curr Alzheimer Res 10:316–323
96. Arispe N, Rojas E, Pollard HB (1993) Alzheimer disease amyloid ß protein forms calcium channels in bilayer membranes: blockade by tromethamine and aluminum. Proc Natl Acad Sci USA 90:567–571
97. Rhee SK, Quist AP, Lal R (1998) Amyloid ß protein-(1-42) forms calcium-permeable, Zn^{2+}-sensitive channel. J Biol Chem 273:13379–13382
98. Durell SR, Guy HR, Arispe N, Rojas E, Pollard HB (1994) Theoretical models of the ion channel structure of amyloid ß-protein. Biophys J 67:2137–3145
99. Strodel B, Lee JW, Whittleston CS, Wales DJ (2010) Transmembrane structures for Alzheimer's Aβ(1-42) oligomers. J Am Chem Soc 132:13300–13312
100. Voelker MJ, Barz B, Urbanc B (2017) Fully atomistic Aβ40 and Aβ42 oligomers in water: observation of pore like conformations. J Chem Theory Comput 13:4567–4583
101. Kawahara M, Arispe N, Kuroda Y, Rojas E (1997) Alzheimer's disease amyloid ß-protein forms Zn^{2+}-sensitive, cation-selective channels across excised membrane patches from hypothalamic neurons. Biophys J 73:67–75
102. Kawahara M, Arispe N, Kuroda Y, Rojas E (2000) Alzheimer's ß-amyloid, human islet amylin and prion protein fragment evoke intracellular free-calcium elevations by a common mechanism in a hypothalamic GnRH neuronal cell-line. J Biol Chem 275:14077–14083
103. Kawahara M, Kuroda Y (2000) Molecular mechanism of neurodegeneration induced by Alzheimer's ß-amyloid protein: channel formation and disruption of calcium homeostasis. Brain Res Bull 53:389–397
104. Kawahara M, Kuroda Y (2001) Intracellular calcium changes in neuronal cells induced by Alzheimer's ß-amyloid protein are blocked by estradiol and cholesterol. Cell Mol Neurobiol 21:1–13
105. Negishi-Kato M, Kawahara M (2008) Neurosteroids block the increase in intracellular calcium level induced by Alzheimer's ß-amyloid protein in long-term cultured rat

hippocampal neurons. Neuropsychiatr Dis Treat 4:209–218
106. Arispe N, Diaz JC, Simakova O (2007) Aß ion channels. Prospects for treating Alzheimer's disease with Aß channel blockers. Biochim Biophys Acta 1768:1952–1965
107. Parodi J, Ormeño D, Ochoa-de la Paz LD (2015) Amyloid pore-channel hypothesis: effect of ethanol on aggregation state using frog oocytes for an Alzheimer's disease study. BMB Rep 48:13–18
108. Arispe N, Pollard HB, Rojas E (1996) Zn^{2+} interactions with Alzheimer's amyloid ß protein calcium channels. Proc Natl Acad Sci USA 93:1710–1715
109. Prusiner SB (2013) Biology and genetics of prions causing neurodegeneration. Annu Rev Genet 47:601–623
110. Jackson GS, Murray I, Hosszu LL, Gibbs N, Waltho JP, Clarke AR, Collinge J (2001) Location and properties of metal-binding sites on the human prion protein. Proc Natl Acad Sci U S A 98:8531–8535
111. Brown DR, Qin K, Herms JW, Madlung A, Manson J, Strome R, Fraser PE, Kruck T, von Bohlen A, Schulz-Schaeffer W, Giese A, Westaway D, Kretzschmar H (1997) The cellular prion protein binds copper in vivo. Nature 390:684–687
112. Huang S, Chen L, Bladen C, Stys PK, Zamponi GW (2018) Differential modulation of NMDA and AMPA receptors by cellular prion protein and copper ions. Mol Brain 11: 62
113. Kawahara M, Koyama H, Nagata T, Sadakane Y (2011) Zinc, copper, and carnosine attenuate neurotoxicity of prion fragment PrP106-126. Metallomics 3:726–734
114. Nguyen XTA, Tran TH, Cojoc D, Legname G (2019) Copper binding regulates cellular prion protein function. Mol Neurobiol 56:6121–6133
115. Schmitt-Ulms G, Ehsani S, Watts JC, Westaway D, Wille H (2009) Evolutionary descent of prion genes from the ZIP family of metal ion transporters. PLoS One 4:e7208
116. Watt NT, Griffiths HH, Hooper NM (2013) Neuronal zinc regulation and the prion protein. Prion 7:203–208
117. Salzano G, Giachin G, Legname G (2019) Structural consequences of copper binding to the prion protein. Cells 8:770
118. Siggs OM, Cruite JT, Du X, Rutschmann S, Masliah E, Beutler B, Oldstone MBA (2012) Disruption of copper homeostasis due to a mutation of Atp7a delays the onset of prion disease. Proc Natl Acad Sci U S A 109:13733–13738
119. Spevacek AR, Evans EGB, Miller JL, Meyer HC, Pelton JG, Millhauser GL (2013) Zinc drives a tertiary fold in the prion protein with familial disease mutation sites at the interface. Structure 21:236–246
120. Singh A, Haldar S, Horback K, Tom C, Zhou L, Meyerson H, Singh N (2013) Prion protein regulates iron transport by functioning as a ferrireductase. J Alzheimers Dis 35:541–552
121. Tripathi AK, Haldar S, Qian J, Beserra A, Suda S, Singh A, Hopfer U, Chen SG, Garrick MD, Turner JR, Knutson MD, Singh N (2015) Prion protein functions as a ferrireductase partner for ZIP14 and DMT1. Free Radic Biol Med 84:322–330
122. Rogers JT, Cahill CM (2020) Iron-responsive-like elements and neurodegenerative ferroptosis. Learn Mem 27:395–413
123. Brown DR (2011) Prions and manganese: a maddening beast. Metallomics 3:229–238
124. White SN, O'Rourke KI, Gidlewski T, VerCauteren KC, Mousel MR, Phillips GE, Spraker TR (2010) Increased risk of chronic wasting disease in Rocky Mountain elk associated with decreased magnesium and increased manganese in brain tissue. Can J Vet Res 74:50–53
125. Slivarichová D, Mitrová E, Ursínyová M, Uhnáková I, Koscová S, Wsólová L (2011) Geographic accumulation of Creutzfeldt-Jakob disease in Slovakia-environmental metal imbalance as a possible cofactor. Cent Eur J Public Health 19:158–164
126. Singh A, Kong Q, Luo X, Petersen RB, Meyerson H, Singh N (2009) Prion protein (PrP) knock-out mice show altered iron metabolism: a functional role for PrP in iron uptake and transport. PLoS One 4:e6115
127. Forloni G, Chiesa R, Bugiani O, Salmona M, Tagliavini F (2019) Review: PrP 106-126—25 years after. Neuropathol. Appl. Neurobiol 45:430–440
128. Lin MC, Mirzabekov T, Kagan BL (1997) Channel formation by a neurotoxic prion protein fragment. J Biol Chem 272:44–47
129. Kourie JI, Culverson A (2000) Prion peptide fragment PrP[106-126] forms distinct cation channel types. J Neurosci Res 62:120–133
130. Paulis D, Maras B, Schininà ME, di Francesco L, Principe S, Galeno R, Abdel-Haq H, Cardone F, Florio T, Pocchiari M et al. (2011) The pathological prion protein forms ionic conductance in lipid bilayer. Neurochem Int 59:168–174
131. Bernal-Conde LD, Ramos-Acevedo R, Reyes-Hernández MA, Balbuena-Olvera AJ, Morales-Moreno ID, Argüero-Sánchez R, Schüle B, Guerra-Crespo M (2020) Alpha-synuclein physiology and pathology:

a perspective on cellular structures and organelles. Front Neurosci 13:1399
132. Brookes AJ, St Clair D (1994) Synuclein proteins and Alzheimer's disease. Trends Neurosci 14:404–405
133. Carboni E, Lingor P (2015) Insights on the interaction of alpha-synuclein and metals in the pathophysiology of Parkinson's disease. Metallomics 7:395–404
134. Uversky VN, Li J, Fink AL (2001) Metal-triggered structural transformations, aggregation, and fibrillation of human alpha-synuclein. A possible molecular NK between Parkinson's disease and heavy metal exposure. J Biol Chem 276:44284–44296
135. Wang H, Mörman C, Sternke-Hoffmann R, Huang CY, Prota A, Ma P, Luo J (2022) Cu2+ ions modulate the interaction between α-synuclein and lipid membranes. J Inorg Biochem 236:111945
136. Miotto MC, Binolfi A, Zweckstetter M, Griesinger C, Fernández CO (2014) Bioinorganic chemistry of synucleinopathies: deciphering the binding features of Met motifs and His-50 in AS-Cu(I) interactions. J Inorg Biochem 141:208–211
137. Davies P, Moualla D, Brown DR (2011) Alpha-synuclein is a cellular ferrireductase. PLoS One 6:e15814
138. McDowall JS, Ntai I, Honeychurch KC et al. (2017) Alpha-synuclein ferrireductase activity is detectible in vivo, is altered in Parkinson's disease and increases the neurotoxicity of DOPAL. Mol Cell Neurosci 85:1–11
139. Cahill CM, Lahiri DK, Huang X, Rogers JT (2009) Amyloid precursor protein and alpha synuclein translation, implications for iron and inflammation in neurodegenerative diseases. Biochim Biophys Acta 1790:615–628
140. Wang TY, Ma Z, Wang C, Liu C, Yan DY, Deng Y, Liu W, Xu ZF, Xu B (2018) Manganese-induced alpha-synuclein over-ex-pression impairs synaptic vesicle fusion by disrupting the Rab3 cycle in primary cultured neurons. Toxicol Lett 285:34–42
141. Lashuel HA, Hartley D, Petre BM, Walz T, Lansbury JrPT (2002) Neurodegenerative disease: amyloid pores from pathogenic mutations. Nature 418:291
142. Ghio S, Camilleri A, Caruana M, Ruf VC, Schmidt F, Leonov A, Ryazanov S, Griesinger C, Cauchi RJ, Kamp F, Giese A, Vassallo N (2019) Cardiolipin promotes pore-forming activity of alpha-synuclein oligomers in mitochondrial membranes. ACS Chem Neurosci 10:3815–3829
143. Choi DW (2020) Excitotoxicity: still hammering the ischemic brain in 2020. Front Neurosci 14:579953
144. Koh JY, Suh SW, Gwag BJ, He YY, Hsu CY, Choi DW (1996) The role of zinc in selective neuronal death after transient global cerebral ischemia. Science 272:1013–1016
145. Calderone A, Jover T, Mashiko T, Noh KM, Tanaka H, Bennett MV, Zukin RS (2004) Late calcium EDTA rescues hippocampal CA1 neurons from global ischemia-induced death. J Neurosci 24:9903–9913
146. Kawahara M, Mizuno D, Koyama H, Konoha K, Ohkawara S, Sadakane Y (2014) Disruption of zinc homeostasis and the pathogenesis of senile dementia. Metallomics 6:209–219
147. Tanaka KI, Kawahara M (2017) Copper enhances zinc-induced neurotoxicity and the endoplasmic reticulum stress response in a neuronal model of vascular dementia. Front Neurosci 11:58
148. Nakano Y, Shimoda M, Okudomi S, Kawaraya S, Kawahara M, Tanaka KI (2020) Seleno-L-methionine suppresses copper-enhanced zinc-induced neuronal cell death via induction of glutathione peroxidase. Metallomics 12:1693–1701
149. Zheng W, Xin N, Chi ZH, Zhao BL, Zhang J, Li JX, Wang ZY (2009) Divalent metal transporter 1 is involved in amyloid precursor protein processing and Abeta generation. FASEB J 23:4207–4217
150. Koh JY, Lee SJ (2020) Metallothionein-3 as a multifunctional player in the control of cellular processes and diseases. Mol Brain 2020(13):116
151. Boldyrev AA, Aldini G, Derave W (2013) Physiology and pathophysiology of carnosine. Physiol Rev 93:1803–1845
152. Bakardjiev A (1998) Carnosine and beta-alanine release is stimulated by glutamatergic receptors in cultured rat oligodendrocytes. Glia 24:346–351
153. Junceda S, Cruz-Alonso M, Fernandez B, Pereiro R, Martínez-Pinilla E, Navarro A (2024) Iron dysregulation in Alzheimer's disease: LA-ICP-MS bioimaging of the distribution of iron and ferroportin in the CA1 region of the human hippocampus. Biomolecules 14:295
154. Liang RF, Li WQ, Wang H, Wang JX, Niu Q (2013) Impact of sub-chronic aluminium-maltolate exposure on catabolism of amyloid precursor protein in rats. Biomed Environ Sci 26:445–452
155. Li XB, Zhang ZU, Yin LH, Schluesener HJ (2012) The profile of β-amyloid precursor protein expression of rats induced by aluminium. Environ Toxicol Pharmacol 33:135–140
156. Woods NK, Padmanabhan J (2012) Neuronal calcium signaling and Alzheimer's disease. Adv Exp Med Biol 740:1193–1217

157. Li GZ, Liu F, Xu C, Li JY, Xu YJ (2018) Selenium and Zinc against Aβ25-35-induced cytotoxicity and tau phosphorylation in PC12 cells and inhibits γ-cleavage of APP. Biol Trace Elem Res 184:442–449
158. Purro SA, Nicoll AJ, Collinge J (2018) Prion protein as a toxic acceptor of amyloid ß oligomers. Biol Psychiatry 83:358–368
159. Roberts HL, Schneider BL, Brown DR (2017) α-Synuclein increases ß-amyloid secretion by promoting β-/γ-secretase processing of APP. PLoS One 12:e0171925
160. Xu J, Xu G, Fang J (2022) Association between serum copper and stroke risk factors in adults: evidence from the National Health and Nutrition Examination Survey, 2011–2016. Biol Trace Elem Res 200:1089–1094
161. Zhang M, Li W, Wang Y, Wang T, Ma M, Tian C (2020) Association between the change of serum copper and ischemic stroke: a systematic review and meta-analysis. J Mol Neurosci 70:475–480
162. Faux NG, Ritchie CW, Gunn A, Rembach A, Tsatsanis A, Bedo J, Harrison J, Lannfelt L, Blennow K, Zetterberg H, Ingelsson M, Masters CL, Tanzi RE, Cummings JL, Herd CM, Bush AI (2010) PBT2 rapidly improves cognition in Alzheimer's disease: additional phase II analyses. J Alzheimers Dis 20:509–516
163. Sampson EL, Jenagaratnam L, McShane R (2014) Metal protein attenuating compounds for the treatment of Alzheimer's dementia. Cochrane Database Syst Rev 2:CD005380
164. Davenward S, Bentham P, Wright J, Crome P, Job D, Polwart A, Exley C (2013) Silicon-rich mineral water as a non-invasive test of the 'aluminum hypothesis' in Alzheimer's disease. J Alzheimers Dis 33:423–430
165. McLachlan DRC, Dalton AJ, Kruck TP, Bell MY, Smith WL, Kalow W, Andrews DF (1991) Intramuscular desferrioxamine in patients with Alzheimer's disease. Lancet 337:1304–1308
166. Farr AC, Xiong MP (2021) Challenges and opportunities of deferoxamine delivery for treatment of Alzheimer's disease, Parkin-son's disease, and intracerebral hemorrhage. Mol Pharm 18:593–609
167. Levi S, Volonté MA (2023) Iron chelation in early Parkinson's disease. Lancet Neurol 22:290–291
168. Thapa K, Khan H, Kanojia N, Singh TG, Kaur A, Kaur G (2022) Therapeutic insights on ferroptosis in Parkinson's disease. Eur J Pharmacol 930:175133
169. Bareggi SR, Braida D, Pollera C, Bondiolotti G, Formentin E, Puricelli M, Poli G, Ponti W, Sala M (2009) Effects of clioquinol on memory impairment and the neurochemical modifications induced by scrapie infection in golden hamsters. Brain Res 1280:195–200
170. Sigurdsson EM, Brown DR, Alim MA, Scholtzova H, Carp R, Meeker HC, Prelli F, Frangione B, Wisniewski T (2003) Copper chelation delays the onset of prion disease. J Biol Chem 278:46199–46202
171. Corona C, Frazzini V, Silvestri E, Lattanzio R, La Sorda R, Piantelli M, Canzoniero LM, Ciavardelli D, Rizzarelli E, Sensi SL (2011) Effects of dietary supplementation of carnosine on mitochondrial dysfunction, amyloid pathology, and cognitive deficits in 3xTg-AD mice. PLoS One 6:e17971
172. Davis CK, Laud PJ, Bahor Z, Rajanikant GK, Majid A (2016) Systematic review and stratified meta-analysis of the efficacy of car-nosine in animal models of ischemic stroke. J Cereb Blood Flow Metab 36:1686–1694
173. Kawahara M, Konoha K, Nagata T, Sadakane Y (2013) JP5382633
174. Kawahara M, Kato-Negishi M, Tanaka K-I (2017) Cross talk between neurometals and amyloidogenic proteins at the synapse and the pathogenesis of neurodegenerative diseases. Metallomics 9:619–633
175. Kawahara M, Kato-Negishi M, Tanaka KI (2021) Neurometals in the Pathogenesis of Prion Diseases. Int J Mol Sci 22:1267
176. Kawahara M, Tanaka K-I, Kato-Negishi M (2018) Zinc, Carnosine, and neurodegenerative diseases. Nutrients 10:147

Check for updates

Chapter 10

From Cause to Treatment: Brain Metallomics in Neuropsychiatric Disorders

Dorota Bederska-Łojewska, Justyna Turek and Bernadeta Szewczyk

Abstract

This chapter provides a comprehensive overview of the role of trace elements in the pathogenesis of neuropsychiatric disorders (NPDs), including mood disorders, psychotic disorders, neurodevelopmental and neurodegenerative diseases. Emphasis is placed on the significance of metal dyshomeostasis-particularly of zinc, iron, copper, calcium, magnesium, manganese, selenium, lithium, chromium, molybdenum, potassium, and sodium—in brain metabolism, neurotransmission, oxidative stress, and neuroinflammation. The chapter discusses how these metals influence neuronal function through mechanisms involving neurotransmitter synthesis, myelination, mitochondrial activity, and blood–brain barrier integrity. Both deficiency and excess of essential elements may lead to or exacerbate conditions such as Alzheimer's disease, Parkinson's disease, schizophrenia, depression, autism spectrum disease, attention-deficit/hyperactivity disorder, and Tourette syndrome. Detailed attention is given to these metals due to their wide-ranging effects on synaptic plasticity, regulation of glutamate and GABA, modulation of membrane potential and excitability, influence on astrocyte function, and the neurotoxic effects of excess metals like manganese and chromium, as well as their potential therapeutic applications. Current treatments partially rely on ion channel modulation, though newer strategies are emerging, including the use of metal chelators and supplementation in specific deficiencies. Understanding the brain metallome offers promising avenues for targeted interventions in complex neuropsychiatric disorders.

Keywords Neuropsychiatric disorders, Metal dyshomeostasis, Brain metallome, Trace elements and mental health, Neurodevelopmental and neurodegenerative disorders, Metal-related brain neurotransmitter regulation

1 Introduction

Current knowledge based on experimental evidence at the cellular and biochemical level shows the significant role of metal dyshomeostasis in the onset of different neuropsychiatric disorders (NPDs):

- mood disorders (major depressive disorder - MDD, bipolar disorder - BD)
- obsessive-compulsive disorder (OCD)

Andreas M. Grabrucker (ed.), *Brain Metallomics*, Neuromethods, vol. 237,
https://doi.org/10.1007/978-1-0716-5468-2_10,
© The Author(s), under exclusive license to Springer Science+Business Media, LLC, part of Springer Nature 2026

- psychotic disorders (schizophrenia - SZ)
- neurodevelopmental disorders (autism spectrum disorder - ASD, attention-deficit/hyperactivity disorder - ADHD, Tourette Syndrome - TS)
- sleep-wake disorders (insomnia disorder, narcolepsy)
- neurodegenerative disease (Alzheimer's disease - AD, Lewy body dementia - DLB, Parkinson's disease - PD, Huntington's disease - HD, multiple sclerosis - MS, vascular dementia - VaD)

This chapter emphasizes the enormously significant role of metals such as zinc (Zn), iron (Fe), copper (Cu), calcium (Ca), magnesium (Mg), manganese (Mn), selenium (Se), lithium (Li), chromium (Cr), molybdenum (Mo), potassium (K), and sodium (Na) in nervous system dysfunction, including oligodendrocytes, astrocytes, microglia, neurons, cerebrospinal fluid, mitochondrial and blood–brain barrier (BBB) function. Metals are constituents of organic and inorganic compounds, influencing a myriad of processes within the nervous system: its proper development, enzymatic activities, cognitive function, mitochondrial function, myelination, BBB integrity, neurotransmission, and adequate composition of cerebrospinal fluid. Within the organism, a series of precise mechanisms regulate the level of the elements in cells and the extracellular space. Studies have shown that metal imbalance in the brain is related to the onset and progression of neuropsychiatric disorders. Brain disorders' pathophysiology can result from a deficiency or too high a concentration of essential elements. Despite the correct overall concentration of a given element in the food, its distribution or functionality may become aberrant because of metal-binding proteins (ferritin, transferrin, hepcidin, hephaestin, ceruloplasmin, albumin, metallothionein) regulating the localization and concentration of intracellular free metals, interactions among some ions or therapeutic drugs. Genetic variations in proteins regulating metal ions can also affect metabolism and utilization. Iron-genetic disorders are the most diffused. NBIA is a neurodegeneration with brain iron accumulation caused by mutations in iron transport and storage proteins [1]. Some reports on metal dyshomeostasis are not consistent. Many different elements in the brain can either increase, decrease, or remain constant in concentration but change in their chemical form under pathological conditions.

2 Zinc (Zn)

Zinc is one of the key elements in the central nervous system (CNS). It is found in significant amounts in essential brain structures such as cortical areas, the hippocampus, amygdala nuclei, and the olfactory bulb [2]. Dyshomeostasis of zinc levels plays a key role in the brain's physiological and pathological functions. Many indications in recent years suggest that zinc-level disorders can lead to numerous neurological diseases, including brain damage and exacerbations of neurological conditions. Major disorders caused by zinc changes in the brain include AD and depression [3–5].

Brain homeostasis of zinc is based on the activity of various cellular processes and the transformation of signaling pathways. Zinc levels in the brain are essential in initiating mechanisms such as learning, memory, and synaptic plasticity [6]. Zinc is stored in the brain in various forms: free zinc ions, vesicular zinc, and protein-bound zinc [7]. Understanding the physiology of how these connections work is crucial in understanding the molecular basis of diseases caused by zinc dyshomeostasis in the brain. In the CNS, the largest amount of zinc is found in protein-bound form, about 85% of zinc. Another location of zinc is in synaptic vesicles. Vesicular zinc is primarily localized in synaptic vesicles at the axon terminals of glutamatergic neurons, particularly in the cortex and hippocampus. It modulates the activity of *N*-methyl-D-aspartate (NMDA), Alpha-amino-3-hydroxy-5-methyl-4-isoxazolepropionic acid receptor (AMPA), γ-aminobutyric acid receptor type A ($GABA_A$) $_A$, and G protein-coupled receptor 39 (GPR39) receptors [8]. The inability of zinc to diffuse freely across membranes is regulated by functional zinc-binding proteins such as metallothioneins (MT) and two families of zinc transporters: SLC39A (ZIP), which facilitates zinc influx into the cytoplasm and SLC30A (ZnT), which mediates zinc efflux and sequestration. Currently, 10 proteins of the ZnT family and 14 of the ZIP family are known. However, in the case of MT, four isoforms have been described: MT-I, MT-II, MT-III, and MT-IV. Abnormalities in the protein transporters contribute to neurological disorders.

One of the most described roles for zinc in the brain is the modulation of NMDA receptors. Zinc acts as an allosteric modulator of NMDA receptor activity. The primary function of zinc, derived from vesicles, is to reduce NMDA receptor currents. Particularly in the hippocampus (a region that contains a large pool of vesicular zinc), zinc has an inhibitory effect on NMDA receptors (postsynaptic GluN2A), which has attracted attention due to the involvement of these receptors in synaptic plasticity processes [9]. Numerous preclinical and clinical studies show a link between zinc deficiency in the body and comorbid depressive symptoms. Animals exposed to a chronic dietary zinc deficit of <3 mg/kg show depressive symptoms in behavioral tests [10, 11]. In studies linking zinc deficit and the stress model, animals show treatment resistance to both the classic antidepressants imipramine and ketamine [12]. However, the mechanisms are not fully understood. What is known is that zinc deficiency affects the induction of inflammation and increased oxidative stress in the body [13]. Proteomic data indicate alterations in the prefrontal cortex, including mutations in zinc transport (upregulated: ZnT1, ZnT4, ZnT6, ZnT7, downregulated: ZnT3), monoamine transport, and mitochondrial function [14]. Few data show the effect of zinc deficiency and chronic stress on the alteration of the morphology of dendritic spines [12]. In this model, organs performing immunological functions in the body are also subject to pathological changes, which could explain an indirect mechanism in the induction of global inflammation for the body [15]. In a subsequent study, the authors demonstrated a reduced potential for antidepressant action in a rodent model of bulbectomy-induced depression when combined with a zinc-deficient diet [16]. Behavioral changes were reversed by drug administration in combination with zinc. This provides further evidence to explain the mechanism of treatment-resistant depression in which zinc, or more specifically, zinc deficiency, may be implicated.

Only a few clinical studies have investigated the effect of dietary zinc deficiency in patients suffering from depression. It is known that depressed patients have reduced levels of zinc in serum [17]. Zinc levels normalize during remission. Several studies suggests the effectiveness of therapies in which antidepressants are used in combination with zinc [18]. These data show that changes in serum zinc levels in patients and after treatment can be an indicator of depression status and remission. Reduced serum zinc levels in patients, on the other hand, may be a marker of drug-resistant depression.

Meta-analyses also show an association between declining zinc levels and ADHD and OCD [19, 20]. However, further research is needed to clarify the role of zinc in these disorders. Available data-show a positive effect of using a zinc complex along with psychostimulants in the treatment of ADHD. In this combination, zinc allowed for a reduction in stimulant dosage, a promising alternative, while long-term use of psychostimulants may carry adverse side effects [21].

In SZ, many studies show reduced serum zinc levels in patients [22]. Zinc dyshomeostasis in SZ affects neurotransmitter dysfunction, inflammation induction, and, most importantly, glutamatergic dysfunction and neurodegeneration. The use of antipsychotics, in combination with zinc, resulted in an improvement in the scale of negative and positive symptoms compared to the placebo groups and to the groups in which only risperidone was used [23].

AD is classified as a neurodegenerative disease. In this disease entity, the status of zinc levels in the body may be an important factor predisposing to the formation of neurodegenerative changes in the brains of patients. The presence of two central brain lesions forms the basis for the diagnosis of AD. These are beta-amyloid protein plaques on the outside of neurons and intraneuronal tau protein. These degenerations occur in the hippocampus, amygdala, and neocortex [24]. Elevated levels of zinc in brain tissue are noted in patients. This level can be twice as high compared to healthy individuals, which characterizes the severity of the disease [25]. Many changes involve zinc transporters (ZnT1, ZnT3, ZnT4, ZnT6, ZnT10, and ZIP1). Zinc transporters have been linked to the development of amyloid plaques in preclinical and clinical studies. Overexpression and altered localization of the transporters contribute to elevated zinc concentrations [26, 27]. This leads to the deposition of pathological changes in the brain, as does excessive zinc supplementation. Excess zinc ions cause hyperphosphorylation of tau proteins and inhibition of extracellular signal-regulated kinase (ERK) and glycogen synthase kinase (GSK-3), contributing to decreased stability of microtubules in neurons, affecting their stabilization and morphological abnormalities [28]. Another mechanism involves disturbances in MT. The excessive activation of MT-I and MT-II, stimulated by free radicals and proinflammatory cytokines, is one of the changes observed in AD. MT activation may be a defensive cellular response, confirming the neuroprotective role of MT in the CNS [29].

The above example shows that zinc deficiency can cause numerous dysfunctions, and an excess of this element can be harmful. Zinc in high concentrations has a cytotoxic effect on brain tissues. The neurotoxicity of high levels of zinc has also been linked to the occurrence of seizures, PD, or MS [30]. Due to excessive dietary zinc intake, copper deficiency can occur, leading to reduced copper absorption by the body. The ratio of the two elements is essential in maintaining overall health.

3 Iron (Fe)

The brain is an organ that requires a high amount of energy to function correctly. An adequate supply of iron is essential in covering the brain's energy consumption. Iron in the nervous system has many functions, such as a cofactor in physiological processes (for cytochromes and iron-sulfur complexes), myelination, metabolism, or biosynthesis of neurotransmitters [31]. Iron undergoes several transformations before entering the brain. In the gut, dietary Fe^{3+} is reduced to Fe^{2+} by duodenal cytochrome b (Dcytb) and then transported into enterocytes via divalent metal transporter 1 (DMT1). Ferroportin then exports Fe^{2+} into the bloodstream, where it is reoxidized to Fe^{3+} by hephaestin or ceruloplasmin and binds to transferrin. Iron enters the brain through the BBB via transferrin receptor-mediated transport [32]. The amount of transferrin in the brain is much lower than in the periphery, and neurons primarily acquire iron from the labile iron pool rather than transferrin-bound iron. Another important source of iron in the brain is ferritin, which serves as an iron storage protein. Ferritin can be transported across the BBB or synthesized locally by endothelial cells, microglia, and astrocytes [33].

As we age, iron tends to be gradually deposited in the brain. High iron levels, along with increasing oxidative stress, can cause neurodegeneration in the brain. Depending on the structures occupied by increased iron aggregation, the most common diseases involved are AD and PD [34]. Increased iron aggregation in neurons, astrocytes, and microglia in the hippocampus, cortex, and substantia nigra are characteristic of AD and PD [35]. These iron-occupied structures cause permanent damage to cells, increasing their susceptibility to inflammation, toxins, apoptosis, and reactive oxygen species (ROS) production [34]. Metal chelators represent a future therapeutic strategy for treating neurodegenerative diseases, AD and PD [36].

Some studies show a link between iron deficiency and comorbid ASD and ADHD in children. Iron deficiency affects memory, attention, emotional behavior, and cognitive and motor skills [37]. Children with ADHD had significantly lower ferritin levels compared to controls. Additionally, iron-deficient children are burdened with more severe ADHD symptoms compared to ADHD children with balanced iron levels [38].

Iron is also a cofactor of tyrosine hydroxylase, and its deficiency disrupts dopamine metabolism in mesolimbic structures, causing behavioral disorders [39]. In patients with SZ, serum iron levels are variable and may depend on the severity of positive or negative symptoms [40]. In one study, it was shown that patients with a first episode of psychosis had iron deficiency and low serum ferritin levels, which also correlated with significantly more severe negative symptoms [41]. However, iron is also a cofactor for the downstream hydroxylases of tryptophan and phenylalanine, and these hydroxylases are essential for synthesizing serotonin and norepinephrine [31]. Therefore, it is believed that iron deficiency may be a presymptomatic factor in the development of depressive and anxiety disorders [42]. In preclinical studies, iron oxide therapy alleviated rodent depressive symptoms in behavioral tests [43].

Among disorders of iron homeostasis in the brain, one of the rare genetic diseases is NBIA. Abnormal iron aggregation in the brain contributes to progressive brain damage. Symptoms of this disease can appear early in life [44]. Varied symptoms of this disease include dystonia, muscle rigidity, spasticity, ataxia, disorientation, and visual changes. Treatment of this disease is limited only to reducing symptoms in the musculoskeletal system.

4 Copper (Cu)

Copper, zinc, and iron are among the body's most abundant trace metals. The presence of this element is significant for the proper functioning of the body. Numerous studies show a substantial effect of this metal on the functioning of neurons. Both deficiency and excess of this element are associated with neurodegenerative disorders such as AD and mental disorders such as SZ, BD, and depression [45]. As mentioned above, proper metabolism and absorption of copper requires an appropriate balance of copper with zinc and manganese [46]. Copper, as a cofactor of many enzymes, plays a vital role in the functioning of the CNS. These enzymes are responsible for, among others, the synthesis of catecholamines, protection against ROS, and activation of neuropeptides and hormones. These include: oxidoreductases, dopamine beta-hydroxylase (DBH), superoxide dismutase (SOD), cyclooxygenase (COX), and others [47]. Another essential aspect is the bioavailability of this element. It depends, among others, on absorption in the gastrointestinal tract, transport in the portal blood, activity of hepatocytes, peripheral tissues, and copper uptake in the nervous system through the BBB and blood-cerebrospinal fluid barrier (BCB). Many genes related to copper absorption and transport participate in this regulation. This group includes numerous ATPases, and copper transport through the BBB, ATP7A, and ATP7B play major roles in this process [48]. ATP7A is involved in copper absorption, while ATP7B is responsible for copper excretion [49].

One rare but characteristic genetic disease caused by mutations in the ATP7B gene is Wilson's disease (WD) [50]. WD is a disorder of copper metabolism, occurring both in hepatocytes and neurons, leading to pathological accumulation of copper in tissues [51]. The main clinical change observed in patients is a diseased liver. Neurological symptoms have not been fully described, but it is known that the structures damaged in people with WD include the frontal cortex, hypothalamus, and cerebellum. Abnormalities in the functioning of ATP7B cause abnormal conversion of dopamine to noradrenaline in the basal ganglia, which in turn contributes to the development of PD symptoms such as rigidity and tremors, as well as concentration disorders. In this disease, there are also disorders of a psychological origin. The broad spectrum of symptoms includes insomnia, depression, anxiety, personality changes, and even hallucinations. WD therapy consists of adding zinc to the diet to disrupt the uptake of copper from the gastrointestinal tract [52, 53]. Among the genetic diseases disrupting copper homeostasis, one can also distinguish Menkes disease (MD). Mutations in the ATP7A gene lead to neuronal degeneration, mental retardation, and disruption of copper efflux through the BBB [54].

Other disorders caused by excess copper include the previously mentioned PD and AD [55]. Copper levels in the brains of AD patients are significantly elevated. Copper binds to beta-amyloid, contributing to increased ROS production and exerting a destructive effect on neurons. Copper chelators can be used to remove copper aggregating with beta-amyloid, as they can selectively extract copper from these complexes [56]. Limited data show a relationship between the effects of copper on neuropsychiatric diseases, including depression. However, copper is involved in several processes that are key to initiating symptoms of low mood. Among these mechanisms, it is worth noting that copper is a component of enzymes involved in the metabolism of catecholamines. Copper plays a vital role in the process of reducing oxidative stress.

Additionally, copper has catalytic properties for some antioxidant enzymes. As is known, increased oxidative stress contributes to the occurrence of inflammation, which in turn is a factor predisposing to the occurrence of depression or the intensification of its symptoms [57]. In clinical studies, no changes in the serum copper level were observed in patients with depression or during remission [58]. However, in patients with high copper levels and depression, memory problems were observed. Preclinical data in an animal stress model show that high levels of copper reduce the growth of dendritic spines in the hippocampus [59]. Limited data are available on copper deficiency, but in several diseases, such as HD and amyotrophic lateral sclerosis, copper may be involved in developing these diseases. Several scientific reports show conflicting results regarding copper in SZ. However, most results conclude that this reduced copper level is due to disrupted copper transport by both ATP7A and ATP7B [60]. Further studies are needed to help understand the impact of copper levels on the pathogenesis of these diseases.

5 Calcium (Ca)

Calcium plays a crucial role in the physiology of the nervous system, including neuronal plasticity. The influx of calcium ions in neurons triggers the presynaptic release of neurotransmitters and, therefore, influences the transmission of many signaling pathways, the regulation of gene expression, membrane excitability, synaptic transmission, the development of dendrites, glial cells, synaptogenesis, and also energy production [61]. These processes mainly affect memory and learning. The regulation of neurotransmitter release by calcium influences forms of synaptic plasticity such as long-term potentiation (LTP) and long-term depression (LTD) [62]. The regulation of intracellular and extracellular calcium concentrations is mediated by voltage-dependent calcium channels (VDCCs), NMDA receptors, and specific Transient Receptor Potential (TRP) channels, along with calcium pumps and exchangers that maintain homeostasis [61]. Intracellular calcium stores such as mitochondria also release calcium ions via receptors located in the endoplasmic reticulum (ER), Ryanodine receptor (RyR), Inositol triphosphate receptor (IP3R), or via sodium-dependent exchange with calcium ions [63]. Membrane ATPases bind to a group of characteristic buffering proteins (calmodulin, calcineurin) and participate in normal calcium homeostasis [64].

VDCCs are activated in response to membrane depolarization. They are responsible for signal processing and regulating intracellular processes in the neuron. Mutations in these channels are associated with numerous diseases and brain disorders. VDCC channels can be divided into several classes: L, P/Q, N, R, and T [65]. Depending on the channel class, their location in the brain and function are diverse. For example, L-type VDCC channels are the most numerous in the mammalian brain. They are located pre- and postsynaptically, forming clusters on dendritic shafts and spines [66]. L-type channels participate in the integration of synaptic signals and initiate the release of neurotransmitters. Structures such as the hippocampus and amygdala nuclei participate in initiating LTP independently of the NMDA receptor [67, 68]. Their significant role is emphasized by the fact that they are involved in several neurodegenerative diseases.

Deregulation of calcium currents is observed during aging, including neuronal degeneration [67]. Blocking L-type channels was associated with improved memory and learning in old mice [69]. These data correlate with the results in patients with dementia. Additionally, blocking these channels reduced the symptoms of PD in a mouse model [70]. Increased expression of these channels is noted in patients with AD. Calcium channel blockers (CCBs) may have a beneficial effect on reducing the occurrence of some psychiatric and neurological disorders. They act on subunits in the L-type channel and may be a promising form of alleviating the symptoms of AD and PD [71]. T-type channels are mainly located in the amygdala, cerebellum, hippocampus, olfactory bulb, and hypothalamus. They participate in both physiological and pathological forms of neuronal rhythmicity [72]. Mutations in one type of this channel are associated with ASD [73]. P/Q and N-type channels are involved in the rapid neuronal transmission of glutamate (Glu), acetylcholine (ACh), and γ-aminobutyric acid (GABA) [61]. Mutations in these channels are responsible for various types of ataxia—congenital or episodic [74]. Another disorder is the genetic disorder—Timothy's syndrome, affecting neurological and behavioral development, including symptoms of ASD [75]. Many studies indicate a broad involvement of P/Q channels in the development of ASD. Other disorders in which mutations in these channels are involved include SZ, depression, and BP [65].

Ligand-gated channels are another group of channels directly or indirectly involved in calcium ion transmission. These include ionotropic glutamate receptors (NMDA, AMPA, kainate), and other nicotinic acetylcholine receptors (nAChR), serotonin 5-HT_3, ATP-gated, and Purinoceptor (P2X). Since glutamate is the most abundant neurotransmitter in the CNS, disruptions in glutamate transmission are associated with numerous neuropsychiatric and neurodegenerative disorders [76]. NMDA receptor activity is associated with many chronic neurodegenerative diseases such as AD, HD, and depression [76]. These channels also participate in the excitotoxic process due to increased calcium ion influx, leading to oxidative stress and neuronal damage [77].

Mitochondria are an essential organelle that plays a regulatory role for calcium ions in the nervous system [63]. Mitochondrial calcium regulates metabolism in the mitochondria, including ATP production and programming cell death. Energy is necessary to maintain proper brain function, which is why disorders in mitochondrial metabolism can lead to impaired brain function [78]. Increased levels of calcium in the cytosol cause its increased uptake by mitochondria. Consequently, mitochondrial calcium overload can play a key role in diseases such as PD, AD, and DLB dementias [79]. Hypercalcemia can be distinguished among calcium metabolism disorders in the body. This is a condition that most often results from primary hyperparathyroidism (PHPT), which affects the distribution of neurotransmitters, usually leading to manic, depressive, and anxiety symptoms [80]. Parathyroidectomy was associated with a lower risk of these neuropsychiatric disorders in patients with PHPT [81].

6 Magnesium (Mg)

In the nervous system, magnesium participates in maintaining ionic homeostasis [82]. Magnesium ions also control synaptic activity and signal transduction by controlling the influx of calcium ions. Magnesium controls calcium influx in two ways: it can inhibit the release of calcium from cytosolic stores and regulate VDCC [83]. In the CNS, magnesium ions affect ionotropic receptors, agonistically on $GABA_AR$, and as a blocker of NMDAR, which is essential in synaptic plasticity [84]. Magnesium also plays a protective role for the BBB and mitochondria—delaying the process of apoptosis and counteracting oxidative stress. Magnesium deficiency contributes to the development of neuroinflammation and neurodegeneration in the brain as a result of initiating a chronic inflammatory state. Reduced magnesium concentration activates numerous cytokines and inflammatory mediators, including Nuclear Factor kappa-light-chain-enhancer of activated B cells (NF-kB), Interleukin 1 (Il-1), Interleukin 6 (Il-6), and Tumor Necrosis Factor alpha (TNF-α) [85].

Numerous in vivo studies show a relationship between magnesium deficiency and chronic inflammation manifested by an increase in proinflammatory cytokines, acute-phase proteins, and an increase in substance P and tachykinin in the CNS and peripheral nervous system (PNS), which maintain inflammation in the body [85]. Low magnesium levels may be responsible for the excitotoxic process. The lack of magnesium ions may increase the activity of NMDA receptors, which contributes to increased Glu signaling [86]. This mechanism may contribute to the development of many neurological and psychiatric disorders, including AD, PD, depression, and anxiety. In anxiety disorders, magnesium is involved in several processes influencing the development of these disorders, including the activity of NMDA receptors and the regulation of the hypothalamic-pituitary adrenal axis (HPA) [87]. There are conflicting data showing the effect of magnesium supplementation on anxiety disorders. Some of them show the potential effect of magnesium supplementation alone and in combination with vitamin B6 on reducing anxiety symptoms and anxiety associated with premenstrual syndrome (PMS) [88]. In the case of depression and anxiety coexisting with depression, supplementing magnesium deficiency has an alleviating effect on depressive states [89]. This effect may involve several mechanisms, such as the protective effect of magnesium in excitotoxicity, interaction with the HPA system, and reduction of inflammation. Magnesium levels in the serum of patients are decreased, but they increase in response to antidepressant treatment, which may be a marker of therapeutic success in patients [87]. However, these data are not entirely consistent, and other studies do not show a correlation between magnesium supplementation and improved mood in patients with depression. Behavioral studies in animals show positive anxiolytic and antidepressant effects of magnesium [90, 91]. Therefore, routes of magnesium administration should be further investigated as a potential target in the treatment of depression.

In PD, where excitotoxicity may play a significant role, contributing to the impairment of dopaminergic pathways, the potential role of magnesium is not sufficiently described. However, there is evidence that higher magnesium concentrations reduced the risk of PD [92]. In preclinical studies in a model of 1-methyl-4-phenyl-1,2,3,6-tetrahydropyridine (MPTP)-induced dopamine neuron damage, magnesium administration ameliorated motor deficits in rodents [93]. However, these limited evidence indicate that further research is needed to better understand the potential association of magnesium therapy with PD. Hypomagnesemia in AD may impair the function of neurons already burdened with pathological forms of beta-amyloid or tau protein in AD. Postmortem tissues of patients are deficient in magnesium in structures such as the hippocampus compared with healthy individuals [94]. As in PD, little data show the effect of magnesium supplementation on a milder course of AD. Preclinical studies provide more evidence that magnesium supplementation has a positive effect. In (amyloid precursors protein) APP mutant rodents, magnesium supplementation has been shown to have several benefits, including reduction of ROS, reduction of synaptic damage in the hippocampus, improvement of spatial memory, and restoration of NMDAR signaling [90, 95, 96]. Magnesium may be one of the alternative treatment options to reduce the symptoms of AD and dementia when detected at an early stage. However, as with other conditions, additional research is still needed.

7 Manganese (Mn)

Manganese toxicity (commonly referred to as *manganism*) is associated with a wide range of neurological disturbances. Elevated manganese levels in the brain have been linked to neurodegenerative conditions such as PD, DLB, AD, and HD. Some studies also suggest possible associations with neurodevelopmental and psychiatric disorders, including ADHD and Tourette syndrome (TS), though these links are less firmly established. Additional symptoms reported in manganese toxicity include cognitive decline, memory impairment, emotional instability, hallucinations, compulsive or aggressive behavior, depression, anxiety, fatigue, insomnia, loss of appetite, apathy, headache, reduced libido, and impaired learning ability [97–100]. Research indicates that manganese not only accumulates in the basal ganglia (striatum, globus pallidus, and substantia nigra) but also in the hippocampus, frontal cortex, and cerebellum following exposure [100, 101]. This suggests that manganese could affect various brain functions [100].

The molecular basis of manganese-induced neurotoxicity is intricate, likely involving various pathways and multiple types of neural cells in the brain. It was shown that the developing brain is more sensitive to manganese accumulation than the adult brain [98]. Moreover, iron deficiency significantly increases manganese absorption, likely due to competition for the divalent metal transporter DMT1 (Divalent Metal Transporter 1), which facilitates cellular uptake of both Fe^{2+} and Mn^{2+} [102]. Manganese penetrates the brain through direct intra-axonal uptake at presynaptic nerve endings, the choroid plexus, or the olfactory epithelium and olfactory nerves [97].

Manganese enters the CNS via direct intra-axonal transport at presynaptic nerve terminals, passage through the choroid plexus, and via the olfactory epithelium and associated neural pathways [103]. Although the extent of manganese accumulation in mitochondria is debated, exposure to elevated manganese levels has been shown to impair mitochondrial respiratory chain function. For example, manganese may inhibit complex I, leading to increased production of ROS [104, 105]. Other research on striatal neurons treated with manganese demonstrated a dose-dependent reduction in mitochondrial membrane potential and complex II activity, suggesting that manganese can induce apoptotic-like neuronal death due to mitochondrial dysfunction and overproduction of reactive nitrogen and ROS [100, 106]. Mitochondria naturally undergo continuous fusion and fission, which is essential for maintaining their shape and function. These processes are regulated by proteins encoded in nuclear DNA, such as OPA1, MFN2 (fusion), and DRP1 (fission). Manganese has been shown to disrupt this balance by reducing OPA1 and MFN2 levels and increasing DRP1 expression, particularly in astrocytes, thereby contributing to mitochondrial fragmentation and dysfunction [107]. Growing evidence also suggests that manganese promotes α-synuclein accumulation, which may be a risk factor for PD pathogenesis due to its toxic effect [100]. While the exact mechanisms are still under investigation, manganese has been shown to activate immune responses in the brain, including releasing IL-1β, IL-6, TNF-α, and proinflammatory chemokines by microglia [108–110]. Astrocytic activation has also been observed, particularly in the substantia nigra (pars reticulata), in manganese-exposed mice [111]. Dysregulation of this structure may be linked to locomotor and emotional disturbances. Astrocytes exposed to increased levels of manganese activate p38 (mitogen-activated protein kinase) and NF-κB (nuclear factor kappa-light-chain-enhancer of activated B cells), further worsening neuroinflammation, which may expand in surrounding tissue [108]. Several studies indicate that manganese disrupts glutamatergic neurotransmission through mechanisms that are not fully understood. Research on manganese's impact on glutamatergic functions, especially involving NMDA receptors, reveals inconsistent results across various experimental conditions, highlighting the receptors' changes [100]. Moreover, elevated levels of manganese in the brain alter the expression of main transporter proteins associated with Glu and GABA turnover. Manganese exposure inhibits glutamine synthetase, particularly in the striatum and globus pallidus (GS, Mn-dependent enzyme predominantly located in astrocytes), which facilitates the conversion of the neurotransmitter glutamate into glutamine (Gln) [112]. Manganese intoxication is associated with impaired neurotransmission of dopamine, norepinephrine, and serotonin. Biochemical analysis showed decreased norepinephrine and serotonin tissue levels with increased dopamine metabolism in the striatum [113]. Furthermore, manganese may intensify dopamine-induced oxidative damage in dopaminergic neurons, as dopamine can oxidize to form quinones and free radicals, which are neurotoxic and contribute to neurodegenerative mechanisms in cells [114].

According to the literature, the treatment of hypermanganesemia, after removal of the source of exposure, is chelation therapy with intravenous disodium calcium edetate. However, this approach is primarily effective while manganese remains in circulation, as $CaNa_2EDTA$ does not readily cross the BBB, limiting its efficacy in cases of CNS toxicity. While some patients exhibit significant progress, others experience less positive results. Certain individuals show no response to treatment, encounter adverse side effects, or continue to decline despite therapeutic interventions. Patients with inherited manganese transport disorders exhibit varying degrees of clinical benefit in response to levodopa treatment [108].

8 Selenium (Se)

Selenium is a ubiquitous element involved in a broad spectrum of biochemical functions in the animal's CNS. Despite being an essential animal micronutrient, selenium has a narrow safety margin, becoming highly toxic when concentrations exceed certain thresholds. Considering these factors, there is a consensus on the various positive impacts of selenium on human health, with a proposed U-shaped non-linear association between selenium levels and health benefits. In general, individuals with initially low selenium concentrations may benefit from supplementation, whereas those with sufficient or elevated levels could experience adverse effects [115]. Se deficiency disrupts oxidative defense, which can contribute to the development or progression of PD, AD, OCD, SZ, depressive disorder, and neuropsychiatric comorbidities frequently observed in ASD, such as anxiety, OCD, or mood disorders [116–121]. SZ, although influenced by multiple environmental and genetic risk factors, may be strongly affected by oxidative stress, which can be induced by lower Se concentration. Selenium deficiency was observed in patients with SZ, depression, and anxiety [115, 122]. Damage to the nervous system due to free radicals includes brain lipid peroxidation, protein carboxylation, DNA damage, and disruption of several neurotransmitter systems, resulting in dysregulation of dopaminergic, glutamatergic, serotonergic, and GABAergic neurotransmission.

Glutathione peroxidases (GPx), which are selenium-dependent enzymes, use reduced glutathione (GSH) as a substrate and are often deficient in patients with SZ. Indirect evidence has indicated that dysfunction of the mitochondrial selenoprotein thioredoxin reductase 2 (TrxR2, essential in redox homeostasis) may also play a role in the development of SZ. Experimental evidence suggests that Se deficiency may enhance dopamine release in the striatum and elevate the turnover rate of dopamine in the substantia nigra, prefrontal cortex, and hippocampus. In contrast, supplementing the diet with selenium decreases the dopamine-degrading enzyme monoamine oxidase (MAO) activity, which is also responsible for producing peroxides [115, 123]. A recent study showed that levels of selenium were correlated with the severity of anorexia nervosa and that its deficiency was linked to a higher risk of suicidality in affected individuals. Neurobiological alterations in the dopamine system may influence the observed variations [124]. In one study involving 21 schizophrenic patients with initially low serum selenium levels, the dose of 60 μg of selenium daily, derived from selenium-enriched yeast, brought functional improvement. After 3 months, patients exhibited increased appetite and improved memory [122]. Animal studies also show changes in the functioning of brain glia depending on the level of selenium. The neuroprotective role of selenium supplementation in lipopolysaccharide (LPS)-induced neuroinflammation was presented by [125]. Se improved integrity of the BBB and suppressed microglia activation after LPS-challenged animals. Mice lacking selenoprotein R (encoded by the MsrB1 gene) exhibit altered synaptic plasticity and increased astrogliosis. Dysfunction of Msr proteins is strongly associated with the development of neurodegenerative diseases. The lack of MsrA in the mouse AD model significantly increases the level of amyloid-beta (Aβ) [126]. Several other studies on stressed rodents have shown promising results in selenium supplementation, preventing the development of depressive-like behaviors in chronically stressed animals [127]. Selenium supplementation has lowered plasma corticosterone levels and diminished lipid peroxidation by regulating oxidative pathways and neuroendocrine functions. It also reduced anxiety-like behavior and normalized the activity of superoxide dismutase and catalase in the mouse hippocampus and prefrontal cortex [123]. Selenium plays a crucial role in thyroid hormone metabolism; it is hypothesized that selenium imbalance in thyroid disorders may contribute to the development of anxiety, OCD, or depressive disorder [123, 128].

Although selenium is an essential trace element for proper brain function, its uptake by the neuronal tissue should be strictly regulated to prevent toxicity due to its narrow safety margin, which was mentioned earlier. This should be considered, especially during the loss of BBB integrity, which promotes increased brain penetration to metabolites, inorganic ions, and proteins. The brain is strongly protected from selenium deficiency and accumulates selenium under its deficiency condition immediately after administration [129]. Exposure to high levels of selenium may disrupt glutamatergic, dopaminergic, serotonergic, and cholinergic systems and dysregulate signaling molecules that control and regulate cognitive, behavioral, and neuro-endocrine functions and may contribute to the development of several neurodegenerative and neuropsychiatric diseases [130].

Selenium deficiency can be completed through supplementation. Most dietary forms of selenium are efficiently absorbed, with organic forms such as selenomethionine and selenocysteine showing the highest bioavailability. This effect can be enhanced by administering it with proteins, vitamins E and A. Moreover, there is a question if a selenium-based drug design may have a therapeutic potential. Ebselen was the first synthetic organic selenium compound developed as a glutathione peroxidase mimetic. It was considered as a potential pharmacological agent. While the exact mechanisms behind its neuroprotective effects remain unclear, they are believed to be linked to its antioxidant and anti-inflammatory properties [131].

9 Lithium (Li)

Although lithium is not officially considered to be a micronutrient for human health, it is widely used in psychopharmacology, particularly for BD treatment [132]. Lithium is also used in treatment-resistant depression to enhance the effectiveness of standard antidepressant therapies, and it has shown benefits in cases of increased impulsiveness and agitation [132, 133]. While not considered a first-line treatment, lithium has occasionally been used as an adjunct therapy in treatment-resistant OCD, especially when standard treatments like SSRIs and cognitive-behavioral therapy have not been fully effective [134].

Clinical and epidemiological data suggest that lithium may reduce suicidal behavior, potentially through both its mood-stabilizing properties and mechanisms independent of mood regulation. A postmortem biochemical study reported that lithium levels in the ocular fluids (aqueous humor) were lower in individuals who died by suicide compared to non-suicidal controls, suggesting that even a low dose of lithium may have a protective, anti-suicidal effect [135, 136]. Lithium's well-established mood-stabilizing effects contribute to the reduction in affective episodes among responsive patients, thereby decreasing the risk of suicide. Alterations in corticotropin-releasing hormone and dysfunctions in the noradrenergic and serotonergic systems have been linked to suicidal ideation due to reduced levels of cerebral tryptophan and serotonin.

Lithium may help regulate the HPA is by modulating CRH expression and downstream signaling. Its serotonin-agonistic properties mitigate serotonin deficiency and may influence serotonin metabolism [132, 136]. Upregulation of platelet serotonin 2_A receptors in individuals with psychiatric disorders is associated with changes in suicidal behavior. Lithium may enhance the activity of glutamate and GABA systems and increase $GABA_b$ receptors in the hippocampus. Some studies suggest that lithium may influence glutamine synthetase activity and brain glutamine levels, potentially reducing suicidal tendencies [136]. Intracellular accumulation of lithium can partially replace sodium ions, which may contribute to the modulation of intracellular calcium signaling. This action can reduce calcium influx and promote the uptake of key neurotransmitters such as norepinephrine, serotonin, and dopamine. Lithium has been observed to regulate the activity of various neurotransmitters, including glutamate, dopamine, serotonin, gamma-aminobutyric acid (GABA), acetylcholine, and glycine. Moreover, lithium enhances the expression of the BDNF (brain-derived neurotrophic factor) and its receptor. By modulating the expression of clock genes (*TIMELESS*, *ARNTL1*, *PER3*, *NR1D1*, *CLOCK*), lithium can potentially normalize circadian rhythms and suppress impulsive behavior. Its inhibitory effects on adenylyl cyclase and modification of inositol content might also play a role in modulating suicidal behavior; patients with BD treated with lithium exhibit increased hippocampal volumes, even after short-term use [132, 136].

10 Chromium (Cr)

Trivalent chromium (Cr(III)) was historically considered an essential trace element, believed to play a role in insulin-mediated regulation of carbohydrate, lipid, and protein metabolism. Despite this, chromium's precise mechanisms and function in metabolic pathways remain unconfirmed. Consequently, the European Food Safety Authority (EFSA) removed chromium from the list of essential micronutrients in its 2014 "Dietary Reference Values for Nutrients Summary Report" [137]. As the conception of chromium's role in the organism constantly changes, this decision remains controversial. Chromium has shown potential in the treatment of atypical depression. Proposed mechanisms include the downregulation of 5HT2A receptors, changes in monoamine neurotransmission, and improvement in insulin sensitivity [138].

Unfortunately, excessive chromium exposure due to its absorption through the nasal membrane among people who are environmentally or occupationally exposed to chromium dust can induce brain injury. Animal studies have demonstrated that sub-chronic chromium exposure increases serum and brain concentrations of tryptophan, corticosterone, serotonin, noradrenaline, and pineal melatonin [139]. Additional findings show that chromium accumulates in the hypothalamus, and prolonged exposure results in more severe neuropathological changes. Mice exposed to chromium for 28 days displayed more significant neuronal damage than those exposed for 14 days. Observed changes included gliosis, neuronal morphological alterations, synaptic degeneration, BBB disruption, reduced neuronal counts in the hypothalamus, impaired Nrf2-mediated antioxidant signaling, and activation of the NF-κB inflammatory pathway [140]. Other toxicological research has reported increased oxidative stress biomarkers such as MDA (malondialdehyde), proinflammatory cytokine-IL-1β, and a lower concentration of brain catalase and GSH. The adverse effects correlate with chromium accumulation in neural tissue [141]. Additional animal research has shown that chromium exposure impairs acetylcholinesterase activity and contributes to neurodegeneration via oxidative stress mechanisms. In humans, chromium exposure has been associated with various neurological and psychiatric conditions, social memory deficits, depression, SZ, ASD, olfactory impairments, and polyneuropathy. More research is required to determine chromium-induced neurotoxicity's scope and underlying mechanisms [142].

Currently, there is no effective therapy for chromium toxicity. Severe chromium poisoning typically results in death despite medical intervention. Management of acute chromium exposure primarily involves supportive and symptomatic care. Therapeutic strategies may include venovenous hemofiltration, plasmapheresis, administration of a chelating agent—dimercaptopropane sulfonic acid—and antioxidants such as N-acetylcysteine and ascorbic acid [143, 144].

11 Molybdenum (Mo)

Molybdenum is a trace element that plays an important role in the nervous system, primarily through the activity of molybdenum-dependent enzymes belonging to two families—SO (sulfite oxidase) and XO (xanthine oxidase). The SO family includes enzymes such as sulfite oxidase and the amidoxime-reducing component (mARC). Disturbances in this enzyme's activity lead to sulfide accumulation, which leads to oxidative damage to neurons, particularly harmful to the developing brain. The XO family comprises two closely related enzymes: xanthine dehydrogenase/oxidase (XDH/XO) and aldehyde oxidase (AOX), which share significant homology. Xanthine oxidase (XO) is key to purine metabolism, where it converts hypoxanthine to xanthine and then xanthine to uric acid. The impaired function of this enzyme leads to dysregulation of purine metabolism, which disrupts neurotransmitter balance and increases the risk of neuroinflammatory responses in the brain. Through these enzyme systems, the molybdenum helps regulate key metabolic pathways vital for maintaining neuronal health [145]. Molybdenum deficiency in humans is very rare. Clinical manifestations have included irritability, tachycardia, tachypnea, and night blindness [146]. Severe symptoms of molybdenum deficiency are associated with a rare genetic disorder affecting molybdopterin production, known as molybdenum cofactor deficiency (MoCD). This deficiency results in the buildup of harmful metabolites, especially sulfite, which can lead to neurological damage [147, 148].

The molecular basis of neurodegeneration in this disease is the accumulation of sulfites, which initially build up in mitochondria, leading to an increase in the production of ROS and decreased ATP synthesis. This mechanism is linked to the inhibition of glutamate dehydrogenase by sulfite, which in the brain reduces the availability of α-ketoglutarate and other compounds of the Krebs cycle [149]. Inhibition of glutamate dehydrogenase also affects glutamate metabolism, which, in excess, exhibits neurotoxic effects. Since glutamate is a precursor of the inhibitory neurotransmitter GABA, disturbances in its metabolism lead to an imbalance between these two neurotransmitters, which may explain the observed neuronal damage. At a later stage, sulfite crosses the outer mitochondrial membrane, accumulates in the cytosol, and eventually in the plasma. Sulfite disrupts disulfide bonds, primarily in membrane and extracellular proteins, affecting their activity. Sulfite likely reacts with cystine, forming the secondary metabolite S-sulfocysteine (SSC) [145]. SSC is structurally similar to glutamate, binds to NMDA receptors, and is considered the primary agent responsible for seizure development and subsequent brain damage in MoCD [150]. Studies in rats have demonstrated that subcutaneous SSC administration induces brain damage similar to that caused by glutamate and other excitatory amino acids. The formation of SSC is accompanied by cystine depletion. In the brain, glial cells take up cystine, reduce it, and incorporate it into glutathione, the key antioxidant in nervous tissue [151].

Studies suggest that molybdenum deficiency may be linked to the development of neurodegenerative diseases, such as PD and AD, due to its impact on oxidative stress and neurotoxicity [152]. Studies conducted on sheep, whose purine metabolism is similar to that of humans, suggest that molybdenum deficiency allows certain dietary purines to accumulate in the CNS. This purine buildup negatively impacts astrocyte function, which is the primary consequence of purine overload. Neurodegeneration is likely a secondary consequence of this disruption [152]. According to recent findings by [153], molybdenum may provide a protective benefit against SZ, as serum molybdenum levels were significantly lower in patients with SZ compared to healthy controls. This leads to the hypothesis that the active form of molybdenum could exert a neuroprotective effect and that disruptions in molybdenum metabolism may contribute to the pathophysiology of SZ [153]. While the data suggest a possible association between low serum molybdenum levels and SZ, further studies are needed to determine whether this relationship is causal or a secondary effect of the disease.

Botchway et al. (2023) [154] highlight the necessity of investigating the role of trace biometals in AD. Their review explores molybdenum's potential role as a candidate for therapeutic intervention and a valuable diagnostic marker for the disease. Molybdenum is involved in the formation of uric acid, and some studies have investigated the relationship between serum uric acid and cognition. They noted a correlation between increased serum uric acid and a lower risk of developing AD and improvement in cognitive decline, but this hypothesis is still being debated [154–157]. Certain molybdenum-containing compounds have been shown to inhibit β-amyloid aggregation and regulate oxidative stress [154, 158–161].

Although molybdenum is highly toxic to some animals, it poses minimal toxicity risk to humans. However, case reports from regions such as Armenia, where soil molybdenum levels are unusually high, describe symptoms such as joint pain, gout-like manifestations, hyperuricosuria, and headaches, particularly in individuals with occupational exposure [146].

12 Potassium and Sodium

Potassium and sodium are the main cations in neurons and are crucial in maintaining membrane potential, essential for proper nerve impulse conduction and metabolic processes. Potassium channels, sodium channels, and the sodium-potassium pump are key components in the depolarization and repolarization of nerve cells. These channels are distributed in both excitatory and inhibitory neurons [162]. Disturbances in sodium and potassium homeostasis have been observed in some neurodegenerative and psychiatric disorders. In a post-mortem study comparing brain samples from 12 neurologically healthy individuals and 16 patients with AD, significant increases in sodium levels were reported in the frontal and parietal cortex (by 25% and 20%, respectively), as well as increased potassium levels in the cerebellum (by 15%). Since no differences were detected in the elemental composition of cerebrospinal fluid, the changes were interpreted as reflecting intracellular ionic shifts [163]. These findings are consistent with earlier reports using in vivo sodium MRI, which showed slightly elevated sodium levels in the hippocampal formations of AD patients [164]. What is interesting is that 20% of acutely hospitalized psychiatric patients showed hypokalemia [165], and similar findings have been reported in one-third of individuals with SZ [166].

Dysfunctions of various potassium channels are implicated in the pathogenesis of both AD and PD. For example, upregulation of the voltage-gated potassium channel Kv3.4 and dysregulation of Kv3.1 have been linked to altered synaptic activity that may underlie neurodegeneration in AD [167]. In addition, the calcium-activated potassium channel KCa3.1 blockade has been shown to exert neuroprotective and anti-inflammatory effects, suggesting therapeutic potential for AD, cerebrovascular disease, and traumatic brain injury [168].

Concerning sodium, it has been demonstrated that pathological increases in intracellular sodium impair the function of sodium-dependent transporters in astrocytes. Under physiological conditions, these transporters mediate the uptake of neurotransmitters such as GABA and glycine. However, these transporters may reverse direction in sodium-overloaded cells, releasing neurotransmitters into the extracellular space. Sodium accumulation also impairs glutamate uptake, leading to extracellular glutamate buildup, which is neurotoxic—especially during conditions such as cerebral ischemia. Elevated intracellular sodium levels can also cause astrocyte swelling, a phenomenon seen in stroke, seizures, and hepatic encephalopathy. More generally, sodium imbalance contributes to cerebral edema and disruption of ionic [163].

Mutations in genes encoding sodium channels, such as *SCN1A*, *SCN2A*, *SCN3A*, and *ATP1A2*, have been identified in individuals with ASD, familial hemiplegic migraine, and epileptic seizures. Mutations in *ATP1A2* disrupt Na^+/K^+-ATPase activity, contributing to ion imbalance, increased seizure susceptibility, and other neurological symptoms [168, 169]. Sodium dysregulation is also observed in cerebrovascular diseases. Hyponatremia frequently follows subarachnoid hemorrhage and may exacerbate symptoms such as lethargy, confusion, seizures, and even coma [170]. In vascular dementia (VaD), postmortem brain tissue analysis has revealed increased cerebral sodium levels across multiple brain regions (including the basal ganglia, cingulate gyrus, frontal gyrus, middle temporal gyrus, occipital cortex, and hippocampus), accompanied by potassium reductions in cingulate gyrus and hippocampus and a changed Na^+/K^+ ratio, suggesting impaired Na^+/K^+-ATPase function [169, 171].

Autoimmune processes also contribute to neuropsychiatric symptoms through disruption of ion channel function. Autoantibodies targeting macromolecular complexes containing potassium channels (e.g., LGI1 and Caspr2 within the VGKC complex) can trigger rapid-onset psychiatric symptoms such as irritability, panic attacks, OCD, sleep disturbances, hallucinations, personality changes, and cognitive decline [169, 172, 173].

Ion channels are essential to psychiatric disorders, and some current treatments target them. Commonly used psychiatric medications modulate ion channel function and thereby affect neurotransmitters such as dopamine, serotonin, glutamate, and GABA. However, these therapies often show limited efficacy and are associated with adverse effects, prompting ongoing development of more selective ion channel modulators [169, 174, 175].

Conflict of Interest

The authors have no conflicts of interest to declare that are relevant to the content of this chapter.

References

1. Lachowicz JI, Lecca LI, Meloni F, Campagna M (2021) Metals and metal-nanoparticles in human pathologies: from exposure to therapy. Molecules 26:6639. https://doi.org/10.3390/molecules26216639
2. Frederickson CJ, Suh SW, Silva D et al (2000) Importance of zinc in the central nervous system: the zinc-containing neuron. J Nutr 130:1471S–1483S. https://doi.org/10.1093/jn/130.5.1471S
3. Szewczyk B (2013) Zinc homeostasis and neurodegenerative disorders. Front Aging Neurosci 5. https://doi.org/10.3389/fnagi.2013.00033
4. Li Z, Liu Y, Wei R et al. (2023) The important role of zinc in neurological diseases. Biomolecules 13:28. https://doi.org/10.3390/biom13010028
5. Ranjbar E, Kasaei MS, Mohammad-Shirazi M et al. (2013) Effects of zinc supplementation in patients with major depression: a randomized clinical trial. Iran J Psychiatry 8:73–79
6. Sindreu C, Storm DR (2011) Modulation of neuronal signal transduction and

memory formation by synaptic zinc. Front Behav Neurosci 5:68. https://doi.org/10.3389/fnbeh.2011.00068
7. Takeda A (2001) Zinc homeostasis and functions of zinc in the brain. Biometals 14:343–351. https://doi.org/10.1023/A:1012982123386
8. Młyniec K, Budziszewska B, Reczyński W et al. (2013) The role of the GPR39 receptor in zinc deficient-animal model of depression. Behav Brain Res 238:30–35. https://doi.org/10.1016/j.bbr.2012.10.020
9. Amico-Ruvio SA, Murthy SE, Smith TP, Popescu GK (2011) Zinc effects on NMDA receptor gating kinetics. Biophys J 100:1910–1918. https://doi.org/10.1016/j.bpj.2011.02.042
10. Doboszewska U, Sowa-Kućma M, Młyniec K et al. (2015) Zinc deficiency in rats is associated with up-regulation of hippocampal NMDA receptor. Prog Neuro-Psychopharmacol Biol Psychiatry 56:254–263. https://doi.org/10.1016/j.pnpbp.2014.09.013
11. Takeda A, Tamano H, Nishio R, Murakami T (2016) Behavioral abnormality induced by enhanced hypothalamo-pituitary-adrenocortical axis activity under dietary zinc deficiency and its usefulness as a model. Int J Mol Sci 17:1149. https://doi.org/10.3390/ijms17071149
12. Pochwat B, Misztak P, Masternak J, et al. (2022) Combined hyperforin and lanicemine treatment instead of ketamine or imipramine restores behavioral deficits induced by chronic restraint stress and dietary zinc restriction in mice. Front Pharmacol 13, 933364. https://doi.org/10.3389/fphar.2022.933364
13. Doboszewska U, Szewczyk B, Sowa-Kućma M et al. (2016) Alterations of bio-elements, oxidative, and inflammatory status in the zinc deficiency model in rats. Neurotox Res 29:143–154. https://doi.org/10.1007/s12640-015-9571-7
14. Gąsior Ł, Pochwat B, Zaręba-Kozioł M et al. (2025) Proteomics analysis in rats reveals convergent mechanisms between major depressive disorder and dietary zinc deficiency. Pharmacol Rep 77:145–157. https://doi.org/10.1007/s43440-024-00681-7
15. Bederska-Łojewska D, Szczepanik K, Turek J et al. (2024) Dietary zinc restriction and chronic restraint stress affect mice physiology, immune organ morphology, and liver function. Nutrients 16:3934. https://doi.org/10.3390/nu16223934
16. Rafało-Ulińska A, Pochwat B, Misztak P et al. (2022) Zinc deficiency blunts the effectiveness of antidepressants in the olfactory bulbectomy model of depression in rats. Nutrients 14:2746. https://doi.org/10.3390/nu14132746
17. Swardfager W, Herrmann N, Mazereeuw G et al. (2013) Zinc in depression: a meta-analysis. Biol Psychiatry 74:872–878. https://doi.org/10.1016/j.biopsych.2013.05.008
18. Siwek M, Dudek D, Schlegel-Zawadzka M et al. (2010) Serum zinc level in depressed patients during zinc supplementation of imipramine treatment. J Affect Disord 126:447–452. https://doi.org/10.1016/j.jad.2010.04.024
19. Shohag H, Ullah A, Qusar S et al. (2012) Alterations of serum zinc, copper, manganese, iron, calcium, and magnesium concentrations and the complexity of interelement relations in patients with obsessive–compulsive disorder. Biol Trace Elem Res 148:275–280. https://doi.org/10.1007/s12011-012-9371-3
20. Ghoreishy SM, Ebrahimi Mousavi S, Asoudeh F, Mohammadi H (2021) Zinc status in attention-deficit/hyperactivity disorder: a systematic review and meta-analysis of observational studies. Sci Rep 11: 14612. https://doi.org/10.1038/s41598-021-94124-5
21. Arnold LE, DiSilvestro RA, Bozzolo D et al. (2011) Zinc for attention-deficit/hyperactivity disorder: placebo-controlled double-blind pilot trial alone and combined with amphetamine. J Child Adolesc Psychopharmacol 21:1–19. https://doi.org/10.1089/cap.2010.0073
22. Joe P, Petrilli M, Malaspina D, Weissman J (2018) Zinc in schizophrenia: a meta-analysis. Gen Hosp Psychiatry 53:19–24. https://doi.org/10.1016/j.genhosppsych.2018.04.004
23. Mortazavi M, Farzin D, Zarhghami M, et al. (2015) Efficacy of zinc sulfate as an add-on therapy to risperidone versus risperidone alone in patients with schizophrenia: a double-blind randomized placebo-controlled trial. Iran J Psychiatry Behav Sci 9, e853. https://doi.org/10.17795/ijpbs-853
24. Watt NT, Whitehouse IJ, Hooper NM (2011) The role of zinc in Alzheimer's disease. Int J Alzheimer's Dis 2011:971021. https://doi.org/10.4061/2011/971021
25. Wang Z-X, Tan L, Wang H-F et al. (2015) Serum iron, zinc, and copper levels in patients with alzheimer's disease: a replication study and meta-analyses. J Alzheimer's Dis 47:565–581. https://doi.org/10.3233/JAD-143108
26. Lyubartseva G, Smith JL, Markesbery WR, Lovell MA (2010) Alterations of zinc transporter proteins znt-1, znt-4 and znt-6 in preclinical alzheimer's disease brain. Brain Pathol 20:343–350. https://doi.org/10.1111/j.1750-3639.2009.00283.x

27. Xu Y, Xiao G, Liu L, Lang M (2019) Zinc transporters in Alzheimer's disease. Mol Brain 12: 106. https://doi.org/10.1186/s13041-019-0528-2
28. Zhang H-L, Wang X-C, Liu R (2022) Zinc in regulating protein kinases and phosphatases in neurodegenerative diseases. Biomolecules 12:785. https://doi.org/10.3390/biom12060785
29. Juárez-Rebollar D, Rios C, Nava-Ruíz C, Méndez-Armenta M (2017) Metallothionein in brain disorders. Oxid Med Cell Longev 2017:5828056. https://doi.org/10.1155/2017/5828056
30. Schoofs H, Schmit J, Rink L (2024) Zinc toxicity: understanding the limits. Molecules 29:3130. https://doi.org/10.3390/molecules29133130
31. Nnah IC, Wessling-Resnick M (2018) Brain iron homeostasis: a focus on microglial iron. Pharmaceuticals 11:129. https://doi.org/10.3390/ph11040129
32. Moos T, Nielsen TR, Skjørringe T, Morgan EH (2007) Iron trafficking inside the brain. J Neurochem 103:1730–1740. https://doi.org/10.1111/j.1471-4159.2007.04976.x
33. McCarthy RC, Kosman DJ (2015) Iron transport across the blood–brain barrier: development, neurovascular regulation and cerebral amyloid angiopathy. Cell Mol Life Sci 72:709–727. https://doi.org/10.1007/s00018-014-1771-4
34. Zecca L, Youdim MBH, Riederer P et al. (2004) Iron, brain ageing and neurodegenerative disorders. Nat Rev Neurosci 5:863–873. https://doi.org/10.1038/nrn1537
35. Chen L, Shen Q, Liu Y et al. (2025) Homeostasis and metabolism of iron and other metal ions in neurodegenerative diseases. Sig Transduct Target Ther 10:1–48. https://doi.org/10.1038/s41392-024-02071-0
36. Schreiner OD, Schreiner TG (2023) Iron chelators as a therapeutic option for Alzheimer's disease—a mini-review. Front Aging 4, 1234958. https://doi.org/10.3389/fragi.2023.1234958
37. Pivina L, Semenova Y, Doşa MD et al. (2019) Iron deficiency, cognitive functions, and neurobehavioral disorders in children. J Mol Neurosci 68:1–10. https://doi.org/10.1007/s12031-019-01276-1
38. Tseng P-T, Cheng Y-S, Yen C-F et al. (2018) Peripheral iron levels in children with attention-deficit hyperactivity disorder: a systematic review and meta-analysis. Sci Rep 8: 788. https://doi.org/10.1038/s41598-017-19096-x
39. Kim J, Wessling-Resnick M (2014) Iron and mechanisms of emotional behavior. J Nutr Biochem 25:1101–1107. https://doi.org/10.1016/j.jnutbio.2014.07.003
40. Cao B, Yan L, Ma J et al. (2019) Comparison of serum essential trace metals between patients with schizophrenia and healthy controls. J Trace Elem Med Biol 51:79–85. https://doi.org/10.1016/j.jtemb.2018.10.009
41. Kim S-W, Stewart R, Park W-Y et al. (2018) Latent iron deficiency as a marker of negative symptoms in patients with first-episode schizophrenia spectrum disorder. Nutrients 10:1707. https://doi.org/10.3390/nu10111707
42. Berthou C, Iliou JP, Barba D (2022) Iron, neuro-bioavailability and depression. eJHaem 3:263–275. https://doi.org/10.1002/jha2.321
43. Saeidienik F, Shahraki MR, Fanaei H, Badini F (2018) The effects of iron oxide nanoparticles administration on depression symptoms induced by LPS in male wistar rats. Basic Clin Neurosci 9:209–216. https://doi.org/10.29252/NIRP.BCN.9.3.209
44. Levi S, Tiranti V (2019) Neurodegeneration with brain iron accumulation disorders: valuable models aimed at understanding the pathogenesis of iron deposition. Pharmaceuticals 12:27. https://doi.org/10.3390/ph12010027
45. Desai V, Kaler SG (2008) Role of copper in human neurological disorders. Am J Clin Nutr 88:855S–858S. https://doi.org/10.1093/ajcn/88.3.855S
46. Osredkar J (2011) Copper and Zinc, biological role and significance of copper/zinc imbalance. J Clin Toxicol 3(2161):0495. https://doi.org/10.4172/2161-0495.S3-001
47. Manchia M, Paribello P, Pinna M, Faa G (2024) The role of copper overload in modulating neuropsychiatric symptoms. Int J Mol Sci 25:6487. https://doi.org/10.3390/ijms25126487
48. Telianidis J, Hung YH, Materia S, La Fontaine S (2013) Role of the P-Type ATPases, ATP7A and ATP7B in brain copper homeostasis. Front Aging Neurosci 5:44. https://doi.org/10.3389/fnagi.2013.00044
49. Scassellati C, Bonvicini C, Benussi L et al. (2020) Neurodevelopmental disorders: metallomics studies for the identification of potential biomarkers associated to diagnosis and treatment. J Trace Elem Med Biol 60:126499. https://doi.org/10.1016/j.jtemb.2020.126499
50. Kitzberger R, Madl C, Ferenci P (2005) Wilson disease. Metab Brain Dis 20:295–302. https://doi.org/10.1007/s11011-005-7910-8

51. Dzieżyc-Jaworska K, Litwin T, Członkowska A (2019) Clinical manifestations of Wilson disease in organs other than the liver and brain. Ann Transl Med 7:S62. https://doi.org/10.21037/atm.2019.03.30
52. Abuduxikuer K, Wang J-S (2014) Zinc mono-therapy in pre-symptomatic chinese children with wilson disease: a single center, retrospective study. PLoS One 9:e86168. https://doi.org/10.1371/journal.pone.0086168
53. Avan A, Członkowska A, Gaskin S et al. (2022) The role of Zinc in the treatment of Wilson's disease. Int J Mol Sci 23:9316. https://doi.org/10.3390/ijms23169316
54. Hordyjewska A, Popiołek Ł, Kocot J (2014) The many "faces" of copper in medicine and treatment. Biometals 27:611–621. https://doi.org/10.1007/s10534-014-9736-5
55. Park J-H, Lee D-W, Park KS (2014) Elevated serum copper and ceruloplasmin levels in Alzheimer's disease. Asia-Pacific Psychiatry 6: 38–45. https://doi.org/10.1111/appy.12077
56. An Y, Li S, Huang X et al. (2022) The role of Copper homeostasis in brain disease. Int J Mol Sci 23:13850. https://doi.org/10.3390/ijms232213850
57. Chen J, Song W, Zhang W (2023) The emerging role of copper in depression. Front Neurosci 17, 1230404. https://doi.org/10.3389/fnins.2023.1230404
58. Styczeń K, Sowa-Kućma M, Siwek M et al. (2016) Study of the serum Copper levels in patients with major depressive disorder. Biol Trace Elem Res 174:287–293. https://doi.org/10.1007/s12011-016-0720-5
59. Liu X, Lin C, Wang S et al. (2023) Effects of high levels of Copper on the depression-related memory disorders. J Gerontol A 78:611–618. https://doi.org/10.1093/gerona/glac222
60. Schoonover KE, Queern, Stacy L, Lapi,Suzanne E, Roberts RC (2020) Impaired copper transport in schizophrenia results in a copper-deficient brain state: A new side to the dysbindin story. W J Biol Psychiatry 21:13–28. https://doi.org/10.1080/15622975.2018.1523562
61. Kawamoto EM, Vivar C, Camandola S (2012) Physiology and pathology of calcium signaling in the brain. Front Pharmacol 3:22359. https://doi.org/10.3389/fphar.2012.00061
62. Chindemi G, Abdellah M, Amsalem O et al. (2022) A calcium-based plasticity model for predicting long-term potentiation and depression in the neocortex. Nat Commun 13:3038. https://doi.org/10.1038/s41467-022-30214-w
63. Giorgi C, Marchi S, Pinton P (2018) The machineries, regulation and cellular functions of mitochondrial calcium. Nat Rev Mol Cell Biol 19:713–730. https://doi.org/10.1038/s41580-018-0052-8
64. Zaidi A, Barŕon L, Sharov VS et al. (2003) Oxidative inactivation of purified plasma membrane Ca2+-ATPase by hydrogen peroxide and protection by calmodulin. Biochemistry 42:12001–12010. https://doi.org/10.1021/bi034565u
65. Nanou E, Catterall WA (2018) Calcium channels, synaptic plasticity, and neuropsychiatric disease. Neuron 98:466–481. https://doi.org/10.1016/j.neuron.2018.03.017
66. Obermair GJ, Szabo Z, Bourinet E, Flucher BE (2004) Differential targeting of the L-type Ca2+ channel α1C (CaV1.2) to synaptic and extrasynaptic compartments in hippocampal neurons. Eur J Neurosci 19:2109–2122. https://doi.org/10.1111/j.0953-816X.2004.03272.x
67. Chan CS, Gertler TS, Surmeier DJ (2009) Calcium homeostasis, selective vulnerability and Parkinson's disease. Trends Neurosci 32:249–256. https://doi.org/10.1016/j.tins.2009.01.006
68. Iacobucci GJ, Popescu GK (2024) Calcium- and calmodulin-dependent inhibition of NMDA receptor currents. Biophys J 123:277–293. https://doi.org/10.1016/j.bpj.2023.12.018
69. Deisseroth K, Heist EK, Tsien RW (1998) Translocation of calmodulin to the nucleus supports CREB phosphorylation in hippocampal neurons. Nature 392:198–202. https://doi.org/10.1038/32448
70. Chan CS, Guzman JN, Ilijic E et al. (2007) 'Rejuvenation' protects neurons in mouse models of Parkinson's disease. Nature 447:1081–1086. https://doi.org/10.1038/nature05865
71. Colbourne L, Harrison PJ (2022) Brain-penetrant calcium channel blockers are associated with a reduced incidence of neuropsychiatric disorders. Mol Psychiatry 27:3904–3912. https://doi.org/10.1038/s41380-022-01615-6
72. Iftinca MC, Zamponi GW (2009) Regulation of neuronal T-type calcium channels. Trends Pharmacol Sci 30:32–40. https://doi.org/10.1016/j.tips.2008.10.004
73. Splawski I, Yoo DS, Stotz SC et al. (2006) CACNA1H mutations in autism spectrum disorders *. J Biol Chem 281:22085–22091. https://doi.org/10.1074/jbc.M603316200
74. Cain SM, Snutch TP (2011) Voltage-gated calcium channels and disease. BioFactors

37:197–205. https://doi.org/10.1002/biof.158

75. Dick IE, Joshi-Mukherjee R, Yang W, Yue DT (2016) Arrhythmogenesis in Timothy syndrome is associated with defects in Ca2+-dependent inactivation. Nat Commun 7: 10370. https://doi.org/10.1038/ncomms10370
76. Foster TC (2007) Calcium homeostasis and modulation of synaptic plasticity in the aged brain. Aging Cell 6:319–325. https://doi.org/10.1111/j.1474-9726.2007.00283.x
77. Zhou X, Hollern D, Liao J et al. (2013) NMDA receptor-mediated excitotoxicity depends on the coactivation of synaptic and extrasynaptic receptors. Cell Death Dis 4:e560–e560. https://doi.org/10.1038/cddis.2013.82
78. Norat P, Soldozy S, Sokolowski JD et al. (2020) Mitochondrial dysfunction in neurological disorders: Exploring mitochondrial transplantation. NPJ Regen Med 5:1–9. https://doi.org/10.1038/s41536-020-00107-x
79. Verma M, Wills Z, Chu CT (2018) Excitatory dendritic mitochondrial calcium toxicity: implications for Parkinson's and other neurodegenerative diseases. Front Neurosci 12:523. https://doi.org/10.3389/fnins.2018.00523
80. Serdenes R, Lewis M, Chandrasekhara S (2021) A clinical review of the psychiatric sequelae of primary hyperparathyroidism. Cureus 13(8), e19078. https://doi.org/10.7759/cureus.19078
81. Song Z, Balachandra S, Wu C et al. (2025) Risk of neuropsychiatric disorders in primary hyperparathyroidism: parathyroidectomy versus nonoperative management. World J Surg 49:106–114. https://doi.org/10.1002/wjs.12285
82. Kirkland AE, Sarlo GL, Holton KF (2018) The role of magnesium in neurological disorders. Nutrients 10:730. https://doi.org/10.3390/nu10060730
83. Maier JAM, Locatelli L, Fedele G et al. (2022) Magnesium and the brain: a focus on neuroinflammation and neurodegeneration. Int J Mol Sci 24:223. https://doi.org/10.3390/ijms24010223
84. Stroebel D, Casado M, Paoletti P (2018) Triheteromeric NMDA receptors: from structure to synaptic physiology. Curr Opin Physiol 2:1–12. https://doi.org/10.1016/j.cophys.2017.12.004
85. Maier JA, Castiglioni S, Locatelli L et al. (2021) Magnesium and inflammation: advances and perspectives. Semin Cell Dev Biol 115:37–44. https://doi.org/10.1016/j.semcdb.2020.11.002
86. Vink R, Nechifor M (2011) Magnesium in the central nervous system. University of Adelaide Press, Adelaide (AU)
87. Boyle NB, Lawton C, Dye L (2017) The effects of magnesium supplementation on subjective anxiety and stress—a systematic review. Nutrients 9:429. https://doi.org/10.3390/nu9050429
88. Khine K, Rosenstein DL, Elin RJ et al. (2006) Magnesium (Mg) retention and mood effects after intravenous Mg infusion in premenstrual dysphoric disorder. Biol Psychiatry 59:327–333. https://doi.org/10.1016/j.biopsych.2005.07.022
89. Eby GA, Eby KL (2006) Rapid recovery from major depression using magnesium treatment. Med Hypotheses 67:362–370. https://doi.org/10.1016/j.mehy.2006.01.047
90. Poleszak E, Szewczyk B, Kędzierska E et al. (2004) Antidepressant- and anxiolytic-like activity of magnesium in mice. Pharmacol Biochem Behav 78:7–12. https://doi.org/10.1016/j.pbb.2004.01.006
91. Jiang W, Chen Q, Li P et al. (2017) Magnesium Isoglycyrrhizinate attenuates lipopolysaccharide-induced depressive-like behavior in mice. Biomed Pharmacother 86:177–184. https://doi.org/10.1016/j.biopha.2016.12.033
92. Miyake Y, Tanaka K, Fukushima W et al. (2011) Dietary intake of metals and risk of Parkinson's disease: a case-control study in Japan. J Neurol Sci 306:98–102. https://doi.org/10.1016/j.jns.2011.03.035
93. Shen Y, Dai L, Tian H et al. (2019) Treatment of magnesium-l-threonate elevates the magnesium level in the cerebrospinal fluid and attenuates motor deficits and dopamine neuron loss in a mouse model of Parkinson's disease. Neuropsychiatric Dis Treat 15:3143–3153. https://doi.org/10.2147/NDT.S230688
94. Veronese N, Zurlo A, Solmi M et al. (2016) Magnesium status in Alzheimer's disease: a systematic review. Am J Alzheimers Dis Other Demen 31:208–213. https://doi.org/10.1177/1533317515602674
95. Xiong Y, Ruan Y-T, Zhao J et al. (2022) Magnesium-L-threonate exhibited a neuroprotective effect against oxidative stress damage in HT22 cells and Alzheimer's disease mouse model. World J Psychiatry 12:410. https://doi.org/10.5498/wjp.v12.i3.410
96. Huang Y, Huang X, Zhang L et al Magnesium boosts the memory restorative effect of environmental enrichment in Alzheimer's disease mice. https://doi.org/10.1111/cns.12775
97. Budinger D, Barral S, Soo AKS, Kurian MA The emerging role of manganese dysregulation in neurological disease : emerging evidence.

Lancet Neurol. 20:956–968. https://doi.org/10.1016/S1474-4422(21)00238-6
98. Dobson AW, Erikson KM, Aschner M (2004) Manganese neurotoxicity. Ann N Y Acad Sci 1012:115–128. https://doi.org/10.1196/annals.1306.009
99. Nitin R, Bowman AB (2018) Chapter Three - connections between manganese neurotoxicity and neurological disease. In: Aschner M, Costa LG (Eds.) Advances in Neurotoxicology. Academic Press, 87–113
100. Pajarillo E, Nyarko-Danquah I, Adinew G et al. (2021) Neurotoxicity mechanisms of manganese in the central nervous system. Adv Neurotoxicol 5:215–238. https://doi.org/10.1016/bs.ant.2020.11.003
101. Calne DB, Chu NS, Huang CC et al. (1994) Manganism and idiopathic parkinsonism: similarities and differences. Neurology 44:1583–1586. https://doi.org/10.1212/wnl.44.9.1583
102. Kim J, Li Y, Buckett PD et al. (2012) Iron-responsive olfactory uptake of manganese improves motor function deficits associated with iron deficiency. PLoS One 7:e33533. https://doi.org/10.1371/journal.pone.0033533
103. Park JH, Hogrebe M, Fobker M et al. (2018) SLC39A8 deficiency: biochemical correction and major clinical improvement by manganese therapy. Genet Med 20:259–268. https://doi.org/10.1038/gim.2017.106
104. Brenneman KA, Cattley RC, Ali SF, Dorman DC (1999) Manganese-induced developmental neurotoxicity in the CD rat: is oxidative damage a mechanism of action?. Neurotoxicology 20:477–487
105. Malecki EA (2001) Manganese toxicity is associated with mitochondrial dysfunction and DNA fragmentation in rat primary striatal neurons. Brain Res Bull 55:225–228. https://doi.org/10.1016/s0361-9230(01)00456-7
106. Roth JA, Feng L, Walowitz J, Browne RW (2000) Manganese-induced rat pheochromocytoma (PC12) cell death is independent of caspase activation. J Neurosci Res 61:162–171
107. Alaimo A, Gorojod RM, Miglietta EA et al. (2013) Manganese induces mitochondrial dynamics impairment and apoptotic cell death: a study in human Gli36 cells. Neurosci Lett 554:76–81. https://doi.org/10.1016/j.neulet.2013.08.061
108. Budinger D, Barral S, Soo AKS, Kurian MA (2021) The role of manganese dysregulation in neurological disease: emerging evidence. Lancet Neurol 20:956–968. https://doi.org/10.1016/S1474-4422(21)00238-6
109. Dodd CA, Filipov NM (2011) Manganese potentiates LPS-induced heme-oxygenase 1 in microglia but not dopaminergic cells: role in controlling microglial hydrogen peroxide and inflammatory cytokine output. Neurotoxicology 32:683–692. https://doi.org/10.1016/j.neuro.2011.09.002
110. Kim J, Pajarillo E, Rizor A et al. (2019) LRRK2 kinase plays a critical role in manganese-induced inflammation and apoptosis in microglia. PLoS One 14:e0210248. https://doi.org/10.1371/journal.pone.0210248
111. Saritha K, Celia DA, Shahryar HK, Nikolay FM (2014) Brain deposition and neurotoxicity of manganese in adult mice exposed via the drinking water. Arch Toxicol 88:47–64. https://doi.org/10.1007/s00204-013-1088-3
112. Horning KJ, Caito SW, Tipps KG et al. (2015) Manganese is essential for neuronal health. Annu Rev Nutr 35:71–108. https://doi.org/10.1146/annurev-nutr-071714-034419
113. Bouabid S, Delaville C, Deurwaerdère PD et al. (2014) Manganese-induced atypical Parkinsonism is associated with altered basal ganglia activity and changes in tissue levels of monoamines in the rat. PLoS One 9:e98952. https://doi.org/10.1371/journal.pone.0098952
114. Nyarko-Danquah I, Pajarillo E, Digman A et al. (2020) Manganese accumulation in the brain via various transporters and its neurotoxicity mechanisms. Molecules 25:5880. https://doi.org/10.3390/molecules25245880
115. Barchielli G, Capperucci A, Tanini D (2022) The role of selenium in pathologies: an updated review. Antioxidants (Basel) 11:251. https://doi.org/10.3390/antiox11020251
116. Dominiak A, Wilkaniec A, Wroczyński P, Adamczyk A (2016) Selenium in the therapy of neurological diseases. where is it going?. Curr Neuropharmacol 14:282–299
117. Długosz A, Wróblewski M, Błaszak B, Szulc J (2025) The role of nutrition, oxidative stress, and trace elements in the pathophysiology of autism spectrum disorders. Int J Mol Sci 26:808. https://doi.org/10.3390/ijms26020808
118. Skalny AV, Skalnaya MG, Bjørklund G et al. (2018) Selenium and autism spectrum disorder. In: Michalke B (Ed.)Eds. Selenium. Springer International Publishing, Cham, 193–210
119. ROS and Brain Diseases: The Good, the Bad, and the Ugly - Popa-Wagner - 2013 - Oxidative Medicine and Cellular Longevity - Wiley Online Library. https://onlinelibrary.wiley.com/doi/full/10.1155/2013/963520. Accessed 27 Mar 2025
120. Zhou J, Zhang W, Cao Z et al. (2023) Association of selenium levels with neurodegenerative disease: a systemic review and

meta-analysis. Nutrients 15:3706. https://doi.org/10.3390/nu15173706
121. Pfaender S, Grabrucker AM (2014) Characterization of biometal profiles in neurological disorders†. Metallomics 6:960–977. https://doi.org/10.1039/c4mt00008k
122. Li Z, Liu Y, Li X et al. (2018) Association of elements with schizophrenia and intervention of selenium supplements. Biol Trace Elem Res 183:16–21. https://doi.org/10.1007/s12011-017-1105-0
123. Paltian JJ, dos Reis AS, de Oliveira RL et al. (2020) The anxiolytic effect of a promising quinoline containing selenium with the contribution of the serotonergic and GABAergic pathways: modulation of parameters associated with anxiety in mice. Behav Brain Res 393:112797. https://doi.org/10.1016/j.bbr.2020.112797
124. Strumila R, Lengvenyte A, Olie E et al. (2022) Selenium deficiency is associated with disease severity, disrupted reward processing, and increased suicide risk in patients with Anorexia Nervosa. Psychoneuroendocrinology 140:105723. https://doi.org/10.1016/j.psyneuen.2022.105723
125. Liang X, Xue Z, Zheng Y et al. (2023) Selenium supplementation enhanced the expression of selenoproteins in hippocampus and played a neuroprotective role in LPS-induced neuroinflammation. Int J Biol Macromol 234:123740. https://doi.org/10.1016/j.ijbiomac.2023.123740
126. Shi T, Song J, You G et al. (2021) The function of selenium in central nervous system: lessons from MsrB1 knockout mouse models. Molecules 26:1372. https://doi.org/10.3390/molecules26051372
127. Torres DJ, Alfulaij N, Berry MJ (2021) Stress and the brain: an emerging role for selenium. Front Neurosci 15, 666601. https://doi.org/10.3389/fnins.2021.666601
128. Ozdemir E, Cetinkaya S, Ersan S et al. (2009) Serum selenium and plasma malondialdehyde levels and antioxidant enzyme activities in patients with obsessive–compulsive disorder. Prog Neuropsychopharmacol Biol Psychiatry 33:62–65. https://doi.org/10.1016/j.pnpbp.2008.10.004
129. Solovyev N, Drobyshev E, Blume B, Michalke B (2021) Selenium at the neural barriers: a review. Front Neurosci 15:630016. https://doi.org/10.3389/fnins.2021.630016
130. Naderi M, Puar P, Zonouzi-Marand M et al. (2021) A comprehensive review on the neuropathophysiology of selenium. Sci Total Environ 767:144329. https://doi.org/10.1016/j.scitotenv.2020.144329
131. Dominiak A, Wilkaniec A, Wroczyńsk P, Adamczyk A (2016) Selenium in the therapy of neurological diseases. where is it going?. Curr Neuropharmacol 14:282–299. https://doi.org/10.2174/1570159X14666151223100011
132. Szklarska D, Rzymski P (2019) Is lithium a micronutrient? from biological activity and epidemiological observation to food fortification. Biol Trace Elem Res 189:18–27. https://doi.org/10.1007/s12011-018-1455-2
133. Nelson JC, Pikalov A, Berman RM (2008) Augmentation treatment in major depressive disorder: focus on aripiprazole. Neuropsychiatr Dis Treat 4:937–948
134. Soleimani R, Jalali M, Keshtkar A, Jalali S (2017) Protocol for a systematic review and meta-analysis of lithium, anticonvulsive or atypical antipsychotic drugs for treatment of refractory obsessive-compulsive disorder. J Med Life 10:208–215
135. Ando S, Suzuki H, Matsukawa T et al. (2022) Comparison of lithium levels between suicide and non-suicide fatalities: cross-sectional study. Transl Psychiatry 12:1–5. https://doi.org/10.1038/s41398-022-02238-9
136. Sarai SK, Mekala HM, Lippmann S (2018) Lithium suicide prevention: a brief review and reminder. Innov Clin Neurosci 15:30–32
137. Authority (EFSA) EFS (2017) Dietary reference values for nutrients summary report. EFSA Support Publ 14:e15121E. https://doi.org/10.2903/sp.efsa.2017.e15121
138. Davidson JRT, Abraham K, Connor KM, McLeod MN (2003) Effectiveness of chromium in atypical depression: a placebo-controlled trial. Biol Psychiatry 53:261–264. https://doi.org/10.1016/S0006-3223(02)01500-7
139. Franklin M, Odontiadis J (2004) Effects of treatment with chromium picolinate on peripheral amino acid availability and brain monoamine function in the rat. Pharmacopsychiatry 36:176–180. https://doi.org/10.1055/s-2003-43046
140. Ding J, Sun B, Gao Y et al. (2024) Evidence for chromium crosses blood brain barrier from the hypothalamus in chromium mice model. Ecotoxicol Environ Saf 273:116179. https://doi.org/10.1016/j.ecoenv.2024.116179
141. Salama A, Hegazy R, Hassan A (2016) Intranasal chromium induces acute brain and lung injuries in rats: assessment of different potential hazardous effects of environmental and occupational exposure to chromium and introduction of a novel pharmacological and toxicological animal model. PLoS One 11:e0168688. https://doi.org/10.1371/journal.pone.0168688

142. Reif BM, Murray BP (2024) Chromium toxicity. In: StatPearls. StatPearls Publishing, Treasure Island (FL)
143. Lin C-C, Wu M-L, Yang C-C et al. (2009) Acute severe chromium poisoning after dermal exposure to hexavalent chromium. J Chin Med Assoc 72:219. https://doi.org/10.1016/S1726-4901(09)70059-0
144. Ueno D, Shiino Y, Fujimoto H et al. (2022) Severe chemical corneal injury from hexavalent chromium exposure: a case report. Toxicol Commun 6:58–61. https://doi.org/10.1080/24734306.2022.2058253
145. Schwarz G, Belaidi AA (2013) Molybdenum in human health and disease. Interrelat Between Essent Met Ions Hum Dis 13:415–450. https://doi.org/10.1007/978-94-007-7500-8_13
146. Novotny JA (2011) Molybdenum nutriture in humans. J Evid Based Complementary Altern Med 16:164–168. https://doi.org/10.1177/2156587211406732
147. Reiss J, Johnson JL (2003) Mutations in the molybdenum cofactor biosynthetic genes MOCS1, MOCS2, and GEPH. Hum Mutat 21:569–576. https://doi.org/10.1002/humu.10223
148. Reiss J, Hahnewald R (2011) Molybdenum cofactor deficiency: mutations in GPHN, MOCS1, and MOCS2. Human Mutation 32:10–18. https://doi.org/10.1002/humu.21390
149. Zhang X, Vincent AS, Halliwell B, Wong KP (2004) A mechanism of sulfite neurotoxicity: direct inhibition of glutamate dehydrogenase *. J Biol Chem 279:43035–43045. https://doi.org/10.1074/jbc.M402759200
150. Kumar A, Dejanovic B, Hetsch F et al. (2017) S-sulfocysteine/NMDA receptor–dependent signaling underlies neurodegeneration in molybdenum cofactor deficiency. J Clin Invest 127:4365–4378. https://doi.org/10.1172/JCI89885
151. Abbas A-K, Xia W, Tranberg M et al. (2008) S-sulfo-cysteine is an endogenous amino acid in neonatal rat brain but an unlikely mediator of cysteine neurotoxicity. Neurochem Res 33:301–307. https://doi.org/10.1007/s11064-007-9441-7
152. Bourke CA (2018) Astrocyte dysfunction following molybdenum-associated purine loading could initiate Parkinson's disease with dementia. NPJ Parkinson's Dis 4:1–4. https://doi.org/10.1038/s41531-018-0045-5
153. Cao B, Yan L, Ma J et al. (2019) Comparison of serum essential trace metals between patients with schizophrenia and healthy controls. J Trace Elem Med Biol 51:79–85. https://doi.org/10.1016/j.jtemb.2018.10.009
154. Botchway BOA, Liu X, Zhou Y, Fang M (2023) Biometals in Alzheimer disease: emerging therapeutic and diagnostic potential of molybdenum and iodine. J Transl Med 21: 351. https://doi.org/10.1186/s12967-023-04220-5
155. Liu M, Wang J, Zeng J, He Y (2017) Relationship between serum uric acid level and mild cognitive impairment in Chinese community elderly. BMC Neurol 17: 146. https://doi.org/10.1186/s12883-017-0929-8
156. Euser SM, Hofman A, Westendorp RGJ, Breteler MMB (2009) Serum uric acid and cognitive function and dementia. Brain 132:377–382. https://doi.org/10.1093/brain/awn316
157. Scheepers LEJM, Jacobsson LTH, Kern S et al. (2019) Urate and risk of Alzheimer's disease and vascular dementia: a population-based study. Alzheimer's Dement 15:754–763. https://doi.org/10.1016/j.jalz.2019.01.014
158. Coelho FC, Cerchiaro G, Araújo SES et al. (2022) Is there a connection between the metabolism of copper, sulfur, and molybdenum in Alzheimer's disease? new insights on disease etiology. Int J Mol Sci 23:7935. https://doi.org/10.3390/ijms23147935
159. Han Q, Cai S, Yang L et al. (2017) Molybdenum disulfide nanoparticles as multifunctional inhibitors against Alzheimer's disease. ACS Appl Mater Interfaces 9:21116–21123. https://doi.org/10.1021/acsami.7b03816
160. Manna S, Karmakar P, Kisan B et al. (2021) Fibril-induced neurodegenerative disorders in an Aβ-mutant Drosophila model: therapeutic targeting using ammonium molybdate. Chem Commun 57:8488–8491. https://doi.org/10.1039/D1CC03752H
161. Kong L, Zhou X, Shi G, Yu Y (2020) Molybdenum disulfide nanosheets-based fluorescent "off-to-on" probe for targeted monitoring and inhibition of β-amyloid oligomers. Analyst 145:6369–6377. https://doi.org/10.1039/D0AN00019A
162. Kuang Q, Purhonen P, Hebert H (2015) Structure of potassium channels. Cell Mol Life Sci 72:3677–3693. https://doi.org/10.1007/s00018-015-1948-5
163. Vitvitsky VM, Garg SK, Keep RF et al. (2012) Na+ and K+ ion imbalances in Alzheimer's disease. Biochim Biophys Acta Mol Basis Dis 1822:1671–1681. https://doi.org/10.1016/j.bbadis.2012.07.004
164. Lam MH, Chau SW, Wing Y (2009) High prevalence of hypokalemia in acute psychiatric inpatients. Gen Hosp Psychiatry

31:262–265. https://doi.org/10.1016/j.genhosppsych.2009.02.007
165. Hatta K, Takahashi T, Nakamura H et al. (1998) Abnormal physiological conditions in acute schizophrenic patients on emergency admission: dehydration, hypokalemia, leukocytosis and elevated serum muscle enzymes. Eur Arch Psychiatry Clin Neurosci 248:180–188. https://doi.org/10.1007/s004060050036
166. Tian C, Zhu R, Zhu L et al. (2014) Potassium channels: structures, diseases, and modulators. Chem Biol Drug Des 83:1–26. https://doi.org/10.1111/cbdd.12237
167. Maezawa I, Jenkins DP, Jin BE, Wulff H (2012) Microglial KCa3.1 channels as a potential therapeutic target for Alzheimer's disease. Int J Alzheimers Dis 2012:868972. https://doi.org/10.1155/2012/868972
168. Rose CR, Verkhratsky A (2016) Principles of sodium homeostasis and sodium signalling in astroglia. Glia 64:1611–1627. https://doi.org/10.1002/glia.22964
169. Imbrici P, Conte Camerino D, Tricarico D (2013) Major channels involved in neuropsychiatric disorders and therapeutic perspectives. Front Genet 4:76. https://doi.org/10.3389/fgene.2013.00076
170. Betjes MGH (2002) Hyponatremia in acute brain disease: the cerebral salt wasting syndrome. Eur J Intern Med 13:9–14. https://doi.org/10.1016/S0953-6205(01)00192-3
171. Philbert SA, Xu J, Church SJ, et al (2022) Pan-cerebral sodium elevations in vascular dementia: eEvidence for disturbed brain-sodium homeostasis. Front Aging Neurosci 14, 926463. https://doi.org/10.3389/fnagi.2022.926463
172. Irani SR, Vincent A (2011) Autoimmune encephalitis—new awareness, challenging questions. Discov Med 11:449–458
173. Somers KJ, Lennon VA, Rundell JR, et al. (2011) Psychiatric manifestations of voltage-gated potassiumchannel complex autoimmunity. JNP 23:425–433. https://doi.org/10.1176/jnp.23.4.jnp425
174. El-Mallakh RS, Huff MO (2001) Mood stabilizers and ion regulation. Harv Rev Psychiatry 9:23–32. https://doi.org/10.1080/10673220127873
175. Aldana BI, Sitges M (2012) Sertraline inhibits pre-synaptic Na+ channel-mediated responses in hippocampus-isolated nerve endings. J Neurochem 121:197–205. https://doi.org/10.1111/j.1471-4159.2012.07674.x

Check for updates

Chapter 11

Brain Metallomics in Neurodevelopmental Disorders: From Pathophysiology to Therapeutics

Asma Ahmadani, Rukhsana Nawaz, Sandeep B. Subramanya, Maria Cristina D'Adamo, Amita Attlee and Mauro Pessia

Abstract

Neurodevelopmental disorders (NDDs) are conditions in which the brain's functioning is affected during developmental stages. They are characterized by challenges in cognition, communication, behavior, and motor skills at an early age. Metal dyshomeostasis, an imbalance of essential and toxic metals, is often a significant hallmark in those with NDDs. This chapter focuses on the emerging field of brain metallomics, which explores the complex roles of biometals in NDDs. It provides an overview of essential metals and their physiological contributions to neural development and function. Then it extends the discussion toward toxic metals, highlighting their neurotoxic effects and the possible links with altered neurodevelopmental trajectories.

Furthermore, it reviews the growing evidence associating the metal dyshomeostasis with the etiopathogenesis of NDDs by examining cellular mechanisms such as oxidative stress, neuronal dysfunction, and synaptic disruption. Finally, therapeutic strategies are discussed in detail, including dietary interventions and lifestyle changes, glutathione-mediated detoxification, the role of natural compounds, and probiotics in restoring metal homeostasis and brain health. Special attention is given to the role of zinc in supplementation and chelation therapy.

By synthesizing insights from basic science and clinical research, the chapter underscores the diagnostic and therapeutic potential of targeting metal imbalances in NDDs. It concludes with future directions for research and therapeutic innovation in the context of brain metallomics.

Keywords Brain metallomics, Neurodevelopmental disorders (NDDs), Metal dyshomeostasis, Autism spectrum disorder (ASD), Therapeutic interventions

1 Introduction

Neurodevelopmental disorders (NDDs) are defined as a group of conditions with onset in the developmental period, inducing deficits that produce impairments of functioning, and are characterized by wide genetic and clinical variability. According to DSM-5, NDDs comprise intellectual disability (ID); communication disorders; autism spectrum disorder (ASD); attention-deficit/hyperactivity

Andreas M. Grabrucker (ed.), *Brain Metallomics*, Neuromethods, vol. 237,
https://doi.org/10.1007/978-1-0716-5468-2_11,
© The Author(s), under exclusive license to Springer Science+Business Media, LLC, part of Springer Nature 2026

disorder (ADHD); neurodevelopmental motor disorders, including Tic disorders; and specific learning disorders [1].

Genetic factors, prenatal exposure to toxins, oxidative stress, and neurodevelopmental abnormalities are among the key contributors, underscoring the heterogeneous origins of the condition. However, the precise interplay among these factors remains complex and not yet fully understood.

Metals are ubiquitous chemical elements that play essential roles in brain physiology, contributing directly to enzymatic activity, learning and memory processes, mitochondrial function, myelination, synaptogenesis and synaptic plasticity, neurotransmission, and the regulation of neuroinflammation. The disruption of any of these mechanisms or the absorption of toxic metals generates a metal homeostasis imbalance in the brain, contributing to abnormal central nervous system (CNS) development [2].

For humans, at least 20 chemical elements are essential for specific brain functions [3]. Several essential metals play critical roles in neurodevelopment, particularly copper (Cu), iron (Fe), and manganese (Mn), which are classic transition metals due to their variable oxidation states. While essential, zinc (Zn) and magnesium (Mg) are not classified as transition metals, and Mg is an alkaline earth metal. In contrast, toxic or nonessential elements such as lead (Pb), cadmium (Cd), aluminum (Al), and arsenic (As) fall under non-transition metals (or metalloids, in the case of As). They exert neurotoxic effects through mechanisms like oxidative stress, synaptic dysfunction, and disruption of essential metal homeostasis. Although mercury (Hg) is chemically a transition metal, it is often grouped with non-essential toxic metals in neurotoxicology due to its biological behavior and d^{10} electron configuration in its common oxidation state.

For instance, Mg supports nucleic acid activity, and Na and K are vital for neuronal cell signaling activity. Fe is the most abundant transition metal in the brain, followed by Zn. Fe is the key component involved in oxygen transport, while Zn is found predominantly in the Zn finger transcription factors in the brain [4]. Dyshomeostasis of these metals inside humans may negatively affect the brain and overall health. The high heterogeneity in phenotypic expression in the NDDs could reflect differences in timing during the brain development period when exposure to risk factors occurs. For the exposures to metal and trace elements, differences exist across populations of pregnant women based on geography in the profile of trace element metal exposures and in the timing of those exposures, thus implying their impact on the brain at different points in development [5]. ASD is one of the most common NDDs, with growing worldwide prevalence in the past few decades, and is characterized by persistent deficits in social communication and interaction across contexts, restricted and repetitive patterns of behavior. Symptoms generally manifest in the first two years of life (American Psychiatric Association, 2022). Environmental factors have been associated with the etiology of ASD in recent times. The involvement of toxic metals in generating reactive oxygen species (ROS) and their epigenetic effects has been implicated in ASD [6]. However, studies predominantly focus on peripheral markers rather than the brain. For example, alterations in the levels of metals and/or trace metals in serum, hair, or nails are reported, correlating with the features or severity of ASD [5]. Zn dysregulation has been frequently reported in individuals with ASD. A meta-analysis confirmed reduced blood Zn levels in ASD compared to typically developing individuals, reinforcing its potential involvement in ASD pathophysiology [7]. However, inconsistencies in Zn findings may stem from geographic factors, differences in dietary Zn intake, and biological limitations in using plasma or serum Zn as reliable indicators due to fluctuations from fasting, circadian rhythms, and albumin-binding capacity [8, 9] Additionally, a consistently lower Zn to Cu ratio in ASD suggests a state of Cu toxicity, which has been associated with increased ASD severity [10].

A low Zn/Cu ratio has been proposed as a potential biomarker for ASD, given that low Zn and high Cu levels contribute to a dyshomeostasis and a combined impact on neurodevelopment [11]. Moreover, Zn functions as a key regulator of gut homeostasis, including stabilizing microbial balance and maintaining intestinal integrity, which is essential for a balanced gut–brain axis. Studies suggest that prenatal Zn deficiency may cause ASD-like behaviors in offspring of animal models, potentially driven by microbiota-mediated inflammatory pathways that impact the early brain development [12, 13]. Furthermore, Cu plays an essential role in maintaining the oxidation and reduction of Cu ion balance of the brain [14], and is an important cofactor in electron transfer reactions [15].

However, due to its powerful redox capacity, high Cu levels can cause neurotoxicity from oxidative stress, resulting in neurodegeneration within the brain [16]. Fe is essential for brain functions such as neuronal development, myelination, the synthesis and catabolism of neurotransmitters, electron transport, and respiration [17]. Dysregulation of Fe metabolism genes may represent risk factors linked to ASD development [18]. While Bener et al. reported significantly reduced serum Fe levels in children with ASD compared to healthy controls [19], excess Fe could increase the levels of ROS, contributing to the oxidative damage of DNA, proteins, and lipids, which eventually lead to cell death, also known as *ferroptosis* [20]. In recent years, an inverse U-shaped association has been reported between Mn exposure and neurodevelopmental outcomes in infants and children, indicating that both low and high levels of Mn may have adverse effects [21]. Similar patterns are expected to be observed in individuals with ASD and ADHD [22]. Mg is an important player in neurodevelopment and neuronal maturation, modulating the pathogenetic mechanisms of both ASD and ADHD development. Adequate Mg levels support plasticity and interconnectivity [23]. Similarly, Na homeostasis is vital for brain function, as it supports electrical neurotransmission by inward transportation of Na ions across membranes [24].

Similarly, K ions, along with Na ions, are important to maintaining this neuronal excitability in the brain [25]. Genetic modification and the resulting dysfunction of K and Na channels have been identified as a cause of various NDDs [26]. Notably, some of these channel types are altered by trace elements, highlighting the importance of micronutrient homeostasis in the pathophysiology of these conditions [27]. A protein complex responsible for regulating the K and Na ion gradient across the cell membranes is known as Na/K-ATPase (NKA) [25]. Dysregulation of NKA activity may potentially be associated with the development of several brain disorders, including mood disorders, depression, epilepsy, and ASD [28, 29]. In addition to the imbalance of essential metals, the presence of toxic metals may exacerbate the condition, and they are often reported as neurodevelopmental toxins. Associations between exposures to toxic metals during the prenatal phase and the risk of developing neurological defects, learning disabilities, and behavioral abnormalities have been found [30]. Toxic metals impact neurodevelopment by triggering inflammatory and oxidative responses that interfere with neuronal balance, axonal transport, neurotransmitter production, and synaptic function [31]. The impact of five such heavy metals, including Pb, Hg, As, Cd, and often Al, is frequently documented in associations with ASD [32]. Among the several techniques used to study metals, inductively coupled plasma mass spectrometry is the most common way to identify metals bound to proteins [33].

Additionally, electrospray ionization mass spectrometry assesses the structure, function, and overall dynamics of metal–protein complexes [34, 35]. For metal distribution in biological samples and for analyzing metal–protein interaction, X-ray absorption spectroscopy, using synchrotron radiation, is particularly effective [36]. Recent developments in genome sequencing techniques have generated extensive protein databases and bioinformatic tools for identifying metalloproteins in genetic sequences [37]. Overall, these studies are of essential importance for the analysis and assessment of metals present in the human system.

A growing body of evidence suggests that re-establishing metal homeostasis may offer promising therapeutic potential in ASD and other NDDs, aiming to correct the imbalance, inflammation, and synaptic dysfunction; an emerging track in precision neuropsychiatry [38–40].

Therefore, this chapter synthesizes contemporary insights into brain metallomics, an emerging interdisciplinary field that promotes understanding of biometals and their biological functions. Metallomics encompasses the study of metalloproteins, metalloenzymes, and metal-binding chaperones, collectively termed *metallomes* [41], and investigates how these components contribute to neural function and dysfunction. Although the associations of metal dyshomeostasis in the neurogenerative disorders are recognized, this chapter focuses on the physiological roles and dysregulation of essential and nonessential metals in the brain in the context of NDDs, with particular emphasis on ASD. By bridging molecular mechanisms with clinical applications, this chapter aims to elucidate the central role of metal dyshomeostasis and neurotoxicity in NDD etiopathogenesis and to discuss plausible therapeutic strategies that target these disturbances for improved diagnosis and treatment outcomes.

2 Essential Metals in Neurodevelopment

2.1 Iron (Fe)

Fe is the most abundant trace element in the human body, playing a vital role in oxygen transport, energy metabolism, and numerous enzymatic processes. Fe plays a unique role in various human physiological and pathological processes, including neurodevelopment, neural differentiation, immunological modulation, embryogenesis, antioxidation, apoptosis, and gene regulation [42–44]. In the brain, Fe is found in the highest concentrations [45] and is mainly located in the extrapyramidal region [46]. This element is required for numerous vital processes such as energy production, myelination, neurotransmitter synthesis and synaptic activity [47]. Given the critical nature of these processes, the Fe inside the brain is tightly regulated. Although different types of brain cells have distinct Fe uptake mechanisms, neurons acquire Fe primarily through the transferrin (Tf)-transferrin receptor 1 (TfR1)-dependent pathway and the non-transferrin-bound Fe uptake pathway from brain interstitial fluid [48]. Most of the Fe is found tightly bound to heme molecules inside the red blood cells, present in the vast capillary network of the brain [45]. Fe is essential for the transportation of oxygen and the development and maturation of neurons. However, excessive Fe can catalyze the formation of ROS, leading to oxidative stress and neuronal damage by catalyzing the *Fenton* reaction, generating highly reactive ·OH from hydrogen peroxide (H_2O_2) in neuronal cells [49]. Notably, Fe plays a crucial role in the biosynthesis of catecholamine-based neurotransmitters. Several studies have reported a strong association between Fe and the dopaminergic system, with Fe imbalance being linked to anxiety-like behaviors and neurodegeneration [50]. Toxicity due to Fe deficiency may pose mental health risks in individuals and has been associated with an increased incidence of psychiatric disorders, including ADHD, ASD, mood disorders, and other developmental disorders [51]. Previous studies have shown that Fe metabolism regulates the function of many enzymes and proteins in the brain, which are directly associated with cognition, social–emotional development, and sensory functioning [52, 53]. Overall, Fe is a vital human body component and plays a critical role in neurodevelopment. Therefore, structural or functional changes in Fe metabolism and their dysregulation may represent risk factors linked to ASD development. A growing body of evidence has identified novel genetic markers related to brain Fe metabolism, which may contribute to the early diagnosis of ASD [18]. An earlier study indicated that serum Fe levels were significantly reduced in ASD children compared to healthy control children [19]. However, using inverse-variance-weighted (IVW) methods it was shown that only transferrin was potentially associated with ASD risk; whereas, serum Fe might not play a direct role in the pathological changes observed in ASD [54]. Conversely, age-related accumulation of Fe in specific brain regions has been closely linked to oxidative stress and neuronal injury, thereby playing a role in the pathogenesis of neurodegenerative diseases [55].

2.2 Calcium (Ca)

Ca plays a vital role in several neuronal functions and is a pivotal secondary messenger in signal transduction pathways within the brain [56, 57]. Ca has been shown to regulate gene expression, membrane excitability, and dendrite development [56]. Additionally, synaptic transmission is tightly controlled by the modulation of neurotransmitter release at the presynaptic terminal. This process is triggered by the influx of calcium ions (Ca^{2+}) through voltage-gated calcium channels in response to action potentials. A synaptic vesicle fusion within the membrane occurs, and neurotransmitters are released into the synaptic cleft. Another crucial role of Ca is in long-term potentiation (LTP) and long-term depression (LTD) forms of synaptic plasticity, which are essential for learning and memory [56, 57]. Notably, Mg is a key regulatory player of Ca flux through neuronal membranes [58]. By regulating voltage-dependent channels and inhibiting Ca release from cytosolic stores, Mg controls Ca influx, acting as a natural antagonist, creating a Mg block under resting conditions [59]. This functional property of Mg is prominent in association with N-methyl-D-aspartate (NMDA) receptor activity [60].

Finally, Ca is involved in gene expression, mitochondrial function, and cell survival, ensuring proper neuronal metabolism and homeostasis. Therefore, dysregulated Ca signaling is not only implicated in NDDs like ASD and ID but also neurodegenerative diseases such as Alzheimer's disease (AD) and Parkinson's disease (PD), where excessive Ca influx can trigger excitotoxicity, oxidative stress, and apoptosis, leading to learning and memory impairments. Thus, maintaining Ca homeostasis is critical for optimal brain function and neuronal health.

Glutamatergic Systems as a Molecular Hub Linking Brain Metallomic Dyshomeostasis to NDD

The NMDA receptor at synapses on glutamatergic neurons and gamma-aminobutyric acid (GABA)-ergic interneurons is critical in mediating excitatory neurotransmission and regulating inhibition. It is essential for maintaining synaptic plasticity and proper neurodevelopment [60]. The NMDA receptor plays a central role in excitatory neurotransmission and synaptic plasticity, which are important for neurodevelopment. Several essential and toxic metals interplay at this NMDA receptor and can have a downstream effect, implicated in NDDs (Fig. 1). NMDA receptor channel are highly permeable to Ca and its controlled influx is essential for downstream signaling. Receptor's activation requires glutamate, a co-agonist (glycine or D-serine), and membrane depolarization to lift the Mg block. Once activated , it permits the influx of Ca and Na, triggering signaling pathways underlying learning, memory, and neuronal maturation.

Notably, Mg acts as a physiological blocker of the NMDA channel, occupying the pore at resting membrane potential and being displaced upon depolarization (see Fig. 2). Low Mg levels reduce this block, potentially increasing Ca influx and excitotoxicity [61]. The Mg block is lifted once sufficient membrane depolarization is achieved, allowing Ca influx into the postsynaptic neuron. This triggers a cascade of intracellular events, including activation of Ca/calmodulin-dependent protein kinase II (CaMKII), cAMP response element-binding protein, and other downstream effectors that control synaptic strength and promote LTP. Importantly, indirect activation of NMDA receptors requires glutamate binding to α -amino-3-hydroxy-5-methyl-4-isoxazolepropionic acid (AMPA) receptors, which leads to Na influx and membrane depolarization that removes the Mg block [56, 62]. Thus, the calibrated gating effect of

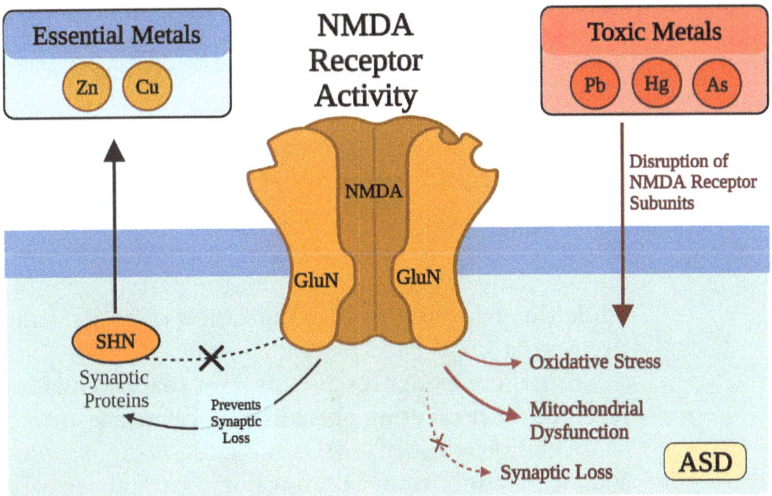

Fig. 1 Essential metals maintain NMDA receptor stability and synaptic plasticity while toxic metals perturb NMDA receptor subunit and synaptic protein, causing oxidative stress, mitochondrial dysfunction, and synaptic loss, which are potentially implicated in ASD

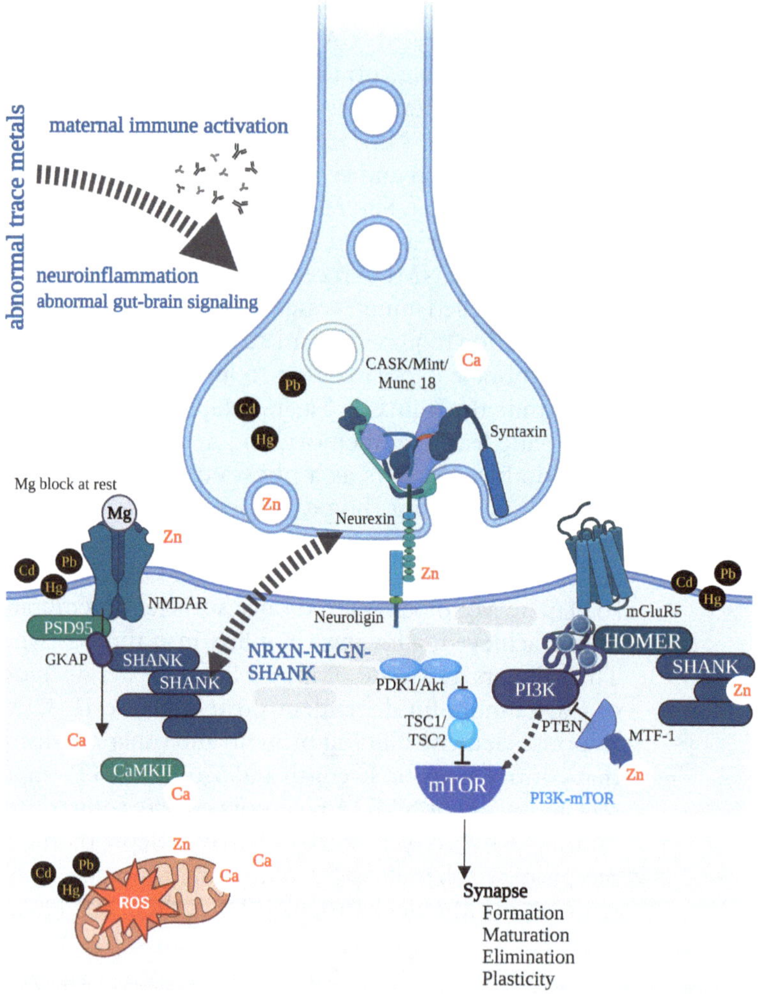

Fig. 2 Schematic illustration of how essential and toxic metals influence NMDA receptor signaling and downstream synaptic processes implicated in ASD

Mg on NMDA receptors is critical for activity-dependent synaptic modifications, allowing the CNS to adapt to environmental stimuli and encode new information. This is a necessary process, especially during neurodevelopment, when the patterns of synaptic activity guide the maturation and stabilization of cortical and subcortical circuits [61].

Emerging evidences also suggest that modulation of NMDA receptor function using pharmacological means, including Mg supplementation or use of NMDA antagonists/agonists, may have therapeutic potential in ameliorating cognitive and behavioral symptoms in certain neurodevelopmental conditions [26, 58]. Both hypo- and hyperactivity of the receptor are implicated in human pathologies. Hyperactivity causes excessive Ca influx and is associated with excitotoxicity, oxidative stress, epilepsy and ASD [60]. Conversely, hypofunction affects synaptic plasticity and cognition and is associated with cognitive deficits in conditions such as schizophrenia and ID [60]. Overall, dysregulated Ca entry through NMDA has been implicated in the pathophysiology of ASD, ADHD, and other NDDs [61]. During early brain development, NMDA receptor expression is especially dynamic, and disruptions in its function, whether by genetic factors, inflammation, or exposure to environmental metals, can exert lasting impacts on synaptic maturation and neural circuitry formation. Indeed, various metals modulate NMDA receptors: Zn inhibits their activity allosterically via GluN2A subunit. At nanomolar concentrations, Zn selectively inhibits receptors containing the GluN2A subunit via high-affinity, noncompetitive binding to an allosteric site. Zn exerts a biphasic and subunit-specific effect on NMDA receptors: at low concentrations, it inhibits NMDA receptors containing the GluN2A subunit, while at higher concentrations, it can have broader inhibitory effects, including partial inhibition of GluN2B-containing receptors. Zn deficiency may decrease this inhibition, resulting in neuronal hyperexcitability, while excess Zn can impair synaptic plasticity and learning [63]. This dual action highlights the complex regulatory role of Zn in synaptic signaling, where both deficiency and excess can lead to neurodevelopmental disturbances. Additionally, Cu modulates NMDA function, though less specifically than Zn, and contributes to oxidative stress, especially when NMDA is overactivated [64].

Pb mimics Ca and disrupts NMDA function, impairing synaptic plasticity and long-term potentiation. Early-life Pb exposure is strongly associated with cognitive deficits through NMDA-mediated pathways [63]. The toxic metal As indirectly influences NMDA activity by inducing oxidative stress and altering intracellular signaling, which may affect NMDA expression or downstream pathways, contributing to developmental delays [63]. Lastly, Hg, particularly in its organic form, induces glutamate-mediated excitotoxicity. NMDA overactivation following Hg exposure has been reported in developing cortical neurons and adult rat frontal cortex, possibly impacting dopamine release in the striatum [63]. Together, these metal–NMDA interactions highlight how even subtle imbalances in these elements can disrupt neurodevelopmental processes and potentially contribute to the pathophysiology of NDDs.

Figure 2 summarizes the molecular interplay between essential (e.g., Zn, Mg, and Ca) and toxic (e.g., Cd, Pb, and Hg) metals at the excitatory synapse, particularly around the NMDAR and associated pathways involved in neurodevelopment.

Zn acts at several levels: it modulates NMDAR activity, stabilizes SH3 and multiple ankyrin repeat domains protein (SHANK) scaffolding proteins, regulates phosphoinositide 3-kinase (PI3K)–mTOR signaling, and supports presynaptic machinery involved in synaptic vesicle release. Ca contributes to signal transduction through CaMKII activation and other downstream events. Mg is a physiological blocker of NMDAR, preventing excessive calcium influx at resting membrane potential, and it is a critical control mechanism in maintaining synaptic stability and preventing excitotoxicity.

In contrast, toxic metals (Cd, Pb, and Hg) disrupt these finely tuned systems. They displace essential metal cofactors, interfere with synaptic receptor function, damage mitochondrial integrity (increasing ROS), and destabilize critical protein complexes like neurexin–neuroligin–SHANK. These effects collectively impair the PI3K–protein kinase 3–mTOR axis and other pathways essential for synaptic formation, maturation, pruning, and plasticity, all commonly dysregulated in ASD.

Additional contributing factors such as maternal immune activation, neuroinflammation, and altered gut–brain axis signaling further impact metal homeostasis and synaptic health, amplifying neurodevelopmental risk.

2.3 Zinc (Zn)

Zn is an essential micronutrient that enters the brain parenchyma via the blood–brain barrier (BBB) and brain cerebrospinal fluid barrier (BCB). Zn transport across the BBB and BCB is tightly regulated by a coordinated network of Zn transporters (such as ZIPs and ZnTs), metallothionein, and ion channels that maintain brain Zn homeostasis [65]. In the adult brain, Zn is more abundant in white than gray matter [46]. The highest concentrations are found within the limbic system, particularly in the hippocampus and amygdala, and in the cerebral cortex, regions enriched with Zn-containing glutamatergic neurons [66]. Notably, these brain structures regulate key physiological functions such as learning, memory formation, emotional processing, and stress responses involved in the pathogenesis of NDD. Within the brain, a substantial proportion of Zn is utilized by Zn-dependent proteins, supporting the functions in both neurons and glial cells [66]. Zn-containing glutamatergic neurons store Zn within synaptic vesicles, which is co-released with glutamate into the synaptic cleft during exocytosis, where Zn concentrations can transiently reach approximately 100 μM, and in some cases, up to 300 μM [67]. At these concentrations, Zn modulates voltage-gated K channels expressed in several brain regions, including the hippocampus [68]. Zn release is tightly coupled to the electrical activity of glutamatergic neurons, acting as a potent neuromodulator of glutamatergic transmission through complex, receptor-specific mechanisms [67]. Indeed, once released into the synaptic cleft, Zn can modulate various glutamate receptor subtypes, influencing excitatory neurotransmission, synaptic plasticity, and neurotoxicity [66]. Zn is an allosteric inhibitor of the NMDA receptor activity [63] (see Sect. 2.2.1). At higher concentrations, Zn can also block the ion channel pore. Zn also inhibits AMPA receptor currents, specifically those that lack the GluA2 subunit, that are Ca permeable [69]. The effects of Zn in Kainate receptors are not well documented, but have been reported to include inhibitory and facilitatory actions, depending on the subunit's configuration [70]. Zn plays an important modulatory role in excitatory neurotransmission and γ-aminobutyric acid (GABA)-mediated inhibitory signaling [71, 72]. Zn regulates the GABA receptor activity, based on concentration and subunit type. Binding to extracellular allosteric sites, Zn can inhibit GABA-mediated chloride currents at micromolar concentrations. The amount of inhibition depends mainly on the receptor's subunit composition. For example, those receptors containing α1 or α2 subunits are more susceptible to Zn inhibition than those with the δ or ε subunits [73]. Notably, Zn does not uniformly suppress GABA receptors. In some settings, even low concentrations may enhance GABAergic currents. Zn regulation of GABAergic transmission differs between synaptic and extrasynaptic receptors. Synaptic GABA receptors (mediating phasic inhibition) seem more susceptible to Zn inhibition [73]. On the other hand, extra-synaptic receptors (mediating tonic inhibition) such as those containing α4βδ subunits are comparatively Zn-insensitive [74]. These differences in sensitivity suggest that Zn can modulate neuronal excitability by selectively regulating components involved in controlling the inhibitory tone. Through Zn modulation of GABAergic neurotransmission, this element influences neural excitability, oscillatory behavior, and seizure thresholds. An altered inhibitory tone associated with Zn homeostasis dysregulation has been linked to various neuropathological conditions, including epilepsy, depression, anxiety, and neurodegeneration [75]. Hyperexcitability caused by excessive synaptic Zn release during seizures may suppress $GABA_A$ receptor function [76]. Conversely, Zn deficiency has been associated with altered neuronal excitability, resulting from disruptions in GABAergic signaling, a feature commonly reported in individuals with ASD [77].

Inadequate Zn intake can alter cerebral Zn homeostasis, potentially contributing to cognitive deficits such as impaired learning and memory [76]. Conversely, excessive neuronal activity may deplete vesicular Zn stores, thereby increasing seizure vulnerability. Fluctuations in synaptically released Zn can directly impact neurotransmission within Zn-enriched synapses. Thus, maintaining optimal Zn concentrations in presynaptic vesicles is essential for the proper function of Zn-containing glutamatergic neurons [76]. Recent studies using ASD mouse models, including Shank3 and BTBR T+ tf/J (BTBR) mice, have shown that dietary Zn supplementation, ranging from 60 ppm in drinking water to 150 ppm in chow, can significantly influence synaptic transmission and behavior, supporting the role of Zn in glutamatergic signaling and its dysregulation in ASD [78, 79]. Zn deficiency induces oxidative stress and promotes neuronal apoptosis by disrupting mitochondrial function and activating intrinsic apoptotic pathways [80]. Previous studies have shown that dysregulation of Zn level is associated with dysfunctional antioxidant enzyme [81, 82], increased levels of oxidants in cells, particularly of H_2O_2 [83], decreased glutathione (GSH) [84], increased oxidation of macromolecules such as protein, lipid, and DNA, causing tissue damage [85]. Overall, mounting evidence highlights the importance of Zn in maintaining proper neuronal and synaptic functions and supports the notion that Zn dyshomeostasis could be a critical risk factor for ASD [27]. A mounting body of evidence has shown that Zn deficiency is associated with altered expression and function in gene products such as SHANK3 and *Neurexin* (*NRXN1*). These disruptions often impair synaptic signaling and neuronal connectivity pathways, thereby contributing to the risk of ASD [86]. Moreover, Zn controls the metallothionine expression, which helps maintain Zn homeostasis and protects against oxidative stress. Disruptions in these Zn–metallothionein complexes can affect the Zn balance, contributing to the susceptibility of NDDs, including ASD [87]. Recent evidence indicates that dysregulation in Zn homeostasis has a potential role in the etiopathophysiology and ASD-related cognitive and behavioral impairments [27, 88].

2.4 Copper (Cu)

Cu plays a critical role in brain function. After Fe and Zn, its abundance in the brain accounts for 9% of the total body's Cu, making it the highest among all organs [45, 89]. Cu is a cofactor for key enzymes such as ceruloplasmin, superoxide dismutase (SOD3), cytochrome c oxidase, and dopamine β-hydroxylase.

Ceruloplasmin functions as a ferroxidase, playing a major role in Fe homeostasis and preventing Fe-mediated oxidative damage. SOD3 provides antioxidant defense by catalyzing the dismutation of superoxide radicals [89]. Cytochrome c oxidase (Complex IV), now reclassified as a translocase, catalyzes the final step of the mitochondrial electron transport chain by transferring electrons to oxygen, thereby driving adenosine triphosphate (ATP) synthesis essential for neuronal energy metabolism. Dopamine β-hydroxylase is a key enzyme in catecholamine biosynthesis, converting dopamine into norepinephrine (noradrenaline), a neurotransmitter crucial for brain function, including attention, arousal, stress response, and mood regulation. Peptidylglycine α-amidating monooxygenase (PAM) is a Cu-dependent enzyme critical for the post-translational modification of numerous neuropeptides and peptide hormones, including oxytocin, which has been implicated in ASD pathophysiology. The maturation of neuropeptides relies on this Cu-dependent enzymatic activity. PAM's function is sensitive to Cu availability; therefore, Cu deficiency can diminish its activity, impairing neuropeptide processing and disrupting various physiological processes. During development, Cu supports myelination, the process by which oligodendrocytes insulate axons, modulating the propagation of action potentials. Cu is required to form and maintain neuronal connections [90], modulate neurotransmitter synthesis [91], and fine-tune synaptic transmission and plasticity by modulating receptors like GABA, NMDA, and AMPA to influence learning and memory. Cu modulates the receptors' function by direct extracellular binding. In NMDA receptors, it serves as a non-competitive inhibitor (see Sect. 2.2.1). Cu plays a dual role in managing inflammation in the brain by regulating pro-inflammatory and anti-inflammatory responses, influencing cytokine production, activating signaling pathways, and distinct immune responses [89].

Furthermore, Cu plays a key role in regulating cell death by influencing autophagy, activating kinase signaling pathways, and triggering cuproptosis [89]. Similar to Zn, the blood-brain barrier (BBB) and the blood-cerebrospinal fluid barrier (BCB) tightly regulate the concentrations of Cu in the brain [89]. Astrocytes, in direct contact with the BBB, serve as key regulators of Cu uptake and distribution, storing Cu and buffering fluctuations in its levels. Cu exists in three oxidation states (Cu^0, Cu^+, and Cu^{2+}), and its high reactivity necessitates a tightly controlled distribution in the brain. Cu concentrations are the highest in the dentate nucleus, substantia nigra, locus coeruleus, and cerebellum [46]. Recent research highlights that individuals with ASD exhibit dysregulated locus coeruleus–norepinephrine (LC–NE) activity, characterized by elevated tonic LC–NE activation (larger resting pupil diameter) linked to enhanced visual search efficiency and sensory hypersensitivity, coupled with reduced phasic responsiveness (diminished pupil dilation/event-related potentials) associated with attentional flexibility deficits [92]. Altered LC connectivity with somatosensory and visual regions further implicates noradrenergic dysfunction in atypical sensory processing and repetitive behaviors, offering potential therapeutic targets for neuromodulation [93, 94].

Notably, cerebellar dysfunction is frequently observed in ASD [95]. The cerebellum is particularly susceptible to Cu toxicity, as evidenced in Wilson's disease, a genetic disorder characterized by Cu accumulation, where early symptoms such as tremors and impaired coordination highlight this region's vulnerability. However, direct evidence linking Cu dyshomeostasis with LC and cerebellar dysfunction, as well as ASD, remains limited, making this an important and compelling area for further investigation.

The primary Cu transporter, CTR1, is responsible for the uptake of approximately 70% of total Cu and its import into the brain [89]. As for all other trace metals, their deficiency or excess can seriously impair brain functions. Excessive Cu can cause neurotoxicity, leading to oxidative stress, inflammation, and neuronal damage [96]. Moreover, excess Cu catalyzes the *Fenton* reaction, producing ·OH that attacks biological macromolecules, including proteins, lipids, and nuclear material, leading to damage [97]. Excess Cu toxicity can affect the human brain moderately or severely. It may be involved in the development of brain disorders, including depression, fear, cognitive impairment, nervousness, Alzheimer's disease (AD), and behavioral disorders [98]. Cu intoxication was found to be high in children diagnosed with ASD [99]. A recent study reported significantly higher Cu levels in the hair of children with ASD, with levels increasing alongside symptom severity, especially in boys [100]. Complementing these findings, recent evidence supports a link between Cu levels and ASD symptomatology [101]. Indeed, this case–control study found a significant positive association between Cu concentration and the core symptoms of ASD in Chinese children [101]. These findings highlight the need for future studies with larger sample sizes to determine the optimal Cu concentration that supports healthy brain function and possibly mitigates ASD-related impairments.

2.5 Magnesium (Mg)

Mg plays fundamental roles in the CNS. In the brain, Mg ensures proper nerve signaling and preservation of BBB integrity [58]. It is a key player in neurotransmission and neuromuscular conduction, a key modulator of synaptic activity and neuronal excitability [102]. As previously stated, Mg helps regulate intracellular Ca homeostasis [58]. Moreover, it acts as a physiological antagonist of NMDA receptors in a neuroprotective role. Mg functions as a voltage-dependent block under resting membrane potentials, preventing Ca influx. This protective regulatory mechanism from excitotoxicity is characterized by glutamatergic stimulation, Ca overload, oxidative stress, and finally neuronal death [102]. Since NMDA is both a ligand and a voltage-gated ion channel, it requires presynaptic glutamate release and postsynaptic depolarization (see Sect. 2.2.1).

Mg influence extends beyond the well-documented regulation of NMDA receptors and synaptic plasticity, including modulation of inwardly-rectifying K (Kir) channels. Kir channels are critical for maintaining resting membrane potential and stabilizing neuronal activity. These channels exhibit "*inward rectification*," allowing K influx more readily than efflux. Cytoplasmic Mg regulates this property by binding to specific sites within the pore region of the channel, where it blocks the outward flux of K ions, thereby functioning as a voltage-dependent gatekeeper. Under physiological conditions, intracellular Mg concentrations modulate Kir channel activity, ensuring balanced neuronal excitability. *Loss-of-function* and *gain-of-function* mutations in Kir channels have been implicated in the pathophysiology of ASD and NDD [103–106]. Given that Kir channel activity is sensitive to intracellular Mg levels, it remains an open and compelling question whether dietary fluctuations in Mg intake, modulating intracellular Mg concentrations, could influence ASD phenotypes through dysregulated Kir channel function [26, 95].

Beyond its ion-channel modulation, Mg also reduces oxidative stress, stabilizes mitochondrial function, and prevents the release of pro-inflammatory and vasoactive molecules [58]. Its anti-inflammatory properties are particularly relevant given that chronic low-grade inflammation is increasingly recognized as a hallmark of NDDs and numerous neurological disorders. Mg deficiency has been linked to heightened inflammatory responses mediated by upregulation of nitric oxide synthase, cytokine production, and prostaglandin synthesis [58]. While the consequences of Mg deficiency have been relatively well-studied, research into the effects of Mg excess, particularly in neurodevelopment, is still emerging. Some studies suggest that elevated Mg levels may exert neuroprotective effects by dampening excitotoxic cascades and attenuating neuroinflammation through improved Ca regulation [107]. However, there is also evidence that excessive Mg concentrations during critical periods of brain development may negatively influence cognitive, motor, or language acquisition, potentially contributing to neurodevelopmental delays [108]. Thus, the role of Mg in NDDs appears to be double-edged, with both deficiency and excess posing risks to neural health. The bidirectional impact of Mg underscores the need for tightly regulated Mg homeostasis during development. Furthermore, research is warranted to delineate the mechanistic pathways through which Mg imbalances contribute to the onset and progression of NDDs and to explore its potential as a therapeutic target in precision pediatric neuropsychiatry.

2.6 Manganese (Mn)

While not as densely concentrated in the brain as Fe, Zn, and Cu, manganese (Mn) is mainly found in the basal ganglia and cerebellum [109]. Mn enters neurons through divalent metal transporter 1 (DMT1) and Zip8/14 [110]. In the brain, it is crucial for activating several proteins required for brain functioning, including being a part of metalloenzymes, for example, manganese superoxide dismutase (MnSOD), found at the mitochondria, which helps in the detoxification of superoxide free radicals [109]. The metal transporter Solute Carrier Family 39 Member 8 (SLC39A8) regulates cellular Mn uptake and has been linked to Mn deficiency in the brain when mutated. Mutations in *SLC39A8* can lead to cerebral atrophy, dystonia, and seizures [111].

Additionally, it plays a role in protein glycosylation, supports immune function, and produces cell energy [109]. Fluctuations in dietary Mn concentrations may disrupt Mn homeostasis in the brain, potentially influencing epileptic susceptibility [112]. It can induce oxidative stress through two putative mechanisms, leading to neurotoxicity via dopamine oxidation and disruption of normal GSH and metallothionein function [113]. Like Cu, excess Mn also catalyzes the *Fenton* reaction, generating high levels of reactive ·OH from H_2O_2, and activates the *Haber–Weiss* reaction to produce ·OH radicals from superoxide anions [114].

Mn is naturally enriched in whole grains, nuts, leafy vegetables, and tea. Although dietary Mn toxicity is rare, excessive intake through supplementation [115] or contaminated water may be harmful, particularly in areas with high Mn content in groundwater. An increased risk of cognitive and behavioral impairments in children has been observed [116, 117]. Several studies have reported that elevated Mn levels are significantly associated with ASD and ADHD, and may be directly linked to impulsive and self-destructive behavior in children [118, 119]. In vitro and in vivo studies have shown that excess Mn has neurotoxic effects, which can adversely affect brain function and induce neuronal oxidative stress, neuroinflammation, apoptosis, and disturbances in neurotransmitter metabolism [120]. Animal studies have shown that Mn exposure is linked to dopaminergic dysfunction, resulting in impulsive and distractible behaviors. However, large-scale human studies are needed to directly assess the relationship between Mn accumulation and neurological alterations in individuals with ASD and ADHD [22].

2.7 Sodium and Potassium (Na and K)

Na and K are essential for body fluid homeostasis and critical to neuronal excitability and signaling. The human body has highly sophisticated mechanisms for detecting Na and K levels in body fluids, and even minor fluctuations trigger highly effective corrective responses to maintain electrolyte balance [121].

Disruptions in Na/K homeostasis' sensory and regulatory mechanisms can lead to imbalanced levels, which have been implicated in developing various neurological disorders. NKA pumps and ion channels work in concert to maintain their physiological gradients across the cell membranes [24] [122](Hille, 2001). Therefore, dysregulation of NKA activity may lead to the development of various brain disorders, including mood disorders, depression, neuronal hyperexcitability, epilepsy, and ASD [28, 29].

A study reported a significant decline in peripheral blood mononuclear cells (PBMC) NKA activity in children with ASD, compared to neurotypical controls, underpinning the importance of NKA activity in neuropsychiatric disorders [123]. The expression of the Na leak channel (NCA) has been shown to influence neural function and synapse development [124]. Several neurodegenerative and psychiatric disorders are associated with dysregulation of NCA through NLF-1, including ADHD and depression [125]. Moreover, mutations in Na channel genes (*SCN1A*, *SCN2A*, and *SCN8A*) have been strongly associated with NDDs such as epileptic encephalopathies, IDs, and movement disorders. As a result, the disrupted channel gating leads to *loss-of-function* deficits in inhibitory interneurons or *gain-of-function* hyperexcitability in excitatory circuits. One such example is Dravet syndrome, a severe, early-onset epilepsy, hallmarks of which include frequent, prolonged seizures, developmental delays, speech and language impairment, and motor difficulties. De novo *loss-of-function* mutations in the *SCN1A* gene are often reported in such cases.

These mutations disrupt inhibitory interneuron function, leading to brain hyperexcitability and treatment-resistant seizures [126].

Conversely, astrocytes help restore the homeostatic gradients of K ions in the brain by redistributing excess K from regions of high concentration to those with lower concentrations [127]. Disruption of this process, known as *K siphoning*, due to mutations in the Kir4.1 channels *(KCNJ10)*, highly expressed in astrocytes, has been implicated in ASD and epilepsy [104–106]. Like Na, several K channel mutations (*KCNC3*, *KCNA4*, and *KCNK18*) impair repolarization and neuronal excitability, leading to spinocerebellar ataxia, epilepsy, and cognitive defects [26,128–131].

Overall, dysregulation of these ion transport systems aggravates developmental delays and seizures through altered neuronal homeostasis. Recent reviews analyze the pathophysiological mechanisms linking Na and K *channelopathies* to NDDs, and several other studies highlight emerging strategies targeting these ion channels [132, 133].

3 Toxic Metals and Neurodevelopment

Heavy metals are not usually found to play any essential roles in our bodies. These have generally been acquired through environmental exposures via the air we breathe, the food and water we consume, and even through skin contact. They come from industrial pollution, vehicle emissions, contaminated water, certain foods, and everyday cosmetics [134, 135]. The most common route of exposure to heavy metals is through the drinking water [136, 137]. Drinking water in various countries has been impacted by heavy metal contamination due to industrial activities, corrosion of pipes, and environmental factors [136]. Such environmental exposures largely seem to affect middle to low-income countries and can contribute to more than 25% of deaths/diseases worldwide, and affect child neurodevelopment [137].

Heavy metals are often reported as neurodevelopmental toxins, and the link between prenatal heavy metal exposure and risk of developing neurological defects, learning disabilities, and behavioral abnormalities has been found [30]. Excessive metal exposure harms the nervous system. Once inside the mother's circulation, they can cross the placenta and reach the baby's brain through still-developing brain barriers [30]. The developing brain is more vulnerable to the detrimental impacts of heavy metal toxicity than an adult [138]. Therefore, metals are the main environmental toxicants linked to abnormal neurodevelopment [138]. These metals activate microglia and astrocytes, triggering inflammatory and oxidative responses that worsen their toxic effects. This disruption can interfere with neuronal balance, axonal transport, neurotransmitter production, and synaptic function, contributing to neurodegenerative diseases like AD and PD, as well as neurodevelopmental conditions like ASD [31]. Five heavy metals, Pb, Hg, Al, As, and Cd, have been identified in the literature as having significant effects on children's neurodevelopment, as discussed below.

3.1 Lead (Pb)

Pb is a neurotoxic compound with a half-life of around 30 days in the blood, while in the brain it is 2–3 years [139]. Pb at early stages (*e.g.*, in infants) can cross the BBB and even be present in the breastmilk, leading to early exposure in fetal and post-natal developmental stages [140]. The impact of Pb exposure is particularly severe during developmental stages due to the immaturity of the BBB and the rapid pace of neural development, which increases susceptibility. Notably, Pb exposure has also been linked to cognitive impairments in both children and adults, such as IDs, learning difficulties, and behavioral disorders in all age groups [141, 142]. The impact is mainly caused by the production of ROS, which causes cellular damage by depleting antioxidant defenses [143–145]. Pb also causes alterations to DNA and impairs chromosomal integrity [146]. Additionally, Pb interferes with proteins, especially those with metal-binding sites, impairing their function [147]. Pb exposure is suggested to cause NMDA receptor inhibition that impairs synapse development, which may have lasting impacts on downstream signaling (see Sect. 2.2.1) [148]. Among animal and cell line studies, Pb was shown to block voltage-activated and receptor-operated Ca channels in invertebrate and mammalian neurons [149]. Most of these alterations in the structure and functions of the CNS caused by Pb are irreversible [140]. A study found that prenatal Pb exposure during late pregnancy (>28 weeks) was significantly associated with lower mental development index scores in infants at 6 months of age, following co-exposure to Hg [150]. Additionally, Pb exposure negatively impacted the psychomotor development index in infants when maternal Hg levels exceeded the 50th percentile, indicating an adverse effect of two well-known heavy metals and their combined exposures in the prenatal phase [150].

Pb at both higher and lower concentrations may cause neurotoxicity. A meta-analysis revealed significantly higher levels of Pb in the hair and blood samples of children with ASD compared to healthy controls [151]. However, a previous study observed lower concentrations of Pb in urine samples of ASD individuals, which affected their excretion ability [152].

Additionally, insufficient antioxidant levels in individuals with ASD may interfere with the body's ability to excrete Pb. This may impair the ability of Pb to bind to antioxidants, resulting in reduced urinary excretion of Pb [153], increased accumulation in the body, and heightened neurotoxic effects, particularly during early brain development. Since Pb can cross the placenta and flow through the breastmilk, fetuses and young children are at high risk [140]. Moreover, Pb exposure has been linked to oxidative stress, disruptions of the synaptic signaling pathways, and, thereby, possibly associated with altered neurodevelopmental outcomes and ASD etiology [153]. Therefore, effective measures to minimize daily Pb exposure in vulnerable populations should be urgently implemented.

3.2 Mercury (Hg)

Elemental Hg and organic methylmercury (MeHg) are two well-known forms of neurotoxic Hg, found in the environment [154]. Elemental Hg can accumulate in the CNS and possibly cross the BBB following its industrial exposure, dental amalgams, mercury vapor, etc. MeHg is consumed via contaminated fish and seafood, and in *thimerosal*, it is a preservative used in some vaccines [154].

Elemental Hg's direct contact with skin or inhalation of vapors can result in organ failures and permanent damage throughout the CNS. While MeHg, once ingested, travels via the gastrointestinal tract, through the blood, BBB, and placental barriers, which can pose risks to fetal neurodevelopment. Cerebral palsy, IDs, and sensory impairments can result from fetal exposure to Hg [155]. Ultimately, among the known NDDs, ASD is closely linked to Hg toxicity, as several studies have reported elevated Hg levels in blood, hair, and teeth of individuals with ASD [156–159]. Being highly reactive, Hg has also been reported to disrupt cellular function, inhibit protein and enzyme function, induce epigenetic modifications, and cause cellular death [154]. The mechanisms causing neurotoxicity are similar to those of Pb. Inside the human body, Hg increases ROS formation, leading to neuronal damage as a direct effect, and indirectly through disruption in the intracellular Ca homeostasis [160]. Hg also interferes with neurotransmitter systems, including dopamine, serotonin, and acetylcholine, essential for mood regulation, cognition, and motor function [161]. In a study using rodent models, exposure to MeHg during early development was shown to impair gut–brain axis communication. This disruption was associated with altered neurotransmitter levels, reduced expression of tight junction (TJ) proteins, increased intestinal permeability ("leaky gut"), immune dysregulation, and chronic inflammation—factors increasingly implicated in the pathophysiology of ASD [161].

Hg exposure profoundly affects the composition and function of the gut microbiota. Studies in animal models have demonstrated that both forms of Hg reduce overall microbial diversity and induce notable shifts in the relative abundance of major bacterial phyla, such as increasing Firmicutes and Bacteroidetes while decreasing Proteobacteria and Verrucomicrobia in the case of inorganic Hg, and elevating Proteobacteria and Bacteroidetes with a reduction in Desulfobacterota following MeHg exposure [162].

At the genus level, Hg exposure is associated with decreased beneficial groups like *Lactobacillaceae* and elevation of potentially pathogenic bacteria, such as *Enterobacteriaceae* and *Pseudomonadaceae*. A pilot study observed that children with ASD in China reported significant positive correlations between hair Hg concentrations and relative abundances of gut bacteria of *Parabacteroides* and *Oscillospira* genera, indicating a dysbiosis in these commensals with increasing Hg levels [163]. Furthermore, Hg-induced dysbiosis disrupts production of microbial metabolites, including neurotransmitters like glutamate and GABA, thereby potentially affecting the gut–brain axis and contributing to neurotoxicity [164]. Further research is required to clarify the exact molecular mechanisms. However, these findings underscore the two-way relationship between Hg and the gut microbiota and emphasize the vital role of the gut–brain axis in facilitating the neurotoxic effects of Hg and its potential involvement in NDDs.

3.3 Cadmium (Cd)

Although a naturally occurring transition metal, Cd is rare and found in the Earth's crust in minute concentrations. However, human activities have significantly increased the environmental impact of Cd. It is mainly produced during mining, smelting, fossil fuel combustion, and via manufacturing plants of batteries, paints, plastics, etc.

Cd can also enter soil and water through phosphate fertilizers and sewage sludge, accumulating in agricultural land and, subsequently, in crops. As a result, humans can ingest Cd primarily by consuming contaminated food, especially vegetables, cereals, rice, shellfish, and organ meats from animals exposed to Cd-rich environments. Tobacco smoke is another primary source, as the tobacco plant accumulates Cd from the soil, and smokers inhale significant amounts with each cigarette. In some cases, Cd can also be ingested via drinking water, particularly if plumbing materials contain Cd or water sources are contaminated. Thus, environmental contamination from industrial and agricultural sources and lifestyle factors such as smoking contribute to human Cd ingestion [165].

Cd can accumulate inside the body as it has a long biological half-life of up to 23.5 years, posing significant risks of neurotoxic effects on the CNS and peripheral nervous system (PNS) [166]. Its ability to cross the BBB causes substantial damage to its integrity and weakens TJs [167]. Thereafter, Cd gains entry to the CNS and disrupts Ca and Zn homeostasis by competitively entering neurons via voltage-gated Ca channels and Zn transporters [166]. Once inside, Cd affects the ATP synthesis, disrupting the mitochondrial respiratory processes, and causing an increase in ROS production [166, 168, 169]. Cd exposure also interferes with neurotransmitter systems by altering the function of signaling proteins and causing asynchrony in neurotransmitter release [166]. Prenatal and early childhood exposure to Cd has been associated with deteriorations in cognitive functions, including reduced concentration and learning disabilities [166]. These findings suggest that even low-level Cd exposure, previously considered safe, can adversely affect children's neurodevelopment due to its bioaccumulation [167].

Studies report structural abnormalities in the forebrain, cerebellum, and midbrain in Cd-exposed rats [170, 171]. Moreover, short-and long-term Cd exposure has been associated with behavioral changes, such as reduced locomotor activity, memory dysfunction, elevated anxiety, aggression, and sleep-related issues [171, 172]. The neurotoxic impact of Cd exposure can be transferred from mothers to their offspring as well, evidenced by animal studies reporting decreased locomotor activity, disrupted neurotransmitter expression, and impaired neuromotor maturation, following maternal exposure [173]. Similar effects were found in humans, where higher maternal urinary Cd during pregnancy (>0.8 μg/L) was associated with lower cognition in 4-year-old children [174]. Another study reported negative correlations of maternal urinary Cd levels (5th to 95th percentile: 0.18–2.0 μg/L, median: 0.63 μg/L) with verbal function, visual perception, and IQ scores in 5-year-old children [175].

Mainly, Cd exposure is considered a modifiable risk factor in ASD. Children with ASD have been reported to have higher Cd levels compared to controls [11]. Studies have confirmed elevated hair Cd concentrations in children with ASD compared to controls (38% and 8%) [11, 176]. A similar trend was observed in blood samples from children with ASD, reporting a higher Cd concentration in their blood plasma. Consumption of seafood may be one of the dietary factors responsible for the elevated concentrations in that group [177]. In contrast, only a modest increase of 1.7% in blood Cd levels was observed in children with ASD in Romania [178]. More recently, it was reported that the risk of developing ASD in children was 78% higher among those whose mothers had detectable levels of urinary Cd during early pregnancy, underscoring the critical impact of early-life exposure to toxic metals on neurodevelopmental outcomes [179].

Recent evidence associates the neurotoxic effects of Cd with its ability to influence the gut microbiota, stressing an important secondary but indirect pathway via the gut–brain axis [180]. Several animal studies have demonstrated a significantly altered gut microbial ecosystem following Cd exposure, resulting in reduced diversity, altered composition in dominant phyla such as *Firmicutes*, *Bacteroidetes*, and *Proteobacteria* [180–183]. To explore further, researchers identified 21 persistently dysregulated microbial taxa in Cd-exposed mice, which were significantly associated with poor performance on the Novel Object Location (NOL) test. To further investigate the association between gut microbiome alterations and Cd-induced cognitive deficits, researchers identified 21 persistently dysregulated microbial taxa in Cd-exposed mice significantly associated with decreased performance on the NOL test.

Eight of these taxa strongly correlated with behavioral outcomes (adjusted $p<0.1$). Notably, Cd exposure led to a persistent reduction in beneficial bacteria such as *UBA660* and members of *Erysipelotrichaceae*, commonly found in the human gut microbiota, while increasing the relative abundance of *Flintia butyricum*,

Faecalibacterium prausnitzii, *Flavonifractor*, *and Escherichia*, *Enterobacteriaceae*, and *Proteobacteria* [180].

Functionally, Cd-induced dysbiosis impairs microbial metabolism, including reduced synthesis of short-chain fatty acids such as butyrate, vital for sustaining gut barrier function and regulating systemic inflammation [180, 181]. Experimental models further show that individuals with disrupted microbiota exhibit enhanced Cd toxicity and neuroinflammatory responses, implying a synergistic role of gut dysbiosis in amplifying Cd-induced neurodevelopmental deficits [181]. Gut microbiota alterations have been observed to precede cognitive and behavioral impairments in some Cd-exposed models, underscoring the role of the microbiome in mediating early-life neurotoxicity [184].

Given the similarity between Cd-induced microbial alterations and those observed in certain NDDs, including ASD, these findings suggest that environmental metal exposure may disrupt microbiome-mediated pathways critical for brain development, warranting further investigation.

3.4 Aluminum (Al)

The potential role of Al in NDDs, particularly in ASD, remains a subject of active investigation and scientific discussion. While some studies propose a possible association between Al exposure and ASD susceptibility, others report no significant correlation. Al is a well-documented neurotoxin capable of crossing the BBB, accumulating in neural tissue, disrupting cellular homeostasis, and inducing neuroinflammation. These processes could contribute to neurodegenerative pathways, cognitive deficits, and behavioral abnormalities [185]. Of particular interest to researchers is Al's use as an adjuvant in certain vaccines, which has raised concerns regarding its potential to induce prolonged neuroinflammatory responses linked to neurological sequelae. Investigations into the hypothesized connection between Al-containing adjuvants and ASD have yielded inconclusive or conflicting outcomes, with methodological variations complicating cross-study comparisons [186]. Extensive toxicological research indicates that typical Al exposure levels from vaccines, environmental sources, and dietary intake fall substantially below regulatory safety thresholds [187]. Despite public anxieties surrounding this issue, rigorous large-scale epidemiological studies and systematic meta-analyses have consistently failed to identify convincing evidence of causality between Al exposure and ASD pathogenesis. The scientific consensus maintains that current exposure levels in developed nations pose minimal risk to neurotypical populations. However, ongoing research explores potential subpopulation susceptibilities and molecular mechanisms, such as epigenetic modifications or BBB permeability variations, that might theoretically mediate Al-related neurotoxicity in atypical developmental contexts.

3.5 Arsenic (As)

As a naturally occurring metalloid, it poses a significant environmental and public health concern due to its widespread presence in drinking water and air. Naturally, arsenic is found in rocks, minerals, and soil across the Earth's crust. It can leach into groundwater as water flows through arsenic-rich geological formations, a process particularly prevalent in regions such as Bangladesh, India, China, and parts of the United States. However, anthropogenic activities have markedly intensified As contamination. Mining operations (notably gold mining), metal smelting, and the historical use of As-based pesticides, fertilizers, and wood preservatives contribute to elevated As levels in soil and water. Runoff from agricultural fields and leaching from industrial waste sites further exacerbate this contamination, allowing As to infiltrate surface and groundwater supplies. In the atmosphere, As adheres to particulate matter released during fossil fuel combustion, or smelting, and other industrial processes, eventually depositing onto land and water.

Additionally, As can enter the food chain through crops irrigated with contaminated water or grown in As-laden soils [188]. Organic As is ingested through the consumption of fish and seafood [189]. While dietary exposure tends to be less significant than drinking water, the cumulative environmental burden underscores arsenic's pervasive presence and potential neurodevelopmental toxicity, particularly during critical periods of brain maturation. As poses a significant global health concern, as a substantial portion of the world's population lives in regions where As concentrations exceed established safety limits [189, 190]. In both forms (organic and inorganic), As is absorbed into the bloodstream, can cross the BBB, and can accumulate in the brain, causing long-lasting effects [189, 191]. Its effect is rapid, right after exposure, within hours or may take up to a few weeks, often presenting with permanent effects such as ID and motor and memory alterations [191, 192]. As' potential to cause neuronal damage is similar to that of Pb and Hg, due to its ability to induce oxidative stress, via the ROS generation [190]. Some of As' negative impacts include mitochondrial dysfunction, inflammation, cellular DNA damage, and genotoxicity [190, 191]. Moreover, As can lead to endoplasmic reticulum stress response caused by protein misfolding in cells, and could result in cell death if not reversed [193]. Studies have shown that long-term exposure to As, particularly during the developmental phase, is a known cause of behavioral changes, cognitive impairments, and learning deficits [190, 191, 193]. Children exposed to As during prenatal or early childhood years are at increased risk for neurodevelopmental deficits, including lowered intelligence and altered behavior. A meta-analysis testing the potential effects on neurodevelopment and behavioral disorders in children exposed to As suggests that a 50% increase in urinary As levels is linked to a 0.4-point drop in IQ among children aged 5–15 [189].

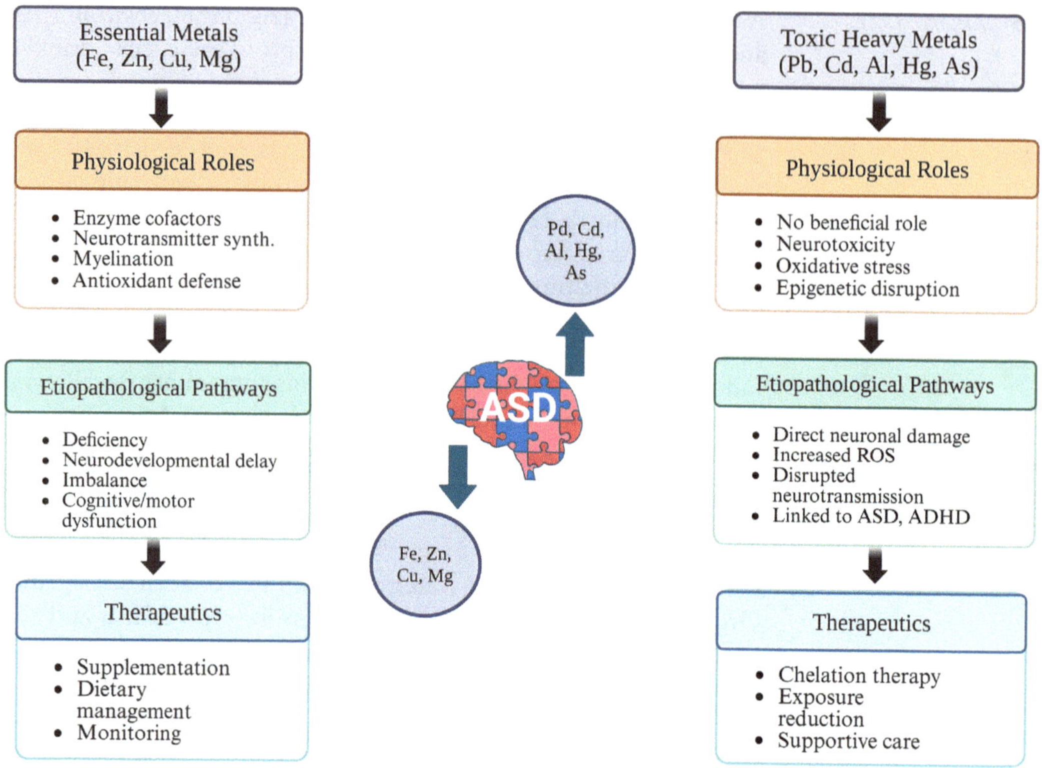

Fig. 3 A flowchart illustrates the contrasting roles of essential and toxic heavy metals in ASD

In summary, essential metals play indispensable and dynamic roles in brain function and neurodevelopment, whereas toxic metals can disrupt these processes and contribute to neurodevelopmental impairments. A summary is provided in Fig. 3. Table 1 summarizes the role of key essential metals, while Table 2 outlines the impact of toxic metals on brain function and neurodevelopment.

4 Therapeutic Approaches: Targeting Metal Dysregulation

To reiterate, metals are present in small quantities in the human body, yet are crucial for developing and maintaining many biological processes, such as metabolism, maintaining protein structures and functions, catalyzing enzyme activities, regulating signal transduction, and maintaining genetic stability. Disrupted metal homeostasis (deficiency, excess, and misallocation of metals) plays an important role as a driver of neurodevelopment. It can lead to oxidative stress, disruption of metal-dependent proteins, and neuroinflammation, although the studies may yield inconsistent outcomes [194].

Table 1
Role of essential metals in brain function and neurodevelopment

Metal	Brain distribution	Key functions	Effects of deficiency	Effects of excess	Transport and regulation
Zinc (Zn)	High in the hippocampus, amygdala, and white matter [76]	Neuromodulation, synaptic communication, antioxidant, enzyme regulation	Cognitive impairments, oxidative stress, and neuronal apoptosis	Seizure susceptibility, tissue damage, and oxidative stress	Crosses BBB and BCB, involved in Zn-containing glutamatergic neurons
Copper (Cu)	High in the substantia nigra, locus coeruleus, dentate nucleus, cerebellum	Neurotransmitter synthesis, oxidative stress regulation, and inflammation control	Impaired neurotransmission, oxidative stress	Neurotoxicity, inflammation, oxidative stress, and NMDA receptor modulation	Crosses BBB and BCB, the CTR1 transporter regulates Cu uptake
Iron (Fe)	High in extrapyramidal regions, red blood cells in the brain capillaries	Oxygen transport, neurotransmitter synthesis, myelination, and synaptic activity	Impaired neuron development, energy deficiency	Oxidative stress, neurodegeneration	Regulated by Tf–TfR1 and NTBI pathways
Manganese (Mn)	Basal ganglia, cerebellum	Antioxidant function, brain development, protein glycosylation	Disrupted Mn homeostasis, altered neural function	Neurotoxicity, oxidative stress, dopamine oxidation	Transported via DMT1 and Zip8/14
Magnesium (Mg)	Throughout the brain, essential for BBB integrity	Neurotransmission, NMDA receptor modulation, and oxidative stress prevention	Increased glutamatergic activity, excitotoxicity, and inflammation	Overactive NMDA signaling, neuroinflammation	Regulates Ca homeostasis, blocks excess Ca influx

Metal	Brain distribution	Key functions	Effects of deficiency	Effects of excess	Transport and regulation
Sodium (Na)	Essential for body fluid homeostasis	Electrical neurotransmission, neuronal excitability	Neuronal dysfunction, impaired neurotransmitter transport	Hyperexcitability, disrupted neural activity	Maintained by Na/K-ATPase channels
Potassium (K)	Works with Na for neuronal excitability	Action potential regulation, neuronal repolarization	Hyperexcitability, impaired neuron function	Neuronal dysfunction, excessive depolarization	Regulated by astrocytes

Table 2
Role of toxic metals in brain function and neurodevelopment

Metal	Source of exposure	Mechanism of neurotoxicity	Effects on neurodevelopment	Association with ASD
Lead (Pb)	Contaminated water, air pollution, breastmilk, old paint, and industrial sources	Crosses BBB and placenta, induces ROS, impairs antioxidant defense, causes DNA damage, inhibits NMDA receptors, blocks Ca channels	Cognitive impairments, IDs, behavioral disorders, irreversible CNS damage	Elevated Pb in ASD children; linked with cognitive/psychomotor impairments, especially in co-exposure with Hg
Mercury (Hg)	Fish (MeHg), dental amalgams, Thimerosal in vaccines, vapor inhalation	Crosses BBB/placenta, induces ROS, disrupts Ca homeostasis, interferes with neurotransmitters (DA, 5-HT, ACh), alters gut–brain axis	CP, IDs, sensory impairments, developmental delays	Elevated Hg in blood/hair of ASD individuals; gut–brain alterations seen in rodent models
Cadmium (Cd)	Smoking, industrial fumes, contaminated food/water, seafood	Disrupts BBB, alters Ca/Zn balance, enters neurons via Ca channels, affects mitochondria/ATP synthesis, increases ROS	Cognitive dysfunction, motor issues, behavioral problems (anxiety, sleep issues)	Higher Cd levels in ASD children's hair, blood, and urine; prenatal exposure linked with ASD risk
Aluminum (Al)	Food, water, cookware, vaccine adjuvants	Accumulates brain tissue, induces inflammation, has potential for mitochondrial dysfunction, and generates ROS	Neurodegeneration, behavioral/cognitive decline	Detected in the brain tissues of individuals with ASD; public concern exists, but no definitive causal link has been established
Arsenic (As)	Contaminated water, air/dust pollution, seafood (organic form)	Crosses BBB, generates ROS, causes cytotoxicity, mitochondrial dysfunction, DNA damage, and inflammation	Memory/motor deficits, ID, neurodegeneration	Suggested risk factor: exposure during pregnancy associated with developmental and cognitive deficits

Therapeutic approaches targeting the restoration of metal homeostasis hold considerable potential in managing NDD-associated issues. By addressing the underlying cause of metal dyshomeostasis, therapeutic approaches can go beyond symptomatic relief to potentially slow or modify disease progression, especially in neurodevelopmental and neurodegenerative conditions [195, 196]. Targeted therapies, such as diet intervention, metal chelators, and chaperones, aim to restore the balance of metal ions in the body. These interventions have been shown to reduce the generation and aggregation of toxic proteins, mitigate synaptic dysfunction, and decrease oxidative stress and neuroinflammation [197, 198].

Metals found in nature accumulate in the environment because of human activities, often through inhalation, ingestion, or skin contact [199, 200]. While non-occupational exposures can cause toxic effects through food intake, polluted air, water, and cigarette smoke, inducing pathological changes [201], occupational metal toxicity can result in serious physical and mental health issues, such as neurological damage associated with Pb and Hg, and respiratory problems related to Cr and Nickel [202], cancer with As and Cd [203] and reproductive problems with Pb and Hg [204]. Advocacy for regulations, policies, prevention and mitigation strategies to address occupational health issues may include personal protection through wearing masks, gloves, and protective clothing, and use of ventilation systems, enclosures, and substitution of hazardous materials [205, 206].

4.1 Dietary and Lifestyle Interventions Targeting Metal Dysregulation in NDDs

Nutritional strategies are crucial in maintaining metal homeostasis, particularly in individuals with NDDs. Dietary interventions aim to correct imbalances, mitigate toxic metal accumulation, and optimize neurological function implicated in NDDs such as ASD.

Evidence suggested that a healthy, well-balanced, nutritious diet comprising whole, unprocessed foods such as fruits, vegetables, whole grains, lean proteins, and healthy fats can protect and alleviate the consequences of environmental toxicants in humans [207]. Plant foods, including vegetables and fruits, as well as whole food components, are important dietary sources of antioxidants and can be used to decrease overall toxicity, promoting beneficial effects against metal toxicity in humans [207, 208]. It is advisable to avoid fish high in Hg and choose options with lower Hg content, such as salmon, sardines, and anchovies. Additionally, avoid consuming foods stored in Pb-soldered cans or prepared using Pb-glazed ceramics [209]. Moreover, test the water for metal contamination, especially if people live in areas with old pipes or nearby industrial sites, and also limit use of pesticides and herbicides, which can contain metals like As and Pb [210].

Besides, implementing health-promoting behaviors can improve body detoxification through natural mechanisms, which can help reduce the accumulation of toxic metals such as Cd, Hg and Pb. Proper hydration is crucial to eliminate toxic metal levels through renal excretion; thus, people are encouraged to drink sufficient, clean water [211]. Meditation, yoga, and deep breathing exercises are counted as stress reduction therapies to control physiological responses in the body and may indirectly support and maintain detoxification processes [212, 213]. Moreover, adequate sleep and regular physical activity may help to enhance circulation and promote endogenous detoxification pathways [214].

Detoxification Through Glutathione

GSH plays a vital role as a significant antioxidant in metal elimination through a metal-binding complex [215]. GSH comprises a tripeptide (glutamate, cysteine, and glycine) and protects against oxidative stress and heavy metal toxicity [215–217]. It directly chelates metals such as Hg, As, Cd and Pb, neutralizing their harmful effects [216]. Additionally, GSH supports phase II liver detoxification, which conjugates with toxic compounds to render them water-soluble for excretion [218]. Diet plays a pivotal role in maintaining optimal GSH levels, with sulfur-rich foods such as garlic, onions, cruciferous vegetables (broccoli, Brussels sprouts, cabbage), and protein and omega-3 sources like eggs and fish providing essential precursors for GSH synthesis [215]. Other dietary components, including vitamin B, C, E, Se and polyphenols (green tea, juices), further enhance GSH activity and regeneration [215]. By incorporating these nutrient-dense foods, individuals can naturally support GSH production and strengthen the body's ability to detoxify harmful metals effectively.

Anti-Neuroinflammatory Properties of Omega-3 Fatty Acids

An imbalance in dietary fats, notably an excessive intake of polyunsaturated omega-6 fatty acids and a disrupted omega-6 to omega-3 ratio, as commonly seen in Western diets, contributes to neuroinflammation, heavy metal accumulation, and the progression of brain-related conditions [219]. To elaborate, elevated Hg, Al and As levels were linked to a high omega-6 to omega-3 ratio, contributing to cognitive decline, neuroinflammation, mood disorders, and impaired brain function. A personalized diet and supplement plan was found to lower this ratio, enhancing omega-3 levels, promoting heavy metal detoxification, and significantly improving symptoms [219]. Another study highlighted omega-3 fatty acids as potent antiapoptotic, antioxidant, and anti-inflammatory agents, beneficial in treating brain, liver, kidney and heart diseases. Furthermore, these also mitigated Pb-induced toxicity, protecting cholinergic neurotransmission, supporting brain development, and reversing learning and memory impairment [220]. Moreover, omega-3 fatty acids significantly mitigated oxidative stress and inflammation in the brains of Cd-exposed rats [221]. The dietary sources of omega-3 fatty acids include fatty fish (salmon, sardines), fish oil, flaxseeds, and walnuts

[222], and their regular inclusion in the diet can mitigate oxidative stress induced by metal imbalances and promote neuronal health.

Probiotics and Gut Health in Metal Metabolism

A healthy gut microbial community is crucial for maintaining intestinal homeostasis and gut–brain axis balance; however, the negative impact of heavy metals on the microbial community can lead to dysbiosis [223]. Disruptions in gut microbial ecology can influence microbial metabolism, potentially altering the physical gut barrier and bioavailability of heavy metals, which impacts their absorption, detoxification, and overall metabolism in the body [223]. Probiotics have been reported to reduce heavy metal absorption by elevating the heavy metal sequestration in the intestines, detoxification, maintaining the gut barrier integrity, and altering the metal transporter expression in the gut [224, 225]. Probiotic-rich foods such as yogurt, kefir, and fermented vegetables enhance gut microbiota diversity, such as *Lactobacillus, Bifidobacterium, Streptococcus*, and *Enterococcus species* [226]. Furthermore, developing research focuses more on the protective benefits of gut microbes against inflammation and oxidative stress and enlightens heavy metal–induced gut dysbiosis [227]. These dietary interventions may help to improve metal balance and neurological outcome in individuals with NDDs.

4.2 Metal Supplementation

Nutritional supplementation may play a protective role. Antioxidants such as vitamin C and essential trace elements like Zn and Se have been demonstrated to mitigate metal-induced oxidative stress. They may contribute to reducing metal burden when included as part of a comprehensive dietary plan. However, the use of such supplements should be guided by a qualified healthcare professional to ensure safety and efficacy [228].

Zinc Supplementation

Given the role of Zn in neurodevelopment and detoxification, its supplementation has been investigated as a possible intervention for Zn deficiency, and/or an elevated Cu/Zn ratio that has been observed not only in ASD but also in other NDDs, particularly ADHD, where they are associated with greater symptom severity [229], reiterating that Zn dysregulation may play a broader role across multiple NDDs.

A study reported significant improvements in ASD symptoms following Zn and vitamin B6 therapy. To specify, the severity of symptoms decreased in individuals with ASD in terms of awareness, receptive language, focus and attention, hyperactivity, tip-toeing, eye contact, sound sensitivity, tactile sensitivity, and seizures. However, in individuals with Asperger's Syndrome, no improvements were observed [75]. Furthermore, this study found that individuals with ASD had lower plasma Zn levels and significantly higher Cu levels than neurotypical controls. After Zn and vitamin B6 supplementation, Cu levels normalized in individuals with ASD but not those with Asperger's. This suggests a possible interaction between Zn,

Cu, and neurotransmitter regulation, further supporting the role of Zn in ASD symptom management.

Other research has shown that Zn supplementation in individuals with ASD improved behavioral symptoms, including enhanced social interaction, reduced hyperactivity, and better attention regulation. These findings align with Zn's role in:

Combating oxidative stress: As discussed earlier, Zn protects protein sulfhydryl groups from oxidative damage, competes with redox-active metals to reduce their harmful effects, and serves as a cofactor for antioxidant enzymes like Cu/Zn SOD [230–232]. Zn acts as an indirect antioxidant agent by inhibiting the oxidation of DNA, RNA, and protein molecules and inhibiting the inflammatory response, eventually resulting in the downregulation of ROS production and thus enhancing the body's ability to counteract free radical damage [232].

Detoxification of heavy metals: The presence of Zn may help dilute the effect of other toxic metals, such as Cd, Pb, As, Hg, as follows:

Support in Cd toxicity: Several studies have shown that Zn can reduce Cd toxicity [166, 233]. Dietary and serum Zn levels in U.S. adults are linked to Cd exposure, influencing its absorption and accumulation [234]. They found that Zn supplement dose and duration were inversely associated with blood Cd and serum Zn. In adjusted regression models, a 10% increase in serum Zn was associated with a 2% decrease in blood Cd and a 4% increase in urinary Cd, suggesting a role for Zn in Cd detoxification.

Support in Pb toxicity: Recent studies show that Zn supplementation reduces Pb accumulation in tissues by competing with Pb for absorption and restores key antioxidant enzymes [235]. In rats, Zn co-exposure mitigated Pb-induced delta aminolaevulinic acid dehydratase enzyme (ALAD) inhibition and enhanced catalase activity [235]. Additionally, mRNA dysregulation in antioxidant and antiapoptotic enzymes encoding genes was minimized by Zn supplementation in rats [235]. Similarly, in mice, dietary Zn lowered Pb levels in blood and testes and improved Cu/Zn–SOD activity, supporting Zn's dual role in Pb competition and oxidative protection [236]. While some studies have focused on other organs, these protective effects are also relevant to the CNS. By reducing Pb deposition and oxidative stress, Zn may help prevent Pb-induced neurotoxicity and support better neurological function.

Support in As toxicity: Chronic low-level As exposure and Zn deficiency pose significant public health concerns, impacting the CNS [237]. Studies have demonstrated Zn's protective effects against As-induced apoptosis in neuronal cells [238, 239], while Zn deficiency with As exposure exacerbated neurotoxicity in zebrafish embryos, disrupting Zn homeostasis, antioxidants, and neuronal genes [237]. In rats, Zn (20 mg/kg/d) co-exposure with As (5 mg/kg/d) reversed oxidative stress markers [240]. Similarly, Zn and vitamins C and E mitigated As-induced oxidative stress and neurotransmitter imbalances in developing rat brains [241, 242]. Zn modulates As-induced neurotoxicity via multiple protective mechanisms. Another study reported that Zn (5 mg/kg) co-exposure reversed oxidative stress markers, with N-acetylcysteine being more effective than Zn alone during As exposure [243]. Zn reversed As-induced oxidative stress without altering baseline oxidative status, suggesting metallothionein involvement [244], although Kreppel et al, found no metallothionein role in Zn-mediated neuroprotection [239].

Human studies indicate a complex interaction. Zeneli et al. reported a positive correlation between blood As and serum Zn levels in exposed workers, suggesting an alteration in the antioxidant defense system [245], while Kordas et al. found no effect of Zn status on As metabolism in Mexican schoolchildren, indicating limited application of Zn supplementation on As toxicity [246]. However, Skalny et al. observed that Zn supplementation reduced As concentration in rats in 7 days, while its presence in hair and serum remained relatively stable [247]. Similarly, Zn supplementation reduced As accumulation in the liver and brain of rats, supporting its neuroprotective potential [248]. Zn counteracts As-induced neurotoxicity by restoring oxidative balance, modulating apoptosis, and potentially regulating metallothionine expression.

Support in Hg toxicity: Zn's protective mechanism against Hg toxicity via metallothionein formation and antioxidant activity has been reported [249]. Additionally, Zn has been shown to mitigate Hg-induced nephrotoxicity in rats, likely through Hg-metallothionine binding [250]. Similarly, in female rats, Zn reduced renal oxidative stress and inhibited Hg accumulation, suggesting a protective role of Zn against inorganic Hg toxicity [251]. Furthermore, Zn protection against Hg-induced biomolecular changes in rat pups was unaffected by Zn pre-treatment [252]. While Hg-induced apoptosis was reported in rat pheochromocytoma cell line (PC12 cells) via ROS generation, Zn mitigated the toxicity via its antioxidant properties, and restored antioxidant GSH levels, contributing to the protection of the cells following Hg administration [253].

A combined therapy with Zn, Selenium, and dithiothreitol chelator or N-acetylcystine chelator counteracted dimethyl Hg-induced tissue damage in rat liver, kidney, and brain [254, 255]. Similarly, combining Zn and N-acetylcystine completely prevented liver and blood Hg accumulation in rats due to increased hepatic metallothionein levels, suggesting promising protective effects against Hg toxicity [256].

While it is established that Zn supplementation does have a protective role against Hg toxicity, limited cellular-level investigations have been conducted on Zn–Hg interactions in the CNS. Further studies are needed to elucidate the precise neuroprotective mechanisms of Zn against Hg toxicity.

Protective role in Zn finger motifs: Zn-binding proteins are abundant in the proteome and are crucial in various biological processes [257]. Many of these proteins contain Zn-binding motifs, specialized structural elements that coordinate Zn ions [258]. Approximately 5% of human proteins consist of Zn finger domains [258], which facilitate the regulation of pre- and post-translational modifications [258, 259]. Zn finger domains involve various cellular and molecular functions [259]. Toxic metals, including Cd, Co, Pb and As, may interrupt Zn finger proteins by oxidizing the cysteine residues, which are important for their structure [259]. Perturbation in Zn finger motifs may inhibit the activity of the DNA repair protein and cause DNA damage [260]. These damages may need more concern in NDDs such as ASD [261–263].

In-vitro studies indicate that adequate or supplemental Zn can counteract the toxic effects of metal binding to Zn finger motifs and protein function by exerting its protective effects through competition with toxic metals for absorption and transport [260]. In the context of As-induced inhibition of DNA repair, supplemental Zn largely restores protein Zn levels and DNA repair capacity in both cell models and in vivo [260]. The critical balance between Zn and toxic metals is evident in studies showing that metal toxicity worsens under Zn-deficient conditions [234, 264]. Overall, these findings suggest that Zn can mitigate the harmful effects of toxic metals on Zn-binding proteins.

Regulation of metallothionein: Zn influences metallothionein expression, a metal-binding protein family that isolates heavy metals and prevents their toxic interactions with cellular components [265, 266]. Increased metallothionein levels have been linked to enhanced detoxification and reduced oxidative damage in metal-exposed individuals [27]. In a study of a cohort of children with ASD, a 3-month supplementation therapy with Zn improved the genetic expression and serum concentration of metallothionine 1, and increased cognitive-motor performance among these children, indicating its role as a therapeutic agent in ASD [266]. Metallothionein also helps neutralize ROS by oxidizing its abundant cysteine residues [233, 265]. Previous studies have shown that some toxic metals, like Chromium (Cr) and As, have low binding affinity with metallothionein [232]. Moreover, Zn deficiency may cause reduced metallothionein and increased Cr toxic levels [267]. In contrast, Zn supplementation protects against toxic metals by promoting metallothionein production and maintaining redox balance [244].

Restoring glutamatergic and GABAergic neurotransmission: Glutamate and GABA are the brain's primary excitatory and inhibitory neurotransmitters, balancing neural activity [268]. Glutamate promotes excitation that helps in learning and memory, while GABA promotes inhibitory neurotrasmission, reducing excessive neuronal activity and supporting relaxation and focused attention [269]. Since

ASD is often associated with an imbalance between excitation (glutamate) and inhibition (GABA), Zn may help restore this balance [27, 270]. This relationship may also exist in some of the individuals with Fragile-X syndrome [271]. Zn plays a role in synaptic plasticity, which is critical for learning and memory. It supports brain-derived neurotrophic factor (BDNF), essential for neuronal growth and function [272]. Zn supplementation may support neurotransmitter balance, synaptic function, and neuroprotection, potentially benefiting individuals with ASD [27].

Metal transporters: Many metal transporters, such as Zrt/Irt-like proteins (ZIP) and Zn transporters (ZnT), primarily transport Zn but can also interact with other metals like Cd and Pb [273, 274]. Some transporters, such as ZIP14 and ZIP8, are specifically identified as Zn transporters, although they also mediate the uptake and transport Mn, Cd, Cu, and Fe [275, 276]. Moreover, Zn can also regulate the expression of transporter genes, help reduce the absorption of toxic metals, and protect against metal toxicity [232, 275]. Co-exposure of Zn and Se revealed that Zn can lower As levels while As can alter the expression of Zn transporter in different tissues such as liver and kidneys [277], which suggested that metal distribution and elimination can depend on transporter expression and mechanism [230]. Although Zn–metal interactions at the transporter level and their complexity are poorly explained, evidence shows that Zn can reduce metal accumulation and lower their toxicity in tissues.

Iron Supplementation

Fe deficiency is associated with impaired cognitive function and overactive performance in children diagnosed with ASD and ADHD [278–280]. A controlled trial study reported noticeable improvement in ADHD symptoms after increasing the dose of ferrous sulfate (80 mg/day) in comparison with the control group [280]. Fe deficiency is associated with impaired function of monoamine transporters, causing disproportion in neurotransmitter activity, thus highlighting the significance of an appropriate level of brain Fe to support monoamine-dependent cognitive functions in ADHD [280].

Magnesium Supplementation

Clinical studies suggest that Mg supplementation, particularly in combination with vitamin B6, may ameliorate behavioral and cognitive symptoms in some children with ASD and ADHD [281, 282]. As a natural antagonist of NMDA receptors, Mg may also play a protective role in reducing seizure susceptibility by modulating excitatory neurotransmission [102].

4.3 Chelation Therapy

One therapeutic approach to treat metal overload disorders, such as hemochromatosis (Fe) and Wilson's disease (Cu), involves the use of metal chelators, including desferrioxamine, penicillamine, and trientine. Chelating agents can be administered intravenously or orally to bind excess metals, remove them from overloaded tissues, and facilitate their excretion through urine. Although several chelating agents are approved for systemic metal intoxication or overload, there is currently no approved chelator specifically designed to efficiently penetrate the BBB and selectively restore brain metal homeostasis. [283].

These chelating agents typically form ring-like structures incorporating metal cations through coordinate bonding with electron-donating atoms such as sulfur, nitrogen, and oxygen [284]. The route of metal excretion, either via bile through the liver or urine through the kidney, depends mainly on the physicochemical properties of the chelating agent, particularly its lipophilicity or hydrophilicity [284]. Some of the common pharmaceutical chelating agents include dimercaprol (DMPA), dimercaptopropane sulfonate (DMPS), and dimercaptosuccinic acid (DMSA).

However, in our body, metal-binding proteins, such as metallothioneins, serve as potent chelators of heavy metals and play a crucial role in the body's natural defense against these toxic elements [285]. Similarly, GSH functions as a key chelator, contributing to cellular detoxification, metal transport, and excretion, while also serving as a biomarker for toxic metal accumulation [284].

Mehrandish et al. [285] reviewed natural compounds for chelation therapy that can offer potential therapeutic benefits with fewer side effects than conventional chemical chelators.

Studies suggest garlic can prevent Cd-induced kidney damage [286] and mitigate oxidative damage caused by Pb exposure in rats [287]. Milk thistle (*Silybum marianum*), a flowering plant in the aster or daisy family, native to Mediterranean regions, contains silymarin, which has been studied for its ability to chelate metals and protect against heavy metal-induced oxidative damage [288]. Cilantro (*Coriandrum sativum*) has demonstrated potential in mobilizing heavy metals such as Hg from body tissues and facilitating excretion [289]. *Ginkgo biloba* exhibits neuroprotective effects that may help counteract heavy metal-induced neurotoxicity via metal chelating activity [285]. Turmeric (*Curcuma longa*) contains curcumin that exhibits metal-binding properties attributed to reducing metal-induced toxicity [290]. Green algae (*Chlorophyta*) have high chlorophyll content as natural chelators, binding to heavy metals and promoting their elimination [285].

Sulfur-containing amino acids, including taurine and methionine, are critical in metal chelation [291]. These amino acids are particularly abundant in excitable tissues and help reduce oxidative stress caused by heavy metal exposure [291]. Some professional specialists use taurine supplements to enhance metal detection before hair analysis [291]. Alpha-lipoic acid (ALA) is a potent antioxidant that functions as a natural metal chelator to eliminate toxic metals [292]. Studies suggested that dietary fiber intake may reduce Hg levels in the brain and bloodstream [285]. High fiber foods such as bran from grains and fruits serve as natural chelators and can help to reduce metals' enterohepatic circulation and control gut flora [293]. N-Acetyl-Cysteine (NAC) is a main precursor in GSH synthesis and serves as a chelator for harmful metals [294]. In the presence of vitamin C and E, GSH boosts metal-binding capacity [285, 294]. In conclusion, integrating natural chelators through diet and/or supplementation may improve the detoxification processes in the body and help reduce the deleterious effects of toxic metals.

4.4 Metal-Based Drug Discovery

Metallodrug discovery is an interdisciplinary field that combines pharmaceutical or medicinal chemistry to develop metal-based therapeutic agents to treat specific disorders. [295].

These metal complexes manifest unique features by coordinating with one or more ligands, including a strong binding tendency for biomolecules, catalytic effect, and oxidoreduction activity, such as oxaliplatin, carboplatin, gallium nitrate, and cisplatin [296].

Metal complexes specifically target particular biological molecules and pathways [297]. Advanced synthetic strategies are used to develop new metal complexes with unique properties to target the biomechanism, such as inhibiting DNA replication and transcription [40, 298].

As the field continues to evolve, the development of precise detection methods and the implementation of integrative studies are crucial for advancing our understanding of how metal imbalances contribute to the onset and progression of ASD and other NDDs. This knowledge will be instrumental in guiding the development of effective preventive and therapeutic strategies to advance detection methods, which will help to understand the role of metal disturbance in the onset of ASD and NDDs.

5 Conclusion and Future Directions

In summary, NDDs such as ASD are on the rise globally and are influenced by a complex interplay of genetic and environmental factors, including disruptions in metal homeostasis. Essential metals like Zn, Cu, and Fe are vital for proper brain development, synaptic functioning, and gut–brain communication. In contrast, accumulating nonessential heavy metals such as Pb, Hg, Cd, Al, and As can pose significant neurotoxic threats. Dysregulation of these metals, whether through deficiency, excess, or environmental exposure, has been increasingly associated with ASD pathophysiology, gut microbiota alterations, and immune responses. As the field advances, precise detection methods and integrative studies are essential to deepen our understanding of how metal imbalances contribute to the onset and progression of ASD and other NDDs, to establish preventative and therapeutic strategies.

Acknowledgments

All figures in the paper were created using BioRender The study was supported by the ASPIRE grant (21M149-AARE20-260) and the START-UP grant (G00003323), which was funded by the Office of the Provost for Research and Graduate Studies, United Arab Emirates University.

Author Contributions Asma Ahmadani and Rukhsana Nawaz: Conceptualization, Writing—Original Draft, Writing— Review & Editing. Sandeep Subramanya and Maria Cristina D'Adamo: Writing—Review & Editing. Amita Attlee and Mauro Pessia: Conceptualization, Writing—Review & Editing, Supervision.

Competing Interests The authors declare no competing interests.

References

1. Diagnostic and Statistical Manual of Mental Disorders | Psychiatry Online [Internet]. American Psychiatric Association; 2022 [cited 2025 May 21]. Available from https://psychiatryonline.org/doi/book/10.1176/appi.books.9780890425787
2. Scassellati C, Bonvicini C, Benussi L, Ghidoni R, Squitti R (2020) Neurodevelopmental disorders: Metallomics studies for the identification of potential biomarkers associated to diagnosis and treatment. J Trace Elem Med Biol 60:126499
3. Maret W (2016) The metals in the biological periodic system of the elements: concepts and conjectures. Int J Mol Sci 17(1):66
4. Que EL, Domaille DW, Chang CJ (2008) Metals in neurobiology: probing their chemistry and biology with molecular imaging. Chem Rev 108(5):1517–1549
5. Cory-Slechta DA, Sobolewski M, Oberdörster G (2020) Air pollution-related brain metal dyshomeostasis as a potential risk factor for neurodevelopmental disorders and neurodegenerative diseases. Atmosphere 11(10):1098
6. Amadi CN, Orish CN, Frazzoli C, Orisakwe OE (2022) Association of autism with toxic metals: a systematic review of case-control studies. Pharmacol Biochem Behav 212:173313
7. Saghazadeh A, Ahangari N, Hendi K, Saleh F, Rezaei N (2017) Status of essential elements in autism spectrum disorder: systematic review and meta-analysis. Rev Neurosci 28(7):783–809
8. Livingstone C (2015) Zinc: physiology, deficiency, and parenteral nutrition. Nutr Clin Pract 30(3):371–382
9. Wieringa FT, Dijkhuizen MA, Fiorentino M, Laillou A, Berger J (2015) Determination of zinc status in humans: which indicator should we use?. Nutrients. 7(5):3252–3263
10. Feng J, Shan L, Miao C, Xue Y, Yue X, Jia F (2023) The association of vitamin A, zinc and copper levels with clinical symptoms in children with autism spectrum disorders in Jilin Province, China. BMC Pediatr 23(1):173
11. Baj J, Flieger W, Flieger M, Forma A, Sitarz E, Skórzyńska-Dziduszko K et al. (2021) Autism spectrum disorder: trace elements imbalances and the pathogenesis and severity of autistic symptoms. Neurosci Biobehav Rev 129:117–132
12. Sauer AK, Malijauskaite S, Meleady P, Boeckers TM, McGourty K, Grabrucker AM (2021) Zinc is a key regulator of gastrointestinal development, microbiota composition and inflammation with relevance for autism spectrum disorders. Cell Mol Life Sci 79(1):46
13. Sauer AK, Grabrucker AM (2019) Zinc deficiency during pregnancy leads to altered microbiome and elevated inflammatory markers in mice. Front Neurosci 13:1295
14. Rihel J (2018) Copper on the brain. Nat Chem Biol 14(7):638–639
15. Burkhead JL, Gogolin Reynolds KA, Abdel-Ghany SE, Cohu CM, Pilon M (2009) Copper homeostasis. New Phytol 182(4):799–816
16. Zlatic S, Comstra HS, Gokhale A, Petris MJ, Faundez V (2015) Molecular basis of neurodegeneration and neurodevelopmental defects in Menkes disease. Neurobiol Dis 81:154–161
17. Levi S, Ripamonti M, Moro AS, Cozzi A (2024) Iron imbalance in neurodegeneration. Mol Psychiatry 29(4):1139–1152

18. Rabaya S, Nairat S, Bader K, Herzallah MM, Darwish HM (2023) Iron metabolism in autism spectrum disorder; inference through single nucleotide polymorphisms in key iron metabolism genes. J Neurol Sci 453:120817
19. Bener A, Khattab AO, Bhugra D, Hoffmann GF (2017) Iron and vitamin D levels among autism spectrum disorders children. Ann Afr Med 16(4):186–191
20. Sousa L, Oliveira MM, Pessôa MTC, Barbosa LA (2020) Iron overload: effects on cellular biochemistry. Clin Chim Acta 504:180–189
21. Balachandran RC, Mukhopadhyay S, McBride D, Veevers J, Harrison FE, Aschner M et al. (2020) Brain manganese and the balance between essential roles and neurotoxicity. J Biol Chem 295(19):6312–6329
22. Aschner M, Martins AC, Oliveira-Paula GH, Skalny AV, Zaitseva IP, Bowman AB, et al (2024) Manganese in autism spectrum disorder and attention deficit hyperactivity disorder: the state of the art. Curr Res Toxicol 6, 100170
23. Skalny AV, Mazaletskaya AL, Ajsuvakova OP, Bjørklund G, Skalnaya MG, Chernova LN et al. (2020) Magnesium status in children with attention-deficit/hyperactivity disorder and/or autism spectrum disorder. J Kor Acad Child Adolesc Psychiatry 31(1):41
24. Rose CR, Verkhratsky A (2016) Principles of sodium homeostasis and sodium signalling in astroglia. Glia. 64(10):1611–1627
25. Huang S, Dong W, Lin X, Bian J (2024) Na+ /K+-ATPase: ion pump, signal transducer, or cytoprotective protein, and novel biological functions. Neural Regen Res 19(12):2684
26. D'Adamo MC, Liantonio A, Conte E, Pessia M, Imbrici P (2020) Ion channels involvement in neurodevelopmental disorders. Neuroscience 440:337–359
27. Ahmadani A, Kittana M, Al-Marzooq F, Subramanya S, D'Adamo MC, Attlee A, et al (2025) Zinc ion dyshomeostasis in autism spectrum disorder. Nutr Res Rev 13:1–65
28. de Lores Arnaiz GR, Ordieres MGL (2014) Brain Na+, K+-ATPase activity in aging and disease. Int J Biomed Sci 10(2):85
29. Ghezzo A, Visconti P, Abruzzo PM, Bolotta A, Ferreri C, Gobbi G et al. (2013) Oxidative stress and erythrocyte membrane alterations in children with autism: correlation with clinical features. PLoS One 8(6):e66418
30. Gorini F, Muratori F, Morales MA (2014) The role of heavy metal pollution in neurobehavioral disorders: a focus on autism. Rev J Autism Dev Disord 1(4):354–372
31. Ijomone OK, Ukwubile II, Aneke VO, Olajide TS, Inyang HO, Omotosho OI et al. (2025) Glial Perturbation in metal neurotoxicity: implications for brain disorders. Neuroglia 6(1):4
32. Ding M, Shi S, Qie S, Li J, Xi X (2023) Association between heavy metals exposure (cadmium, lead, arsenic, mercury) and child autistic disorder: a systematic review and meta-analysis. Front Pediatr [cited 2025 Jun 10] 11. Available from https://www.frontiersin.org/journals/pediatrics/articles/10.3389/fped.2023.1169733/full
33. Gunn AP, Roberts BR (2017) Analysis of trace elements and metalloproteins in fractionated human brain samples using size exclusion inductively coupled mass spectrometry. In: White AR (ed) Metals in the brain [Internet]. Springer New York, NY, pp 119–25 [cited 2025 May 28] (Neuromethods; vol 124). Available from http://link.springer.com/10.1007/978-1-4939-6918-0_7
34. Garcia JS, de Magalhães CS, Arruda MAZ (2006) Trends in metal-binding and metalloprotein analysis. Talanta 69(1):1–15
35. White AR, editor (2017) Metals in the Brain [Internet]. Springer, New York, NY [cited 2025 May 21]. (Neuromethods; vol 124). Available from http://link.springer.com/10.1007/978-1-4939-6918-0
36. Collingwood JF, Adams F (2017) X-ray microscopy for detection of metals in the brain. In: White AR (ed). Metals in the brain [Internet]. Springer, New York, NY, pp 7–32 [cited 2025 May 28]. (Neuromethods; vol. 124). Available from http://link.springer.com/10.1007/978-1-4939-6918-0_2
37. Qin Y, Gee KR, Jin Q, Lai B, Qian C, Colvin RA (2017) Current methods used to probe and quantify intracellular total and free Zn(II) dynamics, and subcellular distribution in cultured neurons. In: White AR (ed). Metals in the brain [Internet]. Springer, New York, NY, pp 195–224 [cited 2025 May 28]. (Neuromethods; vol 124). Available from http://link.springer.com/10.1007/978-1-4939-6918-0_11
38. Bjorklund G (2013) The role of zinc and copper in autism spectrum disorders. Acta Neurobiol Exp 73(2):225–236
39. Chauhan A, Chauhan V (2006) Oxidative stress in autism. Pathophysiology. 13(3):171–181
40. Khan HY, Ansari MF, Tabassum S, Arjmand F (2024) A review on the recent advances of interaction studies of anticancer metal-based drugs with therapeutic targets, DNA and RNAs. Drug Discov Today 29(7):104055
41. Haraguchi H (2017) Metallomics: the history over the last decade and a future outlook. Metallomics 9(8):1001–1013
42. Haschka D, Hoffmann A, Weiss G (2021) Iron in immune cell function and host defense. Semin Cell Dev Biol 115:27–36
43. McCann S, Perapoch Amadó M, Moore SE (2020) The role of iron in brain development: a systematic review. Nutrients. 12(7):2001

44. Chang S, Wang P, Han Y, Ma Q, Liu Z, Zhong S et al. (2023) Ferrodifferentiation regulates neurodevelopment via ROS generation. Sci China Life Sci 66(8):1841–1857
45. DeBenedictis CA, Raab A, Ducie E, Howley S, Feldmann J, Grabrucker AM (2020) Concentrations of essential trace metals in the brain of animal species—a comparative study. Brain Sciences 10(7):460
46. Gh Popescu BF, Nichol H (2011) Mapping brain metals to evaluate therapies for neurodegenerative disease. CNS Neurosci Therap 17(4):256–268
47. D'Mello SR, Kindy MC (2020) Overdosing on iron: elevated iron and degenerative brain disorders. Exp Biol Med (Maywood) 245(16):1444–1473
48. Yu P, Chang YZ (2019) Brain IRon Metabolism and Regulation. In: Chang YZ, (ed). Brain Iron metabolism and CNS diseases [Internet]. Springer, Singapore, pp 33–44 [cited 2025 Mar 2]. (Advances in Experimental Medicine and Biology; vol. 1173). Available from http://link.springer.com/10.1007/978-981-13-9589-5_3
49. Galaris D, Barbouti A, Pantopoulos K (2019) Iron homeostasis and oxidative stress: an intimate relationship. Biochim Biophys Acta 1866(12):118535
50. Hare DJ, Ayton S, Bush AI, Lei P (2013) A delicate balance: Iron metabolism and diseases of the brain. Front Aging Neurosci [Internet] [cited 2025 Mar 2] 5. Available from https://www.frontiersin.org/journals/aging-neuroscience/articles/10.3389/fnagi.2013.00034/full
51. Bener A, Kamal M, Bener H, Bhugra D (2014) Higher prevalence of iron deficiency as strong predictor of attention deficit hyperactivity disorder in children. Ann Med Health Sci Res 4(Suppl 3):S291–S297
52. Walsh JA, Creighton SE, Rutherford MD (2016) Emotion perception or social cognitive complexity: what drives face processing deficits in autism spectrum disorder?. J Autism Dev Disord 46(2):615–623
53. Johnson CP, Myers SM (2007) American Academy of Pediatrics Council on Children With Disabilities. Identification and evaluation of children with autism spectrum disorders. Pediatrics. 120(5):1183–1215
54. Chen L, Guo X, Hou C, Tang P, Zhang X, Chong L, et al (2022) The causal association between iron status and the risk of autism: A Mendelian randomization study. Front Nutr [Internet] [cited 2025 Apr 19] 9. Available from https://www.frontiersin.orghttps://www.frontiersin.org/journals/nutrition/articles/10.3389/fnut.2022.957600/full
55. Gao Q, Zhou Y, Chen Y, Hu W, Jin W, Zhou C et al. (2025) Role of iron in brain development, aging, and neurodegenerative diseases. Ann Med 57(1):2472871
56. Kawamoto EM, Vivar C, Camandola S (2012) Physiology and pathology of calcium signaling in the brain. Front Pharmacol [Internet] [cited 2025 Feb 6] 3. Available from https://www.frontiersin.org/journals/pharmacology/articles/10.3389/fphar.2012.00061/full
57. Siesjö BK (2008) Calcium in the brain under physiological and pathological conditions. Eur Neurol 30(Suppl. 2):3–9
58. Maier JAM, Locatelli L, Fedele G, Cazzaniga A, Mazur A (2023) Magnesium and the brain: a focus on neuroinflammation and neurodegeneration. Int J Mol Sci 24(1):223
59. Bkaily G, Jacques D (2023) Calcium homeostasis, transporters, and blockers in health and diseases of the cardiovascular system. Int J Mol Sci 24(10):8803
60. Chen S, Xu D, Fan L, Fang Z, Wang X, Li M (2022) Roles of N-methyl-D-aspartate receptors (NMDARs) in epilepsy. Front Mol Neurosci [Internet] [cited 2025 May 29] 14. Available from https://www.frontiersin.org/journals/molecular-neuroscience/articles/10.3389/fnmol.2021.797253/full
61. Tumdam R, Hussein Y, Garin-Shkolnik T, Stern S (2024) NMDA receptors in neurodevelopmental disorders: pathophysiology and disease models. Int J Mol Sci 25(22):12366
62. Blanke ML, VanDongen AMJ (2009) Activation mechanisms of the NMDA receptor. In: Van Dongen AM, (ed.) Biology of the NMDA receptor [Internet]. CRC Press/Taylor & Francis, Boca Raton (FL) [cited 2025 Feb 6] (Frontiers in Neuroscience). Available from http://www.ncbi.nlm.nih.gov/books/NBK5274/
63. Carmona A, Roudeau S, Ortega R (2021) Molecular mechanisms of environmental metal neurotoxicity: a focus on the interactions of metals with synapse structure and function. Toxics 9(9):198
64. Li Z, Wang G, Zhong S, Liao X, Lai S, Shan Y et al. (2020) Alleviation of cognitive deficits and high copper levels by an NMDA receptor antagonist in a rat depression model. Compr. Psychiatry 102:152200
65. Chen B, Yu P, Chan WN, Xie F, Zhang Y, Liang L et al. (2024) Cellular zinc metabolism and zinc signaling: from biological functions to diseases and therapeutic targets. Sig Transduct Target Ther 9(1):1–41
66. Benarroch E (2023) What are the functions of zinc in the nervous system?. Neurology 101(16):714–720

67. Wang Y, Song Y, Zhang L, Huang X (2024) The paradoxical role of zinc on microglia. J Trace Elem Med Biol 83:127380
68. Imbrici P, D'Adamo MC, Cusimano A, Pessia M (2007) Episodic ataxia type 1 mutation F184C alters Zn^{2+} -induced modulation of the human K^+ channel Kv1.4-Kv1.1/Kvβ1.1. Am J Physiol 292(2):C778–C787
69. Carrillo E, Bhatia NK, Akimzhanov AM, Jayaraman V (2020) Activity dependent inhibition of AMPA receptors by Zn2+. J Neurosci 40(45):8629–8636
70. Mott DD, Benveniste M, Dingledine RJ (2008) pH-dependent inhibition of kainate receptors by zinc. J Neurosci 28(7):1659–1671
71. Krall RF, Tzounopoulos T, Aizenman E (2021) The function and regulation of zinc in the brain. Neuroscience 457:235–258
72. Wang B, Fang T, Chen H (2023) Zinc and central nervous system disorders. Nutrients 15(9):2140
73. Hosie AM, Dunne EL, Harvey RJ, Smart TG (2003) Zinc-mediated inhibition of GABAA receptors: discrete binding sites underlie subtype specificity. Nat Neurosci 6(4):362–369
74. Menzikov SA, Zaichenko DM, Moskovtsev AA, Morozov SG, Kubatiev AA (2023) Zinc Inhibits the GABAAR/ATPase during postnatal rat development: the role of cysteine residue. Int J Mol Sci 24(3):2764
75. Russo AJ, deVito R (2011) Analysis of copper and zinc plasma concentration and the efficacy of zinc therapy in individuals with Asperger's syndrome, pervasive developmental disorder not otherwise specified (PDD-NOS) and autism. Biomark Insights 6:BMI.S7286
76. Takeda A (2000) Movement of zinc and its functional significance in the brain. Brain Res Rev 34(3):137–148
77. Sauer AK, Hagmeyer S, Grabrucker AM (2022) Prenatal zinc deficient mice as a model for autism spectrum disorders. Int J Mol Sci 23(11):6082
78. Fourie C, Vyas Y, Lee K, Jung Y, Garner CC, Montgomery JM (2018) Dietary zinc supplementation prevents autism related behaviors and striatal synaptic dysfunction in Shank3 exon 13–16 mutant mice. Front Cell Neurosci [cited 2025 Jun 21] 12. Available from https://www.frontiersin.org/journals/cellular-neuroscience/articles/10.3389/fncel.2018.00374/full
79. Zhang L, Xu X, Ma L, Wang X, Jin M, Li L et al. (2023) Zinc water prevents autism-like behaviors in the BTBR mice. Biol Trace Elem Res 201(10):4779–4792
80. Adamo AM, Oteiza PI (2010) Zinc deficiency and neurodevelopment: the case of neurons. BioFactors. 36(2):117–124
81. Oteiza PL, Olin KL, Fraga CG, Keen CL (1996) Oxidant defense systems in testes from zinc-deficient rats. Exp Biol Med 213(1):85–91
82. Virgili F, Canali R, Figus E, Vignolini F, Nobili F, Mengheri E (1999) Intestinal damage induced by zinc deficiency is associated with enhanced CuZn superoxide dismutase activity in rats: effect of dexamethasone or thyroxine treatment. Free Radic Biol Med 26(9–10):1194–1201
83. Zago MP, Mackenzie GG, Adamo AM, Keen CL, Oteiza PI (2005) Differential modulation of MAP kinases by zinc deficiency in IMR-32 cells: role of H_2O_2. Antioxidants Redox Signal 7(11–12):1773–1782
84. Bagchi D, Vuchetich PJ, Bagchi M, Tran MX, Krohn RL, Ray SD et al. (1998) Protective effects of zinc salts on TPA-induced hepatic and brain lipid peroxidation, glutathione depletion, DNA damage and peritoneal macrophage activation in mice. General Pharmacol 30(1):43–50
85. Oteiza PI, Olin KL, Fraga CG, Keen CL (1995) Zinc deficiency causes oxidative damage to proteins, lipids and DNA in rat testes. J Nutr 125(4):823–829
86. Fourie C, Vyas Y, Lee K, Jung Y, Garner CC, Montgomery JM (2018) Dietary zinc supplementation prevents autism related behaviors and striatal synaptic dysfunction in Shank3 exon 13-16 mutant mice. Front Cell Neurosci 12:374
87. Juárez-Rebollar D, Rios C, Nava-Ruíz C, Méndez-Armenta M (2017) Metallothionein in brain disorders. Oxid Med Cell Longev 2017:5828056
88. Mohamed FRR, Irfan MR, Raja A, Sattar MA, Firdous RM, Sharma A et al. (2025) The role of zinc dysregulation in autism spectrum disorder: a systematic review of genetic and environmental influences and therapeutic potential. Res Autism 124:202579
89. An Y, Li S, Huang X, Chen X, Shan H, Zhang M (2022) The role of copper homeostasis in brain disease. Int J Mol Sci 23(22):13850
90. Opazo CM, Greenough MA, Bush AI (2014) Copper: from neurotransmission to neuroproteostasis. Front Aging Neurosci 6:143
91. Chen P, Miah MR, Aschner M (2016) Metals and neurodegeneration. F1000Res 5:Rev-366. Mar 17F1000 Faculty
92. Keehn B, Kadlaskar G, Bergmann S, McNally Keehn R, Francis A (2021) Attentional Disengagement and the Locus Coeruleus—norepinephrine system in children with autism spectrum disorder. Front Integr Neurosci [cited 2025 May 21] 15. Available from https://www.frontiersin.org/journals/

integrative-neuroscience/articles/10.3389/fnint.2021.716447/full
93. Huang Y, Yu S, Wilson G, Park J, Cheng M, Kong X et al. (2021) Altered extended locus coeruleus and ventral tegmental area networks in boys with autism spectrum disorders: a resting-state functional connectivity study. Neuropsychiatr Dis Treat 17:1207–1216
94. London EB (2018) Neuromodulation and a reconceptualization of autism spectrum disorders: using the locus coeruleus functioning as an exemplar. Front Neurol [cited 2025 May 21] 9. Available from https://www.frontiersin.org/journals/neurology/articles/10.3389/fneur.2018.01120/full
95. Guglielmi L, Servettini I, Caramia M, Catacuzzeno L, Franciolini F, D'Adamo MC et al. (2015) Update on the implication of potassium channels in autism: K(+) channelautism spectrum disorder. Front Cell Neurosci 9:34
96. Lu Q, Zhang Y, Zhao C, Zhang H, Pu Y, Yin L (2022) Copper induces oxidative stress and apoptosis of hippocampal neuron via pCREB/BDNF/ and Nrf2/HO-1/NQO1 pathway. J Appl Toxicol 42(4):694–705
97. Halliwell B, Gutteridge JM (1990) [1] Role of free radicals and catalytic metal ions in human disease: an overview. Methods Enzymol 186:1–85
98. Madsen E, Gitlin JD (2007) Copper and iron disorders of the brain. Annu Rev Neurosci 30(1):317–337
99. Faber S, Zinn GM, Kern IIJC, Skip Kingston HM (2009) The plasma zinc/serum copper ratio as a biomarker in children with autism spectrum disorders. Biomarkers 14(3):171–180
100. Zhou X, Xia X, Li L, Ye Y, Chen Q, Ke M, et al (2025) Evaluation of heavy metals and essential minerals in the hair of children with autism spectrum disorder and their association with symptom severity. Biol Trace Elem Res [cited 2025 Apr 19]. Available from https://link.springer.com/10.1007/s12011-025-04588-z
101. Zhang Y, Xie F, Li S, Li Y, Yang L, Wang Z et al. (2025) Associations of serum manganese, zinc, copper, and selenium concentrations with autism spectrum disorders in chinese children: a case–control study. Autism Res 18(2):427–436
102. Kirkland AE, Sarlo GL, Holton KF (2018) The role of magnesium in neurological disorders. Nutrients 10(6):730
103. Ambrosini E, Sicca F, Brignone MS, D'Adamo MC, Napolitano C, Servettini I et al. (2014) Genetically induced dysfunctions of Kir2.1 channels: implications for short QT3 syndrome and autism–epilepsy phenotype. Human Mol Genetics 23(18):4875–4886
104. Hasan S, Balobaid A, Grottesi A, Dabbagh O, Cenciarini M, Rawashdeh R et al. (2017) Lethal digenic mutations in the K^+ channels Kir4.1 (*KCNJ10*) and SLACK (*KCNT1*) associated with severe-disabling seizures and neurodevelopmental delay. J Neurophysiol 118(4):2402–2411
105. Sicca F, Imbrici P, D'Adamo MC, Moro F, Bonatti F, Brovedani P et al. (2011) Autism with seizures and intellectual disability: possible causative role of gain-of-function of the inwardly-rectifying K+ channel Kir4. 1. Neurobiol Dis 43(1):239–247
106. Sicca F, Ambrosini E, Marchese M, Sforna L, Servettini I, Valvo G et al. (2016) Gain-of-function defects of astrocytic Kir4.1 channels in children with autism spectrum disorders and epilepsy. Sci Rep 6(1):34325
107. Patel V, Akimbekov NS, Grant WB, Dean C, Fang X, Razzaque MS (2024) Neuroprotective effects of magnesium: Implications for neuroinflammation and cognitive decline. Front Endocrinol 15:1406455
108. Yang Y, Long Y, Yuan J, Zha Y (2023) U-shaped association of serum magnesium with mild cognitive impairment among hemodialysis patients: a multicenter study. Renal Fail 45(1):2231084
109. Tuschl K, Mills PB, Clayton PT (2013) Chapter twelve—manganese and the brain. In: Bhatia KP, Schneider SA (eds). International review of neurobiology. Academic Press, pp 277–312 [cited 2025 Feb 5]. (Metal Related Neurodegenerative Disease; vol. 110). Available from https://www.sciencedirect.com/science/article/pii/B9780124105027000132
110. Gavin CE, Gunter KK, Gunter T (1999) Manganese and calcium transport in mitochondria: implications for manganese toxicity. Neurotoxicology 20(2–3):445–453
111. Riley LG, Cowley MJ, Gayevskiy V, Roscioli T, Thorburn DR, Prelog K et al. (2017) A SLC39A8 variant causes manganese deficiency, and glycosylation and mitochondrial disorders. J Inherit Metab Dis 40(2):261–269
112. Liu W, Xu J, Zhang L, Li F, Zhang L, Tai Z et al. (2023) Research progress on correlations between trace element levels and epilepsy. Front Cell Dev Biol 11:1167626
113. Erikson KM, Dobson AW, Dorman DC, Aschner M (2004) Manganese exposure and induced oxidative stress in the rat brain. Sci Total Env 334:409–416
114. Gunter KK, Aschner M, Miller LM, Eliseev R, Salter J, Anderson K et al. (2005) Determining the oxidation states of manganese in PC12

and nerve growth factor-induced PC12 cells. Free Radic Biol Med 39(2):164–181
115. Peres TV, Schettinger MRC, Chen P, Carvalho F, Avila DS, Bowman AB et al. (2016) Manganese-induced neurotoxicity: a review of its behavioral consequences and neuroprotective strategies. BMC Pharmacol Toxicol 17(1):57
116. Bouchard MF, Sauvé S, Barbeau B, Legrand M, Brodeur MÈ, Bouffard T et al. (2011) Intellectual impairment in school-age children exposed to manganese from drinking water. Environ Health Perspect 119(1):138–143
117. Khan K, Factor-Litvak P, Wasserman GA, Liu X, Ahmed E, Parvez F et al. (2011) Manganese exposure from drinking water and children's classroom behavior in Bangladesh. Environ Health Perspect 119(10):1501–1506
118. Hawari I, Eskandar MB, Alzeer S (2020) The Role of lead, manganese, and zinc in autism spectrum disorders (ASDs) and attention-deficient hyperactivity disorder (ADHD): a case-control study on syrian children affected by the syrian crisis. Biol Trace Elem Res 197(1):107–114
119. Skalny AV, Mazaletskaya AL, Ajsuvakova OP, Bjørklund G, Skalnaya MG, Notova SV et al. (2020) Hair trace element concentrations in autism spectrum disorder (ASD) and attention deficit/hyperactivity disorder (ADHD). J Trace Elem Med Biol 61:126539
120. Tinkov AA, Paoliello MM, Mazilina AN, Skalny AV, Martins AC, Voskresenskaya ON et al. (2021) Molecular targets of manganese-induced neurotoxicity: a five-year update. Int J Mol Sci 22(9):4646
121. Antunes-Rodrigues J, De Castro M, Elias LLK, Valença MM, McCANN SM (2004) Neuroendocrine control of body fluid metabolism. Physiol Rev 84(1):169–208
122. Sunderland, M. A. (2001). Ion channels of excitable membranes. Sinauer: Sunderland, MA.
123. Bolotta A, Visconti P, Fedrizzi G, Ghezzo A, Marini M, Manunta P et al. (2018) Na^+, K^+ -ATPase activity in children with autism spectrum disorder: searching for the reason(s) of its decrease in blood cells. Autism Res 11(10):1388–1403
124. Cochet-Bissuel M, Lory P, Monteil A (2014) The sodium leak channel, NALCN, in health and disease. Front Cell Neurosci 8:132
125. Xie L, Gao S, Alcaire SM, Aoyagi K, Wang Y, Griffin JK et al. (2013) NLF-1 delivers a sodium leak channel to regulate neuronal excitability and modulate rhythmic locomotion. Neuron 77(6):1069–1082
126. Samanta D (2025) A comprehensive review of evolving treatment strategies for Dravet syndrome: Insights from randomized trials, meta-analyses, real-world evidence, and emerging therapeutic approaches. Epilepsy Behav [cited 2025 May 22] 162. Available from https://www.epilepsybehavior.com/article/S1525-5050(24)00553-5/abstract
127. Kofuji P, Newman EA (2009) Regulation of potassium by glial cells in the centralnervous system. In: Haydon PG, Parpura V (eds). Astrocytes in (patho)physiology of the nervous system. Springer, Boston, MA, USA, pp 151–75 [cited 2025 Mar 6]. Available from https://doi.org/10.1007/978-0-387-79492-1_6
128. D'Adamo MC, Catacuzzeno L, Di Giovanni G, Franciolini F, Pessia M (2013) K+ channelepsy: progress in the neurobiology of potassium channels and epilepsy. Front Cell Neurosci 7:134
129. Imbrici P, Nematian-Ardestani E, Hasan S, Pessia M, Tucker SJ, D'Adamo MC (2020) Altered functional properties of a missense variant in the TRESK K+ channel (KCNK18) associated with migraine and intellectual disability. Pflugers Arch 472(7):923–930
130. Kaya N, Alsagob M, D'Adamo MC, Al-Bakheet A, Hasan S, Muccioli M et al. (2016) KCNA4 deficiency leads to a syndrome of abnormal striatum, congenital cataract and intellectual disability. J Med Genetics 53(11):786–792
131. Pavinato L, Nematian-Ardestani E, Zonta A, De Rubeis S, Buxbaum J, Mancini C et al. (2021) KCNK18 biallelic variants associated with intellectual disability and neurodevelopmental disorders alter TRESK channel activity. Int J Mol Sci 22(11):6064
132. Meisler MH, Hill SF, Yu W (2021) Sodium channelopathies in neurodevelopmental disorders. Nat Rev Neurosci 22(3):152–166
133. Smith RS, Walsh CA (2020) Ion channel functions in early brain development. Trends Neurosci 43(2):103–114
134. Jyothi NR (2020). Heavy metal sources and their effects on human health. In: M. K. Nazal & H. Zhao (Eds.), Heavy metals: Their environmental impacts and mitigation. IntechOpen. 1–12. https://doi.org/10.5772/intechopen.95370
135. Mahurpawar M (2015) Effects of heavy metals on human health. Int J Res Granthaalayah 530(516):1–7
136. Chowdhury S, Mazumder MAJ, Al-Attas O, Husain T (2016) Heavy metals in drinking water: Occurrences, implications, and future needs in developing countries. Sci Total Env 569–570:476–488
137. Heng YY, Asad I, Coleman B, Menard L, Benki-Nugent S, Were FH et al. (2022) Heavy metals and neurodevelopment of children in low and middle-income countries: a systematic review. PLoS ONE 17(3):e0265536

138. Ijomone OM, Olung NF, Akingbade GT, Okoh COA, Aschner M (2020) Environmental influence on neurodevelopmental disorders: potential association of heavy metal exposure and autism. J Trace Elem Med Biol 62:126638
139. Collin MS, Venkatraman SK, Vijayakumar N, Kanimozhi V, Arbaaz SM, Stacey RS et al. (2022) Bioaccumulation of lead (Pb) and its effects on human: a review. J Hazard Mater Adv 7:100094
140. Ramírez Ortega D, González Esquivel DF, Blanco Ayala T, Pineda B, Gómez Manzo S, Marcial Quino J et al. (2021) Cognitive impairment induced by lead exposure during lifespan: mechanisms of lead neurotoxicity. Toxics 9(2):23
141. Glass TA, Bandeen-Roche K, McAtee M, Bolla K, Todd AC, Schwartz BS (2009) Neighborhood psychosocial hazards and the association of cumulative lead dose with cognitive function in older adults. Am J Epidemiol 169(6):683–692
142. Organization WH (2023) Exposure to lead: a major public health concern. Preventing disease through healthy environments. World Health Organization 7 p
143. Hsu PC, Liu MY, Hsu CC, Chen LY, Guo YL (1997) Lead exposure causes generation of reactive oxygen species and functional impairment in rat sperm. Toxicology. 122(1–2):133–143
144. Lopes ACBA, Peixe TS, Mesas AE, Paoliello MMB (2016) Lead exposure and oxidative stress: a systematic review. In: De Voogt P, (ed). Reviews of environmental contamination and toxicology volume 236. Springer International Publishing, Cham, p 193–238 [cited 2025 Feb 13]. (Reviews of Environmental Contamination and Toxicology; vol. 236). Available from http://link.springer.com/10.1007/978-3-319-20013-2_3
145. Patrick L (2006) Lead toxicity part II: the role of free radical damage and the use of antioxidants in the pathology and treatment of lead toxicity. Altern Med Rev [cited 2025 Feb 13] 11(2). Available from https://search.ebscohost.com/login.aspx?direct=true&profile=ehost&scope=site&authtype=crawler&jrnl=10895159&AN=106321740&h=204XUqlix%2F7uCdvYRFauX4WFT7%2Bq3kvN8CP1qHbkhfQb8yQEWE0j244We7pHuS8gqmyEv1lJygQ7Wrwk%2BSoz%2Bg%3D%3D&crl=c
146. Pottier G, Viau M, Ricoul M, Shim G, Bellamy M, Cuceu C et al. (2013) Lead Exposure Induces Telomere Instability in Human Cells. PLoS ONE 8(6):e67501
147. De Souza ID, De Andrade AS, Dalmolin RJS (2018) Lead-interacting proteins and their implication in lead poisoning. Critic Rev Toxicol 48(5):375–386
148. Neal AP, Worley PF, Guilarte TR (2011) Lead exposure during synaptogenesis alters NMDA receptor targeting via NMDA receptor inhibition. NeuroToxicology 32(2):281–289
149. Audesirk G, Audesirk T (1993) The effects of inorganic lead on voltage-sensitive calcium channels differ among cell types and among channel subtypes. Neurotoxicology 14(2–3):259–265
150. Shah-Kulkarni S, Lee S, Jeong KS, Hong YC, Park H, Ha M et al. (2020) Prenatal exposure to mixtures of heavy metals and neurodevelopment in infants at 6 months. Env Res 182:109122
151. Ding M, Shi S, Qie S, Li J, Xi X (2023) Association between heavy metals exposure (cadmium, lead, arsenic, mercury) and child autistic disorder: a systematic review and meta-analysis. Front Pediatr 11:1169733
152. Nakhaee S, Amirabadizadeh A, Farnia V, Ali Azadi N, Mansouri B, Radmehr F (2023) Association between biological lead concentrations and autism spectrum disorder (ASD) in children: a systematic review and meta-analysis. Biol Trace Elem Res 201(4):1567–1581
153. Kern JK, Grannemann BD, Trivedi MH, Adams JB (2007) Sulfhydryl-reactive metals in autism. J Toxicol Environ Health A 70(8):715–721
154. Oliveira CS, Nogara PA, Ardisson-Araújo DMP, Aschner M, Rocha JBT, Dórea JG (2018) Chapter two—neurodevelopmental effects of mercury. In: Aschner M, Costa LG, (eds). Advances in neurotoxicology. Academic Press, pp 27–86 [cited 2025 Feb 13]. (Linking Environmental Exposure to Neurodevelopmental Disorders; vol. 2). Available from https://www.sciencedirect.com/science/article/pii/S2468748018300055
155. Dack K, Fell M, Taylor CM, Havdahl A, Lewis SJ (2022) Prenatal mercury exposure and neurodevelopment up to the age of 5 years: a systematic review. Int J Env Res Public Health 19(4):1976
156. Adams JB, Romdalvik J, Ramanujam VMS, Legator MS (2007) Mercury, lead, and zinc in baby teeth of children with autism versus controls. J Toxicol Environ Health A 70(12):1046–1051
157. Jafari T, Rostampour N, Fallah AA, Hesami A (2017) The association between mercury levels and autism spectrum disorders: a systematic review and meta-analysis. J Trace Elem Med Biol 44:289–297
158. Saghazadeh A, Rezaei N (2017) Systematic review and meta-analysis links autism and toxic metals and highlights the impact of country development status: higher blood

and erythrocyte levels for mercury and lead, and higher hair antimony, cadmium, lead, and mercury. Progress Neuro-Psychopharmacol Biol Psychiatry 79:340–368
159. Stojsavljević A, Lakićević N, Pavlović S (2023) Mercury and autism spectrum disorder: exploring the link through comprehensive review and meta-analysis. Biomedicines 11(12):3344
160. Abbott LC, Nigussie F (2021) Mercury toxicity and neurogenesis in the mammalian brain. Int J Mol Sci 22(14):7520
161. Wang W, Chen F, Zhang L, Wen F, Yu Q, Li P et al. (2023) Neurotransmitter disturbances caused by methylmercury exposure: microbiota-gut-brain interaction. Sci Total Env 873:162358
162. Porru S, Esplugues A, Llop S, Delgado-Saborit JM (2024) The effects of heavy metal exposure on brain and gut microbiota: a systematic review of animal studies. Env Pollut 123732
163. Zhai Q, Cen S, Jiang J, Zhao J, Zhang H, Chen W (2019) Disturbance of trace element and gut microbiota profiles as indicators of autism spectrum disorder: a pilot study of Chinese children. Environ Res 171:501–509
164. O'Grady K, Grabrucker AM (2025) Metal dyshomeostasis as a driver of gut pathology in autism spectrum disorders. J Neurochem 169(3):e70041
165. Agency for Toxic Substances and Disease Registry (ATSDR) (2012) Toxicological profile for cadmium [Internet]. U.S. Department of Health and Human Services (Public Health Service). Available from https://www.atsdr.cdc.gov/toxprofiles/tp5.pdf
166. Arruebarrena MA, Hawe CT, Lee YM, Branco RC (2023) Mechanisms of cadmium neurotoxicity. Int J Mol Sci 24(23):16558
167. Ruczaj A, Brzóska MM (2023) Environmental exposure of the general population to cadmium as a risk factor of the damage to the nervous system: a critical review of current data. J Appl Toxicol 43(1):66–88
168. López E, Arce C, Oset-Gasque MJ, Cañadas S, González MP (2006) Cadmium induces reactive oxygen species generation and lipid peroxidation in cortical neurons in culture. Free Radic Biol Med 40(6):940–951
169. Yuan Y, Bian JC, Liu XZ, Zhang Y, Sun Y, Liu ZP (2012) Oxidative stress and apoptotic changes of rat cerebral cortical neurons exposed to cadmium in vitr. Biomed Env Sci 25(2):172–181
170. Ibiwoye MO, Matthews Q, Travers K, Foster JD (2019) Association of acute, high-dose cadmium exposure with alterations in vascular endothelial barrier antigen expression and astrocyte morphology in the developing rat central nervous system. J Compar Pathol 172:37–47
171. Pulido G, Treviño S, Brambila E, Vazquez-Roque R, Moreno-Rodriguez A, Peña Rosas U et al. (2019) The administration of cadmium for 2, 3 and 4 months causes a loss of recognition memory, promotes neuronal hypotrophy and apoptosis in the hippocampus of rats. Neurochem Res 44(2):485–497
172. Chouit Z, Djellal D, Haddad S, Hanfer M, Hachemi M, Lakroun Z et al. (2021) Potentiation of the apoptotic signaling pathway in both the striatum and hippocampus and neurobehavioral impairment in rats exposed chronically to a low–dose of cadmium. Environ Sci Pollut Res 28(3):3307–3317
173. Tian J, Hu J, Liu D, Yin J, Chen M, Zhou L et al. (2021) Cadmium chloride-induced transgenerational neurotoxicity in zebrafish development. Env Toxicol Pharmacol 81:103545
174. Kippler M, Bottai M, Georgiou V, Koutra K, Chalkiadaki G, Kampouri M et al. (2016) Impact of prenatal exposure to cadmium on cognitive development at preschool age and the importance of selenium and iodine. Eur J Epidemiol 31(11):1123–1134
175. Kippler M, Tofail F, Hamadani JD, Gardner RM, Grantham-McGregor SM, Bottai M et al. (2012) Early-life cadmium exposure and child development in 5-year-old girls and boys: a cohort study in rural Bangladesh. Environ Health Perspect 120(10):1462–1468
176. Yasuda H, Yasuda Y, Tsutsui T (2013) Estimation of autistic children by metallomics analysis. Sci Rep 3(1):1199
177. Qin Yyan, Jian B, Wu C, Jiang Czi, Kang Y, Zhou Jxiu et al. (2018) A comparison of blood metal levels in autism spectrum disorder and unaffected children in Shenzhen of China and factors involved in bioaccumulation of metals. Environ Sci Pollut Res 25(18):17950–17956
178. Hessabi M, Rahbar MH, Dobrescu I, Bach MA, Kobylinska L, Bressler J et al. (2019) Concentrations of lead, mercury, arsenic, cadmium, manganese, and aluminum in blood of romanian children suspected of having autism spectrum disorder. Int J Environ Res Public Health 16(13):2303
179. Dou JF, Schmidt RJ, Volk HE, Nitta MM, Feinberg JI, Newschaffer CJ, et al (2023) Exposure to heavy metals in utero and autism spectrum disorder at age 3: a meta-analysis of two longitudinal cohorts of siblings of children with autism. medRxiv. https://doi.org/10.1101/2023.11.21.23298827
180. Wang H, Lim JJ, Gu H, Xia Z, Cui JY (2025) Cadmium-induced gut dysbiosis precedes the

onset of hippocampus-dependent learning and memory deficits in mice]. bioRxiv [cited 2025 May 24]. pp 2025.04.07.647608. Available from https://www.biorxiv.org/content/10.1101/2025.04.07.647608v1
181. Nehzomi ZS, Shirani K (2025) The gut microbiota: a key player in cadmium toxicity - implications for disease, interventions, and combined toxicant exposures. J Trace Elem Med Biol 88:127570
182. Xie S, Zhang R, Li Z, Liu C, Xiang W, Lu Q et al. (2022) Indispensable role of melatonin, a scavenger of reactive oxygen species (ROS), in the protective effect of *Akkermansia muciniphila* in cadmium-induced intestinal mucosal damage. Free Radic Biol Med 193:447–458
183. Yang J, Chen W, Sun Y, Liu J, Zhang W (2021) Effects of cadmium on organ function, gut microbiota and its metabolomics profile in adolescent rats. Ecotoxicol Env Safety 222:112501
184. Chen Y, Li Y, Fan Y, Chen S, Chen L, Chen Y et al. (2024) Gut microbiota-driven metabolic alterations reveal gut–brain communication in Alzheimer's disease model mice. Gut Microbes 16:2302310
185. Yokel R (2000) The toxicology of aluminum in the brain: a review. NeuroToxicology 21(5):813–828
186. Mold M, Umar D, King A, Exley C (2018) Aluminium in brain tissue in autism. J Trace Elem Med Biol 46:76–82
187. Taylor LE, Swerdfeger AL, Eslick GD (2014) Vaccines are not associated with autism: an evidence-based meta-analysis of case-control and cohort studies. Vaccine 32(29):3623–3629
188. Arsenic Toxicity: Where is Arsenic Found? (2023) Environmental Medicine | ATSDR [cited 2025 May 26]. Available from https://archive.cdc.gov/www_atsdr_cdc_gov/csem/arsenic/where_arsenic.html
189. Rodríguez-Barranco M, Lacasaña M, Aguilar-Garduño C, Alguacil J, Gil F, González-Alzaga B et al. (2013) Association of arsenic, cadmium and manganese exposure with neurodevelopment and behavioural disorders in children: a systematic review and meta-analysis. Sci Total Env 454–455:562–577
190. Thakur M, Rachamalla M, Niyogi S, Datusalia AK, Flora SJS (2021) Molecular mechanism of arsenic-induced neurotoxicity including neuronal dysfunctions. Int J Mol Sci 22(18):10077
191. Vázquez Cervantes GI, González Esquivel DF, Ramírez Ortega D, Blanco Ayala T, Ramos Chávez LA, López-López HE et al. (2023) Mechanisms associated with cognitive and behavioral impairment induced by arsenic exposure. Cells 12(21):2537
192. Vahidnia A, van der Voet GB, de Wolff FA (2007) Arsenic neurotoxicity—a review. Hum Exp Toxicol 26(10):823–832
193. Park HR, Sun R, Panganiban RA, Christiani DC, Lu Q (2020) MicroRNA-124 reduces arsenic-induced endoplasmic reticulum stress and neurotoxicity and is linked with neurodevelopment in children. Scie Rep 10(1):5934
194. Behl S, Mehta S, Pandey MK (2020) Abnormal levels of metal micronutrients and autism spectrum disorder: a perspective review. Front Mol Neurosci [cited 2025 Jun 9] 13. Available from https://www.frontiersin.org/journals/molecular-neuroscience/articles/10.3389/fnmol.2020.586209/full
195. Chen GT, Geschwind DH (2022) Challenges and opportunities for precision medicine in neurodevelopmental disorders. Adv Drug Deliv Rev 191:114564
196. Lamptey RN, Chaulagain B, Trivedi R, Gothwal A, Layek B, Singh J (2022) A review of the common neurodegenerative disorders: current therapeutic approaches and the potential role of nanotherapeutics. Int J Mol Sci 23(3):1851
197. Bruzzese A, Martino EA, Mendicino F, Lucia E, Olivito V, Bova C et al. (2023) Iron chelation therapy. Eur J Haematol 110(5):490–497
198. Rafique S, Idrees M, Nasim A, Akbar H, Athar A (2010) Transition metal complexes as potential therapeutic agents. Biotechnol Mol Biol Rev 5(2):38–45
199. Dockrell ME, Purchase D, Price RG (2023) E-waste and metal contamination in the environment: health effects. In: Trace Metals in the Environment [Internet]. IntechOpen [cited 2025 Mar 12]. Available from https://www.intechopen.com/chapters/1132524
200. Wu Y, Qi M, Yu H, Li G, An T (2025) Assessment of internal exposure risk from metals pollution of occupational and non-occupational populations around a non-ferrous metal smelting plant. J Env Sci 147:62–73
201. De Paula Arrifano G, Crespo-Lopez ME, Lopes-Araújo A, Santos-Sacramento L, Barthelemy JL, De Nazaré CGL et al. (2023) Neurotoxicity and the global worst pollutants: astroglial involvement in arsenic, lead, and mercury intoxication. Neurochem Res 48(4):1047–1065
202. Nemery B (2022) Metals and the respiratory tract—ScienceDirect. Handb Toxicol Metals (Fifth Ed) 1:421–443
203. Khlifi R, Olmedo P, Gil F, Hammami B, Chakroun A, Rebai A et al. (2013) Arsenic, cadmium, chromium and nickel in cancerous and healthy tissues from patients with head and neck cancer. Sci Total Env 452–453:58–67

204. Massányi P, Massányi M, Madeddu R, Stawarz R, Lukáč N (2020) Effects of cadmium, lead, and mercury on the structure and function of reproductive organs. Toxics 8(4):94
205. Ayenimo JG, Yusuf AM, Adekunle AS, Makinde OW (2010) Heavy metal exposure from personal care products. Bull Environ Contam Toxicol 84(1):8–14
206. Vandebroek E, Haufroid V, Smolders E, Hons L, Nemery B (2019) Occupational exposure to metals in shooting ranges: a biomonitoring study. Safety Health Work 10(1):87–94
207. Petriello MC, Newsome B, Hennig B (2014) Influence of nutrition in PCB-induced vascular inflammation. Environ Sci Pollut Res 21(10):6410–6418
208. Petriello MC, Newsome BJ, Dziubla TD, Hilt JZ, Bhattacharyya D, Hennig B (2014) Modulation of persistent organic pollutant toxicity through nutritional intervention: Emerging opportunities in biomedicine and environmental remediation. Sci Total Env 491–492:11–16
209. Ayodeji-Fapohunda GN, Ugwoha E, Nwaichi EO (2023) Nutrition as a therapeutic intervention for metal toxicity. Stress 17:18
210. Adil N, Ashraf K, Munir M, Mohiuddin M, Abbasi A, Riaz U et al. (2023) Pesticides, heavy metals and plasticizers: contamination and risk assessment of drinking-water quality. Sustainability. 15(17):13263
211. Moody EC, Coca SG, Sanders AP (2018) Toxic metals and chronic kidney disease: a systematic review of recent literature. Curr Envir Health Rpt 5(4):453–463
212. Arora S, Bhattacharjee J (2008) Modulation of immune responses in stress by Yoga. Int J Yoga 1(2):45–55
213. Fincham GW, Kartar A, Uthaug MV, Anderson B, Hall L, Nagai Y et al. (2023) High ventilation breathwork practices: an overview of their effects, mechanisms, and considerations for clinical applications. Neurosci Biobehav Rev 155:105453
214. Healthline (2017) [cited 2025 Mar 6]. Heavy Metal Detox Diet: Symptoms and Recommended Foods. Available from https://www.healthline.com/health/heavy-metal-detox
215. Minich DM, Brown BI (2019) A review of dietary (phyto)nutrients for glutathione support. Nutrients 11(9):2073
216. Jozefczak M, Remans T, Vangronsveld J, Cuypers A (2012) Glutathione is a key player in metal-induced oxidative stress defenses. Int J Mol Sci 13(3):3145–3175
217. Deponte M (2017) The incomplete glutathione puzzle: just guessing at numbers and figures?. Antioxid Redox Signal 27(15):1130–1161
218. Potęga A (2022) Glutathione-mediated conjugation of anticancer drugs: an overview of reaction mechanisms and biological significance for drug detoxification and bioactivation. Molecules 27(16):5252
219. Epure A, Anastasoi V, Cheta DM (2022) Correlation between the presence of metals with potential for intoxication, omega 3 deficiency, increased omega 6: omega 3 ratio and their associated symptoms. Roman Med J [cited 2025 Feb 27] 69(2). Available from https://rmj.com.ro/articles/2022.2/RMJ_2022_2_Art-05.pdf
220. Singh PK, Gupta MK, Nath R (2023) Omega-3 fatty acid as a protectant in lead-induced neurotoxicity. In: Treatments, nutraceuticals, supplements, and herbal medicine in neurological disorders. Elsevier, pp 285–301 [cited 2025 Feb 27]. Available from https://www.sciencedirect.com/science/article/pii/B9780323900522000020
221. Alnahdi HS, Sharaf IA (2019) Possible prophylactic effect of omega-3 fatty acids on cadmium-induced neurotoxicity in rats' brains. Environ Sci Pollut Res 26(30):31254–31262
222. Tur JA, Bibiloni MM, Sureda A, Pons A (2012) Dietary sources of omega 3 fatty acids: public health risks and benefits. Br J Nutr 107(S2):S23–S52
223. Bist P, Choudhary S (2022) Impact of heavy metal toxicity on the gut microbiota and its relationship with metabolites and future probiotics strategy: a review. Biol Trace Elem Res 200(12):5328–5350
224. Duan H, Yu L, Tian F, Zhai Q, Fan L, Chen W (2020) Gut microbiota: a target for heavy metal toxicity and a probiotic protective strategy. Sci Total Env 742:140429
225. Arun KB, Madhavan A, Sindhu R, Emmanual S, Binod P, Pugazhendhi A et al. (2021) Probiotics and gut microbiome—prospects and challenges in remediating heavy metal toxicity. J Hazard Mater 420:126676
226. Khushboo, Karnwal A, Malik T (2023) Characterization and selection of probiotic lactic acid bacteria from different dietary sources for development of functional foods. Front Microbiol [cited 2025 Feb 27] 14. Available from https://www.frontiersin.org/journals/microbiology/articles/10.3389/fmicb.2023.1170725/full
227. Peng Z, Liao Y, Yang W, Liu L (2024) Metal(loid)-gut microbiota interactions and microbiota-related protective strategies: a review. Environ Int 192:109017
228. Patil R, Sontakke T, Biradar A, Nalage D (2023) Zinc: an essential trace element for human health and beyond. Food Health 5(3):13

229. Viktorinova A, Ursinyova M, Trebaticka J, Uhnakova I, Durackova Z, Masanova V (2016) Changed plasma levels of zinc and copper to zinc ratio and their possible associations with parent- and teacher-rated symptoms in children with attention-deficit hyperactivity disorder. Biol Trace Elem Res 169(1):1–7
230. Hudson LG, Dashner-Titus EJ, MacKenzie D (2025) Zinc as a mechanism-based strategy for mitigation of metals toxicity. Curr Environ Health Rep 12(1):5
231. Prasad AS, Bao B (2019) Molecular mechanisms of zinc as a pro-antioxidant mediator: clinical therapeutic implications. Antioxidants (Basel) 8(6):164
232. Rahman MM, Hossain KFB, Banik S, Sikder MT, Akter M, Bondad SEC et al. (2019) Selenium and zinc protections against metal-(loids)-induced toxicity and disease manifestations: a review. Ecotoxicol Environ Saf 168:146–163
233. Yu HT, Zhen J, Leng JY, Cai L, Ji HL, Keller BB (2021) Zinc as a countermeasure for cadmium toxicity. Acta Pharmacol Sin 42(3):340–346
234. Vance TM, Chun OK (2015) Zinc intake is associated with lower cadmium burden in US Adults 1, 2, 3. J Nutr 145(12):2741–2748
235. Kataba A, Nakayama SMM, Yohannes YB, Toyomaki H, Nakata H, Ikenaka Y et al. (2021) Effects of zinc on tissue uptake and toxicity of lead in Sprague Dawley rat. J Vet Med Sci 83(11):1674–1685
236. Zhang Z, Yu J, Xie J, Liu D, Fan Y, Ma H et al. (2021) Improvement roles of zinc supplementation in low dose lead induced testicular damage and glycolytic inhibition in mice. Toxicology 462:152933
237. Beaver LM, Truong L, Barton CL, Chase TT, Gonnerman GD, Wong CP et al. (2017) Combinatorial effects of zinc deficiency and arsenic exposure on zebrafish (*Danio rerio*) development. PLoS ONE 12(8):e0183831
238. Milton AG, Zalewski PD, Ratnaike RN (2004) Zinc protects against arsenic-induced apoptosis in a neuronal cell line, measured by DEVD-caspase activity. Biometals. 17(6):707–713
239. Kreppel H, Liu J, Liu Y, Reichl FX, Klassen CD (1994) Zinc-induced arsenite tolerance in Mice1. Toxicol Sci 23(1):32–37
240. Ghabaee DNZ, Amiri FT, Moghaddam AE, Khalatbary AR, Zargari M (2016) Administration of zinc against arsenic-induced nephrotoxicity during gestation and lactation in rat model. J Nephropathol 6(2):74
241. Garcia MTA, Dueñas AH, Pampliega JP (2013) Hematological effects of arsenic in rats after subchronical exposure during pregnancy and lactation: the protective role of antioxidants. Exp Toxicol Pathol 65(5):609–614
242. Herrera A, Pineda J, Antonio MT (2013) Toxic effects of perinatal arsenic exposure on the brain of developing rats and the beneficial role of natural antioxidants. Env Toxicol Pharmacol 36(1):73–79
243. Modi M, Kaul RK, Kannan GM, Flora SJ (2006) Co-administration of zinc and n-acetylcysteine prevents arsenic-induced tissue oxidative stress in male rats. J Trace Elem Med Biol 20(3):197–204
244. Ganger R, Garla R, Mohanty BP, Bansal MP, Garg ML (2016) Protective effects of zinc against acute arsenic toxicity by regulating antioxidant defense system and cumulative metallothionein expression. Biol Trace Elem Res 169(2):218–229
245. Zeneli L, Sekovanić A, Ajvazi M, Kurti L, Daci N (2016) Alterations in antioxidant defense system of workers chronically exposed to arsenic, cadmium and mercury from coal flying ash. Environ Geochem Health 38(1):65–72
246. Kordas K, Roy A, López P, García-Vargas G, Cebrián ME, Vera-Aguilar E et al. (2017) Iron and zinc supplementation does not impact urinary arsenic excretion in Mexican school children. J Pediatr 185:205–210.e1
247. Skalny AA, Tinkov AA, Medvedeva YS, Alchinova IB, Karganov MYu, Ajsuvakova OP et al. (2015) Zinc asparaginate supplementation induces redistribution of toxic trace elements in rat tissues and organs. Interdiscip Toxicol 8(3):131–138
248. Kumar A, Nair P, Malhotra A, Majumdar S, Garg ML, Dhawan DK (2011) Altered uptake and biological half-lives of 65Zn on arsenic exposure—modulation by zinc treatment. Biol Trace Elem Res 144(1):1059–1068
249. Jadán-Piedra C, Chiocchetti GM, Clemente MJ, Vélez D, Devesa V (2018) Dietary compounds as modulators of metals and metalloids toxicity. Crit Rev Food Sci Nutr 58(12):2055–2067
250. Afonne OJ, Orisakwe OE, Obi E, Dioka CE, Ndubuka GI (2002) Nephrotoxic actions of low-dose mercury in mice: protection by zinc. Arch Env Health 57(2):98–102
251. Mesquita M, Pedroso TF, Oliveira CS, Oliveira VA, do Santos RF, Bizzi CA et al. (2016) Effects of zinc against mercury toxicity in female rats 12 and 48 hours after HgCl2 exposure. EXCLI J 15:256–267
252. Franciscato C, Moraes-Silva L, Duarte FA, Oliveira CS, Ineu RP, Flores EMM et al. (2011) Delayed biochemical changes induced by mercury intoxication are prevented by zinc pre-exposure. Ecotoxicol Env Safety 74(3):480–486

253. Hossain KFB, Hosokawa T, Saito T, Kurasaki M (2021) Zinc-pretreatment triggers glutathione and Nrf2-mediated protection against inorganic mercury-induced cytotoxicity and intrinsic apoptosis in PC12 cells. Ecotoxicol Env Safety 207:111320
254. Joshi D, Mittal DK, Bhadauria M, Nirala SK, Shrivastava S, Shukla S (2010) Role of micronutrients against dimethylmercury intoxication in male rats. Env Toxicol Pharmacol 29(2):97–103
255. Deepmala J, Deepak M, Srivastav S, Sangeeta S, Kumar SA, Kumar SS (2013) Protective effect of combined therapy with dithiothreitol, zinc and selenium protects acute mercury induced oxidative injury in rats. J Trace Elem Med Biol 27(3):249–256
256. Oliveira VA, Oliveira CS, Mesquita M, Pedroso TF, Costa LM, Fiuza T, da L et al. (2015) Zinc and *N*-acetylcysteine modify mercury distribution and promote increase in hepatic metallothionein levels. J Trace Elem Med Biol 32:183–188
257. Vilas CK, Emery LE, Denchi EL, Miller KM (2018) Caught with one's zinc fingers in the genome integrity cookie jar. Trends Genet 34(4):313–325
258. Padjasek M, Kocyła A, Kluska K, Kerber O, Tran JB, Krężel A (2020) Structural zinc binding sites shaped for greater works: structure-function relations in classical zinc finger, hook and clasp domains. J Inorg Biochem 204:110955
259. Abbehausen C (2019) Zinc finger domains as therapeutic targets for metal-based compounds—an update. Metallomics. 11(1):15–28
260. Zhou X, Speer RM, Volk L, Hudson LG, Liu KJ (2021) Arsenic co-carcinogenesis: inhibition of DNA repair and interaction with zinc finger proteins. Semin Cancer Biol 76:86–98
261. Bu S, Lv Y, Liu Y, Qiao S, Wang H (2021) Zinc finger proteins in neuro-related diseases progression. Front Neurosci 15:760567
262. Joyce PI, Fratta P, Landman AS, Mcgoldrick P, Wackerhage H, Groves M et al. (2016) Deficiency of the zinc finger protein ZFP106 causes motor and sensory neurodegeneration. Human Mol Genetics 25(2):291–307
263. Li D, Quan Z, Ni J, Li H, Qing H (2023) The many faces of the zinc finger protein 335 in brain development and immune system. Biomed Pharmacother 165:115257
264. House WA, Hart JJ, Norvell WA, Welch RM (2003) Cadmium absorption and retention by rats fed durum wheat. Br J Nutr 89(4):499–508
265. Marreiro D, do N, Cruz KJC, Morais JBS, Beserra JB, Severo JS, de Oliveira ARS (2017) Zinc and oxidative stress: current mechanisms. Antioxidants (Basel) 6(2):24
266. Meguid NA, Bjørklund G, Gebril OH, Doş a MD, Anwar M, Elsaeid A et al. (2019) The role of zinc supplementation on the metallothionein system in children with autism spectrum disorder. Acta Neurol Belg 119(4):577–583
267. Kimura T, Onodera A, Okumura F, Nakanishi T, Itoh N (2015) Chromium (VI)-induced transformation is enhanced by Zn deficiency in BALB/c 3T3 cells. J Toxicol Sci 40(3):383–387
268. Sears SM, Hewett SJ (2021) Influence of glutamate and GABA transport on brain excitatory/inhibitory balance. Exp Biol Med (Maywood) 246(9):1069–1083
269. Pal MM (2021) Glutamate: the master neurotransmitter and its implications in chronic stress and mood disorders. Front Hum Neurosci 15:722323
270. Lee K, Mills Z, Cheung P, Cheyne JE, Montgomery JM (2022) The role of zinc and NMDA receptors in autism spectrum disorders. Pharmaceuticals 16(1):1
271. Siller SS, Broadie K (2012) Matrix metalloproteinases and minocycline: therapeutic avenues for fragile X syndrome. Neural Plast 2012:1–9
272. Kirsten TB, Queiroz-Hazarbassanov N, Bernardi MM, Felicio LF (2015) Prenatal zinc prevents communication impairments and BDNF disturbance in a rat model of autism induced by prenatal lipopolysaccharide exposure. Life Sci 130:12–17
273. Cirovic A, Cirovic A, Yimthiang S, Vesey DA, Satarug S (2024) Modulation of adverse health effects of environmental cadmium exposure by zinc and its transporters. Biomolecules. 14(6):650
274. Maret W (2017) Zinc in cellular regulation: the nature and significance of "Zinc signals. Int J Mol Sci 18(11):2285
275. Aydemir TB, Cousins RJ (2018) The multiple faces of the metal transporter ZIP14 (SLC39A14). J Nutr 148(2):174–184
276. Fujishiro H, Himeno S (2019) New insights into the roles of ZIP8, a cadmium and manganese transporter, and its relation to human diseases. Biol Pharm Bull 42(7):1076–1082
277. Dashner-Titus EJ, Schilz JR, Alvarez SA, Wong CP, Simmons K, Ho E et al. (2023) Zinc supplementation alters tissue distribution of arsenic in *Mus musculus*. Toxicol Appl Pharmacol 478:116709
278. Gunes S, Ekinci O, Celik T (2017) Iron deficiency parameters in autism spectrum disorder: clinical correlates and associated factors. Ital J Pediatr 43(1):86

279. Kaczmarek KA, Dobrzyńska M, Drzymała-Czyż S (2024) Iron, magnesium, zinc and selenium—the most common elemental deficiencies in children with autism spectrum disorder. Res Autism Spectrum Disord 110:102288
280. Konofal E, Lecendreux M, Deron J, Marchand M, Cortese S, Zaïm M et al. (2008) Effects of iron supplementation on attention deficit hyperactivity disorder in children. Pediatr Neurol 38(1):20–26
281. Mousain-Bosc M, Siatka C, Bali JP (2018) Magnesium, hyperactivity and autism in children. [cited 2025 Feb 26]. Available from https://europepmc.org/article/nbk/nbk507249
282. Hariri M, Azadbakht L (2015) Magnesium, iron, and zinc supplementation for the treatment of attention deficit hyperactivity disorder: a systematic review on the recent literature. Int J Prevent Med 6(1):83
283. Kim YJ, Kwon S, Kim MK (2009) Effect of Chlorella vulgaris intake on cadmium detoxification in rats fed cadmium. Nutr Res Pract 3(2):89
284. Sears ME (2013) Chelation: harnessing and enhancing heavy metal detoxification—a review. Sci World J 2013(1):219840
285. Mehrandish R, Rahimian A, Shahriary A (2019) Heavy metals detoxification: a review of herbal compounds for chelation therapy in heavy metals toxicity. J HerbMed Pharmacol 8(2):69–77
286. Tosato M, Ponte F, Franchi S, Menegazzo I, Gabryel-Skrodzka M, Zanoni G et al. (2024) Unveiling the potential of sulfur-rich macrocyclic chelators against cadmium poisoning. Inorg Chem 63(50):23970–23982
287. Manoj Kumar V, Henley AK, Nelson CJ, Indumati O, Prabhakara Rao Y, Rajanna S et al. (2017) Protective effect of Allium sativum (garlic) aqueous extract against lead-induced oxidative stress in the rat brain, liver, and kidney. Environ Sci Pollut Res 24(2):1544–1552
288. Fanoudi S, Alavi MS, Karimi G, Hosseinzadeh H (2020) Milk thistle (Silybum Marianum) as an antidote or a protective agent against natural or chemical toxicities: a review. Drug Chem Toxicol 43(3):240–254
289. Bone K, Morgan M (2006) Herbs and heavy metal detoxification. Pythotherap Perspect 94:1–7
290. Qayoom A, Kazmi SA, Ali SN (2017) Turmeric powder as a natural heavy metal chelating agent: Surface characterisation. Pak J Sci Ind Res Ser A 60(1):1–8
291. Colovic MB, Vasic VM, Djuric DM, Krstic DZ (2018) Sulphur-containing amino acids: protective role against free radicals and heavy metals. Curr Med Chem 25(3):324–335
292. Ou P, Tritschler HJ, Wolff SP (1995) Thioctic (lipoic) acid: a therapeutic metal-chelating antioxidant?. Biochem Pharmacol 50(1):123–126
293. Raji Z, Karim A, Karam A, Khalloufi S (2023) A review on the heavy metal adsorption capacity of dietary fibers derived from agro-based wastes: opportunities and challenges for practical applications in the food industry. Trends Food Sci Technol 137:74–91
294. Rossignol DA (2019) The use of N-acetylcysteine as a chelator for metal toxicity. In: Frye RE, Berk M (eds). The therapeutic use of N-acetylcysteine (NAC) in medicine. Springer, Singapore, pp 169–79 [cited 2025 Feb 27] Available from http://link.springer.com/10.1007/978-981-10-5311-5_10
295. Fricker SP (2007) Metal based drugs: from serendipity to design. Dalton Trans 43:4903–4917
296. Karges J, Stokes RW, Cohen SM (2021) Metal complexes for therapeutic applications. TRECHEM 3(7):523–534
297. Kostova I, Balkansky S (2013) Metal complexes of biologically active ligands as potential antioxidants. Curr Med Chem 20(36):4508–4539
298. Mezgebe K, Mulugeta E (2024) Synthesis and pharmacological activities of Schiff bases with some transition metal complexes: a review. Med Chem Res 33(3):439–463

Chapter 12

Environmental Factors Influencing Brain Metallomics

Anna Błażewicz, Michał Kiełbus, Agostinho Almeida and Julia Wojnicka

Abstract

Neurotoxic metals, including lead, mercury, cadmium, arsenic, and aluminum, have long been linked to the development and progression of various brain disorders, from neurodegenerative diseases such as Alzheimer's disease or Parkinson's disease to cognitive and developmental impairments. While manganese is essential in trace amounts, excessive exposure can also be neurotoxic. All these metals disrupt critical molecular processes in the brain, affecting cell function, oxidative balance, protein folding, and neuronal communication. This chapter examines the molecular mechanisms underlying metal-induced neurotoxicity, focusing on their impact on enzyme activity, metal ion homeostasis, and neuroinflammation. Additionally, it explores human exposure sources and current risk assessment strategies, addressing challenges in quantifying exposure, bioaccumulation, and individual susceptibility. Finally, key gaps and challenges in characterizing the full spectrum of neurotoxicants within human brain metallomics research are discussed.

Keywords Neurotoxic metals, Brain diseases, Molecular mechanisms, Risk assessment

Abbreviations

Aβ	amyloid-beta
ADA	adenosine deaminase
AD	Alzheimer's disease
ADHD	attention-deficit hyperactivity disorder
AChE	acetylcholinesterase
ALS	amyotrophic lateral sclerosis
AMPA	α-amino-3-hydroxy-5-methyl-4-isoxazolepropionic acid
AMPK	adenosine monophosphate-activated protein kinase
APP	amyloid precursor protein
ARC	activity-regulated cytoskeleton-associated protein
AS3MT	arsenic methyltransferase
ASD	autism spectrum disorders
ATP	adenosine triphosphate
Bax	BCL2-associated X apoptosis regulator

Andreas M. Grabrucker (ed.), *Brain Metallomics*, Neuromethods, vol. 237,
https://doi.org/10.1007/978-1-0716-5468-2_12,
© The Author(s), under exclusive license to Springer Science+Business Media, LLC, part of Springer Nature 2026

BBB	blood–brain barrier
Bcl-2	Apoptosis regulator Bcl-2, encoded by BCL2 gene
BDNF	brain-derived neurotrophic factor
Beclin-1	Bcl-2-interacting protein 1
Bim	Bcl-2-interacting mediator of cell death
CaM	calmodulin
CAT	catalase
CD4+ T cell	CD4-positive T cell
CHRM2	acetylcholine receptor M2
ChAT	choline acetyltransferase
CNS	central nervous system
CO	carbon monoxide
COX	cyclooxygenase
CREB	cAMP response element-binding protein
CSF	cerebrospinal fluid
CTIP2	COUP-TF interacting protein 2, aka BCL11C B-cell lymphoma/ leukemia 11B or BCL11 transcription factor B
Daxx	death domain-associated protein 6, encoded by DAXX gene
DNA	deoxyribonucleic acid
Drp-1	dynamin-related protein 1 (or Dnm1L dynamin-like protein 1)
DS	Down's syndrome
EAAC1	excitatory amino acid carrier 1
EAAT1	excitatory amino acid transporter 1
EAAT2	excitatory amino acid transporter 2
EDs	endocrine disruptors
ERK	extracellular signal-regulated kinase
Fas	Fas cell surface death receptor
FasLG	Fas ligand
FDOPA	L-3,4-dihydroxy-6-[^{18}F]fluorophenylalanine
FOXO3a	forkhead box O3a
GABA	gamma-aminobutyric acid
GCR	glucocorticoid receptor
GFAP	glial fibrillary acidic protein
GPX	glutathione peroxidase
GR	glutathione reductase
GSH	reduced glutathione
GST	glutathione-S-transferase
ICP-MS	inductively coupled plasma mass spectrometry
IL	interleukin
IQ	intelligence quotient
JNK	c-Jun N-terminal kinase
LA-ICP-MS	laser ablation ICP-MS
LTP	long-term potentiation
MAPK	mitogen-activated protein kinase
MBP	myelin basic protein
MEK5	mitogen-activated protein kinase kinase 5
MeHg	methylmercury

miRNA	micro ribonucleic acid
Mfn2	mitofusin 2
MIP-1α	macrophage inflammatory protein-1 alpha
MnSOD	manganese-dependent superoxide dismutase
MMP-9	matrix metalloproteinase-9
mRNA	messenger ribonucleic acid
MS	multiple sclerosis
mTOR	mechanistic target of rapamycin
NAC	N-acetylcysteine
NF-κB	nuclear factor kappa-light-chain-enhancer of activated B cells
NLRP3	nucleotide-binding domain leucine-rich-containing family pyrin domain-containing-3
NMDA	N-methyl-D-aspartate
NOS	nitric oxide synthase
NOX	NADPH oxidase family
NQO-1	NAD(P)H quinone dehydrogenase 1
Nrf2	nuclear factor-erythroid 2-related factor 2
NSCs	neural stem cells
NTPDase	nucleoside triphosphate diphosphohydrolase
Opa1	mitochondrial dynamin-like GTPase, encoded by OPA1 gene
PAHs	polycyclic aromatic hydrocarbons
PCBs	polychlorinated biphenyls
PD	Parkinson's disease
PET	positron emission tomography
PHF8	plant homeodomain finger protein 8
PI3K	phosphoinositide 3-kinase
PKB	protein kinase B (Akt)
PINK1	PTEN-induced kinase 1
PKC	protein kinase C
PLP	proteolipid protein
PM	particulate matter
POPs	persistent organic pollutants
PP2A	protein phosphatase 2A
PTEN	phosphatase and tensin homolog
PUMA	p53 upregulated modulator of apoptosis
p53	tumor protein p53
p73	tumor protein p73
PGC1α	peroxisome proliferator-activated receptor gamma coactivator 1 alpha
Ras	rat sarcoma virus protein family
RfD	reference dose
ROS	reactive oxygen species
SAM	S-adenosylmethionine
SAH	S-adenosylhomocysteine
SIRT1	NAD^+-dependent deacetylase
SOX2	SRY (sex determining region Y)-box 2
SOD	superoxide dismutase
Sp1	specificity protein 1

STAT-3	signal transducer and activator of transcription 3
TBR2	T-box brain 2
TDI	tolerable daily intake
TH	tyrosine hydroxylase
TLR4	toll-like receptor 4
TNF-α	tumor necrosis factor alpha
TORC1	target of rapamycin complex 1
TRPCs	transient receptor potential canonical channels family
Trx	thioredoxin
TrxR	thioredoxin reductase family
VEGF	vascular endothelial growth factor
VGCCs	voltage-gated calcium channel family
VOCs	volatile organic compounds
WHO	World Health Organization
Wip1	wild-type p53-induced phosphatase 1
XRF	X-ray fluorescence
xCT	cystine/glutamate antiporter
YY1	Yin-Yang 1 transcription factor
γ-GCS	γ-glutamylcysteine synthetase

1 Introduction

Understanding the complex interactions between the brain and its surrounding environment is a key aspect of modern neuroscience [1]. The brain metallome refers to the profile of essential and nonessential metals present in the brain, which play critical roles in neural function, signaling, and homeostasis [2–7].

Brain metallomics primarily investigates the distribution, content, and role of metals and metalloids in relation to brain function, neurological diseases, and environmental influences. A recently emerging subfield, environmental metallomics, focuses on the impact of metal exposure and associated environmental hazards on brain health [8].

Various environmental factors influence the brain's metallome, thereby affecting neurological processes and overall health outcomes. Key factors include the intake of essential metals such as iron (Fe), zinc (Zn), copper (Cu), manganese (Mn), and magnesium (Mg), which are necessary for enzymatic activity, neurotransmitter synthesis, and neuronal signaling. Any imbalance—whether deficiency or excess—can disrupt brain function. Additionally, vitamins and antioxidants influence metal absorption and metabolism, potentially mitigating oxidative stress caused by metal imbalances. Conversely, exposure to toxic metals such as lead (Pb), mercury (Hg), cadmium (Cd), arsenic (As), and aluminum (Al) can significantly impair brain function and the entire nervous system. Their ability to accumulate in neural tissues can result in developmental delays, cognitive deficits, and behavioral changes. There is substantial evidence that exposure to air pollution, including tobacco smoke, particulate matter (PM), volatile organic compounds, and other environmental pollutants, contribute to neuroinflammatory processes and neurodegeneration. Long-term exposure to pesticides and industrial chemicals (e.g., organophosphates, polychlorinated biphenyls, and other persistent organic pollutants) can also disrupt metal homeostasis in the brain. Certain substances capable of mimicking or disrupting the endocrine system—known as endocrine disruptors (EDs), such as Cd—may influence the uptake, storage, and excretion of metals in the body.

The brain's response to exposure to metals with no known beneficial physiological roles varies with age and developmental stage. For example, prenatal exposure to environmental pollutants, including metals, can significantly affect brain development and cause lasting effects on the metallome and cognitive function. Research indicates that childhood exposure to metals may increase the risk of mental health problems, including anxiety and depression, later in life [9]. As individuals age, the brain's ability to regulate metal homeostasis may decline, contributing to neurodegenerative diseases such as Alzheimer's disease (AD) or Parkinson's disease (PD). Many of the aforementioned environmental factors may exacerbate this process. Heavy metal exposure poses a significant public health risk due to its widespread presence in the environment and its well-documented effects on brain function. These metals can cross the blood–brain barrier (BBB), accumulate in neural tissues, and interfere with essential biochemical pathways, causing neuronal damage and dysfunction [9]. Despite extensive research, many aspects of metal-induced neurotoxicity remain incompletely understood, particularly in terms of synergistic effects, dose–response relationships, and long-term consequences. Advancing knowledge in this field is crucial for regulatory policies, developing preventive strategies, and therapeutic interventions aimed at minimizing human exposure and mitigating neurological risks.

This chapter focuses on issues related to metal neurotoxicity, particularly Mn, Hg, Pb, Cd, As, and Al. It also explores the mechanisms through which metal neurotoxicity contributes to brain pathology and examines exposure to neurotoxic metals to assess health risk levels. Recent findings from epidemiological and in vivo studies are presented that shed light on the potential mechanisms by which metals impair neurological function and contribute to disease pathogenesis. Additionally, we discuss key challenges researchers face in studying the brain metallome and the impact of selected environmental factors on it.

2 Identifying Possible Associations Between Exposure to Neurotoxic Metals and Brain Diseases

This section examines the potential connections between neurotoxic metal exposure and the development of brain diseases. It explores the neurotoxic effects of metals such as Mn, Hg, Pb, Cd, As, and Al, analyzing their proven contribution to neurotoxicity and resulting neurological dysfunction through various molecular mechanisms. Furthermore, we report how metal-induced neurotoxicity may be involved in the onset and progression of neurodegenerative and neurodevelopmental disorders, including AD, PD, and autism spectrum disorders (ASD). With the rising prevalence of neurological disorders—now the primary cause of disability and the second leading cause of death—coupled with increasing environmental contamination, understanding the role of metal exposure in brain diseases is of paramount importance [9, 10]. Industrial activities have greatly increased human exposure to neurotoxic metals, which are extensively used in manufacturing, agriculture, and energy production. Environmental pollution from mining, smelting, and industrial emissions has also led to widespread exposure, disproportionately affecting vulnerable populations [11]. Identifying the mechanistic links between metal exposure and neurological disorders is crucial for developing targeted public health interventions and regulatory measures [12].

2.1 Manganese

Manganese is an essential element at low concentrations, acting as a cofactor for enzymes such as Mn-dependent superoxide dismutase (MnSOD) and arginase. However, excessive Mn exposure can trigger developmental complications and, with chronic exposure, may result in PD-like syndromes, also known as manganism [13]. Occupational Mn exposure (e.g., in mining and welding) has been strongly associated with PD-like syndromes characterized by motor dysfunction, bradykinesia, and rigidity. Welders exposed to airborne Mn particles show a higher prevalence of Parkinsonian symptoms than the general population [14, 15]. Although manganism and PD are recognized as two distinct neurological disorders, Mn exposure may increase the risk of developing PD, particularly in individuals with a genetic predisposition [14, 16]. Chronic exposure to Mn-contaminated drinking water has been linked to significant cognitive impairments, particularly in children and older adults, with elevated Mn levels correlating with reduced intelligence quotient (IQ) scores and memory deficits in children [17]. Prenatal Mn exposure, especially in regions with high environmental contamination, has been associated with attention-deficit hyperactivity disorder (ADHD) and other neurodevelopmental behavioral problems, including internalizing, externalizing, and hyperactivity in children [18].

Exposure to Mn fumes in welders has been linked to early signs of dopaminergic dysfunction in the nigrostriatal system, as evidenced by inhibited ^{18}F-6-fluoro-L-dihydroxyphenylalanine (FDOPA) positron emission tomography (PET) uptake. Asymptomatic welders showed decreased dopamine uptake in the caudate nucleus, with a pattern distinct from idiopathic PD. Their unique pattern of dysfunction (caudate>anterior putamen>posterior putamen) suggests that Mn neurotoxicity differs from the regional dopamine deficits seen in idiopathic PD and may serve as an early marker of neurotoxicity in exposed workers [19]. Another study highlights reduced caudate FDOPA uptake as an indicator of Mn-related dysfunction but questions its direct relevance to clinical Parkinsonism [20].

Elevated Mn exposure has been consistently associated with behavioral and emotional problems in adolescents. Higher blood Mn levels have been linked to aggressive behaviors and emotional dysregulation [21]. Similarly, elevated blood Mn levels have been associated with heightened anxiety, social difficulties, and delinquent behaviors, particularly in children with ADHD [22]. Further evidence suggests that Mn exposure, especially in boys, correlates with disruptive behaviors and increased symptoms of anxiety and depression [23].

Epidemiological evidence strongly links Mn exposure to neurotoxicity, including Parkinsonian syndromes, cognitive decline, and neurodevelopmental disorders. Environmental contamination, mainly through drinking water, exacerbates these risks, while chronic exposure contributes to neurodegeneration. Addressing this issue requires targeted research and interventions focused on protecting vulnerable populations.

2.2 Mercury

Mercury is recognized as a pervasive environmental pollutant, widely present in seafood, exposing all age groups to its toxic effects [24, 25]. The neurotoxicity of Hg depends on its chemical form and route of exposure, with both inorganic and organic compounds contributing to its harmful consequences [26, 27]. Hg poisoning is particularly concerning due to its central nervous system (CNS) toxicity, which has been linked to neurodegenerative diseases, neural stem cell dysfunction, and developmental abnormalities. These risks are especially pronounced during fetal development, a period of heightened susceptibility [27, 28]. For example, maternal exposure to methylHg ((CH3Hg)+, MeHg), a highly neurotoxic chemical form of Hg, as assessed by maternal hair Hg concentrations, is associated with memory deficits in children [29]. Other studies on pregnant women exposed to MeHg have reported developmental and cognitive impairments in their offspring [30]. Hg can cross the placenta, posing a significant risk to fetal development, particularly in pregnant women with amalgam dental fillings [31].

Chronic exposure to MeHg can cause severe CNS damage, characterized by oxidative stress, neurodegeneration, and histological changes, resulting in motor dysfunction, sensory impairments, and paralysis in extreme cases [32, 33]. Research suggests that even low-level Hg exposure may increase the risk of neurodegenerative diseases such as amyotrophic lateral sclerosis (ALS), PD, and AD. For example, studies have observed Hg-induced protein alterations and mitochondrial dysfunction in hippocampal cells, exacerbating oxidative damage and memory deficits [34, 35]. In animal studies, Hg deposits in the spinal cord and brain were associated with hallmark neurodegenerative changes, including axonal atrophy and reduced axon caliber, suggesting that long-term exposure may increase the risk of neurodegeneration in humans [36]. In AD models, Hg has been shown to worsen amyloid-beta (Aβ)-induced memory impairments by inducing mitochondrial dysfunction, oxidative stress, and hippocampal damage, ultimately impairing spatial memory and cognitive function [37].

Dietary Hg exposure has also been linked to later-life neurological disorders, including extrapyramidal motor dysfunction, sensory and cranial nerve impairments, neurocognitive deficits, and mood disorders [38]. Several studies have identified a significant link between elevated Hg levels and the progression of ASD [39, 40]. Elevated Hg concentrations have been detected in the teeth of ASD patients, possibly due to impaired excretion mechanisms exacerbated by prolonged antibiotic use [41]. Although animal studies have confirmed that antibiotic use almost completely inhibits Hg excretion in rats due to changes in gut flora [41], the specific mechanisms by which long-term antibiotic use impairs Hg elimination in humans are not fully understood. Additionally, Hg exposure during critical developmental periods has been associated with increased ASD symptoms, as evidenced by higher Hg levels in the blood of pregnant women and young children, which correlated with the emergence of autistic behaviors. Hg accumulation in endocrine glands, where interactions with sex hormones further compound developmental disruptions, may also contribute to ASD-like manifestations [42]. Studies have also reported increased Hg levels in the hair and blood samples of ASD patients, emphasizing the metal's systemic impact on neurodevelopment [43]. Furthermore, exposure to thimerosal, an Hg-containing preservative used in some vaccines, has been linked to atypical autistic symptoms in certain studies [44].

Research highlights the role of Hg in disrupting redox homeostasis through interactions with cysteine and selenocysteine residues, as well as through microglial and astrocytic activation, which amplify its neurotoxic effects [35].

Even at low levels, Hg exposure poses significant neurotoxic risks, contributing to neurodegenerative diseases, developmental impairments, and cognitive dysfunction, particularly among the most vulnerable populations, such as pregnant women and children. Its widespread presence in the environment underscores the pressing need for stricter exposure regulations and further research into its long-term neurotoxic impact.

2.3 Lead

Lead is a potent neurotoxicant that profoundly affects the CNS, particularly during developmental stages. Once absorbed, Pb transiently accumulates in the brain, which induces morphological and functional disruptions [45]. Pb primarily targets the CNS, accumulating particularly in the hippocampus, resulting in deficits in motor skills, memory, intelligence, language, and emotional regulation [46]. Pb exposure impairs neuronal differentiation, myelination, and synaptogenesis, resulting in structural alterations in brain cells. Functionally, Pb exerts its effects through ion mimicry, competing with Ca, Zn, and Fe for binding sites essential for their biochemical functions. This disrupts synaptic neurotransmission, altering intracellular Ca levels and contributing to mitochondrial dysfunction and oxidative stress [45].

A study indicates that Pb exposure significantly contributes to mental disorders, particularly in children under 6 years old, who bear the most significant burden. These findings emphasize the increased vulnerability of young developing brains to Pb toxicity and underscore the urgent need for targeted interventions to mitigate this health risk [47].

Increasing evidence points to a significant connection between AD and Pb exposure. Early-life exposure to Pb has been suggested as a potential trigger for the most common form of late-onset AD, potentially influencing disease development later in life through epigenetic regulation of genes associated with AD pathogenesis [48, 49]. Population-based studies indicate that higher cumulative Pb levels are associated with decreased memory performance [50]. The studies consistently reported that blood Pb levels are significantly higher in AD patients than in healthy controls [51, 52]. A 2019 cohort study analyzing data from over 8000 participants aged 60+ (between 1999 and 2008) found a positive but not statistically significant association between blood Pb level and AD mortality [53]. Animal studies corroborate these findings by demonstrating the link between AD and the neurotoxic effects of Pb. In primates exposed to Pb during the first 2 months of life, adverse brain alterations, increased amyloidogenesis, and senile plaque deposition were observed at the age of 23, accompanied by upregulation of key amyloid-processing genes [54]. Similarly, rats exposed to Pb early in life exhibited increased amyloid precursor protein (APP) messenger ribonucleic acid (mRNA) expression and Aβ aggregation in adulthood [55].

Environmental factors, such as Pb, have also been linked to an increased risk of PD [56, 57]. Among individuals studied for bone Pb concentrations, those with higher Pb accumulation had more than three times the risk of developing PD compared to those with lower levels [58]. The link between long-term Pb exposure and an increased risk of PD is further supported by evidence showing that individuals with higher lifetime Pb exposure had more than twice the risk of developing PD compared to those with lower exposure [59].

Neurotoxic Pb profoundly affects brain development, especially in young children, Pb to cognitive, behavioral, and neurodegenerative consequences. Growing evidence also links early-life and chronic Pb exposure to an increased risk of neurodegenerative diseases, such as AD and PD, highlighting the urgent need for preventive measures to mitigate its long-term neurological impact.

2.4 Cadmium

Cadmium is a known carcinogen and neurotoxin that infiltrates both the peripheral and CNS [60]. Exposure to Cd disrupts cellular homeostasis, inducing oxidative stress, inhibiting DNA repair mechanisms, and triggering apoptosis, all of which contribute to neuronal degeneration and brain dysfunction [61, 62]. Chronic Cd exposure has been strongly associated with neurodegenerative disorders, including AD, PD, and Huntington's disease, due to its ability to cause widespread structural changes in the brain and impair essential neurological functions [60, 63]. Additionally, Cd toxicity is linked to cerebral hemorrhage, leading to cognitive decline, memory impairment, olfactory dysfunction, and behavioral disturbances such as anxiety and depression in exposed individuals [64–66]. Cd can damage the BBB and increase its permeability, but it can also escape the BBB via the olfactory route, entering the brain through the nasal mucosa and accumulating in the choroid plexus [67]. The underdeveloped BBB in young individuals makes them particularly vulnerable to Cd's neurotoxic effects. Exposure to Cd during pregnancy and early life is associated with adverse outcomes, including reduced birth weight, impaired brain development, and cognitive deficits [68].

Epidemiological studies suggest that prenatal and early-life exposure to Cd negatively affects pregnancy outcomes, neonatal health, and cognitive abilities in children, such as verbal IQ and learning capacity [69, 70]. However, studies on Cd's neurotoxic effects are limited by confounding factors, emphasizing the need for animal models to establish a direct link between Cd exposure and cognitive or olfactory impairments [65].

Exposure to Cd causes substantial histopathological changes in the brain, highlighting its potent neurotoxic effects. Key structural alterations include separation of the pia mater, capillary hyperemia, disruption of mitochondrial cristae, and accumulation of apoptotic cells, as observed in juvenile mice exposed to Cd [71]. Further damage, such as neuronal degeneration, Nissl body dissolution, and disrupted myelination, has been documented in rat models, highlighting widespread neuronal apoptosis and lipid peroxidation [72, 73]. Histological impairments extend across various brain regions, with fetal Cd exposure causing pyknosis in cortical and hippocampal areas and adult rats displaying cortical inflammation and vacuolation [74, 75]. These findings validate animal models of Cd neurotoxicity and emphasize its ability to disrupt cellular integrity and neural function, reflecting the critical role of structural damage in Cd-induced neurotoxicity [75–77]. Moreover, chronic Cd exposure disrupts neural differentiation, cortical organization, and synaptogenesis in human cerebral organoids [78].

Due to its well-established neurotoxic properties, Cd disrupts cellular homeostasis, causing neuronal degeneration and cognitive impairments, particularly in pregnant women and their fetuses. Its strong link to neurodegenerative diseases, cerebral hemorrhage, and structural brain damage underscores the urgent need for further research and preventive strategies to mitigate its neurological effects.

2.5 Arsenic

Arsenic is a well-documented toxic metalloid with carcinogenic properties. Exposure to As can cause mitochondrial oxidative stress, Ca imbalance, disrupted adenosine triphosphate (ATP) production, and altered membrane potential. Early exposure to As is linked to reduced brain weight, as well as a decrease in the number of neurons and glia, which may contribute to neurodegenerative disorders such as AD and ALS [79]. Furthermore, As can readily cross both the placenta and BBB, accumulating in higher concentrations in the fetal brain compared to adults [80, 81]. This underscores the significant concern of As-mediated neurotoxicity, particularly during prenatal development, as the developing brain is far more vulnerable to toxic agents than the adult brain [82].

Multiple epidemiological studies have linked As exposure to neurodevelopmental impairments. One notable study in Japan found that infants aged 6–10 months exposed to As-contaminated milk powder exhibited lower IQ and an increased risk of mental retardation before the age of 60 [83, 84]. A meta-analysis reported significant deficits in verbal IQ, memory, motor function, and attention, indicating that increased levels of As in urine or drinking water correlate with significant reductions in IQ [85]. Similar findings, including associations with intellectual disabilities and memory impairments following As exposure during childhood, have been confirmed by other systematic reviews [86, 87]. While most studies agree that As exposure negatively affects brain development, the findings vary due to differences in study design, exposure assessment, and developmental stages examined. Notably, 13 out of 41 studies, particularly those focusing on early childhood, found no clear association between As exposure and neurodevelopmental impairments [86]. The lack of clear evidence of As-related neurodevelopmental effects in young children may be because such impairments only become apparent over time. This is supported by studies showing no detectable effects in infants but significant cognitive deficits, including lower verbal IQ, in the same population at age 5 [86, 88].

The most recent systematic review of 16 studies examined As exposure from drinking water and its effects on children's neurodevelopment [89]. Among five high-quality studies, only one found a link between As and lower intellectual function, while others showed no significant associations after adjusting for confounders. Three intermediate-quality studies suggested cognitive impairments but lacked methodological rigor. Four low-quality studies were excluded from the review. The same study also reviewed six studies on adults. Two high-quality studies conducted in highly exposed populations reported cognitive decline, but no effects were observed for As exposure below 75μg/L, while the lower-quality studies showed inconsistent findings. The authors concluded that while As may impact cognition at high exposures, evidence at lower levels remains inconclusive, highlighting the need for further research. Finally, the authors found no definitive link between As in drinking water and neurobehavioral effects. This conclusion was mainly attributed to study limitations, underscoring the need for further research in this important area [89].

While high As exposure has been associated with cognitive decline, lower IQ, memory impairments, and neurodevelopmental delays, findings remain inconsistent due to variations in study design, exposure assessment, and developmental stages examined. Recent systematic reviews suggest that while As may contribute to neurotoxicity at high concentrations, evidence at lower levels is inconclusive. This underscores the need for further rigorous research to clarify its long-term neurological effects and to better understand the potential risks associated with As exposure at lower concentrations.

2.6 Aluminum

Aluminum exposure has been shown to have harmful effects on the brain, with excessive exposure potentially contributing to neuronal damage, raising concerns about its long-term impact on human health [90]. Due to its neurotoxic properties, Al has been linked to impaired neurodevelopment and the progression of neurodegenerative disorders [91]. Studies suggest that Al can contribute to oxidative stress, promote the aggregation of misfolded proteins, and interfere with neuronal function, thereby exacerbating conditions such as AD and other brain disorders [92–94]. Additionally, Al levels were found to be significantly higher in the brain tissue of patients with AD, Down's syndrome (DS), and dialysis dementia syndrome compared to age- and gender-matched controls from the same brain regions. In contrast, no significant increase in Al levels was observed in other neurological disorders, including PD, multiple sclerosis (MS), ALS, or schizophrenia [95]. Moreover, Al exposure is associated with increased neuroinflammation and disruption of neuronal signaling, further linking it to the etiopathology of various neurological conditions [96].

Al exposure has been consistently linked to AD in both epidemiological and experimental studies. Elevated Al levels have been observed in the brain, serum, and cerebrospinal fluid (CSF) of AD patients, with some studies suggesting that these levels may serve as early markers of the disease or contribute to its progression [97, 98]. For instance, chronic Al exposure through drinking water has been associated with a twofold higher risk of developing AD, even after controlling for demographic and lifestyle factors [99, 100].

Research highlights that Al accumulates in amyloid plaques and colocalizes with Aβ in senile plaques and cerebral vasculature, promoting Aβ aggregation and tau hyperphosphorylation [101]. This association may stem from the effects of Al accumulation, which include enhanced production and aggregation of Aβ, tau protein hyperphosphorylation, cytoskeletal disruptions, decreased synaptic plasticity, changes in signal transduction, and eventual neuronal death [102]. These molecular processes are pathological hallmarks of AD [103]. Experimental models have shown that Al exposure, particularly in combination with agents like D-galactose, induces AD-like cognitive deficits in rodents, further supporting its role in disease development [104]. Moreover, elevated Al concentrations have also been found in the hippocampal regions of AD patients' brains, with levels exceeding those found in controls by more than fourfold [105, 106]. Importantly, Al accumulation appears to be more pronounced in familial forms of AD, likely influenced by genetic factors such as carrying the ε4 allele of the apolipoprotein E gene (ApoEε4), which exacerbates amyloid pathology [101, 107, 108].

Al has been proposed as a potential risk factor for the development of PD. While age, genetic predisposition, and pesticide exposure are established risk factors, long-term exposure to Al also increases PD risk [109]. Al accumulates in the substantia nigra and Lewy bodies, disrupting the dopaminergic system by impairing dopamine biosynthesis pathways [110, 111]. The zebrafish model has effectively explored Al neurotoxicity, revealing histological changes, oxidative damage, and deregulation of Parkinsonism-related genes [112]. Occupational and environmental exposure to Al is a primary source of increased PD risk [113, 114]. A retrospective study on 37,000 males in Ontario, Canada, found a 30% higher incidence of PD among miners with respiratory exposure to Al, with the risk positively correlating to the duration of exposure to Al-containing dust [114]. Similarly, Martell et al. [115] demonstrated that inhaling finely ground Al dust significantly elevated PD risk. The synergistic toxicity of Al with other metals, such as Mn and Fe, as well as pesticides, has also been linked to accelerated alpha-synuclein fibril formation, a hallmark of PD [116, 117]. Moreover, elevated serum Al levels, along with elemental imbalances, may serve as early diagnostic markers for PD [118]. Additionally, Al neurotoxicity in PD is associated with its ability to facilitate the misfolding and aggregation of α-synuclein into Lewy bodies, which disrupts dopaminergic neurons and induces mitochondrial dysfunction and oxidative stress. This aggregation exacerbates neuronal death, which is central to PD pathology [119].

Evidence also suggests that Al exposure may contribute to neuroinflammatory processes in MS, where Al has been detected at elevated levels in the brain tissue of patients, and localization has been observed both within cells and in extracellular areas. Notably, its association with corpora amylacea—aggregates of cellular waste found in the CNS—points to a potential role of Al in the neurodegenerative processes characteristic of MS [120]. Studies have reported higher Al levels in MS tissues compared to healthy controls, indicating its potential role in the disease's progression [121]. Elevated Al levels in neural tissues are thought to disrupt the BBB, facilitating immune cell infiltration and subsequent damage to white matter. This demyelination contributes to the neurological impairment characteristic of MS [95, 122].

Al exposure has been associated with various neurodegenerative diseases, including AD, PD, and MS, with strong evidence suggesting its role in promoting oxidative stress, protein misfolding, and neuroinflammation. While epidemiological and experimental studies indicate its neurotoxic effects, further research is needed to clarify its precise contribution to disease progression and to establish exposure thresholds for potential health risks.

3 Molecular Mechanisms of Metal-Induced Neurotoxicity

Studying molecular and cellular mechanisms underlying neurotoxicity induced by environmental neurotoxic metals—including Mn, Hg, Pb, Cd, As, and Al—requires a multidisciplinary approach due to its complexity. It examines how these metals disrupt critical molecular processes essential for normal brain function, including synaptic transmission, oxidative stress regulation, mitochondrial function, and neuroinflammation. Particular attention is given to their impact on neurodevelopment and cognitive function, as early-life exposure to these metals has been linked to long-term cognitive deficits, behavioral abnormalities, and an increased susceptibility to neurodegenerative diseases [9].

Understanding how toxic metals affect molecular and cellular pathways in normal and pathological brain states is essential for developing strategies to reduce exposure, mitigate existing damage, and protect brain health—especially in vulnerable populations such as children, pregnant women, and the elderly. The following sections summarize recent findings from in vivo and in vitro studies, comprehensively reviewing cellular and molecular disruptions caused by neurotoxic metals. This discussion aims to bridge the gap between toxicological research and clinical implications, shedding light on potential biomarkers of metal-induced neurotoxicity and opportunities for future research.

3.1 Manganese

Manganese-induced neurotoxicity involves a range of mechanisms, including oxidative stress, mitochondrial dysfunction, glutamate excitotoxicity, neuroinflammation, protein misfolding, and disruptions in autophagy, mitophagy, and endoplasmic reticulum stress [123–126]. These pathological processes collectively contribute to neuronal injury and have been implicated in the development and progression of neurodegenerative diseases such as PD [126]. A deeper understanding of these mechanisms is essential for developing targeted therapeutic strategies to mitigate the harmful neurological effects of Mn exposure.

Mn induces oxidative stress by generating reactive oxygen species (ROS), causing oxidative damage to neural cells, especially in the CNS [127, 128]. This effect is most pronounced in the basal ganglia regions, including the globus pallidus, striatum, and substantia nigra, which are especially vulnerable to Mn-induced oxidative damage [129]. Mitochondrial dysfunction plays a crucial role in this process since Mn disrupts mitochondrial ATP production, especially its trivalent form (Mn^{3+}), enhancing electron leakage and subsequent ROS generation [130]. Beyond directly increasing ROS production, Mn also impairs the cellular antioxidative response, notably by disrupting the synthesis of reduced glutathione (GSH), a key antioxidant. The depletion of GSH amplifies oxidative stress and further compromises cellular defenses [131, 132]. At the transcriptional level, Mn disturbs redox homeostasis primarily through the downregulation of sirtuins, leading to increased acetylation of transcription factors such as forkhead box O3a, a protein encoded by the FOXO3 gene (FOXO3a), and peroxisome proliferator-activated receptor gamma coactivator 1 alpha (PGC1α), a protein encoded by the PPARGC1A gene [133]. Acetylated PGC1α is associated with a decreased nuclear factor-erythroid 2-related factor 2 (Nrf2) expression and the subsequent downregulation of its target genes, which include crucial redox-regulating enzymes such as γ-glutamylcysteine synthetase (γ-GCS), glutathione peroxidase (GPX), glutathione reductase (GR), glutathione synthetase, NAD(P)H quinone dehydrogenase 1 (NQO-1), and heme oxygenase 1 [133]. Additionally, Mn may further impair Nrf2 signaling by altering kelch-like ECH-associated protein 1 gene expression, a key regulator of Nrf2 activity [126, 133]. The increased acetylation of FOXO3a further contributes to oxidative stress by reducing the expression of antioxidant enzymes such as superoxide dismutase (SOD) and catalase (CAT), which are normally upregulated by deacetylated FOXO3. Moreover, acetylated FOXO3a enhances proapoptotic signaling through the activation of B-cell leukemia/lymphoma 2 protein (Bim) and p53 upregulated modulator of apoptosis (PUMA), further exacerbating neuronal damage and oxidative stress [134].

The accumulation of Mn in mitochondria disrupts their homeostasis and bioenergetics, triggering oxidative stress, as mentioned above, ultimately resulting in necrotic or apoptotic cell death [135, 136]. Another consequence of Mn toxicity is mitochondrial fragmentation, resulting from its interference with fission and fusion processes. Key regulatory proteins such as dynamin--related protein 1 (Drp-1), mitochondrial dynamin-like GTPase (Opa1), and mitofusin

2 (Mfn2) are affected, causing imbalanced mitochondrial dynamics that impair function and contribute to neuroinflammation [137–139]. Mitophagy, the selective autophagy mechanism responsible for clearing damaged mitochondria, appears to play a critical protective role against Mn-induced neurotoxicity. Initially, this process helps preserve mitochondria, mitigating damage [140, 141]. However, prolonged Mn exposure disrupts this defense mechanism by inducing S-nitrosylation of phosphatase and tensin homolog (PTEN)-induced kinase 1 (PINK1), an essential protein in the PINK1/Parkin pathway. This disruption increases dysfunctional mitochondria, exacerbating cellular damage [142, 143]. Autophagy, vital for cellular homeostasis, is initially activated following Mn exposure as a protective response against brain tissue damage. However, prolonged exposure impairs this process, resulting in cytotoxicity [144, 145]. In rats, Mn-induced autophagy in the striatum was observed within 4–12h, but its suppression was observed after 1–28 days [146]. Additionally, Mn injections led to mitochondrial vacuolation, swollen endoplasmic reticulum, and dysfunctional lysosomes in the striatum, further underscoring its role in autophagy disruption and brain dysfunction [147].

Mn-induced inflammation background was defined as disrupting the regulatory functions of astrocytes and microglia, key cells responsible for maintaining CNS homeostasis through immune regulation and the production of inflammatory mediators. By impairing mitochondrial bioenergetics in astrocytes, Mn triggers the release of proinflammatory cytokines, causing neuronal damage in adjacent regions [138]. At the same time, Mn activates microglia, promoting the production of proinflammatory cytokines that contribute to dopaminergic neuronal death [148, 149]. In nonhuman primates, Mn exposure further amplifies neuroinflammation by increasing the release of factors such as interleukin-1 beta (IL-1β) and tumor necrosis factor-alpha (TNF-α), and inducible nitric oxide synthase (NOS), intensifying neuronal injury [150]. The nuclear factor kappa-light-chain-enhancer of activated B cells (NF-κB) signaling pathway plays a central role in this process by mediating Mn-induced cytokine release from microglia and amplifying astrocyte activation, which exacerbates inflammation [149]. Additionally, Mn induces a proinflammatory M1 microglial phenotype, characterized by the overexpression of inflammatory molecules such as IL-6 and IL-1β, along with activation of the nucleotide-binding domain, leucine-rich–containing family, pyrin domain–containing-3 (NLRP3) inflammasome, further aggravating neuronal damage [151].

Mn neurotoxicity has also been implicated in the accumulation of AD and PD-related proteins. Mn exposure increases Aβ levels by upregulating β-secretase, APP, and presenilin, particularly in the presence of activated microglia and inflammatory cytokines such as IL-1β and TNF-α [152]. Additionally, Mn promotes tau hyperphosphorylation through demethylation of protein phosphatase 2A (PP2A), leading to oxidative stress, apoptosis, and neuronal damage [153]. Mn nanoparticles further enhance tau aggregation by activating caspase-3 and caspase-9 and increasing the Bax/Bcl-2 ratio, signaling apoptosis pathways [154]. Mn also facilitates α-synuclein misfolding, aggregation, and neurotoxicity by interacting with the tropomyosin kinase B receptor (TrkB), thereby disrupting brain-derived neurotrophic factor (BDNF)/TrkB and N-methyl-D-aspartate (NMDA) receptor signaling [155]. Additionally, Mn enhances α-syn exosomal release, propagating neurotoxic signals to microglia and intensifying neuroinflammation [155]. Furthermore, Mn exposure has been shown to alter micro ribonucleic acid (miRNA) expression, notably by downregulating miR-7 and miR-433, contributing to α -synuclein accumulation and toxicity in the brain [156].

Mn also disrupts neurotransmission by inducing excitotoxicity. It increases extracellular glutamate levels by impairing astrocytic glutamate transporters, particularly human excitatory amino acid transporter 1 (EAAT1) [glutamate aspartate transporter (GLAST) in rodents] and excitatory amino acid transporter 2 (EAAT2) (glutamate transporter-1 in rodents), which are crucial for synaptic glutamate clearance and excitotoxicity prevention. This dysfunction ultimately results in neuronal injury [157]. Studies indicate that Mn reduces glutamate uptake by astrocytes through the transcriptional downregulation of EAAT1 and EAAT2, an effect linked to increased expression of the Yin Yang 1 transcription factor (YY1) in the brain and astrocytes [158, 159]. Additionally, Mn contributes to neurotoxicity by repressing NMDA receptors, further amplifying its detrimental impact on neuronal health [160]. Mn significantly affects neurotransmitter systems by altering gamma-aminobutyric acid (GABA) and dopamine signaling. In welders, thalamic GABA levels were positively correlated with Mn deposition and exposure duration, though findings at lower exposure levels suggest nonlinear associations [161–164]. Animal studies show that Mn decreases GABA concentrations in the basal ganglia, alters glutamate-to-GABA ratios, and disrupts GABA receptor function and enzyme activity, impairing inhibitory neurotransmission and contributing to related behavioral abnormalities [165, 166]. Moreover, Mn exposure reduced norepinephrine and serotonin levels in rats, impairing catecholaminergic function in the prefrontal cortex, which correlates with motor and cognitive deficits [167].

Finally, Mn induces apoptosis through several interconnected mechanisms. It has been shown to disrupt mitochondrial function, causing cytochrome c release and caspase-3 activation, along with changes in Mfn2 expression in the rat striatum, which triggers apoptotic signaling [168]. Mn-induced oxidative stress activates the mitogen-activated protein kinase kinase 5/extracellular signal-regulated kinase 5 (MEK5/ERK5) pathway, causing phosphorylation and inhibition of a key antiapoptotic protein, Bcl-2, thereby promoting apoptosis [169]. Additionally, Mn impairs the cyclic adenosine monophosphate/protein kinase A/mitogen-activated protein kinase/cAMP response element-binding protein (cAMP/PKA/MAPK/CREB) signaling pathway, which reduces the expression of BDNF and Bcl-2, further facilitating neuronal death [170]. Alterations in Ca homeostasis caused by Mn also increase MAPK and CREB phosphorylation, exacerbating apoptotic processes [171]. Moreover, Mn disrupts rat sarcoma virus (Ras) protein/MAPK and phosphoinositide 3-kinase/protein kinase B (PI3K/PKB) signaling pathways, impairing neurotrophin-3/ tropomyosin receptor kinase C (NT3/TrkC) pathways essential for neuronal survival. Activation of cellular tumor protein (p53) further amplifies proapoptotic events by suppressing wild-type p53-induced phosphatase 1 (Wip1) and mouse double minute 2 homolog expression [172, 173]. Mn-induced apoptosis is also associated with the downregulation of tumor protein 73 (p73) and upregulation of KH-type splicing regulatory protein, which regulates proapoptotic genes and interacts with active caspase-3 [174, 175].

Mn-induced neurotoxicity is a highly complex process that involves oxidative stress, mitochondrial dysfunction, excitotoxicity, neuroinflammation, and disruptions in protein homeostasis, all of which contribute to neuronal damage and neurodegeneration. While significant progress has been made in understanding these mechanisms, further research is needed to clarify their interactions and long-term consequences and to identify potential therapeutic targets for mitigating Mn-related neurological disorders.

3.2 Mercury

Mercury exposure disrupts antioxidant defense systems, including GSH, GPX, and thioredoxin reductase (TrxR), thereby inducing oxidative stress in brain cells. MeHg, in particular, has a strong affinity for selenothiol and sulfhydryl groups, so it causes inhibition of these enzymes crucial for maintaining cellular homeostasis, resulting in elevated ROS levels and abnormal oxidative stress [176]. This oxidative stress alters neural stem cell fate, promoting astrogliogenesis over neurogenesis, while higher Hg concentrations impede neuronal migration, disrupting the CNS structure [177, 178]. MethylHg-induced neurotoxicity also significantly impacts mitochondrial function, causing K accumulation, hydrogen ion efflux, and a decreased inner membrane potential, further amplifying ROS production [179]. Additionally, MeHg inhibits the electron transport chain, triggering cytochrome c release and initiating apoptotic cell death [179]. These mitochondrial impairments disrupt cellular energetics, exacerbate oxidative stress, and contribute to neuronal damage.

The oxidative stress induced by Hg exposure not only disrupts cellular homeostasis but also plays a crucial role in the progression of neurodegenerative diseases such as AD and PD, where it is proposed as a key pathological mechanism. In AD, Hg-induced oxidative stress disrupts γ-secretase activity, causing increased production of Aβ, a hallmark of AD pathology [180, 181]. Additionally, Hg exposure suppresses protein kinase C activity, further promoting Aβ accumulation by altering amyloidogenic pathways [182]. The combined effects of elevated oxidative stress and Aβ accumulation contribute to tau hyperphosphorylation, driving the formation of neurofibrillary tangles, another hallmark of AD [182]. In PD, Hg accumulates in the motor cortex, causing increased nitric oxide generation, lipid peroxidation, and disruption of membrane lipids, ultimately triggering apoptosis of glial and neuronal cells [183]. These processes are closely linked to mitochondrial dysfunction, exacerbating oxidative damage and cellular degeneration [184].

Beyond its role in stimulating oxidative stress and the progression of neurodegenerative disorders, Hg also disrupts neuronal communication by interfering with glutamate signaling, a critical pathway in neuron excitotoxicity. A key mechanism of MeHg neurotoxicity involves disrupting glutamate transport, the primary excitatory neurotransmitter in mammals [185]. MeHg enhances glutamate release from presynaptic terminals while inhibiting uptake by astrocytes, causing its accumulation in the synaptic cleft. This accumulation finally results in neuronal excitotoxicity and functional impairment [186]. Astrocytes, which play a crucial role in maintaining the extracellular environment by clearing neurotransmitters and buffering extracellular K, are particularly vulnerable to MeHg accumulation, impairing their neuroprotective functions [187]. Elevated glutamate levels further promote excessive Ca influx into neurons via NMDA receptors, triggering inducible NOS activity, nitric oxide (NO) production, and mitochondrial dysfunction [188].

Neuroinflammation is a hallmark of Hg neurotoxicity, with microglia and astrocytes playing pivotal roles. MeHg activates microglia at subcytotoxic concentrations, triggering the release of proinflammatory cytokines such as IL-6 [189–191]. Reactive astrocytes further exacerbate the inflammation, creating a feedback loop of neuronal injury. Redox-sensitive signaling pathways, such as NF-κ B and signal transducer and activator of transcription 3 (STAT-3), play a crucial role in mediating these inflammatory processes, amplifying CNS damage [192]. At the cellular level, Hg disrupts several processes and organelles, including modulating cytokine production, such as IL-6 and TNF-α. This may contribute to autoimmune responses in the brain, including immunoglobulin G accumulation and cluster of differentiation 4-positive T cell (CD4+ T cell) infiltration [193, 194]. Additionally, Hg-induced activation of vascular endothelial growth factor (VEGF) and IL-6 release from mast cells has been implicated in the breakdown of the BBB, further exacerbating its neurotoxic effects [195].

In addition to driving neuroinflammation, Hg toxicity impairs glucocorticoid receptor (GCR) function, disrupting neuroendocrine signaling and stress response. MeHg binds to redox-sensitive cysteine residues in GCR, causing conformational changes that impair its function [196]. This inhibition disrupts critical pathways controlling development, metabolism, and immune responses, contributing to developmental neurotoxicity [196]. GCR regulation is further affected by interactions with thioredoxin (Trx), highlighting the interconnected nature of Hg's redox and signaling effects [197].

Hg-induced neurotoxicity is driven by oxidative stress, mitochondrial dysfunction, excitotoxicity, and neuroinflammation, all of which contribute to neuronal damage and the progression of neurodegenerative diseases such as AD and PD. By disrupting glutamate signaling, impairing antioxidant defenses, and triggering inflammatory responses, Hg exposure amplifies cellular stress and neurodegenerative mechanisms. Understanding these interconnected pathways is crucial for developing strategies to mitigate Hg's harmful neurological effects.

3.3 Lead

The integrity of the BBB and blood–cerebrospinal fluid barrier (BCB) is essential for protecting the CNS. However, Pb can penetrate and compromise these barriers, significantly increasing its neurotoxic potential. Pb's ability to traverse and disrupt the BBB and BCB highlights its harmful impact on the CNS. Pb enters the brain through multiple mechanisms. Initially, it may cross the BBB by passive diffusion, with pH levels playing a critical role in its transport efficiency [198]. Pb also utilizes divalent metal transporters (DMT1), a mechanism shared with essential metals like Fe^{2+}. Notably, Fe deficiency can enhance Pb uptake via DMT1, further increasing CNS exposure [199, 200].

Additionally, Pb can permeate the brain through non-voltage-gated Ca channels and transient receptor potential canonical (TRPCs) channels, which mediate Ca influx, offering another route for Pb entry into the brain [201]. Pb exposure disrupts tight junctions that maintain the integrity of the BBB and BCB by reducing the levels of critical tight junction proteins (claudin-1 and occludin) and scaffold proteins (ZO1 and ZO2), thereby increasing barrier permeability and enhancing CNS exposure [202, 203]. Furthermore, Pb interferes with essential signaling pathways in barrier cells, such as activating ERK phosphorylation, which lowers Cx43 gap junction hemichannel levels, further impairing barrier function [204]. Additionally, Pb activates the non-receptor tyrosine kinase Src, contributing to the breakdown of tight junction proteins and increasing BBB permeability [203]. Moreover, Pb exposure, in combination with inflammatory cytokines, induces the production of matrix metalloproteinase-9 (MMP-9), which degrades extracellular matrix components and increases BBB permeability [205].

Once Pb crosses the brain barriers, it induces severe neurotoxic effects, primarily by disrupting redox balance and triggering oxidative stress, which accelerates cellular damage and neuronal dysfunction. Pb alters the redox balance and promotes oxidative stress through its pro-oxidative effects on δ-aminolevulinic acid (δ-ALA), lipid peroxidation, and the impairment of antioxidant defense systems. These mechanisms lead to membrane damage, disrupted neuronal signaling, and oxidative damage to cellular biomolecules [45]. Pb disrupts the antioxidant defense system by binding to thiol groups in GSH, depleting its levels and reducing its capacity to neutralize ROS [206]. Additionally, Pb inhibits key enzymes involved in ROS detoxification, including GR, SOD, CAT, and GPX, further impairing the cellular defense against oxidative damage [207]. While epidemiological studies have established a clear link between occupational Pb exposure and oxidative stress, fewer studies have focused on the impact of environmental Pb exposure in children and adults [208, 209].

Pb further exacerbates neurotoxicity by mimicking essential metal ions, disrupting the homeostasis of Ca, Zn, and Fe, all of which are essential for neuronal signaling and function. It exerts neurotoxic effects through ionic mimicry, competing with these essential divalent metals due to its similar charge and binding properties. One key mechanism is Pb's ability to block voltage-gated Ca channels (VGCCs), reducing intracellular Ca levels critical for neurotransmitter release and synaptic function [210]. Additionally, Pb binds to proteins such as calmodulin (CaM) and protein kinase C (PKC), altering their activity and disrupting intracellular signaling and neurotransmitter release [211]. Pb inhibits Ca-dependent processes like vesicle fusion and neurotransmitter release, while high concentrations disrupt synaptic plasticity and impair learning [212]. Moreover, Pb displaces Zn in transcription factors such as specificity protein 1 (Sp1) and early growth response 1 (Egr-1), disrupting gene regulation that is essential for brain development and synaptic plasticity [213]. Pb also affects NMDA receptors, binding to allosteric sites with high affinity for Zn ions, reducing Ca permeability, impairing synaptic plasticity, and disrupting long-term potentiation (LTP), a key process for learning and memory [214]. In addition, Pb increases cellular Fe^{2+} levels by downregulating ferroportin 1, a critical Fe efflux protein, contributing to oxidative stress and cellular damage [215].

In addition to disrupting metal homeostasis, Pb exposure drives neuroinflammation, further amplifying neuronal damage through microglial activation, cytokine release, and oxidative stress. Pb-triggered neuroinflammation is rooted in complex interactions between activated microglia and astrocytes, signaling pathways, and inflammatory cytokines, all contributing to oxidative stress, neuronal injury, and eventual neuron loss [45, 216]. Prolonged exposure to Pb has been linked to microgliosis and astrogliosis, which correspond to the reactive activation and proliferation of microglia and astrocytes in response to toxic stimuli. The results are increased inflammation and potential neuronal damage, as indicated by increased levels of glial markers such as ionized Ca-binding adapter molecule 1 and glial fibrillary acidic protein (GFAP) [217]. Moreover, intermittent Pb exposure, especially from the fetal period through adulthood, induces neuroinflammation, as seen in the hippocampus, characterized by microgliosis, astrogliosis, and presynaptic dysfunction in rodent models [216]. Pb also triggers innate immune responses through toll-like receptor 4 (TLR4) and its downstream signaling pathways, including MyD88 and NF-κB, which amplify inflammatory responses in the CNS. Pb exposure activates MAPK pathways, including p38 MAPK and ERK 1/2 [218]. Pb affects cytokines metabolism, increasing the levels of IL-2, IL-1b, IL-6, IL-4, IL-8, and TNF-α, alongside the expression of inflammatory enzymes such as cyclooxygenases [218–220]. Proinflammatory cytokines, such as IL-1β and TNF-α, activate NADPH oxidase (NOX) expression, driving ROS production, which amplifies inflammation by activating redox-sensitive transcription factors like NF-κB and AP-1 [219]. Additionally, Pb^{2+} depletes GSH, a key antioxidant, increasing microglial susceptibility to oxidative damage while sparing astrocytes [221]. IL-1-associated inflammation driven by chronic Pb exposure promotes the formation of amyloid plaques, a hallmark of AD, by increasing the expression of APP, β-secretase, and presenilin [61]. Furthermore, Pb induces epigenetic modifications, including DNA methylation, histone acetylation, and noncoding RNA regulation, which may mediate long-term neurotoxic effects [222].

Pb's ability to breach the brain barriers and disrupt redox homeostasis triggers a cascade of neurotoxic events, including oxidative stress, ionic dysregulation, and excitotoxicity. These mechanisms not only exacerbate neuroinflammation but also amplify neuronal damage, accelerating the progression of neurodegenerative disorders like AD. Understanding this complex interplay of processes is essential for developing strategies to mitigate Pb-induced neurotoxicity and its long-term impact on brain health.

3.4 Cadmium

Cadmium neurotoxicity is primarily driven by oxidative stress, resulting from an imbalance between ROS production and the brain's antioxidant defense system. The brain's antioxidant defense includes both enzymatic components, such as SOD, CAT, glutathione-S-transferase (GST), GPX, and GR, as well as non-enzymatic antioxidants like GSH, thiol groups, and ascorbic acid. Cd exposure depletes these defense mechanisms, increasing oxidative stress, which contributes to neurodegenerative diseases, memory impairment, and neuronal apoptosis and necrosis [223, 224]. Antioxidants such as N-acetylcysteine (NAC), curcumin, cysteine, quercetin, melatonin, and omega-3 fatty acids have been shown to mitigate Cd-induced oxidative stress by restoring antioxidant enzyme activities and protecting against neuronal damage [75, 77, 225]. Furthermore, plant-based diets rich in foods like nuts, parsley, garlic, ginger, turmeric, green tea, and grapes have demonstrated efficacy in reducing Cd-induced oxidative damage and improving cognitive outcomes in rodents [66, 226]. In addition to oxidative stress, Cd induces neuronal damage by disrupting mitochondrial function. It inhibits the electron transport chain complexes II and III, reducing membrane potential and ATP production, which amplifies ROS generation and further exacerbates oxidative damage [227, 228]. Cd also impairs critical signaling pathways, including c-Jun N-terminal kinase (JNK), ERK1/2, and MAPK/mechanistic target of rapamycin (mTOR), contributing to cell death and potentially linking Cd exposure to neurodegenerative diseases like AD and PD [229–231]. In animal studies, Cd exposure has been shown to induce apoptosis, as evidenced by increased expression of Bax and caspase-3 alongside reduced levels of Bcl-2, a protein associated with cell survival [232].

Cd-induced neurotoxicity involves oxidative stress and disruptions to both mitochondrial and autophagic processes. SIRT1, a NAD^+-dependent deacetylase, is crucial in mitigating oxidative stress by regulating mitochondrial function and cellular health. Cd suppresses SIRT1 activity, increasing ROS production, mitochondrial dysfunction, and neuronal apoptosis. On the other hand, SIRT1 activators like resveratrol can counteract these detrimental effects by promoting mitochondrial integrity and enhancing oxidative metabolism through PGC-1α deacetylation, which supports the maintenance of mitochondrial function [224, 233]. Moreover, Cd interferes with autophagy, causing the accumulation of autophagosomes by impairing the autophagosome–lysosome fusion. This dysfunction contributes to neuronal apoptosis [234, 235]. Protective compounds such as linagliptin have been shown to suppress the phosphoinositide 3-kinase/BCL-2-interacting protein 1 (PI3K/Beclin-1) pathway, reducing autophagic dysregulation. Additionally, linagliptin stimulates the adenosine monophosphate-activated protein kinase (AMPK)/mTOR pathway, which enhances autophagic activity, thereby preventing Cd-induced cognitive deficits and providing a potential strategy for mitigating the neurotoxic effects of Cd exposure [236].

Additionally, Cd exposure significantly interferes with both cholinergic and purinergic systems, leading to neurotransmission deficits and cognitive impairments. One of the major effects of Cd is the disruption of key enzymes, such as acetylcholinesterase (AChE) and sodium–potassium adenosine triphosphatase (Na^+, K^+-ATPase), likely due to oxidative damage and lipid peroxidation caused by ROS. This impairs enzyme structure and function, inhibiting Na^+ K^+-ATPase and thereby disrupting ionic gradients essential for neurotransmission, affecting memory and learning abilities [237, 238]. Cd exposure also decreases the transcription of muscarinic receptors M1, M2, and M4 in the murine brain while sparing M3 and M5 receptors. Furthermore, Cd reduces AChE and choline acetyltransferase (ChAT) levels, disrupting cholinergic neurotransmission and neuronal integrity. These changes in the hippocampus and frontal cortex contribute to cognitive deficits, including impaired learning and memory, which may be mitigated by quercetin [227]. Beyond its effects on the cholinergic system, Cd also disrupts the purinergic system, further impairing neurotransmission by altering key enzymes such as nucleoside triphosphate diphosphohydrolase (NTPDase), 5′-nucleotidase, and adenosine deaminase (ADA), which regulate extracellular ATP and adenosine levels. Compensatory increases in NTPDase and 5′ -nucleotidase activity in the cortex may enhance adenosine production as a neuroprotective response [239, 240]. However, inhibited 5'-nucleotidase and increased ADA activity reduce adenosine levels, exacerbating neuronal damage [241, 242]. In addition to disrupting neurotransmission, Cd exposure may affect neural development. In cerebral organoids, chronic Cd exposure may also impair neural differentiation and cortical layer organization by altering key neuronal markers, such as GFAP, SRY (sex determining region Y)-box 2 (SOX2), neuron-specific class III beta-tubulin, T-box brain 2 (TBR2), and COUP-TF interacting protein 2 (CTIP2). This effect is linked to metal ion imbalance, including Zn and Cu dysregulation and synaptic dysfunction. Excessive metallothionein synthesis and inhibited postsynaptic protein expression further contribute to these developmental disruptions [78].

Cd exposure induces the transcriptional activation of inflammatory genes and the secretion of proinflammatory molecules, such as TNF-α and IL-1β, primarily by activating M1 phenotype microglial cells. This triggers a neuroinflammatory response [243]. The activation of microglia toward the proinflammatory M1 phenotype by Cd exposure can occur via two distinct pathways: through Toll-like receptors, which signal via MyD88, and through TREM2 receptors. Both pathways activate the IRAK kinase family, initiating downstream inflammatory cascades [243]. This elevated inflammatory state contributes to cognitive impairments and cholinergic dysfunction, particularly in the hippocampus and prefrontal cortex, as demonstrated in animal models [244].

Additionally, Cd-induced neuroinflammation is linked to the activation of NF-κB and interferon-gamma, further amplifying the inflammatory response and resulting in neuronal damage. This chronic inflammation may contribute to the long-term cognitive and behavioral deficits observed following Cd exposure [75]. Furthermore, Cd exposure increases AChE activity, exacerbating the disruption of the cholinergic system. Elevated AChE activity in key brain regions like the hippocampus and prefrontal cortex impairs memory and learning processes [61]. Anti-inflammatory interventions, such as extracts from ginger and turmeric, have been shown to reduce inflammatory biomarkers (IL-6, IL-10, and TNF-alpha) and restore AChE activity in rats, suggesting their potential to mitigate Cd-induced neurotoxicity and cognitive decline [244].

Cd-induced neurotoxicity results from a complex interplay of oxidative stress, mitochondrial dysfunction, neuroinflammation, and disruptions in neurotransmission. This complex cascade ultimately results in cognitive impairments and neuronal loss. The widespread impact of Cd on various cellular and molecular pathways highlights its potential contribution to developing neurodegenerative diseases. The growing evidence supporting antioxidant and anti-inflammatory interventions underscores the need for further research into preventive and therapeutic strategies to mitigate Cd-related neurological damage.

3.5 Arsenic

Exposure to As can cause methylation imbalance, mitochondrial oxidative stress, Ca imbalance, disrupted ATP production, and altered membrane potential, while early-life exposure is associated with reduced brain weight and decreased neurons and glia, potentially contributing to neurodegenerative disorders [79, 245]. As methylation in the hippocampus and brain cortex by As methyltransferase (As3MT) disrupts methylation processes in the brain by depleting S-adenosylmethionine (SAM), a critical methyl donor, and increasing levels of S-adenosylhomocysteine (SAH), a known inhibitor of methylation [84, 246]. This imbalance results in DNA hypomethylation and altered posttranslational protein modifications, which can impair cytoskeletal integrity and membrane function [247, 248]. In the developing brain, SAM-dependent methylation is essential for oligodendrocyte maturation and myelin synthesis, particularly through the regulation of myelin basic protein (MBP) and proteolipid protein (PLP) [249, 250]. As exposure reduces an endogenous inhibitor of NOS, dimethylarginine, in the striatum [251], it has been associated with impaired oligodendrogenesis, potentially contributing to cognitive and motor deficits [252]. In the mouse model, brain SAM depletion disrupts mitochondrial GSH synthesis and reduces membrane fluidity, impairing the regulation of oxidative stress and increasing ROS accumulation [253]. As-induced oxidative stress plays a crucial role in its neurotoxicity, primarily through the disruption of the antioxidant defense system and the accumulation of reactive oxygen species. As exposure depletes GSH and inhibits critical enzymes such as SOD, CAT, GPX, and GSR, causing an imbalance between oxidative and reductive cellular processes [253]. Thus, a key consequence of As-induced oxidative stress is mitochondrial dysfunction, which disrupts energy production and leads to neuronal damage by causing membrane potential collapse, swelling, and cytochrome c release, triggering apoptosis [254].

Moreover, oxidative stress caused by As exposure impairs neurogenesis, depletes neural stem cells (NSCs), and inhibits neuronal differentiation, resulting in cognitive deficits and behavioral impairments. In animal models, these effects positively correlate with learning and memory deficits, highlighting the vulnerability of brain regions essential for cognition to oxidative damage [252, 255]. In addition to mitochondrial dysfunction, As exposure disrupts glutamate homeostasis, causing excitotoxicity and further neuronal damage. As-exposed brains show increased extracellular L-glutamate levels due to the overexpression of glutamate transporters such as cystine/glutamate antiporter (xCT) and excitatory amino acid carrier 1 (EAAC1), which impair GSH metabolism and the oxidative stress response [256]. Moreover, As disrupts synaptic function by downregulating the expression of NMDA and α-amino-3-hydroxy-5-methyl-4-isoxazolepropionic acid (AMPA) glutamate receptor subunits, impairing synaptic plasticity and LTP [256, 257]. As exposure also increases extracellular glutamate in the brain, causing excitotoxicity and oxidative stress, which contribute to neuronal apoptosis and impaired LTP [258]. Furthermore, As toxicity disrupts key neurotransmitter systems, including the dopaminergic and cholinergic pathways. As exposure reduces dopamine levels, it impairs tyrosine hydroxylase (TH) expression and alters dopamine receptor signaling, resulting in deficits in motor control and cognitive function [259]. The cholinergic system is affected by decreased expression of muscarinic acetylcholine receptor M2 (CHRM2), ChAT, and AChE, which are crucial for learning and memory processes [260, 261]. Promisingly, antioxidant interventions such as α-lipoic acid, vitamin E, and coenzyme Q10, particularly in combination, have shown potential in mitigating As-induced oxidative damage, emphasizing the need for further research on preventive and therapeutic strategies [258].

Beyond its role in maintaining membrane stability and regulating oxidative stress, As-induced SAM imbalance also disrupts epigenetic processes, altering DNA and histone methylation through depleted methyl donor availability. Chromatin methylation, which plays a critical role in neurodevelopment and memory formation, is altered by As exposure through depletion of methyl donors, such as SAM, and increases in SAH, a methyltransferase inhibitor. This results in altered DNA and histone methylation [247]. Developmental As exposure impairs cognition in mice by inducing DNA hypomethylation and dysregulating hippocampal gene expression. As promotes DNA (cytosine-5)-methyltransferase 3 alpha methylase degradation, causing hypomethylation, particularly in embryonic cells. Metformin restores methylation and gene expression, improving cognitive function and suggesting its potential as a therapeutic intervention [262]. In addition to methylation, As exposure also disrupts acetylation, another key epigenetic mechanism regulating gene expression and neurodevelopment. As affects histone acetylation, with increased histone deacetylase activity linked to impaired learning and memory in animal models [263]. Transgenerational studies in zebrafish revealed persistent histone modifications, suggesting long-term epigenetic consequences of As exposure [264].

As exposure exerts its neurotoxic effects through a combination of oxidative stress, mitochondrial dysfunction, excitotoxicity, and epigenetic alterations, ultimately causing impaired neurogenesis, synaptic dysfunction, and cognitive deficits. The disruption of methylation and acetylation processes further amplifies these effects, contributing to long-term neurological consequences. Given its widespread environmental presence, understanding As's impact on brain function is crucial for developing targeted preventive and therapeutic strategies.

3.6 Aluminum

Excessive Al levels disrupt the mitochondrial membrane's electrical potential, hampering oxidative phosphorylation and ultimately increasing ROS production [265]. Oxidative stress is a key mechanism in Al-induced neurotoxicity [266]. Al exposure increases brain lipid peroxidation [267, 268] and reduces antioxidant enzyme activities, including SOD, CAT, GPX, GR, and GST, impairing the glutathione system by lowering cerebral and cerebellar levels of glutathione [102]. Al also promotes prooxidant activity by increasing inducible and endothelial (constitutive) NOS activity, elevating brain nitric oxide (NO) levels [269–271]. Additionally, Al stimulates NADPH oxidase, increasing superoxide production in brain and liver issues [272]. The findings indicate that Al selectively enhances oxidative damage, reduces antioxidant enzyme activity in key brain areas, and amplifies the neurotoxic effects of 6-hydroxydopamine, a known inducer of PD-like pathology [273]. The oxidative stress induced by Al exposure damages cellular components and acts as a key trigger for apoptotic pathways, causing neuronal loss in vulnerable brain regions. Animal studies have shown that Al induces neural apoptosis in the hippocampus [274] and in the motor cortex and hippocampus astrocytes [275]. Mitochondrial dysfunction plays a central role in Al-induced apoptosis, as cytochrome c release triggers apoptosome formation, causing caspase-9 activation and apoptotic neuronal death [276]. Activation of this apoptotic pathway is further linked to increased cerebral levels of proapoptotic Bax and caspase-3 and downregulation of the antiapoptotic Bcl-2 protein [270]. Additionally, Al activates the Fas cell surface death receptor (Fas)/ Fas ligand (FasLG)-mediated apoptotic pathway, as indicated by elevated levels of Fas, caspase-8, and death domain-associated protein 6 (Daxx), suggesting the involvement of both p53-dependent and independent mechanisms [254, 266, 277]. Moreover, Al-induced apoptosis is further modulated by signaling pathways, with low doses triggering apoptosis via the IL-1β/JNK pathway, while higher doses cause necroptosis, a programmed form of inflammatory cell death [278, 279]. The ERK and p38 MAPK/ERK signaling pathways also play distinct roles in mediating apoptosis and necrosis in neuroblastoma SH–SY5Y cells [280]. In addition to apoptosis, Al disrupts cell cycle regulation by inhibiting neural progenitor cell proliferation, increasing the sub-G1 cell population, and reducing the G2/M phase during neural differentiation [281].

Beyond inducing oxidative stress and apoptosis, Al exposure disrupts Ca homeostasis, further amplifying neuronal damage through excitotoxicity and impaired synaptic signaling [282]. Al interferes with protective Ca-regulating mechanisms by delaying the closure of voltage-dependent Ca channels and inhibiting CaM-dependent Ca^{2+}/Mg^{2+}-ATPase, impairing Ca extrusion and amplifying excitotoxic damage [283, 284]. Studies in Al sulfate-exposed rats revealed dose-dependent Al accumulation in brain tissues, particularly in the hippocampus, accompanied by increased glutamate and decreased GABA levels. This imbalance enhances excitotoxicity, resulting in spongiform neuronal changes, nuclear deformity, and neurofibrillary degeneration [285–287]. Reduced GABA levels due to Al exposure contribute to excitotoxicity, disrupting inhibitory signaling and exacerbating neuronal vulnerability, ultimately linking Ca dysregulation to neuroinflammation [288]. Additionally, they promote the synaptic insertion of GluR2-lacking AMPA receptors, amplifying excitatory neurotransmission and neuronal vulnerability [282].

The persistent excitotoxicity and disrupted inhibitory signaling caused by Al exposure not only contribute to neuronal injury but also activate inflammatory pathways, driving neuroinflammation and further exacerbating neurodegeneration. Al is strongly implicated in neuroinflammation through its ability to upregulate pro-inflammatory signaling pathways. Exposure to Al increases the expression of cytokines such as IL-1β, IL-6, TNF-α, and macrophage inflammatory protein-1 alpha (MIP-1α) while reducing levels of the neuroprotective BDNF in rodent models [289–291]. This pro-inflammatory response is mediated via activation of the NF-κB pathway, a key regulator of inflammation [292]. Al-induced neuroinflammation involves inflammasome activation and IL-1β secretion by microglia alongside systemic inflammation characterized by elevated levels of pro-inflammatory cytokines and microRNAs such as miRNA-9 and miRNA-146a [293, 294]. The role of cyclooxygenases (COX-1 and COX-2) is also significant, with Al exposure increasing their activity and downstream prostaglandin signaling, which is linked to hippocampal inflammation. Treatment with COX inhibitors like ibuprofen or prostaglandin E1 analogs has been shown to mitigate these inflammatory effects and improve cognitive deficits [295, 296]. A key feature of Al-induced neuroinflammation is microgliosis, characterized by the activation and proliferation of microglia in response to injury or toxic insults [297, 298], highlighting the pivotal role of glial cells in mediating inflammatory responses and contributing to neurodegenerative processes [299, 300]. These mechanisms collectively underscore Al's role in promoting neuroinflammation and its contribution to neurodegenerative processes.

Further, Al exposure also profoundly affects synaptic function, as chronic inflammatory signaling disrupts neuroplasticity and contributes to synaptic degeneration. Both presynaptic and postsynaptic mechanisms are affected, with studies showing significant reductions in field excitatory postsynaptic potentials after high-frequency stimulation, indicative of synaptic dysfunction [277]. Structural changes in synapses, such as decreased postsynaptic density thickness, increased synaptic cleft width, and reduced perforated synapses, further highlight Al's impact on neural communication [301]. These changes are compounded by increased membrane rigidity and altered lipid composition in synaptosomes, which reduce the release of synaptic vesicles [302]. Al also inhibits synaptic (Na^+/K^+)ATPase activity, further impairing synaptic function [284]. Structural alterations in neural connections are accompanied by reduced axon branching and fewer dendritic spines, associated with neuronal atrophy and dysregulation of key proteins such as mGluR, PKC, and NMDA receptor subunits [303]. Disruptions in signaling pathways, particularly the PI3K/PKB–mTOR pathway, further exacerbate the impairment of synaptic plasticity [303]. Moreover, Al significantly impairs the production of neurotrophic factors essential for synaptic plasticity and neuronal survival, such as nerve growth factor (NGF) and BDNF [304]. This reduction is primarily driven by epigenetic changes, including histone demethylation (H3K9me2) and the involvement of plant homeodomain finger protein 8 (PHF8), which negatively impact synaptic long-term potentiation [305]. Additional epigenetic alterations, including changes in trimethylation of lysine 4 on histone and trimethylation of lysine 27 on histone H3, further decrease BDNF and early growth response protein 1 expression, exacerbating Al's neurotoxic effects [306]. Further, prenatal exposure to Al has been linked to decreased hippocampal expression of the activity-regulated cytoskeleton-associated protein (ARC), an essential regulator of synaptic plasticity. This effect is mediated by disruptions in BDNF-induced ERK signaling, emphasizing Al's role in developmental neurotoxicity [307, 308]. Additional disruptions in critical signaling pathways, including SIRT1, target of rapamycin complex 1 (TORC1), phosphorylated CREB, and the cAMP–PKA–CREB pathway, highlight the broad and multifaceted impact of Al exposure on synaptic plasticity and neurotrophic support [309, 310].

In addition to impairing synaptic plasticity, Al exposure also plays a crucial role in the pathological accumulation of Aβ and tau hyperphosphorylation, further contributing to neurodegenerative processes associated with AD. Studies have shown that Al exposure in vivo and in vitro increases Aβ accumulation and tau phosphorylation [103]. Al contributes to amyloid pathology in AD by modulating Aβ production, aggregation, and clearance. Exposure to Al-maltolate increases Aβ1–42 expression through the upregulation of APP, β-secretase, and γ-secretase while reducing α-secretase proteins, thereby favoring amyloidogenic pathways [311]. Furthermore, Al decreases neprilysin expression, an enzyme critical for Aβ degradation, and inhibits plasmin activity, disrupting Aβ clearance [312]. Al also inhibits microglial phagocytosis of Aβ 42, linked to TREM2 downregulation, further exacerbating amyloid accumulation [313]. Al increases Aβ production and promotes its aggregation by binding to Aβ peptides and inducing β-sheet formation, making it a stronger aggregation promoter than Cu or Zn [314]. Moreover, Al exposure is associated with tau hyperphosphorylation, mediated by increased glycogen synthase kinase 3 beta activity, which plays a key role in tau pathology in AD [315, 316]. Emerging evidence supports the notion that Al exposure is a potential environmental factor that may exacerbate the effects of known AD risk factors, including the apolipoprotein E epsilon 4 (ApoEε4) allele [107] and aging [317]. However, more research is needed to fully elucidate these interactions and their implications for AD pathogenesis.

Al exposure in ApoEε4-transgenic mice led to significant increases in hippocampal Aβ42 levels, accompanied by downregulation of ApoER2 and low-density lipoprotein receptor-related protein 1 proteins. This disruption impaired synaptic plasticity, resulting in deficits in memory and learning [318]. An epidemiological study conducted on workers in an Al factory identified a 44.2% increase in the risk of cognitive impairment among those with elevated plasma Al levels and the ApoEε4 allele [319]. Similarly, research on drinking water revealed a significant association between exposure to organic monomeric Al and increased AD risk, even after adjusting for confounding factors such as education, family history, and ApoEε 4 status [320].

Al-induced neurotoxicity is driven by oxidative stress, apoptosis, and disrupted Ca homeostasis, resulting in excitotoxicity and neuroinflammation. These mechanisms collectively impair synaptic plasticity, promoting neuronal degeneration and cognitive dysfunction. Additionally, Al exposure accelerates Aβ accumulation and tau hyperphosphorylation, contributing to AD pathology, particularly in individuals carrying the ApoEε4 allele, the strongest genetic risk factor for AD. This highlights Al as a potential environmental risk factor for neurodegenerative diseases Fig. 1

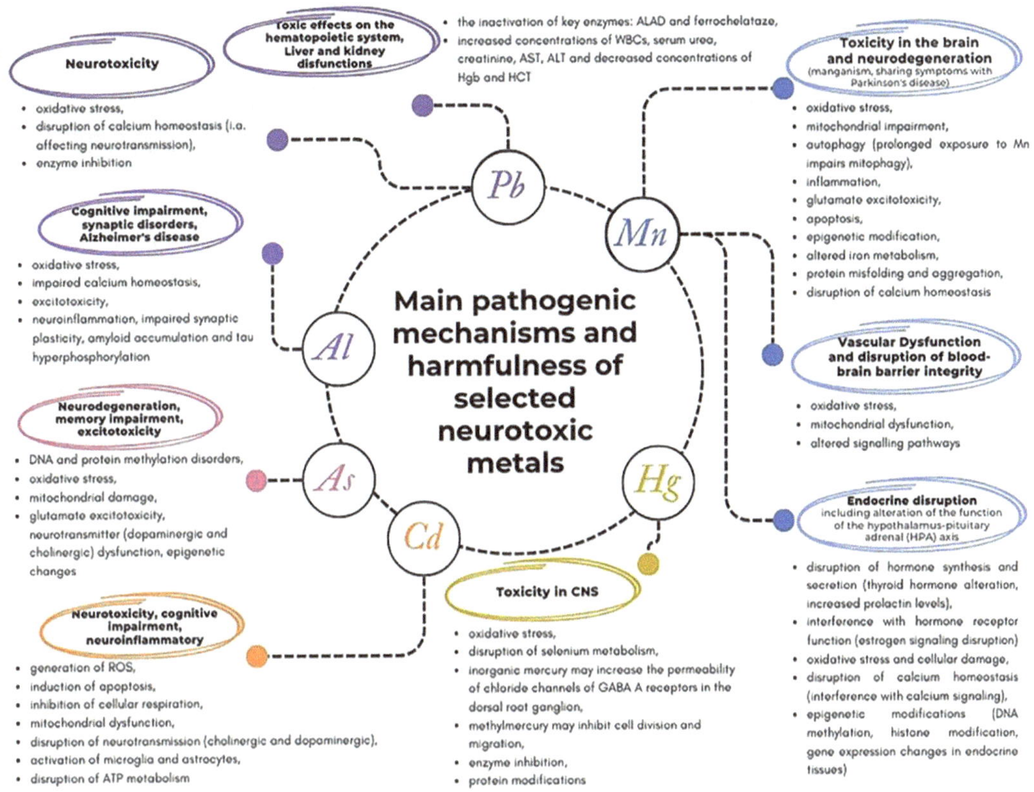

Fig. 1 Summary of the main pathogenic mechanisms and harmfulness of Mn, Hg, Pb, Cd, As, and Al

4 Air, Water, Soil, and Food Pollutants and Brain Metallome

The link between environmental pollutants and the brain's metallome is a broad area of research that studies how exposure to toxic substances affects the content and distribution of metals in the brain and how it affects neurological health. Scientific evidence suggests a link between exposure to environmental pollutants and the risk of developing serious brain diseases in humans [321, 322].

4.1 Air Pollutants

Air pollutants negatively affect the structure and function of the CNS [323]. They include PM, heavy metals (which are associated with PM), and a wide range of gaseous pollutants. Ultrafine (nanoparticles) can carry numerous metal and trace element contaminants [324].

Air pollutants originate from natural sources (such as wildfires, volcanic eruptions, and dust storms) and human activities (such as vehicle emissions, industrial processes, agriculture, and fossil fuel combustion) [325]. Tiny solid particles and liquid droplets suspended in air are defined by their equivalent aerodynamic diameter for air quality regulatory purposes. PM10 refers to particles with a diameter ≤10μm, corresponds to the inhalable fraction of PM ("inhalable particles"), and originates mainly from dust, pollen, and construction activities. PM2.5 refers to fine inhalable particles with a diameter≤2.5μm and originates mainly from vehicle emissions, industrial processes, and wildfires. PM10 affects the respiratory system but does not penetrate as deeply as PM2.5, which can reach the alveoli and even enter the bloodstream [326–328].

Directive 2008/50/EC of the European Parliament and of the Council [329] and Directive (EU) 2024/2881 (recast) [330], which set EU air quality standards, limit PM2.5 (annual average) to 25μg/m^3 and PM10 (annual average) to 40μg/m^3. However, there is discussion about lowering these limits to, respectively, 10μg/m^3 and 20μg/m^3 by 2030, bringing them more in line with the World Health Organization (WHO) guidelines (5μg/m^3 and 15μg/m^3, respectively) [331].

"Heavy metals" is a collective term commonly used to refer to metals of particularly high toxicity (i.e., which are toxic or harmful to living organisms at very low concentrations) [332]. Examples include Pb (present in industrial emissions, particularly from smelters, leaded gasoline, old paints, and batteries), Hg (released from coal-burning power plants and waste incineration, mining, and certain medical products), Cd (released from metal production and waste incineration; present in cigarette smoke and batteries), As (present in groundwater, pesticides, and industrial by-products), Cr (used in electroplating, dyes, and leather tanning), Ni (released from industrial processes such as metal refining, combustion of fossil fuels, especially oil and coal, mining, and waste incineration, as well as from vehicle exhaust and natural sources such as volcanic activity). These metals can accumulate in living tissues, causing environmental and human health risks, including neurological damage [11]. EU regulations limit Pb to 0.5μg/m^3 (1-year average), and target values for the total content in the PM10 fraction, averaged over a calendar year, are set at 5ng/m^3 for Cd, 6ng/m^3 for As, and 20mg/m^3 for Ni [333].

Airborne metals increase oxidative stress, altering the homeostasis of essential brain metals such as Fe and Cu, which are vital for neurotransmitter synthesis [334, 335]. Pb, Hg, and Mn from airborne sources can cross the blood–brain barrier, disrupting brain metal levels. Traffic emissions containing Pb can elevate brain levels, displacing Ca and affecting synaptic plasticity. High levels of Mn from industrial emissions have been linked to PD-like symptoms by interfering with metal-dependent brain enzymes [27, 323, 336, 337].

Gaseous air pollutants include carbon monoxide (CO, resulting from the incomplete combustion of fossil fuels), nitrogen oxides (NO_x, produced by vehicle emissions, power plants, and industrial processes), and sulfur dioxide (SO_2, emitted from the burning of fossil fuels and volcanic eruptions) [338]. Special "groups" of gaseous air pollutants are volatile organic compounds (VOCs), including benzene (from vehicle exhaust, tobacco smoke, and industrial solvents), formaldehyde (released by building materials, household products, and combustion processes), and toluene (from gasoline, adhesives, and paint thinners), and persistent organic pollutants (POPs), including dioxins and furans (by-products of industrial combustion and waste incineration), and polychlorinated biphenyls (PCBs) (released from electrical equipment and industrial processes) [339, 340]. Other gaseous pollutants include ground-level ozone (O_3), a key component of smog, formed by chemical reactions between VOCs and NO_x under the action of sunlight; ammonia (NH_3), emitted from agricultural activities, fertilizers, and animal waste; and radioactive pollutants (e.g., radon gas, originating from the natural radioactive decay of soil and rocks, which can seep into buildings) [339–342]. Table 1 lists the EU air quality standards and the WHO guidelines.

Table 1
EU air quality standards (long-term and short-term averages) and WHO guidelines

Pollutant	Averaging period	Concentration until 2030	Concentration as of 2030	WHO recommendations
Fine particles (PM2.5)	1 year	25μg/m³	10μg/m³	5μg/m³
	1 day	–	25μg/m³(−18d)	15μg/m³(−3d)
Particulate matter (PM10)	1 year	40μg/m³	20μg/m³	15μg/m³
	1 day	50μg/m³(−35d)	45μg/m³(−18d)	45μg/m³
Nitrogen dioxide (NO_2)	1 year	40μg/m³	20μg/m³	10μg/m³
	1 day	–	50μg/m³(−18d)	50μg/m³
	1 hour	200μg/m³(−18h)	200μg/m³(−3h)	200μg/m³
Sulfur dioxide (SO_2)	1 year	–	20μg/m³	–
	1 day	125μg/m³(−3d)	50μg/m³(−18d)	40μg/m³(−3d)
	1 hour	350μg/m³(−24h)	350μg/m³(−3h)	–
Benzene	1 year	5μg/m³	3.4μg/m³	1.7μg/m³
	1 day			
Lead (Pb)	1 year	0.5μg/m³	0.5μg/m³	0.5μg/m³
Arsenic (As)	1 year	6ng/m³	6ng/m³	6.6ng/m³
Cadmium (Cd)	1 year	5ng/m³	5ng/m³	5ng/m³
Nickel (Ni)	1 year	20ng/m³	20ng/m³	25ng/m³
Benzo(a)pyrene[a]	1 year	1ng/m³	1ng/m³	0.12ng/m³
Carbon monoxide (CO)	1 day	–	4mg/m³(−18d)	4mg/m³(−3d)
	8h max.	10mg/m³	10mg/m³	10mg/m³
Ozone	3y avg 8h max[b]	120μg/m³(−25d)	120μg/m³(−18d)	100μg/m³(−3d)
	3y avg 8h max[c]	120μg/m³(−3d)	100μg/m³(−3d)	100μg/m³(−3d)

[a]Polycyclic aromatic hydrocarbons (PAH) are expressed as the concentration of benzo(a)pyrene [b]Target value [c]Long-term objective [14, 239, 330, 343]

4.2 Water Pollutants

Water can be contaminated by a range of pollutants, including microorganisms (bacteria, viruses, and parasites), heavy metals, and numerous organic pollutants, from pesticides, fertilizers, and pharmaceuticals to industrial waste (organic compounds and solvents) and micro/nanoplastics [344–347].

Given the potential adverse effects of any contamination of water intended for human consumption on human health, strict legislation exists (e.g., Directive (EU) 2020/2184 on the quality of water intended for human consumption [348]) that sets out rules to ensure it is "wholesome and clean."

Within the chemical parameters, limits are set for a wide range of metals (parametric values in parentheses), including Sb (10μg/L), As (10μg/L), Cd (5.0μg/L), Cr (25μg/L), Cu (2.0mg/L), Pb (5μ g/L), Hg (1.0μg/L), Ni (20μg/L), Se (20μg/L), U (30μg/L); and also, as "indicator parameters", that is, metals that do not have a direct impact on public health but are important in terms of water quality: Al (200μg/L), Fe (200μg/L), Mn (50μg/L), and Na (200mg/L) [349].

There is strong evidence that As and Pb in drinking water bioaccumulate and disrupt the balance of essential trace elements Zn and Cu, which are crucial for neurotransmitter function, leading to cognitive dysfunction [350–352]. Chronic exposure to methylHg impairs Se metabolism, which is essential for antioxidant enzymes in the brain [353–357]. Hg may create a selenium-deficient state and directly inhibit selenium's role in controlling the intracellular redox environment in organisms [354]. Exposure to As from groundwater has been shown to cause dysregulation of Zn and Cu levels in both cell culture and mouse feeding studies [358].

4.3 Soil Pollutants

Soil pollution primarily affects the brain indirectly through contaminated crops. Heavy metals, namely Cd and Pb from agricultural soil, can accumulate in food crops and, after ingestion, impact the entire human body, including the brain. Persistent organic pollutants (POPs) in soil can disrupt essential metal homeostasis. In particular, they can interact with Mg and Fe, affecting neurodevelopment. Exposure to Cd from contaminated soil has been shown to disrupt Fe homeostasis in the brain, contributing to cognitive decline [359, 360].

4.4 Food Pollutants

Dietary intake is a major route of metal exposure. Bioaccumulation can occur throughout the food chain. Seafood can be a significant source of exposure to As and Hg, especially the highly toxic methyl–Hg species. High consumption of fish containing $(CH_3Hg)^+$ has been shown to disrupt Zn and Cu homeostasis, affecting memory and motor function [357, 361].

There is also concern about food contamination from packaging materials, e.g., the accumulation of Al in the brain, contributing to AD [103].

In order to ensure efficient protection of public health, including the most vulnerable population groups (such as children and pregnant women), an EU regulation [Commission Regulation [EU] 2023/915 [362]) has recently updated the maximum levels of certain contaminants in food. Regarding metallic/metalloid contaminants, this regulation covers the "big four" heavy metals Pb, Cd, Hg, and

Table 2
Maximum levels for Pb, Cd, Hg, inorganic As, and Sn in selected foods.

	Maximum level (mg/kg)				
Food	**Pb**	**Cd**	**Hg**	**(i)As**	**(i)Sn**
Cranberries, currants, elderberries, and strawberry tree fruits	0.20	0.030[a]			
Other fruits	0.10	0.050[a]			
Bulb vegetables	0.10	0.030[a]			
Leafy brassica	0.30	0.10			
Legume vegetables	0.10	0.020			
Cereals	0.20	0.10[a]			
Non-parboiled milled rice (polished or white rice)		0.15		0.15	
Parboiled rice and husked rice				0.25	
Rice flour				0.25	
Rice for food for infants and young children				0.10	
Meat of bovine animals and sheep	0.20	0.050			
Meat of a pig	0.15	0.050			
Meat of poultry	0.10	0.050			
Muscle meat of fish	0.30	0.050[a]	0.50		
Crustaceans, mollusks		0.50, 1.0	0.50		
Cephalopods		1.0	0.30		
Raw milk	0.020				
Salt		0.50	0.10	0.50 (T)	
Wine (from the 2022 fruit harvest onwards)	0.10				
Fruit juices				0.020	
Infant formulae (powder)	0.020	0.010[a]		0.020	
Infant formulae (liquid)	0.010	0.005[a]		0.010	
Food supplements		1.0[a]	0.10		
Canned food					200/50[b]
Canned beverage					100/50[b]

[a]With some exceptions. (i) inorganic. T—total As
[b]Products for babies, infants, young children [364]

As, as well as inorganic Sn in canned food [363]. Table 2 shows the maximum levels of Pb, Cd, Hg, inorganic As, and Sn in selected foods [364].

5 Risk Assessment Studies on Exposure to Neurotoxic Metals Present in the Environment

Metals with neurotoxic potential, widespread in the environment, can potentially accumulate in biological systems and interfere with neurological function [365–368]. The most relevant neurotoxic metals are Pb, Hg, As, Cd, Mn, and Al. Risk assessment studies on neurotoxic metals are essential for understanding their impact on human health and guiding countermeasures. Integrating scientific research with regulatory action can mitigate exposure risks, protect vulnerable populations, and safeguard public health from the neurological hazards of environmental metal contamination.

Current exposure to Pb is mainly due to Pb-based paints (present in older homes and buildings), contaminated water (old pipes containing Pb), contaminated soil (due to the past use of leaded gasoline and industrial waste), industrial emissions (from mining, smelting, and battery manufacturing), consumer products (e.g., toys, cosmetics, and ceramics) imported from countries without strict Pb ban and control rules, and occupational exposure (in industries such as construction, automotive, and recycling) [366]. Chronic Pb exposure can cause cognitive deficits, behavioral problems, and developmental delays, especially in children [367].

Exposure to Hg occurs mainly through consuming seafood and fish, especially large and predatory species such as tuna and swordfish, which accumulate methylHg. This highly toxic Hg species bioaccumulates in the aquatic food chain, posing risks to populations with high fish consumption [368]. Other sources of exposure include contaminated air due to coal combustion (a major source of atmospheric Hg emissions) and industrial processes (chlor-alkali production, waste incineration, and cement manufacturing), consumer products (e.g., fluorescent lamps and batteries), and dental amalgams. In specific regions, artisanal and small-scale gold mining using Hg for gold extraction is of great relevance [369]. Neurological effects of Hg include memory loss, tremors, and cognitive impairment [370].

As exposure occurs mainly through contaminated drinking water (especially groundwater, in certain regions) and food consumption (especially rice, seafood, and some vegetables) [371]. Air contaminated by fossil fuel combustion and waste incineration, some industrial processes (mining, smelting, pesticide production), and treated wood (preserved with As-based chemicals) are other sources of As exposure [372–375]. As exposure has been linked to cancer, cardiovascular disease, and neurotoxicity [81, 148, 376]. Chronic exposure can lead to developmental abnormalities, cognitive dysfunction, and peripheral neuropathy [377, 378].

Exposure to Cd occurs through contaminated food (especially leafy vegetables, cereals, shellfish, and organ meats) and air contaminated by industrial emissions (from mining, smelting, and battery manufacturing) [379]. The soil is polluted mainly by industrial waste and phosphate fertilizers [380]. Occupational exposure occurs in metallurgy, battery manufacturing, and electronics [381]. Tobacco smoke may be the most significant source of Cd exposure for both smokers and those exposed to second-hand smoke [382–385]. Exposure to Cd is associated with kidney and bone damage and neurodegenerative conditions [60, 386]. Cd disrupts the homeostasis of Ca and Zn, which are vital for neuronal function [387].

Exposure to Mn occurs primarily through drinking water (Mn naturally present in groundwater or as a result of environmental contamination) and food (especially nuts, grains, leafy vegetables, and tea) [388, 389]. Air pollution is caused by various industrial activities (mining, welding, and steelmaking) and by the use of Mn-based fuel additives. Occupational exposure occurs in the welding, battery production, and mining sectors. Inhaled Mn is transported directly to the brain [390, 391]. Excessive exposure can lead to "manganism," a neurological disorder with PD-like symptoms, which occurs primarily among industrial workers, such as welders, who are chronically exposed to Mn fumes or dust [392].

Al is ubiquitous and abundant in nature, accounting for about 8% (m/m) of the Earth's crust [393]. Oral exposure is, therefore, very high, occurring mainly through food and beverages (Al-containing additives, kitchenware, and food packaging) and drinking water (both naturally occurring Al and Al used in water treatment processes) [394, 395]. Atmospheric Al is mainly associated with soil-derived particles [396]. Anthropogenic sources include industrial emissions and waste incineration. Occupational exposure occurs in the Al production, welding, and mining industries. Al is also found in pharmaceuticals (antacids and vaccines) and personal care products (e.g., antiperspirants), providing a possibility for dermal exposure [397–399]. The bioavailability of Al is very low (oral bioavailability has been estimated at 0.3% from drinking water and 0.1% from food and beverages), so normal blood levels are very low [400]. However, when introduced directly into the bloodstream, as can happen during chronic hemodialysis, Al is very toxic, especially to bones and the CNS. It can impair synaptic transmission and promote oxidative stress, leading to cognitive decline and memory loss [401, 402]. Excessive accumulation of Al in the brain has been linked to neurodegenerative conditions such as AD by promoting the formation and accumulation of insoluble Aβ and hyperphosphorylated tau [403].

Over the past decade, numerous studies have evaluated the neurotoxic effects of environmental exposure to various heavy metals. Very recent studies highlight that heavy metals can induce neurotoxic effects by triggering epigenetic changes and disrupting the epigenome [404, 405], discuss how long-term exposure to As has been associated with memory loss and cognitive functions in humans, including early indicators of AD [406], and evaluate the neurobehavioral effects of exposure to trace elements, including Pb, Hg, Cd, Mn, and As [407]. Studies by Lucchini et al. revealed that individuals with higher socioeconomic status had better neurocognitive scores than those with lower socioeconomic status, highlighting the combined effect of metal exposure and social stressors on neurocognitive outcomes [408]. Another study assessed the longitudinal impact of environmental chemical mixtures on child neurodevelopment, addressing the complexity and high dimensionality of environmental exposures [409]. A study by Vasconcellos et al. [410] evaluated the health risks of consuming Hg-contaminated fish among Munduruku indigenous communities in the Middle Tapajós region. The findings highlighted significant health risks from high levels of Hg in fish locally consumed.

The risk assessment of Hg through dietary exposure in China was reported by Qing et al. [411]. This study estimated dietary Hg exposure across China, revealing that plant-based foods, especially cereals and vegetables, contributed to 62.7% of total dietary Hg exposure. A study by Weinhouse et al. [412] in communities around the Amarakaeri Communal Reserve found abnormally high levels of Hg in hair and blood samples, identifying Indigenous communities as having an increased risk of Hg exposure, regardless of their proximity to mining activities. A risk assessment due to consumption of Hg-contaminated river fish was recently published by Mufakhir et al. [413]. Focusing on the Katingan River, contaminated by artisanal small-scale gold mining activities, this study assessed the risks of Hg exposure from local fish consumption. The results showed that the estimated daily intake was significantly higher than the safety thresholds, posing health risks to the local population. Deng et al. [414] evaluated Hg exposure risks for populations living near an urban solid waste incinerator in southern China by combining internal (biomonitoring) and external (environmental measurements) exposure assessment methods. This integrated approach provided a comprehensive understanding of exposure sources and health risks.

Recent studies highlight that even low-level exposure to Pb and Hg during critical developmental periods can impair cognitive function, lower IQ, and cause behavioral disorders. Children are more susceptible due to their higher metabolic rate and increased BBB permeability, which is attributed to less mature tight junctions between endothelial cells and a different composition of barrier proteins [415]. In turn, cognitive impairment was found in working adults exposed to Mn in terms of processing speed, attention, working memory, reaction time, and visual attention [416].

It should be noted that mutual physiological interactions between metals complicate single-metal exposure studies. Emerging research suggests that combined exposure to multiple neurotoxic metals may have additive or synergistic effects, increasing the risk of neurodegenerative diseases like AD and PD [417–419]. Co-exposure may result in different outcomes than expected from exposure to individual metals. For example, animal studies have shown that As is more toxic in isolated exposure and that when co-exposed with Mn, there is an antagonistic effect [420].

Understanding the risks associated with neurotoxic metals (like any other health risk factor) requires comprehensive risk assessment studies that quantify exposure levels, identify vulnerable populations, and assess long-term health effects.

Toxic metal exposure assessments include the characterization of metal-specific exposure pathways and routes, chemical forms, and doses in a manner consistent with defining risk thresholds for individual organisms. Factors affecting the threshold dose include the physicochemical properties of the substance, the route and duration of exposure, and the characteristics of the exposed population, such as age, gender, and health status [421].

Risk assessment involves several steps. Firstly, it is necessary to identify the hazard, i.e., the potential danger posed by the environmental pollutant, through the analysis of available scientific studies, epidemiological data, and toxicological profiles. Secondly, quantification of the level (magnitude) of exposure by measuring the concentration of the pollutant in environmental compartments (air, water, soil, food) and biological samples (blood, urine, hair) must take place. Different routes of exposure (inhalation, ingestion, dermal contact) and population-specific factors (e.g., age, occupation, diet) are also considered at this stage [421]. Thirdly, the identification of safe exposure limits, such as the *"reference dose"* (RfD) or *"tolerable daily intake"* (TDI), based on available experimental and human studies, should be done. Finally, hazard, exposure, and dose–response data are integrated to estimate the likelihood and magnitude of health risks. High-risk groups (e.g., children, pregnant women, industrial workers) should be identified, and the likelihood of toxic outcomes should be quantified. Based on the risk characterization, appropriate measures must be suggested and implemented to mitigate the identified risks. Figure 2 presents a structured framework for assessing risks associated with exposure to heavy metals.

A comprehensive overview of methodologies for assessing metal exposure in environmental contexts was provided by Newman et al. [422], who addressed the need for standardized approaches to protect human health and the environment. In general, standardized methodologies for accurately estimating the extent of exposure and associated health risks may be categorized into direct and indirect methods. Direct methods involve measuring pollutants at the point of contact

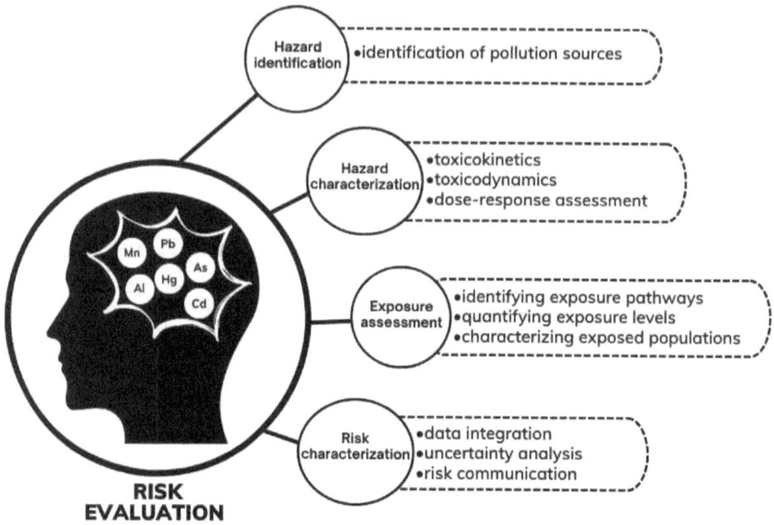

Fig. 2 Structured framework for assessing risks associated with exposure to heavy metals

or inside the human body, providing data on exposure levels (e.g., measuring pollutants in inhaled air or analyzing biological samples such as blood, urine, or tissues to detect the presence of metals). On the other hand, indirect methods estimate exposure by correlating environmental concentrations (contaminant levels in air, water, soil, or food samples) with human activities and time-activity patterns to estimate exposure. Exposure assessments should consider identifying sources of uncertainty and natural variability so that data on people's behavior and characteristics, such as time spent in certain places, dietary habits, and body weight, can be used to model and estimate exposure levels. Both direct and indirect methods are essential for a comprehensive exposure assessment and are often used in a complementary manner [423].

A recent paper by Ogwu et al. [424] delves into the complexity of environmental risk assessment of trace metal pollution. The authors point out that traditional environmental risk assessment methods often fail to address the complex dynamics of trace metals. To improve the accuracy and reliability of these assessments, the authors recommend integrating advanced statistical techniques, including Bayesian modeling, machine learning, and geostatistics.

There are many benefits of integrating statistical models in the context described above. They improve understanding of the mechanisms governing trace metal transport, bioavailability, and environmental impacts. In addition, the ability to predict future contamination levels and pollution patterns is enhanced. Spatial and temporal analysis, in turn, facilitates the assessment of pollution in different locations and times, identifying trends and anomalies. Increasing attention is being given to incorporating big data analytics and machine learning to improve the predictive capabilities of environmental assessment models, enabling real-time monitoring and assessment [425].

6 Environmental Neurotoxicity and the Brain Metallome: Toward a Comprehensive Understanding

Although the significance of metals and some non-metallic elements in brain health and disease is well established, there are still many gaps in creating a complete picture of the brain metallome. This is partly due to the highly dynamic nature of the metallome itself, which is considered a subcategory of the genome, transcriptome, proteome, and metabolome [426]. As sequencing tools and genome analysis capabilities have advanced our understanding of susceptibility to environmental exposures, environmental health scientists are developing new tools to study exposures and their cumulative impact over the lifespan [427]. Researchers expect that brain metallome studies will contribute to a better understanding of the human exposome [427–429].

Mapping the human brain metallome—a comprehensive profile of metal ions and their interactions in the brain—could provide valuable insights for omics research and help fill critical gaps in neuroscience, contributing to advancements in prevention, diagnosis, and therapeutics. Metallomics and environmental neurotoxicology are closely related fields. Metallomics focuses on the comprehensive study of metal species within biological systems, including their distribution, interactions, and functions, whereas environmental neurotoxicology examines how environmental factors, particularly neurotoxic metals, negatively impact the nervous system. Understanding the brain's metallome is essential in this context, as it sheds light on how metals influence neuronal function and contribute to neurotoxicity.

Environmental neurotoxicity research has identified numerous neurotoxicants, including heavy metals, pesticides, and air pollutants, which are linked to conditions like neurodegenerative diseases, developmental delays, and cognitive impairments. However, a broad spectrum of potential neurotoxins remains uncharacterized. This presents a significant challenge in understanding the impact of the environment on the complex function, development, and structure of the nervous system. Therefore, further research is needed to assess the neurological effects of emerging contaminants.

Studies on interactions between toxic and essential metals and metalloids, the synergistic and antagonistic effects of various metal species, their transport through biological fluids and membranes, and metabolic pathway alterations due to the overexpression or inhibition of metallobiomolecules require thorough, comprehensive, and interdisciplinary approaches [430].

Metallomics provides advanced tools and methodologies to study the roles of metals in biological systems. However, most of our knowledge about the health effects of toxic metals comes from studies conducted in cell lines, animals, or populations with relatively high exposure levels. Despite research advancements, several gaps remain in understanding the impact of environmental neurotoxins, such as toxic metals, on the human brain. These gaps are outlined below.

A major limitation is our incomplete understanding of the effects of combined exposures to toxic metals [431, 432]. In reality, individuals are typically exposed to mixtures of neurotoxicants rather than single substances. Understanding how these combinations interact and their cumulative effects on brain health is a significant challenge in environmental neurotoxicity research. However, most studies focus on single pollutants, even though real-world exposure often involves complex mixtures. Therefore, further research is needed to assess the interactive effects of multiple environmental toxins on neurological health.

In metallomics research, significant obstacles to fully understanding the role of metals in brain function include insufficient longitudinal data and small study populations. Long-term studies tracking environmental exposures and their neurological outcomes over decades are scarce, yet such data are crucial for establishing causal relationships between environmental factors and chronic neurological conditions. Additionally, large-scale brain metallomics studies are limited by several challenges. Sampling difficulties arise due to variations in postmortem samples, influenced by factors such as cause of death, time between death and tissue collection, and individual differences (e.g., age, sex, and health status). Moreover, the brain's complexity and the heterogeneous distribution of metals across regions and individuals make obtaining standardized samples from large populations difficult. Furthermore, numerous factors—including age, sex, diet, genetic background, environmental exposure, and disease states (e.g., neurodegenerative diseases)—affect metal content in specific brain regions. This variability complicates efforts to isolate specific influences and assess the impact of metal homeostasis on neurological health across diverse populations.

Large-scale population studies often require long-term follow-up and repeated brain tissue sampling, which is not feasible for most human populations. Without longitudinal studies, tracking how changes in brain metal levels correlate with the progression of neurological and other diseases over time remains challenging. Moreover, brain metallomics research generates vast data, often involving complex interactions between metals, proteins, and other molecules. Therefore, advanced bioinformatics and statistical tools are essential for interpreting these data in large populations.

Another challenge lies in the study of vulnerable populations. Children, the elderly, and marginalized communities may be more susceptible to environmental neurotoxicity. While numerous studies have examined neurotoxin exposure in pregnant women, there is a notable lack of concurrent research involving both mothers and their offspring. For example, many epidemiological studies investigating exposure to persistent organic pollutants (POPs) rarely explore disruptions in the metallome within mother–child pairs. This gap limits our understanding of how prenatal neurotoxin exposure affects maternal and child health outcomes [433, 434].

It is well established that the prenatal environment plays a crucial role in shaping brain development through epigenetic mechanisms [435]. The global rise in neurodevelopmental disorders has heightened interest in the epigenetic modifications induced by toxic metal exposure. Evidence suggests that As, Cd, nickel, Hg, Pb, and chromium can significantly alter the epigenetic landscape of stem cells, affecting DNA methylation modifications, DNA hydroxymethylation, and histone modifications [436, 437]. For instance, Pb exposure has been linked to decreased histone acetylation and methylation, resulting in altered gene expression [438]. These modifications can disrupt normal brain development and function, potentially contributing to neurological disorders.

It is crucial to carefully examine the critical windows of metal exposure, as the effects of Pb and other toxic metals during these periods of brain development (such as in utero, early childhood, or adolescence) on brain plasticity and long-term cognitive function are not yet fully understood. More research is needed to explore how environmental toxins interfere with processes like synaptic pruning, neurogenesis, and myelination. Profiling the metallome during these stages could provide insights into how Pb disrupts brain development at the molecular and cellular levels. Additionally, further studies are required to determine whether childhood exposure to toxic metals leads to lasting changes in synaptic function and neural network organization in adulthood.

Emerging research suggests that heavy metals and environmental pollutants may play a role in the development of ASD through epigenetic pathways [439]. However, large-scale studies over extended periods are necessary to validate these findings and better understand the impact of environmental toxin exposure on the genome.

Gull et al. [440] reported that prenatal exposure to As, Cd, Hg, and Pb among pregnant women and fetuses is associated with IQ deficits, attention-related behaviors, poor academic achievement, reduced birth length, smaller head and chest circumferences, reproductive effects, and lower birth weight as consequences of toxic metal exposure. Issah et al. [441] evaluated the strength and consistency of the association between metal mixtures and pregnancy and birth outcomes, such as low birth weight and preterm birth. As mentioned earlier, the combined effects of multiple metal exposures can differ from those of individual metals; for instance, certain combinations may amplify adverse effects on brain development, while others may mitigate them. Thus, studies examining the effects of exposure to metal mixtures on both mothers and their children are highly desirable. Furthermore, because critical neurodevelopmental processes—such as functional cortical differentiation, synaptogenesis, myelination, and programmed apoptosis [442]—occur in the human CNS during fetal development and the first three years of life, parallel exposure studies involving both mothers and their offspring would enhance our understanding of environmental neurotoxicity.

While associations between environmental pollutants and neurotoxicity have been documented, the underlying biological mechanisms remain poorly understood. Sect. 2 presents studies that have revealed several key mechanisms by which metal exposure leads to pathological effects impacting human brain health. However, the cellular and molecular mechanisms still require further exploration. For example, research on how toxic metals interact with neurotransmitter systems, ion channels, and neuroinflammatory pathways is limited [443].

Undoubtedly, research on gene-environment interactions, combined with machine learning methods applied to multidimensional neuroimaging data, will continue to investigate links between exposure and indicators of structural brain health [444].

Researchers are increasingly integrating animal study findings with human epidemiological data, in vitro models, and advanced computational tools to enhance the predictive value for humans. This integrated approach aims to bridge the gap between animal research and human health outcomes, improving risk assessments associated with metal exposures. Modern toxicology emphasizes in vitro methods and computational approaches for testing chemicals to better understand and predict human hazards. Recreating human brain responses to environmental toxins in animal models is complex, and animal models do not fully replicate the brain's response to many toxins, including metals. Therefore, alternative testing methods, such as organs-on-a-chip, represent the future of modeling the brain's response to toxins and their mixtures [445]. Modern brain-on-a-chip devices can encompass multiple levels of complexity in cell culturing methodologies and may involve brain parenchyma and/or BBB tissues [445]. The potential of such devices for environmental neurotoxicity research is likely to be widely explored. While advancements in alternative testing methods show promise, further development is needed to mimic human neurological responses accurately.

Advancements in research methodologies, increased regulatory attention, and a growing focus on emerging contaminants are shaping the future of environmental neurotoxicity research. One of the most significant recent developments in brain metallomics is the creation of high-resolution, multimetal analytical techniques that enable precise and spatially resolved profiling of metals in brain tissue [446]. This advancement has opened new avenues for understanding the role of metals in brain function, neurodegeneration, and neurodevelopmental disorders. The combination of inductively coupled plasma mass spectrometry (ICP–MS), laser ablation ICP–MS (LA–ICP–MS), and synchrotron X-ray fluorescence (XRF) allows for the detailed mapping of metal distribution at unprecedented spatial resolution. These techniques are of crucial importance in the spatial mapping of metal accumulation. In the past, metal distribution in the brain was assessed in bulk tissue, limiting our ability to observe how metals accumulate in specific brain regions. With LA–ICP–MS and synchrotron XRF, scientists can now visualize metal distribution in intact brain tissue at submicron resolution. This capability allows for identifying specific regions where metal imbalances may contribute to neurological diseases such as AD, PD, and MS [447, 448].

It is important to note that metallomic analysis of brain tissue requires highly specialized and expensive technologies, making it financially prohibitive to scale up for large populations. Additionally, the need for high precision and sensitivity in detecting trace metals means that these analyses demand significant technical expertise, further limiting their widespread use in large-scale studies. Brain metallomics requires interdisciplinary collaboration between neurobiologists, chemists, biochemists, and clinicians. Integrating knowledge from these diverse fields can be challenging in extensive population studies, so smaller, more focused studies are often more feasible.

Metallome profiling may also play a role in neuroprotection studies [449, 450]. Future research will likely explore the potential of using antioxidants, anti-inflammatories, or metal-regulating compounds to protect the brain from the toxic effects of chemical elements and their compounds. These interventions should focus on restoring metal homeostasis and reducing oxidative stress in the brain.

Brain metallomics studies face many other challenges, including determining the threshold exposure concentrations for neurotoxic effects. These thresholds depend on factors such as the specific element, its speciation, the route and duration of exposure, the cumulative effects of neurotoxic contaminants, and, particularly, the age of the exposed individuals [451]. Overcoming the challenges of accurately assessing exposure to neurotoxic metals and metal compounds is difficult because humans are often exposed to these substances in complex and diverse ways (e.g., through air, water, food, and soil pollution). Measuring chronic or low-level exposure over time, especially with contaminants like air pollutants or residues from endocrine-active compounds, presents a significant technical challenge.

To reduce the risk of exposure to neurotoxic metals, stricter guidelines for acceptable metal concentrations in air, water, soil, and food should be enforced. Programs to screen at-risk populations for exposure to neurotoxic metals should be implemented, offering early interventions for affected individuals. Additionally, there is a need to promote cleaner industrial technologies, phase out hazardous materials, and improve waste management practices to reduce environmental pollution. Increasing public awareness of the risks associated with neurotoxin exposure is also essential.

Conflict of Interest

The authors have no conflicts of interest to declare.

References

1. Abondio P, Bruno F (2025) Single-cell pan-omics, environmental neurology, and artificial intelligence: the time for holistic brain health research. Neural Regen Res 20(6):1703–1704. https://doi.org/10.4103/NRR.NRR-D-24-00324
2. de Prado Bert P, Mercader EMH, Pujol J et al (2018) The effects of air pollution on the brain: a review of studies interfacing environmental epidemiology and neuroimaging. Curr Environ Health Rep 5(3):351–364. https://doi.org/10.1007/s40572-018-0209-9
3. Potter NA, Meltzer GY, Avenbuan ON et al (2021) Particulate matter and associated metals: a link with neurotoxicity and mental health. Atmosphere 12:425. https://doi.org/10.3390/atmos12040425
4. Morrel J, Dong M, Rosario MA et al (2025) A systematic review of air pollution exposure and brain structure and function during development. Environ Res 15(275), 121368. https://doi.org/10.1016/j.envres.2025.121368
5. Scorza FA, Chaddad-Neto F, Beltramim L et al (2024) Water pollution and the brain. Clinics (Sao Paulo) 28(79):100424. https://doi.org/10.1016/j.clinsp.2024.100424
6. Spencer PS, Palmer VS (2021) Direct and indirect neurotoxic potential of metal/metalloids in plants and fungi used for food, dietary supplements, and herbal medicine. Toxics 9(3):57. https://doi.org/10.3390/toxics9030057
7. Karri V, Schuhmacher M, Kumar V (2016) Heavy metals (Lead, Cd, As and MeHg) as risk factors for cognitive dysfunction: a general review of metal mixture mechanism in brain. Environ Toxicol Pharmacol 48:203–213. https://doi.org/10.1016/j.etap.2016.09.016
8. Liu L, Hu L, Chen B et al (2024) Chapter 3. Environmetallomics. In: Li Y, Sun H (eds) Applied metallomics: from life sciences to environmental sciences. Wiley-VCH GmbH. https://doi.org/10.1002/9783527840397.ch3
9. Althomali RH, Abbood MA, Saleh EAM et al (2024) Exposure to heavy metals and neurocognitive function in adults: a systematic review. Environ Sci Eur 36:18. https://doi.org/10.1186/s12302-024-00843-7
10. Feigin VL, Vos T, Nichols E et al (2020) The global burden of neurological disorders: translating evidence into policy. Lancet Neurol 19:255–265. https://doi.org/10.1016/S1474-4422(19)30411-9
11. Mitra S, Chakraborty AJ, Tareq AM et al (2022) Impact of heavy metals on the environment and human health: novel therapeutic insights to counter the toxicity. J King Saud Univ 34:101865. https://doi.org/10.1016/j.jksus.2022.101865
12. Landrigan PJ, Lucchini RG, Kotelchuck D et al (2022) Chapter 29—Principles for prevention of the toxic effects of metals. In: Nordberg GF, Costa M (eds) Handbook on the toxicology of metals, 5th edn. Academic Press, pp 685–703
13. Kwakye GF, Paoliello MMB, Mukhopadhyay S et al (2015) Mn-induced Parkinsonism and Parkinson's disease: shared and distinguishable features. Int J Environ Res Public Health 12:7519–7540. https://doi.org/10.3390/ijerph120707519
14. Racette B, McGee-Minnich L, Moerlein S et al (2001) Welding-related Parkinsonism: clinical features, treatment, and pathophysiology. Neurology 56:8–13. https://doi.org/10.1212/WNL.56.1.8

15. Racette BA, Criswell SR, Lundin JI et al (2012) Increased risk of Parkinsonism associated with welding exposure. Neurotoxicology 33:1356–1361. https://doi.org/10.1016/j.neuro.2012.08.011
16. Calne D, Chu N (1996) Manganism and idiopathic Parkinsonism: similarities and differences. Neurology 46:1581–1586. https://doi.org/10.1212/WNL.46.6.1581
17. Bouchard M, Sauvé S, Barbeau B et al (2011) Intellectual impairment in school-age children exposed to Mn from drinking water. Environ Health Perspect 119:138–143. https://doi.org/10.1289/ehp.1002321
18. Mora A, Arora M, Harley K et al (2015) Prenatal and postnatal Mn exposure and neurodevelopment in the CHAMACOS cohort. Environ Res 140:461–468. https://doi.org/10.1016/j.envres.2015.04.001
19. Criswell S, Perlmutter J, Videen T et al (2012) Reduced uptake of [18F] FDOPA PET in asymptomatic welders with occupational Mn exposure. Neurology 79:1278–1284. https://doi.org/10.1212/WNL.0b013e31826aac4a
20. Criswell SR, Nielsen SS, Warden M et al (2018) [18F]FDOPA positron emission tomography in Mn-exposed workers. Neurotoxicology 64:43–49. https://doi.org/10.1016/j.neuro.2017.07.004
21. Yıldız S, Pirinççioğlu A, Arıca E (2023) Evaluation of heavy metal (Lead, Hg, Cd, and Mn) levels in blood, plasma, and urine of adolescents with aggressive behavior. Cureus 15(1):e33902. https://doi.org/10.7759/cureus.33902
22. Hong SB, Kim JW, Choi BS et al (2014) Blood Mn levels in relation to comorbid behavioral and emotional problems in children with attention-deficit/hyperactivity disorder. Psychiatry Res 220(1–2):418–25. https://doi.org/10.1016/j.psychres.2014.05.049
23. Menezes-Filho JA, de Carvalho-Vivas CF, Viana GF et al (2014) Elevated Mn exposure and school-aged children's behavior: a gender-stratified analysis. Neurotoxicology 45:293–300. https://doi.org/10.1016/j.neuro.2013.09.006
24. Bjørklund G, Dadar M, Mutter J et al (2017) The toxicology of Hg: current research and emerging trends. Environ Res 159:545–554. https://doi.org/10.1016/j.envres.2017.08.051
25. Kaur I, Behl T, Aleya L et al (2021) Role of metallic pollutants in neurodegeneration: effects of aluminium, Lead, Hg, and As in mediating brain impairment events and autism spectrum disorder. Environ Sci Pollut Res 28:8989–9001. https://doi.org/10.1007/s11356-020-12255-0
26. Bjørklund G, Aaseth J, Ajsuvakova OP et al (2017) Molecular interaction between Hg and selenium in neurotoxicity. Coord Chem Rev 332:30–37. https://doi.org/10.1016/j.ccr.2016.10.009
27. Branco V, Aschner M, Carvalho C (2021) Chapter seven—neurotoxicity of Hg: an old issue with contemporary significance. In: Aschner M, Costa LG (eds) Advances in neurotoxicology. Academic Press, pp 239–262
28. Raposo RDS, Pinto DV, Moreira R et al (2020) MethylHg impact on adult neurogenesis: is the worst yet to come from recent brazilian environmental disasters?. Front Aging Neurosci 12:591601. https://doi.org/10.3389/fnagi.2020.591601
29. Grandjean P, Weihe P, White RF et al (1998) Cognitive performance of children prenatally exposed to "Safe" levels of methylHg. Environ Res 77:165–172. https://doi.org/10.1006/enrs.1997.3804
30. Wu Y-S, Osman AI, Hosny M et al (2024) The toxicity of hg and its chemical compounds: molecular mechanisms and environmental and human health implications: a comprehensive review. ACS Omega 9:5100–5126. https://doi.org/10.1021/acsomega.3c07047
31. Bose-O'Reilly S, McCarty KM, Steckling N et al (2010) Hg exposure and children's health. Curr Probl Pediatr Adolesc Health Care 40:186–215. https://doi.org/10.1016/j.cppeds.2010.07.002
32. Arrifano GPF, de Oliveira MA, Souza-Monteiro JR et al (2018) Role for apolipoprotein E in neurodegeneration and Hg intoxication. Front Biosci-Elite 10:229–241. https://doi.org/10.2741/E819
33. Novo JP, Martins B, Raposo RS et al (2021) Cellular and molecular mechanisms mediating methylHg neurotoxicity and neuroinflammation. Int J Mol Sci 22:3101. https://doi.org/10.3390/ijms22063101
34. Magnavita N, Sabatelli M, Scoditti E et al (2020) Personalized prevention in Hg-induced amyotrophic lateral sclerosis: a case report. Appl Sci 10:7839. https://doi.org/10.3390/app10217839
35. Bittencourt LO, Chemelo VS, Aragão WAB et al (2021) From molecules to behavior in long-term inorganic Hg intoxication: unraveling proteomic features in cerebellar neurodegeneration of rats. Int J Mol Sci 23:111. https://doi.org/10.3390/ijms23010111
36. Navarro-Sempere A, García M, Rodrigues AS et al (2022) Occurrence of volcanogenic inorganic Hg in wild mice spinal cord: potential health implications. Biol Trace Elem Res 200:2838–2847. https://doi.org/10.1007/s12011-021-02890-0

37. Behzadfar L, Hassani S, Feizpour H et al (2020) Effects of mercuric chloride on spatial memory deficit-induced by beta-amyloid and evaluation of mitochondrial function markers in the hippocampus of rats. Metallomics 12:144–153. https://doi.org/10.1039/c9mt00161a
38. Philibert A, Fillion M, Da Silva J et al (2022) Past Hg exposure and current symptoms of nervous system dysfunction in adults of a First Nation community (Canada). Environ Health 21: 34. https://doi.org/10.1186/s12940-022-00838-y
39. Kern JK, Geier DA, Sykes LK et al (2016) The relationship between Hg and autism: a comprehensive review and discussion. J Trace Elem Med Biol 37:8–24. https://doi.org/10.1016/j.jtemb.2016.06.002
40. Jafari T, Rostampour N, Fallah AA et al (2017) The association between Hg levels and autism spectrum disorders: a systematic review and meta-analysis. J Trace Elem Med Biol 44:289–297. https://doi.org/10.1016/j.jtemb.2017.09.002
41. Adams JB, Romdalvik J, Ramanujam VMS et al (2007) Hg, lead, and zinc in baby teeth of children with autism versus controls. J Toxicol Environ Health A 70:1046–1051. https://doi.org/10.1080/15287390601172080
42. Ryu J, Ha E-H, Kim B-N et al (2017) Associations of prenatal and early childhood Hg exposure with autistic behaviors at 5 years of age: the Mothers and Children's Environmental Health (MOCEH) study. Sci Total Environ 605–606:251–257. https://doi.org/10.1016/j.scitotenv.2017.06.227
43. Yassa HA (2014) Autism: a form of Lead and Hg toxicity. Environ Toxicol Pharmacol 38:1016–1024. https://doi.org/10.1016/j.etap.2014.10.005
44. Geier DA, Kern JK, Homme KG et al (2017) Abnormal brain connectivity spectrum disorders following thimerosal administration: a prospective longitudinal case–control assessment of medical records in the vaccine safety datalink. Dose-Response 15:1559325817690849. https://doi.org/10.1177/1559325817690849
45. Virgolini MB, Aschner M (2021) Molecular mechanisms of Lead neurotoxicity. Adv Neurotoxicol 5:159–213. https://doi.org/10.1016/bs.ant.2020.11.002
46. Mason LH, Harp JP, Han DY (2014) Lead neurotoxicity: neuropsychological effects of lead toxicity. BioMed Res Int 2014:840547. https://doi.org/10.1155/2014/840547
47. Xu T, Lin K, Cao M et al (2023) Patterns of global burden of 13 diseases attributable to Lead exposure, 1990-2019. BMC Public Health 23: 1121. https://doi.org/10.1186/s12889-023-15874-7
48. Bakulski KM, Rozek LS, Dolinoy DC et al (2012) Alzheimer's disease and environmental exposure to lead: the epidemiologic evidence and potential role of epigenetics. Curr Alzheimer Res 9:563–573. https://doi.org/10.2174/156720512800617991
49. Senut M-C, Cingolani P, Sen A et al (2012) Epigenetics of early-life lead exposure and effects on brain development. Epigenomics 4:665–674. https://doi.org/10.2217/epi.12.58
50. Kunert HJ, Wiesmüller GA, Schulze-Röbbecke R et al (2004) Working memory deficiencies in adults associated with low-level Lead exposure: implications of neuropsychological test results. Int J Hyg Environ Health 207:521–530. https://doi.org/10.1078/1438-4639-00323
51. Lee J-Y, Kim J-H, Choi D-W et al (2012) The association of heavy metal of blood and serum in the Alzheimer's diseases. Toxicol Res 28:93–98. https://doi.org/10.5487/TR.2012.28.2.093
52. Fathabadi B, Dehghanifiroozabadi M, Aaseth J et al (2018) Comparison of blood lead levels in patients with Alzheimer's disease and healthy people. Am J Alzheimers Dis Dementias 33:541–547. https://doi.org/10.1177/1533317518794032
53. Horton CJ, Weng H-Y, Wells EM (2019) Association between blood Lead level and subsequent Alzheimer's disease mortality. Environ Epidemiol 3:e045. https://doi.org/10.1097/EE9.0000000000000045
54. Wu J, Basha MR, Brock B et al (2008) Alzheimer's disease (AD)-like pathology in aged monkeys after infantile exposure to environmental metal lead (lead): evidence for a developmental origin and environmental link for AD. J Neurosci 28:3–9. https://doi.org/10.1523/JNEUROSCI.4405-07.2008
55. Basha MdR MuraliM, Siddiqi HK et al (2005) Lead (Lead) exposure and its effect on APP proteolysis and Aβ aggregation. FASEB J 19:2083–2084. https://doi.org/10.1096/fj.05-4375fje
56. Gorell JM, Johnson CC, Rybicki BA et al (1999) Occupational exposure to Mn, copper, lead, iron, Hg and zinc and the risk of Parkinson's disease. Neurotoxicology 20:239–247
57. Tamegart L, Abbaoui A, Makbal R et al (2019) Crocus sativus restores dopaminergic and noradrenergic damages induced by Lead in Meriones shawi: a possible link with Parkinson's disease. Acta Histochem 121:171–181. https://doi.org/10.1016/j.acthis.2018.12.003
58. Weisskopf MG, Weuve J, Nie H et al (2010) Association of cumulative Lead exposure with

Parkinson's disease. Environ Health Perspect 118:1609–1613. https://doi.org/10.1289/ehp.1002339
59. Coon S, Stark A, Peterson E et al (2006) Whole-body lifetime occupational lead exposure and risk of Parkinson's disease. Environ Health Perspect 114:1872–1876. https://doi.org/10.1289/ehp.9102
60. Wang B, Du Y (2013) Cd and its neurotoxic effects. Oxid Med Cell Longev 2013:898034. https://doi.org/10.1155/2013/898034
61. Ashok A, Rai NK, Tripathi S et al (2015) Exposure to As-, Cd-, and Lead-mixture induces Aβ, amyloidogenic APP processing and cognitive impairments via oxidative stress-dependent neuroinflammation in young rats. Toxicol Sci 143:64–80. https://doi.org/10.1093/toxsci/kfu208
62. Ben P, Zhang Z, Zhu Y et al (2016) l-Theanine attenuates Cd-induced neurotoxicity through the inhibition of oxidative damage and tau hyperphosphorylation. Neurotoxicology 57:95–103. https://doi.org/10.1016/j.neuro.2016.09.010
63. Ruczaj A, Brzóska MM (2023) Environmental exposure of the general population to Cd as a risk factor of the damage to the nervous system: a critical review of current data. J Appl Toxicol 43:66–88. https://doi.org/10.1002/jat.4322
64. Ciesielski T, Bellinger DC, Schwartz J et al (2013) Associations between Cd exposure and neurocognitive test scores in a cross-sectional study of US adults. Environ Health 12:13. https://doi.org/10.1186/1476-069X-12-13
65. Wang H, Zhang L, Abel GM et al (2018) Cd exposure impairs cognition and olfactory memory in male C57BL/6 mice. Toxicol Sci 161:87–102. https://doi.org/10.1093/toxsci/kfx202
66. Batool Z, Agha F, Tabassum S et al (2019) Prevention of Cd-induced neurotoxicity in rats by essential nutrients present in nuts. Acta Neurobiol Exp (Warsz) 79:169–183
67. Czarnecki LA, Moberly AH, Turkel DJ et al (2012) Functional rehabilitation of Cd-induced neurotoxicity despite persistent peripheral pathophysiology in the olfactory system. Toxicol Sci 126:534–544. https://doi.org/10.1093/toxsci/kfs030
68. Feng J, Chen S, Wang Y et al (2019) Maternal exposure to Cd impairs cognitive development of male offspring by targeting the Coronin-1a signaling pathway. Chemosphere 225:765–774. https://doi.org/10.1016/j.chemosphere.2019.03.094
69. Al-Saleh I, Shinwari N, Mashhour A et al (2014) Birth outcome measures and maternal exposure to heavy metals (Lead, Cd and Hg) in Saudi Arabian population. Int J Hyg Environ Health 217:205–218. https://doi.org/10.1016/j.ijheh.2013.04.009
70. Jeong KS, Park H, Ha E et al (2015) Performance IQ in children is associated with blood Cd concentration in early pregnancy. J Trace Elem Med Biol 30:107–111. https://doi.org/10.1016/j.jtemb.2014.11.007
71. Yang XF, Fan GY, Liu DY et al (2015) Effect of Cd exposure on the histopathology of cerebral cortex in juvenile mice. Biol Trace Elem Res 165:167–172. https://doi.org/10.1007/s12011-015-0246-2
72. Zhang Y-M, Liu X-Z, Lu H et al (2009) Lipid peroxidation and ultrastructural modifications in brain after perinatal exposure to Lead and/or Cd in rat pups. Biomed Environ Sci 22:423–429. https://doi.org/10.1016/S0895-3988(10)60021-9
73. Afifi OK, Embaby AS (2016) Histological study on the protective role of ascorbic acid on Cd induced cerebral cortical neurotoxicity in adult male albino rats. J Microsc Ultrastruct 4:36–45. https://doi.org/10.1016/j.jmau.2015.10.001
74. Ben Mimouna S, Boughammoura S, Chemek M et al (2018) Disruption of the zinc metabolism in rat fœtal brain after prenatal exposure to Cd. Chem Biol Interact 286:88–95. https://doi.org/10.1016/j.cbi.2018.03.005
75. Almeer RS, Kassab RB, AlBasher GI et al (2019) Royal jelly mitigates Cd-induced neuronal damage in mouse cortex. Mol Biol Rep 46:119–131. https://doi.org/10.1007/s11033-018-4451-x
76. Wang X-Y, Yang H, Wang M-G et al (2017) Trehalose protects against Cd-induced cytotoxicity in primary rat proximal tubular cells via inhibiting apoptosis and restoring autophagic flux. Cell Death Dis 8: e3099. https://doi.org/10.1038/cddis.2017.475
77. Alnahdi HS, Sharaf IA (2019) Possible prophylactic effect of omega-3 fatty acids on Cd-induced neurotoxicity in rats' brains. Environ Sci Pollut Res 26:31254–31262. https://doi.org/10.1007/s11356-019-06259-8
78. Huang Y, Guo X, Lu S et al (2024) Long-term exposure to Cd disrupts neurodevelopment in mature cerebral organoids. Sci Total Environ 912:168923. https://doi.org/10.1016/j.scitotenv.2023.168923
79. Zarazúa S, Bürger S, Delgado JM et al (2011) As affects expression and processing of amyloid precursor protein (APP) in primary neuronal cells overexpressing the Swedish mutation of human APP. Int J Dev Neurosci 29:389–396. https://doi.org/10.1016/j.ijdevneu.2011.03.004

80. Parajuli RPInt J Dev Neurosci, Fujiwara T, Umezaki M et al (2013) Association of cord blood levels of lead, As, and zinc with neurodevelopmental indicators in newborns: a birth cohort study in Chitwan Valley, Nepal. Environ Res 121:45–51. https://doi.org/10.1016/j.envres.2012.10.010
81. Thakur M, Rachamalla M, Niyogi S et al (2021) Molecular mechanism of as-induced neurotoxicity including neuronal dysfunctions. Int J Mol Sci 22:10077. https://doi.org/10.3390/ijms221810077
82. Hong YS, Song KH, Chung JY (2014) Health effects of chronic As exposure. J Prev Med Public Health 47:245–252. https://doi.org/10.3961/jpmph.14.035
83. Dakeishi M, Murata K, Grandjean P (2006) Long-term consequences of As poisoning during infancy due to contaminated milk powder. Environ Health 5:31. https://doi.org/10.1186/1476-069X-5-31
84. Masjosthusmann S, Tigges J, Fritsche E et al (2021) Chapter two—As-mediated developmental neurotoxicity: recent advances in understanding the adverse outcomes and underlying mechanisms. In: Aschner M, Costa LG (eds) Advances in neurotoxicology. Academic Press, pp 35–80
85. Rodríguez-Barranco M, Lacasaña M, Aguilar-Garduño C et al (2013) Association of As, Cd and Mn exposure with neurodevelopment and behavioural disorders in children: a systematic review and meta-analysis. Sci Total Environ 454–455:562–577. https://doi.org/10.1016/j.scitotenv.2013.03.047
86. Tolins M, Ruchirawat M, Landrigan P (2014) The developmental neurotoxicity of As: cognitive and behavioral consequences of early life exposure. Ann Glob Health 80:303–314. https://doi.org/10.1016/j.aogh.2014.09.005
87. Tsuji JS, Garry MR, Perez V et al (2015) Low-level As exposure and developmental neurotoxicity in children: a systematic review and risk assessment. Toxicology 337:91–107. https://doi.org/10.1016/j.tox.2015.09.002
88. Hamadani JD, Tofail F, Nermell B et al (2011) Critical windows of exposure for As-associated impairment of cognitive function in pre-school girls and boys: a population-based cohort study. Int J Epidemiol 40:1593–1604. https://doi.org/10.1093/ije/dyr176
89. Boffetta P, Sambati L, Sassano M (2024) Systematic review of studies on exposure to As in drinking water and cognitive and neurobehavioral effects. Crit Rev Toxicol 54:174–193. https://doi.org/10.1080/10408444.2023.2297751
90. Klotz K, Weistenhöfer W, Neff F et al (2017) The health effects of aluminium exposure. Dtsch Arzteblatt Int 114:653–659. https://doi.org/10.3238/arztebl.2017.0653
91. Inan-Eroglu E, Ayaz A (2018) Is aluminium exposure a risk factor for neurological disorders?. J Res Med Sci 23:51. https://doi.org/10.4103/jrms.JRMS_921_17
92. Kawahara M (2005) Effects of aluminium on the nervous system and its possible link with neurodegenerative diseases. J Alzheimers Dis 8:171–182. https://doi.org/10.3233/jad-2005-8210 discussion 209–215
93. Bondy SC, Camleadell A (2017) Chapter five—aluminium and neurodegenerative diseases. In: Aschner M, Costa LG (eds) Advances in neurotoxicology. Academic Press, pp 131–156
94. Dey M, Singh RK (2022) Neurotoxic effects of aluminium exposure as a potential risk factor for Alzheimer's disease. Pharmacol Rep 74:439–450. https://doi.org/10.1007/s43440-022-00353-4
95. Lukiw WJ, Kruck TPA, Percy ME et al (2019) Aluminium in neurological disease—a 36 year multicenter study. J Alzheimers Dis Parkinsonism 8:457. https://doi.org/10.4172/2161-0460.1000457
96. Zhu X, Hao W, Liu Z et al (2023) Aluminium induces neuroinflammation via P2X7 receptor activating NLRP3 inflammasome pathway. Ecotoxicol Environ Saf 249:114373. https://doi.org/10.1016/j.ecoenv.2022.114373
97. Virk SA, Eslick GD (2015) Aluminium levels in brain, serum, and cerebrospinal fluid are higher in Alzheimer's disease cases than in controls: a series of meta-analyses. J Alzheimers Dis 47:629–638. https://doi.org/10.3233/JAD-150193
98. Xu L, Zhang W, Liu X et al (2018) Circulatory levels of toxic metals (aluminium, Cd, Hg, Lead) in patients with Alzheimer's disease: a quantitative meta-analysis and systematic review. J Alzheimers Dis 62:361–372. https://doi.org/10.3233/JAD-170811
99. McLachlan DRC, Bergeron C, Smith JE et al (1996) Risk for neuropathologically confirmed Alzheimer's disease and residual aluminium in municipal drinking water employing weighted residential histories. Neurology 46:401–405. https://doi.org/10.1212/WNL.46.2.401
100. Rondeau V, Commenges D, Jacqmin-Gadda H et al (2000) Relation between aluminium concentrations in drinking water and Alzheimer's disease: an 8-year follow-up study. Am J Epidemiol 152:59–66. https://doi.org/10.1093/aje/152.1.59
101. Mold MJ, O'Farrell A, Morris B et al (2020) Aluminium and neurofibrillary tangle co-localization in familial Alzheimer's Disease and related neurological disorders. J Alzheimers Dis 78:139–149. https://doi.org/10.3233/JAD-200838

102. Skalny AV, Aschner M, Jiang Y et al (2021) Molecular mechanisms of aluminium neurotoxicity: update on adverse effects and therapeutic strategies. Adv Neurotoxicol 5:1–34. https://doi.org/10.1016/bs.ant.2020.12.001
103. Kandimalla R, Vallamkondu J, Corgiat EB et al (2016) Understanding aspects of aluminium exposure in Alzheimer's disease development. Brain Pathol 26:139–154. https://doi.org/10.1111/bpa.12333
104. Xiao F, Li X-G, Zhang X-Y et al (2011) Combined administration of D-galactose and aluminium induces Alzheimer-like lesions in brain. Neurosci Bull 27:143–155. https://doi.org/10.1007/s12264-011-1028-2
105. Andrási E, Páli N, Molnár Z et al (2005) Brain aluminium, magnesium and phosphorus contents of control and Alzheimer-diseased patients. J Alzheimers Dis 7:273–284. https://doi.org/10.3233/jad-2005-7402
106. Rusina R, Matěj R, Kašparová L et al (2011) Higher aluminium concentration in Alzheimer's disease after Box-Cox data transformation. Neurotox Res 20:329–333. https://doi.org/10.1007/s12640-011-9246-y
107. Michaelson DM (2014) APOE ε4: the most prevalent yet understudied risk factor for Alzheimer's disease. Alzheimers Dement J 10:861–868. https://doi.org/10.1016/j.jalz.2014.06.015
108. Mold M, Linhart C, Gómez-Ramírez J et al (2020) Aluminium and amyloid-β in familial Alzheimer's disease. J Alzheimers Dis 73:1627–1635. https://doi.org/10.3233/JAD-191140
109. Beitz JM (2014) Parkinson's disease: a review. Front Biosci6:65–74. https://doi.org/10.2741/S415
110. Hirsch EC, Brandel JP, Galle P et al (1991) Iron and Aluminium Increase in the Substantia Nigra of patients with Parkinson's disease: an x-ray microanalysis. J Neurochem 56:446–451. https://doi.org/10.1111/j.1471-4159.1991.tb08170.x
111. Erazi H, Ahboucha S, Gamrani H (2011) Chronic exposure to aluminium reduces tyrosine hydroxylase expression in the substantia nigra and locomotor performance in rats. Neurosci Lett 487:8–11. https://doi.org/10.1016/j.neulet.2010.09.053
112. Capriello T, Di Meglio G, De Maio A et al (2022) Aluminium exposure Leads to neurodegeneration and alters the expression of marker genes involved to parkinsonism in zebrafish brain. Chemosphere 307:135752. https://doi.org/10.1016/j.chemosphere.2022.135752
113. Bjorklund G, Stejskal V, Urbina MA et al (2018) Metals and Parkinson's disease: mechanisms and biochemical processes. Curr Med Chem 25:2198–2214. https://doi.org/10.2174/0929867325666171129124616
114. Zeng X, Macleod J, Berriault C et al (2021) Aluminium dust exposure and risk of neurodegenerative diseases in a cohort of male miners in Ontario, Canada. Scand J Work Environ Health 47:531–539. https://doi.org/10.5271/sjweh.3974
115. Martell J, Guidotti TL (2022) Trading one risk for another: consequences of the unauthenticated treatment and prevention of silicosis in Ontario miners in the mcintyre powder aluminium inhalation program. NEW Solut 31:422–433. https://doi.org/10.1177/10482911211037007
116. Gorell JM, Rybicki BA, Cole Johnson C et al (1999) Occupational metal exposures and the risk of Parkinson's disease. Neuroepidemiology 18:303–308. https://doi.org/10.1159/000026225
117. Uversky VN, Li J, Bower K et al (2002) Synergistic effects of pesticides and metals on the fibrillation of α-synuclein: implications for Parkinson's disease. NeuroToxicology 23:527–536. https://doi.org/10.1016/S0161-813X(02)00067-0
118. Ahmed SSSJ, Santosh W (2010) Metallomic profiling and linkage map analysis of early Parkinson's disease: a new insight to aluminium marker for the possible diagnosis. PloS ONE 5:e11252. https://doi.org/10.1371/journal.pone.0011252
119. Raj K, Kaur P, Gupta GD et al (2021) Metals associated neurodegeneration in Parkinson's disease: insight to physiological, pathological mechanisms and management. Neurosci Lett 753:135873. https://doi.org/10.1016/j.neulet.2021.135873
120. Mold M, Chmielecka A, Rodriguez MRR et al (2018) Aluminium in brain tissue in multiple sclerosis. Int J Environ Res Public Health 15:1777. https://doi.org/10.3390/ijerph15081777
121. Exley C, Clarkson E (2020) Aluminium in human brain tissue from donors without neurodegenerative disease: a comparison with Alzheimer's disease, multiple sclerosis and autism. Sci Rep 10: 7770. https://doi.org/10.1038/s41598-020-64734-6
122. Pamphlett R, Buckland ME, Bishop DP (2023) Potentially toxic elements in the brains of people with multiple sclerosis. Sci Rep 13: 655. https://doi.org/10.1038/s41598-022-27169-9
123. Martinez-Finley E, Gavin C, Aschner M et al (2013) Mn neurotoxicity and the role of reactive oxygen species. Free Radic Biol Med 62:65–75. https://doi.org/10.1016/j.freeradbiomed.2013.04.003

124. Harischandra DS, Rokad D, Neal ML et al (2019) Mn promotes the aggregation and prion-like cell-to-cell exosomal transmission of α-synuclein. Sci Signal 12:eaau4543. https://doi.org/10.1126/scisignal.aau4543
125. Pajarillo E, Johnson J, Rizor A et al (2020) Astrocyte-specific deletion of the transcription factor Yin Yang 1 in murine substantia nigra mitigates Mn-induced dopaminergic neurotoxicity. J Biol Chem 295:15662–15676. https://doi.org/10.1074/jbc.RA120.015552
126. Tinkov AA, Paoliello MMB, Mazilina AN et al (2021) Molecular targets of Mn-induced neurotoxicity: a five-year update. Int J Mol Sci 22:4646. https://doi.org/10.3390/ijms22094646
127. Cordova FM, Aguiar AS, Peres TV et al (2013) Mn-exposed developing rats display motor deficits and striatal oxidative stress that are reversed by Trolox. Arch Toxicol 87:1231–1244. https://doi.org/10.1007/s00204-013-1017-5
128. da Silva Santos V, Bisen-Hersh E, Yu Y et al (2014) Anthocyanin-rich acai (*Euterpe oleracea* Mart.) extract attenuates Mn-induced oxidative stress in rat primary astrocyte cultures. J Toxicol Env Health A 77:390–404. https://doi.org/10.1080/15287394.2014.880392
129. Balachandran RC, Mukhopadhyay S, McBride D et al (2020) Brain Mn and the balance between essential roles and neurotoxicity. J Biol Chem 295:6312–6329. https://doi.org/10.1074/jbc.REV119.009453
130. Chen JY, Tsao GC, Zhao Q et al (2001) Differential cytotoxicity of Mn(II) and Mn(III): special reference to mitochondrial [Fe-S] containing enzymes. Toxicol Appl Pharmacol 175:160–168. https://doi.org/10.1006/taap.2001.9245
131. Szpetnar M, Luchowska-Kocot D, Boguszewska-Czubara A et al (2016) The influence of Mn and glutamine intake on antioxidants and neurotransmitter amino acids levels in rats' brain. Neurochem Res 41:2129–2139. https://doi.org/10.1007/s11064-016-1928-7
132. Yang X, Yang H, Wu F et al (2018) Mn inhibits GSH synthesis via downregulation of neuronal EAAC1 and astrocytic xCT to cause oxidative damage in the striatum of mice. Oxid Med Cell Longev 2018:4235695. https://doi.org/10.1155/2018/4235695
133. Moyano P, García JM, García J et al (2020) Mn increases Aβ and Tau protein levels through proteasome 20S and heat shock proteins 90 and 70 alteration, leading to SN56 cholinergic cell death following single and repeated treatment. Ecotoxicol Environ Saf 203:110975. https://doi.org/10.1016/j.ecoenv.2020.110975
134. Zhao X, Liu Y, Zhu G et al (2019) SIRT1 downregulation mediated Mn-induced neuronal apoptosis through activation of FOXO3a-Bim/PUMA axis. Sci Total Environ 646:1047–1055. https://doi.org/10.1016/j.scitotenv.2018.07.363
135. Chen P, Culbreth M, Aschner M (2016) Exposure, epidemiology, and mechanism of the environmental toxicant Mn. Env Sci Pollut Res Int 23:13802–13810. https://doi.org/10.1007/s11356-016-6687-0
136. Gorojod RM, Alaimo A, Porte Alcon S et al (2018) Heme Oxygenase-1 protects astroglia against Mn-induced oxidative injury by regulating mitochondrial quality control. Toxicol Lett 295:357–368. https://doi.org/10.1016/j.toxlet.2018.07.045
137. Alaimo A, Gorojod RM, Beauquis J et al (2014) Deregulation of mitochondria-shaping proteins Opa-1 and Drp-1 in Mn-induced apoptosis. PLoS ONE 9:e91848. https://doi.org/10.1371/journal.pone.0091848
138. Sarkar S, Malovic E, Harischandra DS et al (2018) Mn exposure induces neuroinflammation by impairing mitochondrial dynamics in astrocytes. Neurotoxicology 64:204–218. https://doi.org/10.1016/j.neuro.2017.05.009
139. Morcillo P, Cordero H, Ijomone OM et al (2021) Defective mitochondrial dynamics underlie Mn-induced neurotoxicity. Mol Neurobiol 58:3270–3289. https://doi.org/10.1007/s12035-021-02341-w
140. Roberts RF, Tang MY, Fon EA et al (2016) Defending the mitochondria: The pathways of mitophagy and mitochondrial-derived vesicles. Int J Biochem Cell Biol 79:427–436. https://doi.org/10.1016/j.biocel.2016.07.020
141. Shefa U, Jeong NY, Song IO et al (2019) Mitophagy links oxidative stress conditions and neurodegenerative diseases. Neural Regen Res 14:749–756. https://doi.org/10.4103/1673-5374.249218
142. Zhang HT, Mi L, Wang T et al (2016) PINK1/Parkin-mediated mitophagy play a protective role in Mn induced apoptosis in SH-SY5Y cells. Toxicol Vitro 34:212–219. https://doi.org/10.1016/j.tiv.2016.04.006
143. Liu K, Liu Z, Liu Z et al (2022) Mn induces S-nitrosylation of PINK1 Leading to nerve cell damage by repressing PINK1/Parkin-mediated mitophagy. Sci Total Env 834:155358. https://doi.org/10.1016/j.scitotenv.2022.155358
144. Zhou Q, Fu X, Wang X et al (2018) Autophagy plays a protective role in Mn-induced

toxicity in PC12 cells. Toxicology 394:45–53. https://doi.org/10.1016/j.tox.2017.12.001
145. Porte Alcon S, Gorojod RM, Kotler ML (2020) Kinetic and protective role of autophagy in Mn-exposed BV-2 cells. Biochim Biophys Acta 1867:118787. https://doi.org/10.1016/j.bbamcr.2020.118787
146. Zhang J, Cao R, Cai T et al (2013) The role of autophagy dysregulation in Mn-induced dopaminergic neurodegeneration. Neurotoxicol Res 24:478–490. https://doi.org/10.1007/s12640-013-9392-5
147. Zhang Z, Yan J, Bowman AB et al (2020) Dysregulation of TFEB contributes to Mn-induced autophagic failure and mitochondrial dysfunction in astrocytes. Autophagy 16:1506–1523. https://doi.org/10.1080/15548627.2019.1688488
148. Zhao F, Cai T, Liu M et al (2009) Mn induces dopaminergic neurodegeneration via microglial activation in a rat model of manganism. Toxicol Sci 107:156–164. https://doi.org/10.1093/toxsci/kfn213
149. Tjalkens RB, Popichak KA, Kirkley KA (2017) Inflammatory activation of microglia and astrocytes in Mn neurotoxicity. Adv Neurobiol 18:159–181. https://doi.org/10.1007/978-3-319-60189-2_8
150. Verina T, Kiihl SF, Schneider JS et al (2011) Mn exposure induces microglia activation and dystrophy in the substantia nigra of non-human primates. Neurotoxicology 32:215–226. https://doi.org/10.1016/j.neuro.2010.11.003
151. Kondru N, Manne S, Hepker M et al (2019) Mn exposure augments misfolded α-synuclein-induced proinflammatory M1 microglial phenotype and inflammasome activation. FASEB J 33:505.7--505.7. https://doi.org/10.1096/fasebj.2019.33.1_supplement.505.7
152. Lin G, Li X, Cheng X et al (2020) Mn exposure aggravates β-amyloid pathology by microglial activation. Front Aging Neurosci 12:556008. https://doi.org/10.3389/fnagi.2020.556008
153. Wu B, Cai H, Tang S et al (2020) Methionine-mediated protein phosphatase 2A catalytic subunit (PP2Ac) methylation ameliorates the tauopathy induced by Mn in cell and animal models. Neurotherapeutics 17:1878–1896. https://doi.org/10.1007/s13311-020-00930-6
154. Mehdizadeh P, Fesharaki SSH, Nouri M et al (2019) Tau folding and cytotoxicity of neuroblastoma cells in the presence of Mn oxide nanoparticles: biophysical, molecular dynamics, cellular, and molecular studies. Int J Biol Macromol 125:674–682. https://doi.org/10.1016/j.ijbiomac.2018.11.191
155. Ma Z, Liu K, Li XR et al (2020) Alpha-synuclein is involved in Mn-induced spatial memory and synaptic plasticity impairments via TrkB/Akt/Fyn-mediated phosphorylation of NMDA receptors. Cell Death Dis 11: 834. https://doi.org/10.1038/s41419-020-03051-2
156. Tarale P, Daiwile AP, Sivanesan S et al (2018) Mn exposure: linking down-regulation of miRNA-7 and miRNA-433 with α -synuclein overexpression and risk of idiopathic Parkinson's disease. Toxicol In Vitro 46:94–101. https://doi.org/10.1016/j.tiv.2017.10.003
157. Deng Y, Xu Z, Xu B et al (2012) The protective effects of riluzole on Mn-induced disruption of glutamate transporters and glutamine synthetase in the cultured astrocytes. Biol Trace Elem Res 148:242–249. https://doi.org/10.1007/s12011-012-9365-1
158. Karki P, Webb A, Smith K et al (2014) Yin Yang 1 is a repressor of glutamate transporter EAAT2, and it mediates Mn-induced decrease of EAAT2 expression in astrocytes. Mol Cell Biol 34:1280–1289. https://doi.org/10.1128/MCB.01176-13
159. Pajarillo E, Johnson J, Kim J et al (2018) 17\(\beta\)-estradiol and tamoxifen protect mice from Mn-induced dopaminergic neurotoxicity. Neurotoxicology 65:280–288. https://doi.org/10.1016/j.neuro.2017.11.008
160. Xu B, Xu ZF, Deng Y (2010) Mn exposure alters the expression of N-methyl-D-aspartate receptor subunit mRNAs and proteins in rat striatum. J Biochem Mol Toxicol 24:1–9. https://doi.org/10.1002/jbt.20306
161. Ma RE, Ward EJ, Yeh C-L et al (2018) Thalamic GABA levels and occupational Mn neurotoxicity: association with exposure levels and brain MRI. Neurotoxicology 64:30–42. https://doi.org/10.1016/j.neuro.2017.08.013
162. Casjens S, Dydak U, Dharmadhikari S et al (2018) Association of exposure to Mn and iron with striatal and thalamic GABA and other neurometabolites—neuroimaging results from the WELDOX II study. NeuroToxicology 64:60–67. https://doi.org/10.1016/j.neuro.2017.08.004
163. Edmondson DA, Ma RE, Yeh CL et al (2019) Reversibility of neuroimaging markers influenced by lifetime occupational Mn exposure. Toxicol Sci 172:181–190. https://doi.org/10.1093/toxsci/kfz174
164. Edmondson DA, Yeh CL, Hélie S et al (2020) Whole-brain R1 predicts Mn exposure and biological effects in welders. Arch Toxicol 94:3409–3420. https://doi.org/10.1007/s00204-020-02839-7

165. Ou C-Y, He Y-H, Sun Y et al (2019) Effects of sub-acute Mn exposure on thyroid hormone and glutamine (Gln)/Glutamate (Glu)-γ-aminobutyric acid (GABA) cycle in serum of rats. Int J Environ Res Public Health 16:2157. https://doi.org/10.3390/ijerph16122157
166. Yang X, Tan J, Xu X et al (2020) Prepubertal overexposure to Mn induce precocious puberty through GABAA receptor/nitric oxide pathway in immature female rats. Ecotoxicol Environ Saf 188:109898. https://doi.org/10.1016/j.ecoenv.2019.109898
167. Lasley SM, Fornal CA, Mandal S et al (2020) Early postnatal Mn exposure reduces rat cortical and striatal biogenic amine activity in adulthood. Toxicol Sci 173:144–155. https://doi.org/10.1093/toxsci/kfz208
168. Liu X, Yang J, Lu C et al (2017) Downregulation of Mfn2 participates in Mn-induced neuronal apoptosis in rat striatum and PC12 cells. Neurochem Int 108:40–51. https://doi.org/10.1016/j.neuint.2017.02.008
169. Ding H, Wang F, Su L et al (2020) Involvement of MEK5/ERK5 signalling pathway in Mn-induced cell injury in dopaminergic MN9D cells. J Trace Elem Med Biol 61:126546. https://doi.org/10.1016/j.jtemb.2020.126546
170. Yang Y, Ma S, Wei F et al (2019) Pivotal role of cAMP-PKA-CREB signalling pathway in Mn-induced neurotoxicity in PC12 cells. Environ Toxicol 34:1052–1062. https://doi.org/10.1002/tox.22776
171. Zhu G, Liu Y, Zhi Y et al (2019) PKA- and Ca2+-dependent p38 MAPK/CREB activation protects against Mn-mediated neuronal apoptosis. Toxicol Lett 309:10–19. https://doi.org/10.1016/j.toxlet.2019.04.004
172. Ma X, Han J, Wu Q et al (2015) Involvement of dysregulated Wip1 in Mn-induced p53 signalling and neuronal apoptosis. Toxicol Lett 235:17–27. https://doi.org/10.1016/j.toxlet.2014.12.019
173. Yang Y, Wei F, Wang J et al (2020) Mn modifies neurotrophin-3 (NT3) and its tropomyosin receptor kinase C (TrkC) in the cortex: Implications for Mn-induced neurotoxicity. Food Chem Toxicol 135:110925. https://doi.org/10.1016/j.fct.2019.110925
174. Shi S, Zhao J, Yang L et al (2015) KHSRP participates in Mn-induced neurotoxicity in rat striatum and PC12 cells. J Mol Neurosci 55:454–465. https://doi.org/10.1007/s12031-014-0367-7
175. Kim DS, Jin H, Anantharam V et al (2017) p73 gene in dopaminergic neurons is highly susceptible to Mn neurotoxicity. NeuroToxicology 59:231–239. https://doi.org/10.1016/j.neuro.2016.04.012
176. Branco V, Canário J, Lu J et al (2012) Hg and selenium interaction in vivo: effects on thioredoxin reductase and glutathione peroxidase. Free Radic Biol Med 52:781–793. https://doi.org/10.1016/j.freeradbiomed.2011.12.002
177. Xavier JM, Morgado AL, Rodrigues CM et al (2014) Tauroursodeoxycholic acid increases neural stem cell pool and neuronal conversion by regulating mitochondria-cell cycle retrograde signalling. Cell Cycle Georget Tex 13:3576–3589. https://doi.org/10.4161/15384101.2014.962951
178. Ren X, Zou L, Zhang X et al (2017) Redox signalling mediated by thioredoxin and glutathione systems in the central nervous system. Antioxid Redox Signal 27:989–1010. https://doi.org/10.1089/ars.2016.6925
179. Mori N, Yasutake A, Marumoto M et al (2011) MethylHg inhibits electron transport chain activity and induces cytochrome c release in cerebellum mitochondria. J Toxicol Sci 36:253–259. https://doi.org/10.2131/jts.36.253
180. Stringari J, Meotti FC, Souza DO et al (2006) Postnatal methylHg exposure induces hyperlocomotor activity and cerebellar oxidative stress in mice: dependence on the neurodevelopmental period. Neurochem Res 31:563–569. https://doi.org/10.1007/s11064-006-9051-9
181. Alattia J-R, Kuraishi T, Dimitrov M et al (2011) Hg is a direct and potent γ-secretase inhibitor affecting Notch processing and development in Drosophila. FASEB J 25:2287–2295. https://doi.org/10.1096/fj.10-174078
182. Olivieri G, Brack C, Müller-Spahn F et al (2000) Hg induces cell cytotoxicity and oxidative stress and increases beta-amyloid secretion and tau phosphorylation in SHSY5Y neuroblastoma cells. J Neurochem 74:231–236. https://doi.org/10.1046/j.1471-4159.2000.0740231.x
183. Teixeira FB, Fernandes RM, Farias-Junior PMA et al (2014) Evaluation of the effects of chronic intoxication with inorganic Hg on memory and motor control in rats. Int J Environ Res Public Health 11:9171–9185. https://doi.org/10.3390/ijerph110909171
184. Cariccio VL, Samà A, Bramanti P et al (2019) Hg involvement in neuronal damage and in neurodegenerative diseases. Biol Trace Elem Res 187:341–356. https://doi.org/10.1007/s12011-018-1380-4
185. Fonfría E, Vilaró MT, Babot Z et al (2005) Hg compounds disrupt neuronal glutamate transport in cultured mouse cerebellar granule cells. J Neurosci Res 79:545–553. https://doi.org/10.1002/jnr.20375

186. Yuan Y, Atchison WD (2003) MethylHg differentially affects GABA(A) receptor-mediated spontaneous IPSCs in Purkinje and granule cells of rat cerebellar slices. J Physiol 550:191–204. https://doi.org/10.1113/jphysiol.2003.040543
187. Felix L, Delekate A, Petzold GC et al (2020) Sodium fluctuations in astroglia and their potential impact on astrocyte function. Front Physiol 11:871. https://doi.org/10.3389/fphys.2020.00871
188. Farina M, Aschner M, Rocha JBT (2011) Oxidative stress in MeHg-induced neurotoxicity. Toxicol Appl Pharmacol 256:405–417. https://doi.org/10.1016/j.taap.2011.05.001
189. Ni M, Li X, Yin Z et al (2011) Comparative study on the response of rat primary astrocytes and microglia to methylHg toxicity. Glia 59:810–820. https://doi.org/10.1002/glia.21153
190. Noguchi Y, Shinozaki Y, Fujishita K et al (2013) Astrocytes protect neurons against methylHg via ATP/P2Y(1) receptor-mediated pathways in astrocytes. PloS ONE 8:e57898. https://doi.org/10.1371/journal.pone.0057898
191. Bjørklund G, Peana M, Dadar M et al (2020) Hg-induced autoimmunity: drifting from micro to macro concerns on autoimmune disorders. Clin Immunol 213:108352. https://doi.org/10.1016/j.clim.2020.108352
192. Vilhardt F, Haslund-Vinding J, Jaquet V et al (2017) Microglia antioxidant systems and redox signalling. Br J Pharmacol 174:1719–1732. https://doi.org/10.1111/bph.13426
193. Begam M, Sengupta M (2015) Immunomodulation of intestinal macrophages by Hg involves oxidative damage and rise of pro-inflammatory cytokine release in the fresh water fish *Channa punctatus* Bloch. Fish Shellfish Immunol 45:378–385. https://doi.org/10.1016/j.fsi.2015.04.017
194. Toyama T, Hoshi T, Noguchi T et al (2021) MethylHg induces neuronal cell death by inducing TNF-α expression through the ASK1/p38 signalling pathway in microglia. Sci Rep 11:9832. https://doi.org/10.1038/s41598-021-89210-7
195. Kempuraj D, Asadi S, Zhang B et al (2010) Hg induces inflammatory mediator release from human mast cells. J Neuroinflammation 7:20. https://doi.org/10.1186/1742-2094-7-20
196. Spulber S, Raciti M, Dulko-Smith B et al (2018) MethylHg interferes with glucocorticoid receptor: potential role in the mediation of developmental neurotoxicity. Toxicol Appl Pharmacol 354:94–100. https://doi.org/10.1016/j.taap.2018.02.021
197. Carvalho CML, Chew E-H, Hashemy SI et al (2008) Inhibition of the human thioredoxin system. A molecular mechanism of Hg toxicity. J Biol Chem 283:11913–11923. https://doi.org/10.1074/jbc.M710133200
198. Deane R, Bradbury MW (1990) Transport of Lead-203 at the blood-brain barrier during short cerebrovascular perfusion with saline in the rat. J Neurochem 54:905–914. https://doi.org/10.1111/j.1471-4159.1990.tb02337.x
199. Wang Q, Luo W, Zhang W et al (2011) Involvement of DMT1 +IRE in the transport of Lead in an in vitro BBB model. Toxicol Vitro 25:991–998. https://doi.org/10.1016/j.tiv.2009.11.006
200. An DZ, Ai JT, Fang HJ et al (2015) Influence of iron supplementation on DMT1 (IRE)-induced transport of Lead by brain barrier systems in vivo. Biomed Environ Sci 28:651–659. https://doi.org/10.3967/bes2015.091
201. Bouron A, Kiselyov K, Oberwinkler J (2015) Permeation, regulation and control of expression of TRP channels by trace metal ions. Pflugers Arch 467:1143–1164. https://doi.org/10.1007/s00424-014-1590-3
202. Balbuena P, Li W, Ehrich M (2011) Assessments of tight junction proteins occludin, claudin 5 and scaffold proteins ZO1 and ZO2 in endothelial cells of the rat blood–brain barrier: cellular responses to neurotoxicants malathion and Lead acetate. NeuroToxicology 32:58–67. https://doi.org/10.1016/j.neuro.2010.10.004
203. Song H, Zheng G, Shen X-F et al (2014) Reduction of brain barrier tight junctional proteins by Lead exposure: role of activation of nonreceptor tyrosine kinase Src via chaperon GRP78. Toxicol Sci 138:393–402. https://doi.org/10.1093/toxsci/kfu007
204. Song H, Zheng G, Liu Y et al (2016) Cellular uptake of Lead in the blood-cerebrospinal fluid barrier: novel roles of Connexin 43 hemichannel and its down-regulations via Erk phosphorylation. Toxicol Appl Pharmacol 297:1–11. https://doi.org/10.1016/j.taap.2016.02.021
205. Lahat N, Shapiro S, Froom P et al (2002) Inorganic Lead enhances cytokine-induced elevation of matrix metalloproteinase MMP-9 expression in glial cells. J Neuroimmunol 132:123–128. https://doi.org/10.1016/s0165-5728(02)00323-5
206. Carocci A, Catalano A, Lauria G et al (2016) Lead toxicity, antioxidant defense and environment. Rev Environ Contam Toxicol 238:45–67. https://doi.org/10.1007/398_2015_5003
207. Garza-Lombó C, Posadas Y, Quintanar L et al (2018) Neurotoxicity linked to dysfunctional

metal ion homeostasis and xenobiotic metal exposure: redox signalling and oxidative stress. Antioxid Redox Signal 28:1669–1703. https://doi.org/10.1089/ars.2017.7272
208. Lopes ACBA, Peixe TS, Mesas AE et al (2016) Lead exposure and oxidative stress: a systematic review. Rev Environ Contam Toxicol 236:193–238. https://doi.org/10.1007/978-3-319-20013-2_3
209. Kshirsagar MS, Patil JA, Patil A (2020) Increased blood Lead level induces oxidative stress and alters the antioxidant status of spray painters. J Basic Clin Physiol Pharmacol 31:(2). https://doi.org/10.1515/jbcpp-2018-0229
210. Atchison WD (2003) Effects of toxic environmental contaminants on voltage-gated calcium channel function: from past to present. J Bioenergy Biomembr 35:507–532. https://doi.org/10.1023/B:JOBB.0000008023.11211.13
211. Villalobo A, Ishida H, Vogel HJ et al (2018) Calmodulin as a protein linker and a regulator of adaptor/scaffold proteins. Biochim Biophys Acta 1865:507–521. https://doi.org/10.1016/j.bbamcr.2017.12.004
212. Florea A-M, Taban J, Varghese E et al (2013) Lead (Lead2+) neurotoxicity: Ion-mimicry with calcium (Ca2+) impairs synaptic transmission. A review with animated illustrations of the pre- and post-synaptic effects of Lead. J Local Glob Health Sci 2013:4. https://doi.org/10.5339/jlghs.2013.4
213. Ordemann JM, Austin RN (2016) Lead neurotoxicity: exploring the potential impact of Lead substitution in zinc-finger proteins on mental health. Met Integr Biometal Sci 8:579–588. https://doi.org/10.1039/c5mt00300h
214. Ishida K, Kotake Y, Sanoh S et al (2017) Lead-induced ERK activation is mediated by GluR2 non-containing AMPA receptor in cortical neurons. Biol Pharm Bull 40:303–309. https://doi.org/10.1248/blead.b16-00784
215. Zhou F, Chen Y, Fan G et al (2014) Lead-induced iron overload and attenuated effects of ferroportin 1 overexpression in PC12 cells. Toxicol Vitro 28:1339–1348. https://doi.org/10.1016/j.tiv.2014.07.005
216. Shvachiy L, Amaro-Leal Â, Outeiro TF et al (2023) Intermittent Lead exposure induces behavioral and cardiovascular alterations associated with neuroinflammation. Cells 12:818. https://doi.org/10.3390/cells12050818
217. Shvachiy L, Geraldes V, Amaro-Leal Â et al (2018) Intermittent low-level Lead exposure provokes anxiety, hypertension, autonomic dysfunction and neuroinflammation. Neurotoxicology 69:307–319. https://doi.org/10.1016/j.neuro.2018.08.001
218. Liu JT, Dong MH, Zhang JQ et al (2015) Microglia and astroglia: the role of neuroinflammation in Lead toxicity and neuronal injury in the brain. Neuroimmunol Neuroinflamm 2:131–137. https://doi.org/10.4103/2347-8659.156980
219. Kasten-Jolly J, Heo Y, Lawrence DA (2011) Central nervous system cytokine gene expression: modulation by Lead. J Biochem Mol Toxicol 25:41–54. https://doi.org/10.1002/jbt.20358
220. Harshitha P, Bose K, Dsouza HS (2024) Influence of Lead-induced toxicity on the inflammatory cytokines. Toxicology 503:153771. https://doi.org/10.1016/j.tox.2024.153771
221. Peng J, Zhou F, Wang Y et al (2019) Differential response to Lead toxicity in rat primary microglia and astrocytes. Toxicol Appl Pharmacol 363:64–71. https://doi.org/10.1016/j.taap.2018.11.010
222. Wang T, Zhang J, Xu Y (2020) Epigenetic basis of Lead-induced neurological disorders. Int J Environ Res Public Health 17:4878. https://doi.org/10.3390/ijerph17134878
223. Gonçalves JF, Dressler VL, Assmann CE et al (2021) Chapter three—Cd neurotoxicity: from its analytical aspects to neuronal impairment. In: Aschner M, Costa LG (eds) Advances in neurotoxicology. Academic Press, pp 81–113
224. Arruebarrena MA, Hawe CT, Lee YM et al (2023) Mechanisms of Cd neurotoxicity. Int J Mol Sci 24:16558. https://doi.org/10.3390/ijms242316558
225. Abdalla FH, Schmatz R, Cardoso AM et al (2014) Quercetin protects the impairment of memory and anxiogenic-like behavior in rats exposed to Cd: possible involvement of the acetylcholinesterase and Na(+), K(+)-ATPase activities. Physiol Behav 135:152–167. https://doi.org/10.1016/j.physbeh.2014.06.008
226. Zhai Q, Narbad A, Chen W (2014) Dietary strategies for the treatment of Cd and Lead toxicity. Nutrients 7:552–571. https://doi.org/10.3390/nu7010552
227. Gupta R, Shukla RK, Chandravanshi LP et al (2017) Protective role of quercetin in Cd-induced cholinergic dysfunctions in rat brain by modulating mitochondrial integrity and MAP kinase signalling. Mol Neurobiol 54:4560–4583. https://doi.org/10.1007/s12035-016-9950-y
228. Belyaeva EA (2018) Respiratory complex II in mitochondrial dysfunction-mediated cytotoxicity: insight from Cd. J Trace Elem Med Biol 50:80–92. https://doi.org/10.1016/j.jtemb.2018.06.009
229. Chen L, Xu B, Liu L et al (2011) Cd induction of reactive oxygen species activates the mTOR pathway, leading to neuronal cell death. Free Radic Biol Med 50(5):624e632

230. Monroe RK, Halvorsen SW (2006) Cd blocks receptor-mediated Jak/STAT signaling in neurons by oxidative stress. Free Radic Biol Med 41(3):493e502
231. Xu B, Chen S, Luo Y et al (2011) Calcium signaling is involved in Cd-induced neuronal apoptosis via induction of reactive oxygen species and activation of MAPK/mTOR network. PLoS ONE 6(4):e19052. https://doi.org/10.1371/journal.pone.0019052
232. Almeer RS, AlBasher GI, Alarifi S et al (2019) Royal jelly attenuates Cd-induced nephrotoxicity in male mice. Sci Rep 9(1):5825. https://doi.org/10.1038/s41598-019-42368-7
233. Shati AA (2019) Resveratrol protects against Cd chloride-induced hippocampal neurotoxicity by inhibiting ER stress and GAAD 153 and activating sirtuin 1/AMPK/Akt. Environ Toxicol 34:1340–1353. https://doi.org/10.1002/tox.22835
234. Zhang H, Dong X, Zhao R et al (2019) Cd results in accumulation of autophagosomes-dependent apoptosis through activating Akt-impaired autophagic flux in neuronal cells. Cell Signal 55:26–39. https://doi.org/10.1016/j.cellsig.2018.12.008
235. Zhang L, Xia Q, Zhou Y et al (2019) Endoplasmic reticulum stress and autophagy contribute to Cd-induced cytotoxicity in retinal pigment epithelial cells. Toxicol Lett 311:105–113. https://doi.org/10.1016/j.toxlet.2019.05.001
236. Arab HH, Eid AH, Alsufyani SE et al (2023) Neuroprotective impact of linagliptin against Cd-induced cognitive impairment and neuropathological aberrations: targeting SIRT1/Nrf2 axis, apoptosis, and autophagy. Pharmaceuticals 16:1065. https://doi.org/10.3390/ph16081065
237. Abdalla FH, Cardoso AM, Schmatz R et al (2014) Protective effect of quercetin in ecto-enzymes, cholinesterases, and myeloperoxidase activities in the lymphocytes of rats exposed to Cd. Mol Cell Biochem 396:201–211. https://doi.org/10.1007/s11010-014-2155-7
238. Adefegha SA, Oboh G, Omojokun OS et al (2016) Alterations of Na+/K+-ATPase, cholinergic and antioxidant enzymes activity by protocatechuic acid in Cd-induced neurotoxicity and oxidative stress in Wistar rats. Biomed Pharmacother 83:559–568. https://doi.org/10.1016/j.biopha.2016.07.017
239. Gonçalves JF, Spanevello RM, Fiorenza AM et al (2013) NTPDase and 5′-nucleotidase activities from synaptosomes and platelets of rats exposed to Cd and treated with N-acetylcysteine. Int J Dev Neurosci 31(1):69–74. https://doi.org/10.1016/j.ijdevneu.2012.11.001
240. Schmatz R, Schetinger MRC, Spanevello RM et al (2009) Effects of resveratrol on nucleotide degrading enzymes in streptozotocin-induced diabetic rats. Life Sci 84(11–12):345–350. https://doi.org/10.1016/j.lfs.2008.12.019
241. da Costa P, Gonçalves JF, Baldissarelli J et al (2015) Curcumin attenuates memory deficits and the impairment of cholinergic and purinergic signaling in rats chronically exposed to Cd. Environ Toxicol 32:70–83. https://doi.org/10.1002/tox.22213
242. Abdalla FH, Cardoso AM, Pereira LB et al (2013) Neuroprotective effect of quercetin in ectoenzymes and acetylcholinesterase activities in cerebral cortex synaptosomes of Cd-exposed rats. Mol Cell Biochem 381:1–8. https://doi.org/10.1007/s11010-013-1659-x
243. Martínez-Hernández MI, Acosta-Saavedra LC, Hernández-Kelly LC et al (2023) Microglial activation in metal neurotoxicity: impact in neurodegenerative diseases. BioMed Res Int 2023:7389508. https://doi.org/10.1155/2023/7389508
244. Akinyemi AJ, Adeniyi PA (2018) Effect of essential oils from ginger (*Zingiber officinale*) and turmeric (*Curcuma longa*) rhizomes on some inflammatory biomarkers in Cd induced neurotoxicity in rats. J Toxicol 2018:4109491. https://doi.org/10.1155/2018/4109491
245. Niño SA, Chi-Ahumada E, Carrizales L et al (2022) Life-long As exposure damages the microstructure of the rat hippocampus. Brain Res 1775:147742. https://doi.org/10.1016/j.brainres.2021.147742
246. Chernoff M, Tong L, Demanelis K et al (2020) Genetic determinants of reduced As metabolism efficiency in the 10q24.32 region are associated with reduced AS3MT expression in multiple human tissue types. Toxicol Sci 176:382–395. https://doi.org/10.1093/toxsci/kfaa075
247. Reichard JF, Puga A (2010) Effects of As exposure on DNA methylation and epigenetic gene regulation. Epigenomics 2:87–104. https://doi.org/10.2217/epi.09.45
248. Goicoechea L, Torres S, Fàbrega L et al (2024) S-Adenosyl-l-methionine restores brain mitochondrial membrane fluidity and GSH content improving Niemann-Pick type C disease. Redox Biol 72:103150. https://doi.org/10.1016/j.redox.2024.103150
249. Bianchi R, Calzi F, Savaresi S et al (1999) Biochemical analysis of myelin lipids and proteins in a model of methyl donor pathway deficit: effect of S-adenosylmethionine. Exp Neurol 159:258–266. https://doi.org/10.1006/exnr.1999.7132
250. Rai NK, Ashok A, Rai A et al (2022) Corrigendum to "Exposure to As, Cd and Lead-mixture impairs myelin and axon development

in rat brain, optic nerve and retina" [Toxicology and Applied Pharmacology 273 (2013) 242–258]. Toxicol Appl Pharmacol 450:116165. https://doi.org/10.1016/j.taap.2022.116165
251. Zarazúa S, Ríos R, Delgado JM et al (2010) Decreased arginine methylation and myelin alterations in As exposed rats. Neurotoxicology 31:94–100. https://doi.org/10.1016/j.neuro.2009.10.014
252. Masjosthusmann S, Siebert C, Hübenthal U et al (2019) Arsenite interrupts neurodevelopmental processes of human and rat neural progenitor cells: the role of reactive oxygen species and species-specific antioxidative defense. Chemosphere 235:447–456. https://doi.org/10.1016/j.chemosphere.2019.06.123
253. Hu Y, Li J, Lou B et al (2020) The role of reactive oxygen species in As toxicity. Biomolecules 10:240. https://doi.org/10.3390/biom10020240
254. Keshavarz-Bahaghighat H, Sepand MR, Ghahremani MH et al (2018) Acetyl-L-carnitine attenuates As-induced oxidative stress and hippocampal mitochondrial dysfunction. Biol Trace Elem Res 184:422–435. https://doi.org/10.1007/s12011-017-1210-0
255. Biswas S, Banna HU, Jahan M et al (2020) In vivo evaluation of As-associated behavioral and biochemical alterations in F0 and F1 mice. Chemosphere 245:125619. https://doi.org/10.1016/j.chemosphere.2019.125619
256. Ramos-Chávez LA, Rendón-López CRR, Zepeda A et al (2015) Neurological effects of inorganic As exposure: altered cysteine/glutamate transport, NMDA expression and spatial memory impairment. Front Cell Neurosci 9:21. https://doi.org/10.3389/fncel.2015.00021
257. Nelson-Mora J, Escobar ML, Rodríguez-Durán L et al (2018) Gestational exposure to inorganic As (iAs3+) alters glutamate disposition in the mouse hippocampus and ionotropic glutamate receptor expression Leading to memory impairment. Arch Toxicol 92:1037–1048. https://doi.org/10.1007/s00204-017-2111-x
258. Sharma A, Kshetrimayum C, Sadhu HG et al (2018) As-induced oxidative stress, cholinesterase activity in the brain of Swiss albino mice, and its amelioration by antioxidants Vitamin E and Coenzyme Q10. Environ Sci Pollut Res Int 25:23946–23953. https://doi.org/10.1007/s11356-018-2398-z
259. Chandravanshi LP, Gupta R, Shukla RK (2019) As-induced neurotoxicity by dysfunctioning cholinergic and dopaminergic system in brain of developing rats. Biol Trace Elem Res 189:118–133. https://doi.org/10.1007/s12011-018-1452-5
260. Chandravanshi LP, Gupta R, Shukla RK (2018) Developmental neurotoxicity of As: involvement of oxidative stress and mitochondrial functions. Biol Trace Elem Res 186:185–198. https://doi.org/10.1007/s12011-018-1286-1
261. Vázquez Cervantes GI, González Esquivel DF, Ramírez Ortega D et al (2023) Mechanisms associated with cognitive and behavioral impairment induced by As exposure. Cells 12:2537. https://doi.org/10.3390/cells12212537
262. Yan N, Li Y, Xing Y et al (2022) Developmental As exposure impairs cognition, directly targets DNMT3A, and reduces DNA methylation. EMBO Rep 23:e54147. https://doi.org/10.15252/embr.202154147
263. Tyler CRS, Smoake JJW, Solomon ER et al (2018) Sex-dependent effects of the histone deacetylase inhibitor, sodium valproate, on reversal learning after developmental As exposure. Front Genet 9:200. https://doi.org/10.3389/fgene.2018.00200
264. Valles S, Hernández-Sánchez J, Dipp VR et al (2020) Exposure to low doses of inorganic As induces transgenerational changes on behavioral and epigenetic markers in zebrafish (*Danio rerio*). Toxicol Appl Pharmacol 396:115002. https://doi.org/10.1016/j.taap.2020.115002
265. Iglesias-González J, Sánchez-Iglesias S, Beiras-Iglesias A et al (2017) Effects of aluminium on rat brain mitochondria bioenergetics: an in vitro and in vivo study. Mol Neurobiol 54:563–570. https://doi.org/10.1007/s12035-015-9650-z
266. Kumar V, Gill KD (2014) Oxidative stress and mitochondrial dysfunction in aluminium neurotoxicity and its amelioration: a review. Neurotoxicology 41:154–166. https://doi.org/10.1016/j.neuro.2014.02.004
267. Yuan C-Y, Lee Y-J, Hsu G-SW (2012) Aluminium overload increases oxidative stress in four functional brain areas of neonatal rats. J Biomed Sci 19:51. https://doi.org/10.1186/1423-0127-19-51
268. Ghorbel I, Amara IB, Ktari N et al (2016) Aluminium and acrylamide disrupt cerebellum redox states, cholinergic function and membrane-bound ATPase in adult rats and their offspring. Biol Trace Elem Res 174:335–346. https://doi.org/10.1007/s12011-016-0716-1
269. Bondy SC, Liu D, Guo-Ross S (1998) Aluminium treatment induces nitric oxide synthase in the rat brain. Neurochem Int 33:51–54. https://doi.org/10.1016/s0197-0186(05)80008-7
270. Mesole SB, Alfred OO, Yusuf UA et al (2020) Apoptotic inducement of neuronal cells by

aluminium chloride and the neuroprotective effect of eugenol in wistar rats. Oxid Med Cell Longev 2020:8425643. https://doi.org/10.1155/2020/8425643
271. Mokhemer SA, El-Tahawy NFG, Rifaai RA et al (2020) Comparative study of the effect of chronic aluminium chloride administration on the expression of endothelial nitric oxide synthase in rat brain. Indian J Public Health Res Dev 11:1729–1734. https://doi.org/10.37506/v11/i2/2020/ijphrd/195077
272. Martinez CS, Piagette JT, Escobar AG et al (2017) Aluminium exposure at human dietary levels promotes vascular dysfunction and increases blood pressure in rats: a concerted action of NAD(P)H oxidase and COX-2. Toxicology 390:10–21. https://doi.org/10.1016/j.tox.2017.08.004
273. Sánchez-Iglesias S, Méndez-Alvarez E, Iglesias-González J et al (2009) Brain oxidative stress and selective behaviour of aluminium in specific areas of rat brain: potential effects in a 6-OHDA-induced model of Parkinson's disease. J Neurochem 109:879–888. https://doi.org/10.1111/j.1471-4159.2009.06019.x
274. Çabuş N, Oğuz EO, Tufan AÇ et al (2015) A histological study of toxic effects of aluminium sulfate on rat hippocampus. Biotech Histochem 90:132–139. https://doi.org/10.3109/10520295.2014.965277
275. Laabbar W, Abbaoui A, Elgot A et al (2021) Aluminium induced oxidative stress, astrogliosis and cell death in rat astrocytes, is prevented by curcumin. J Chem Neuroanat 112:101915. https://doi.org/10.1016/j.jchemneu.2020.101915
276. Savory J, Herman MM, Ghribi O (2003) Intracellular mechanisms underlying aluminium-induced apoptosis in rabbit brain. J Inorg Biochem 97:151–154. https://doi.org/10.1016/s0162-0134(03)00258-7
277. Qin X, Li L, Nie X et al (2020) Effects of chronic aluminium lactate exposure on neuronal apoptosis and hippocampal synaptic plasticity in rats. Biol Trace Elem Res 197:571–579. https://doi.org/10.1007/s12011-019-02007-8
278. Hao Y-X, Li M-Q, Zhang J-S et al (2020) Aluminium-induced "mixed" cell death in mice cerebral tissue and potential intervention. Neurotox Res 37:835–846. https://doi.org/10.1007/s12640-019-00123-w
279. Zhang H, Wei M, Lu X et al (2020) Aluminium trichloride caused hippocampal neural cells death and subsequent depression-like behavior in rats via the activation of IL-1β/JNK signalling pathway. Sci Total Environ 715:136942. https://doi.org/10.1016/j.scitotenv.2020.136942
280. Jia X, Zhang Q, Niu Q (2014) [MAPK signalling pathways involved in aluminium-induced apoptosis and necroptosis in SH-SY5Y cells]. Wei Sheng Yan Jiu 43(6):917–922
281. Reichert KP, Schetinger MRC, Pillat MM et al (2019) Aluminium affects neural phenotype determination of embryonic neural progenitor cells. Arch Toxicol 93:2515–2524. https://doi.org/10.1007/s00204-019-02522-6
282. Blaylock RL (2024) Additive aluminium as a cause of induced immunoexcitoxicity resulting in neurodevelopmental and neurodegenerative disorders: a biochemical, pathophysiological, and pharmacological analysis. Surg Neurol Int 15:171. https://doi.org/10.25259/SNI_296_2024
283. LaFerla FM (2002) Calcium dyshomeostasis and intracellular signalling in Alzheimer's disease. Nat Rev Neurosci 3:862–872. https://doi.org/10.1038/nrn960
284. Silva VS, Gonçalves PP (2003) The inhibitory effect of aluminium on the (Na+/K+)ATPase activity of rat brain cortex synaptosomes. J Inorg Biochem 97:143–150. https://doi.org/10.1016/s0162-0134(03)00257-5
285. El-Rahman SSA (2003) Neuropathology of aluminium toxicity in rats (glutamate and GABA impairment). Pharmacol Res 47:189–194. https://doi.org/10.1016/s1043-6618(02)00336-5
286. Stellwagen D, Beattie EC, Seo JY et al (2005) Differential regulation of AMPA receptor and GABA receptor trafficking by tumor necrosis factor-alpha. J Neurosci 25:3219–3228. https://doi.org/10.1523/JNEUROSCI.4486-04.2005
287. Bondy SC (2010) The neurotoxicity of environmental aluminium is still an issue. Neurotoxicology 31:575–581. https://doi.org/10.1016/j.neuro.2010.05.009
288. Wigerblad G, Huie JR, Yin HZ et al (2017) Inflammation-induced GluA1 trafficking and membrane insertion of Ca2+ permeable AMPA receptors in dorsal horn neurons is dependent on spinal tumor necrosis factor, PI3 kinase and protein kinase A. Exp Neurol 293:144–158. https://doi.org/10.1016/j.expneurol.2017.04.004
289. Kasbe P, Jangra A, Lahkar M (2015) Mangiferin ameliorates aluminium chloride-induced cognitive dysfunction via alleviation of hippocampal oxido-nitrosative stress, proinflammatory cytokines and acetylcholinesterase level. J Trace Elem Med 31:107–112. https://doi.org/10.1016/j.jtemb.2015.04.002
290. Cao Z, Yang X, Zhang H et al (2016) Aluminium chloride induces neuroinflammation, loss of neuronal dendritic spine and cognition impairment in developing rat.

Chemosphere 151:289–295. https://doi.org/10.1016/j.chemosphere.2016.02.092
291. Prema A, Justin Thenmozhi A, Manivasagam T et al (2017) Fenugreek seed powder attenuated aluminium chloride-induced tau pathology, oxidative stress, and inflammation in a rat model of Alzheimer's disease. J Alzheimers Dis 60:S209–S220. https://doi.org/10.3233/JAD-161103
292. Zhao Y, Dang M, Zhang W et al (2020) Neuroprotective effects of Syringic acid against aluminium chloride induced oxidative stress mediated neuroinflammation in rat model of Alzheimer's disease. J Funct Foods 71:104009. https://doi.org/10.1016/j.jff.2020.104009
293. Gustin A, Kirchmeyer M, Koncina E et al (2015) NLRP3 inflammasome is expressed and functional in mouse brain microglia but not in astrocytes. PloS One 10:e0130624. https://doi.org/10.1371/journal.pone.0130624
294. Pogue AI, Jaber V, Zhao Y et al (2017) Systemic Inflammation in C57BL/6J mice receiving dietary aluminium sulfate; up-regulation of the pro-inflammatory cytokines IL-6 and TNFα, C-reactive protein (CRP) and miRNA-146a in blood serum. J Alzheimers Dis Parkinsonism 7:403. https://doi.org/10.4172/2161-0460.1000403
295. Guo Y, Lei W, Wang J et al (2016) Misoprostol reverse hippocampal neuron cyclooxygenase-2 downstream signalling imbalance in aluminium-overload rats. Curr Alzheimer Res 13:1006–1016. https://doi.org/10.2174/1567205013666160401114601
296. Jamil A, Mahboob A, Ahmed T (2016) Ibuprofen targets neuronal pentraxins expression and improves cognitive function in mouse model of AlCl3-induced neurotoxicity. Exp Ther Med 11:601–606. https://doi.org/10.3892/etm.2015.2928
297. Gao C, Jiang J, Tan Y et al (2023) Microglia in neurodegenerative diseases: mechanism and potential therapeutic targets. Signal Transduct Target Ther 8:1–37. https://doi.org/10.1038/s41392-023-01588-0
298. Pan B, Li Y, Zhang J et al (2020) Role of mGluR 1 in synaptic plasticity impairment induced by maltol aluminium in rats. Environ Toxicol Pharmacol 78:103406. https://doi.org/10.1016/j.etap.2020.103406
299. Akinrinade ID, Memudu AE, Ogundele OM (2015) Fluoride and aluminium disturb neuronal morphology, transport functions, cholinesterase, lysosomal and cell cycle activities. Pathophysiology 22:105–115. https://doi.org/10.1016/j.pathophys.2015.03.001
300. Mold M, Cottle J, King A et al (2019) Intracellular aluminium in inflammatory and glial cells in cerebral amyloid angiopathy: a case report. Int J Environ Res Public Health 16:1459. https://doi.org/10.3390/ijerph16081459
301. Jing Y, Wang Z, Song Y (2004) Quantitative study of aluminium-induced changes in synaptic ultrastructure in rats. Synapse 52:292–298. https://doi.org/10.1002/syn.20025
302. A-R Ahmed G, Khalil SKH, El Hotaby W et al (2020) ATR-IR and EPR spectroscopy for detecting the alterations in cortical synaptosomes induced by aluminium stress. Spectrochim Acta A 228:117535. https://doi.org/10.1016/j.saa.2019.117535
303. Li H, Xue X, Li L et al (2020) Aluminium-induced synaptic plasticity impairment via PI3K-Akt-mTOR signalling pathway. Neurotox Res 37:996–1008. https://doi.org/10.1007/s12640-020-00165-5
304. Johnson VJ, Sharma RP (2003) Aluminium disrupts the pro-inflammatory cytokine/neurotrophin balance in primary brain rotation-mediated aggregate cultures: possible role in neurodegeneration. Neurotoxicology 24:261–268. https://doi.org/10.1016/S0161-813X(02)00194-8
305. Li H, Xue X, Li Z et al (2020) Aluminium-induced synaptic plasticity injury via the PHF8-H3K9me2-BDNF signalling pathway. Chemosphere 244:125445. https://doi.org/10.1016/j.chemosphere.2019.125445
306. Pan B, Zhou Y, Li H et al (2020) Relationship between occupational aluminium exposure and histone lysine modification through methylation. J Trace Elem Med 61:126551. https://doi.org/10.1016/j.jtemb.2020.126551
307. Chen T-J, Cheng H-M, Wang D-C et al (2011) Nonlethal aluminium maltolate can reduce brain-derived neurotrophic factor-induced Arc expression through interrupting the ERK signalling in SH-SY5Y neuroblastoma cells. Toxicol Lett 200:67–76. https://doi.org/10.1016/j.toxlet.2010.10.016
308. Inohana M, Eguchi A, Nakamura M et al (2018) Developmental exposure to aluminium chloride irreversibly affects postnatal hippocampal neurogenesis involving multiple functions in mice. Toxicol Sci 164:264–277. https://doi.org/10.1093/toxsci/kfy081
309. Zhang L, Jin C, Lu X et al (2014) Aluminium chloride impairs long-term memory and downregulates cAMP-PKA-CREB signalling in rats. Toxicology 323:95–108. https://doi.org/10.1016/j.tox.2014.06.011
310. Yan D, Jin C, Cao Y et al (2017) Effects of aluminium on long-term memory in rats and on SIRT1 mediating the transcription of CREB-dependent gene in hippocampus. Basic Clin Pharmacol Toxicol 121:342–352. https://doi.org/10.1111/bcpt.12798

311. Wang L, Hu J, Zhao Y et al (2014) Effects of aluminium on β-amyloid (1-42) and secretases (APP-cleaving enzymes) in rat brain. Neurochem Res 39:1338–1345. https://doi.org/10.1007/s11064-014-1317-z
312. Luo Y, Niu F, Sun Z et al (2009) Altered expression of Abeta metabolism-associated molecules from D-galactose/AlCl(3) induced mouse brain. Mech Ageing Dev 130:248–252. https://doi.org/10.1016/j.mad.2008.12.005
313. Alexandrov PN, Zhao Y, Jones BM et al (2013) Expression of the phagocytosis-essential protein TREM2 is down-regulated by an aluminium-induced miRNA-34a in a murine microglial cell line. J Inorg Biochem 128:267–269. https://doi.org/10.1016/j.jinorgbio.2013.05.010
314. Zhang Q, Zhang F, Ni Y et al (2019) Effects of aluminium on amyloid-beta aggregation in the context of Alzheimer's disease. Arab J Chem 12:2897–2904. https://doi.org/10.1016/j.arabjc.2015.06.019
315. Chiroma SM, Mohd Moklas MA, Mat Taib CN et al (2018) D-galactose and aluminium chloride induced rat model with cognitive impairments. Biomed Pharmacother 103:1602–1608. https://doi.org/10.1016/j.biopha.2018.04.152
316. Culbreth M, Aschner M (2018) GSK-3β, a double-edged sword in Nrf2 regulation: implications for neurological dysfunction and disease. F1000Research 7:1043. https://doi.org/10.12688/f1000research.15239.1
317. Hou Y, Dan X, Babbar M et al (2019) Ageing as a risk factor for neurodegenerative disease. Nat Rev Neurol 15:565–581. https://doi.org/10.1038/s41582-019-0244-7
318. Liang L, Xing-li X, Ya-qin L et al (2020) Effects of subchronic aluminium exposure on β-amyloid protein and low density lipoprotein family in transgenic mice with human apolipoprotein E4 gene. J Environ Occup Med 37:103–110. https://doi.org/10.13213/j.cnki.jeom.2020.19565
319. Wang S, Xue Y, Zhang J et al (2023) Interaction between aluminium exposure and ApoEε4 gene on cognitive function of in-service workers. Chemosphere 323:138282. https://doi.org/10.1016/j.chemosphere.2023.138282
320. Gauthier E, Fortier I, Courchesne F et al (2000) Aluminium forms in drinking water and risk of Alzheimer's disease. Environ Res 84:234–246. https://doi.org/10.1006/enrs.2000.4101
321. Chin-Chan M, Navarro-Yepes J, Quintanilla-Vega B (2015) Environmental pollutants as risk factors for neurodegenerative disorders: Alzheimer and Parkinson diseases. Front Cell Neurosci 9:124. https://doi.org/10.3389/fncel.2015.00124
322. Zundel CG, Ryan P, Brokamp C et al (2022) Air pollution, depressive and anxiety disorders, and brain effects: a systematic review. Neurotoxicology 93:272–300. https://doi.org/10.1016/j.neuro.2022.10.011
323. Genc S, Zadeoglulari Z, Fuss SH et al (2012) The adverse effects of air pollution on the nervous system. J Toxicol 2012:782462. https://doi.org/10.1155/2012/782462
324. Cory-Slechta DA, Sobolewski M, Oberdörster G (2020) Air pollution-related brain metal dyshomeostasis as a potential risk factor for neurodevelopmental disorders and neurodegenerative diseases. Atmosphere 11:1098. https://doi.org/10.3390/atmos11101098
325. Manisalidis I, Stavropoulou E, Stavropoulos A et al (2020) Environmental and health impacts of air pollution: a review. Front Public Health 20(8):14. https://doi.org/10.3389/fpubh.2020.00014
326. Wright S, Andrews G, Sabir H (2009) A review of heat exchanger fouling in the context of aircraft air-conditioning systems, and the potential for electrostatic filtering. Appl Therm Eng 29(13):2596–2609. https://doi.org/10.1016/j.applthermaleng.2009.01.002
327. https://www.epa.gov/pm-pollution/particulate-matter-pm-basics
328. Xing YF, Xu YH, Shi MH et al (2016) The impact of PM2.5 on the human respiratory system. J Thorac Dis 8(1):E69–74. https://doi.org/10.3978/j.issn.2072-1439.2016.01.19
329. European Parliament & Council (2008) Directive 2008/50/EC on ambient air quality and cleaner air for Europe. Official Journal of the European Union, L152, pp 1–44. https://eur-lex.europa.eu/eli/dir/2008/50/oj/eng. Accessed 13 March 2025
330. European Parliament & Council (2024) Directive (EU) 2024/2881 on ambient air quality and cleaner air for Europe (recast). Official Journal of the European Union, L2881, pp 1–44. Accessed 13 March 2025
331. Wakenhut F (2024) Air quality revision of EU rules. Presentation at the 8th Copernicus Atmosphere Monitoring Service General Assembly, Brussels, Belgium. https://atmosphere.copernicus.eu/sites/default/files/custom-uploads/CAMS%208TH%20GA/Presentations/120624_09_30_sIntro_Francois%20Wakenhut_v1.pdf. Accessed 13 March 2025
332. Ali H, Khan E, Ilahi I (2019) Environmental chemistry and ecotoxicology of hazardous heavy metals: environmental persistence, toxicity, and bioaccumulation. J Chem https://doi.org/10.1155/2019/6730305
333. European Commission. "EU Air Quality Standards." European Commission. https://environment.ec.europa.eu/topics/air/

air-quality/eu-air-quality-standards_en. Accessed 13 March 2025
334. Haywood S (2019) Brain–barrier regulation, metal (Cu, Fe) dyshomeostasis, and neurodegenerative disorders in man and animals. Inorganics 7:108. https://doi.org/10.3390/inorganics7090108
335. Yoo J, Han J, Lim MH (2023) Transition metal ions and neurotransmitters: coordination chemistry and implications for neurodegeneration. RSC Chem Biol 4(8):548–563. https://doi.org/10.1039/d3cb00052d
336. Sanders T, Liu Y, Buchner V et al (2009) Neurotoxic effects and biomarkers of Lead exposure: a review. Rev Environ Health 24(1):15–45. https://doi.org/10.1515/reveh.2009.24.1.15
337. Neal AP, Guilarte TR (2013) Mechanisms of Lead and Mn neurotoxicity. Toxicol Res (Camb) 2(2):99–114. https://doi.org/10.1039/C2TX20064C
338. El Morabet R (2019) Effects of outdoor air pollution on human health. In: Nriagu J (ed) Encyclopedia of environmental health (Second Edition). Elsevier, pp 278–286. https://doi.org/10.1016/B978-0-12-409548-9.11509-X
339. David E, Niculescu VC (2021) Volatile organic compounds (VOCs) as environmental pollutants: occurrence and mitigation using nanomaterials. Int J Environ Res Public Health 18(24):13147. https://doi.org/10.3390/ijerph182413147
340. Kirkok SK, Kibet JK, Kinyanjui TK et al (2020) A review of persistent organic pollutants: dioxins, furans, and their associated nitrogenated analogues. SN Appl Sci 2:1729. https://doi.org/10.1007/s42452-020-03551-y
341. Wyer KE, Kelleghan DB, Blanes-Vidal V et al (2022) Ammonia emissions from agriculture and their contribution to fine particulate matter: a review of implications for human health. J Env Manag 323:116285. https://doi.org/10.1016/j.jenvman.2022.116285
342. Abed TK, Fayad MA, Al-Amiery AA et al (2024) Radon gas emission from home appliances: understanding sources, implications, and mitigation strategies. Results Eng 22:102133. https://doi.org/10.1016/j.rineng.2024.102133
343. World Health Organization (2021) WHO global air quality guidelines: Particulate matter ($PM_{2.5}$ and PM_{10}), ozone, nitrogen dioxide, sulfur dioxide and carbon monoxide [Internet]. World Health Organization, Geneva. https://www.who.int/publications/i/item/9789240034228. Accessed 11 March 2025
344. U.S. Environmental Protection Agency. (n.d.). Potential well water contaminants and their impacts. U.S. Environmental Protection Agency. https://www.epa.gov/privatewells/potential-well-water-contaminants-and-their-impacts. Accessed 14 March 2025
345. Oladimeji TE, Oyedemi M, Emetere ME et al (2024) Review on the impact of heavy metals from industrial wastewater effluent and removal technologies. Heliyon 10(23):e40370. https://doi.org/10.1016/j.heliyon.2024.e40370
346. Saravanan A, Kumar PS, Jeevanantham S et al (2021) Effective water/wastewater treatment methodologies for toxic pollutants removal: Processes and applications towards sustainable development. Chemosphere 280:130595. https://doi.org/10.1016/j.chemosphere.2021.130595
347. Ziani K, Ioniță-Mîndrican CB, Mititelu M et al (2023) Microplastics: a real global threat for environment and food safety: a state of the art review. Nutrients 15(3):617. https://doi.org/10.3390/nu15030617
348. European Parliament & Council (2020) Directive (EU) 2020/2184 on the quality of water intended for human consumption (recast). Off J Eur Union, L437, pp 1–62. https://eur-lex.europa.eu/eli/dir/2020/2184/oj/eng. Accessed 13 March 2025
349. Commission of the European Communities Directive (2020) Directive (EU) 2020/2184 of the European Parliament and of the council of 16 December 2020 on the quality of water intended for human consumption. https://eur-lex.europa.eu/legal-content/EN/TXT/PDF/?uri=CELEX:32003L0040&from=EL. Accessed 13 March 2025
350. Ahmad A, Bhattacharya P (2019) As in drinking water: is 10 μg/L a safe limit?. Curr Pollut Rep 5:1–3. https://doi.org/10.1007/s40726-019-0102-7
351. Fewtrell L, Kay D, MacGill S (2001) A review of the science behind drinking water standards for copper. Int J Env Health Res 11:161–7. https://doi.org/10.1080/09603120020047573
352. World Health Organization (2017) Lead in drinking-water: background document for development of WHO guidelines for drinking-water quality. https://cdn.who.int/media/docs/default-source/wash-documents/wash-chemicals/Lead-background-feb17.pdf. Accessed 13 March 2025
353. Canton SP (1999) Acute aquatic life criteria for selenium. Env Toxicol Chem 18:1425–1432. https://doi.org/10.1002/etc.5620180712

354. Spiller HA (2018) Rethinking Hg: the role of selenium in the pathophysiology of Hg toxicity. Clin Toxicol (Phila) 56(5):313–326. https://doi.org/10.1080/15563650.2017.1400555
355. Ralston NVC, Raymond LJ (2010) Dietary selenium's protective effects against methylHg toxicity. Toxicology 278(1):112–123. https://doi.org/10.1016/j.tox.2010.06.004
356. Fujimura M, Usuki F (2020) MethylHg-mediated oxidative stress and activation of the cellular protective system. Antioxidants (Basel) 9(10):1004. https://doi.org/10.3390/antiox9101004
357. Usuki F, Yamashita A, Fujimura M (2011) Post-transcriptional defects of antioxidant selenoenzymes cause oxidative stress under methylHg exposure. J Biol Chem 286(8):6641–6649
358. Wong CP, Dashner-Titus EJ, Alvarez SC et al (2019) Zinc deficiency and As exposure can act both independently or cooperatively to affect zinc status, oxidative stress, and inflammatory response. Biol Trace Elem Res 191:370–381. https://doi.org/10.1007/s12011-019-1631-z
359. Angon LEAD IslamMS, Kc S et al (2024) Sources, effects and present perspectives of heavy metals contamination: Soil, plants and human food chain. Heliyon 10(7):e28357. https://doi.org/10.1016/j.heliyon.2024.e28357
360. Sankhla SM, Mayurikumari, Nandan M et al (2016) Heavy metal contamination in soil and their toxic effect on human health: a review study. Int J All Res Educ Sci Methods 4:13–19
361. Taylor V, Goodale B, Raab A et al (2017) Human exposure to organic As species from seafood. Sci Total Environ 580:266–282. https://doi.org/10.1016/j.scitotenv.2016.12.113
362. European Commission (2023) Commission Regulation (EU) 2023/915 of 25 April 2023 on maximum levels for certain contaminants in food and repealing Regulation (EC) No 1881/2006. Official Journal of the European Union, L119:103–107. https://eur-lex.europa.eu/eli/reg/2023/915/oj/eng. Accessed 13 March 2025
363. European Union. (n.d.). Maximum levels for certain contaminants in food. EUR-Lex. https://eur-lex.europa.eu/EN/legal-content/summary/maximum-levels-for-certain-contaminants-in-food.html?fromSummary=30. Accessed 14 March 2025
364. European Commission (2006) Commission Regulation (EC) No 1881/2006 of 19 December 2006 Setting Maximum Levels for Certain Contaminants in Foodstuffs. Official Journal of the European Union, no. L 364 (December 20):5–24. https://eur-lex.europa.eu/legal-content/EN/TXT/PDF/?uri=CELEX%3A32006R1881. Accessed 13 March 2025.
365. Jomova K, Valko M (2011) Advances in metal-induced oxidative stress and human disease. Toxicology 283(2-3):65–87. https://doi.org/10.1016/j.tox.2011.03.001
366. Huang H, Guan H, Tian ZQ et al (2024) Exposure sources, intake pathways and accumulation of Lead in human blood. Soil Security 15:100150. https://doi.org/10.1016/j.soisec.2024.100150
367. Maria SMP, Hill BD, Kline J (2018) Lead (Lead) neurotoxicology and cognition. Appl Neuropsychol 8(3):272–293. https://doi.org/10.1080/21622965.2018.1428803
368. Karimi R, Silbernagel S, Fisher NS et al (2014) Elevated blood Hg at recommended seafood consumption rates in adult seafood consumers. Int J Hygiene Env Health 217(7):758–764. https://doi.org/10.1016/j.ijheh.2014.03.007
369. Zhao S, Pudasainee D, Duan Y et al (2019) A review on Hg in coal combustion process: content and occurrence forms in coal, transformation, sampling methods, emission and control technologies. Progr Energy Combust Sci 73:26–64. https://doi.org/10.1016/j.pecs.2019.02.001
370. Paduraru E, Iacob D, Rarinca V et al (2022) Comprehensive review regarding Hg poisoning and its complex involvement in Alzheimer's disease. Int J Mol Sci 23(4):1992. https://doi.org/10.3390/ijms23041992
371. Khosravi-Darani K, Rehman Y, Katsoyiannis IA et al (2022) As exposure via contaminated water and food sources. Water 14:1884. https://doi.org/10.3390/w14121884
372. Wang Y, Yu J, Wang Z et al (2021) A review on As removal from coal combustion: advances, challenges and opportunities. Chem Eng J 414:128785. https://doi.org/10.1016/j.cej.2021.128785
373. Garelick H, Jones H, Dybowska A et al (2009) As pollution sources. Reviews of environmental contamination and toxicology, vol 197. Springer, New York, NY. https://doi.org/10.1007/978-0-387-79284-2_2
374. Sierra-Alvarez R (2009) Removal of copper, chromium and As from preservative-treated wood by chemical extraction-fungal bioleaching. Waste Manag 29(6):1885–1891. https://doi.org/10.1016/j.wasman.2008.12.015
375. Welch K, Higgins I, Oh M et al (1982) As exposure, smoking, and respiratory cancer in copper smelter workers. Arch Env Health 37(6):325–335. https://doi.org/10.1080/00039896.1982.10667586

376. States JC, Srivastava S, Chen Y et al (2009) As and cardiovascular disease. Toxicol Sci 107(2):312–23. https://doi.org/10.1093/toxsci/kfn236
377. Tyler CR, Allan AM (2014) The effects of As exposure on neurological and cognitive dysfunction in human and rodent studies: a review. Curr Environ Health Rep 1(2):132–147. https://doi.org/10.1007/s40572-014-0012-1
378. Prakash PR, Narayan A, Jain S et al (2024) Chronic As poisoning: a sinister cause of peripheral neuropathy in a young couple. J Postgrad Med 70(2):105–108. https://doi.org/10.4103/jpgm.jpgm_708_23
379. Chunhabundit R (2016) Cd exposure and potential health risk from foods in contaminated area, Thailand. Toxicol Res 32(1):65–72. https://doi.org/10.5487/TR.2016.32.1.065
380. Suciu NA, De Vivo R, Rizzati N et al (2022) Cd content in phosphate fertilizer: which potential risk for the environment and human health?. Curr Opin Env Sci Health 30:100392. https://doi.org/10.1016/j.coesh.2022.100392
381. Khan Z, Elahi A, Bukhari DA et al (2022) Cd sources, toxicity, resistance and removal by microorganisms—a potential strategy for Cd eradication. J Saudi Chem Soc 26(6):101569. https://doi.org/10.1016/j.jscs.2022.101569
382. American Toxic Substances and Disease Registry (ATSDR) Toxicological Profile for Cd (2012) American Toxic Substances and Disease Registry. Atlanta, GA, USA, pp 12–105 http://www.atsdr.cdc.gov/toxprofiles/tp5.pdf. Accessed on 31 August 2017
383. Wróblewski K, Wojnicka J, Tutka P et al (2024) Measurements of Cd levels in relation to tobacco dependence and as a function of cytisine administration. Sci Rep 14: 1883. https://doi.org/10.1038/s41598-024-52234-w
384. Richter P, Faroon O, Pappas RS (2017) Cd and Cd/zinc ratios and tobacco-related morbidities. Int J Environ Res Public Health 14(10):1154. https://doi.org/10.3390/ijerph14101154
385. Matt GE, Quintana PJE, Hoh E et al (2021) Tobacco smoke is a likely source of Lead and Cd in settled house dust. J Trace Elem Med Biol 63:126656. https://doi.org/10.1016/j.jtemb.2020.126656
386. Satarug S, Moore MR (2004) Adverse health effects of chronic exposure to low-level Cd in foodstuffs and cigarette smoke. Environ Health Perspect 112(10):1099–103. https://doi.org/10.1289/ehp.6751
387. Cirovic A, Cirovic A, Yimthiang S et al (2024) Modulation of adverse health effects of environmental Cd exposure by zinc and its transporters. Biomolecules 14(6):650. https://doi.org/10.3390/biom14060650
388. Yoon M, Ring C, Van Landingham CB et al (2019) Assessing children's exposure to Mn in drinking water using a LEADPK model. Toxicol Appl Pharmacol 380:114695. https://doi.org/10.1016/j.taap.2019.114695
389. Office of Dietary Supplements, National Institutes of Health. Mn—Health Professional Fact Sheet. U.S. Department of Health & Human Services https://ods.od.nih.gov/factsheets/Mn-HealthProfessional/. Accessed 13 March 2025
390. Lucchini RG, Aschner M, Kim Y et al (2015) Chapter 45—Mn. In: Nordberg GF, Fowler BA, Nordberg M (eds) Handbook on the toxicology of metals (Fourth Edition). Academic Press, pp 975–1011. https://doi.org/10.1016/B978-0-444-59453-2.00045-7.
391. Aschner M, Dos Santos APM, Erikson KM et al (2008) Mn transport into the brain: putative mechanisms. Met Ions Biol Med 10:695–700
392. Horning KJ, Caito SW, Tipps KG et al (2015) Mn is essential for neuronal health. Annu Rev Nutr 35:71–108. https://doi.org/10.1146/annurev-nutr-071714-034419
393. Gad SC (2014) Al. In: Wexler P (ed) Encyclopedia of Toxicology (Third Edition). Academic Press, pp 161_163. https://doi.org/10.1016/B978-0-12-386454-3.00810-1
394. World Health Organization (2003) Aluminium in Drinking-Water: Background Document for Development of WHO Guidelines for Drinking-Water Quality. https://iris.who.int/bitstream/handle/10665/75362/WHO_SDE_WSH_03.04_53_eng.pdf. Accessed 13 March 2025
395. Rahimzadeh MR, Rahimzadeh MR, Kazemi S et al (2022) Al poisoning with emphasis on its mechanism and treatment of intoxication. Emerg Med Int 11(2022):1480553. https://doi.org/10.1155/2022/1480553
396. Hashimoto Y, Sekine Y, Otoshi T (1992) Atmospheric Al from human activities. Atmos Env 26(3):295–300. https://doi.org/10.1016/0957-1272(92)90005-D
397. Bryliński Ł, Kostelecka K, Woliński F et al (2023) Aluminium in the human brain: routes of penetration, toxicity, and resulting complications. Int J Mol Sci 24(8):7228. https://doi.org/10.3390/ijms24087228
398. Krewski D, Yokel RA, Nieboer E et al (2007) Human health risk assessment for aluminium, aluminium oxide, and aluminium hydroxide. Erratum in: J Toxicol Environ Health B Crit Rev. 2008 11(2):147. J Toxicol Environ Health B Crit Rev 10 Suppl 1(Suppl 1):1–269. doi: 10.1080/10937400701597766
399. Affourtit F, Bakker MI, Pronk MEJ (2020) Human Health Risk Assessment of

Aluminium. RIVM Report No. 2020-0001 National Institute for Public Health and the Environment. https://www.rivm.nl/bibliotheek/rapporten/2020-0001.pdf. Accessed 13 March 2025
400. European Food Safety Authority (2011) Statement of EFSA on the evaluation of a new study related to the bioavailability of aluminium in food. EFSA J 9(5):2157. https://doi.org/10.2903/j.efsa.2011.2157
401. Kawahara M, Kato-Negishi M (2011) Link between Al and the pathogenesis of Alzheimer's Disease: the Integration of the Al and Amyloid Cascade Hypotheses. Int J Alzheimers Dis 2011:276393. https://doi.org/10.4061/2011/276393
402. Kawahara M, Tanaka K, Kato-Negishi M (2021) Neurotoxicity of Al and its link to neurodegenerative diseases. Metallomics. Research 1(1)::rev-47–rev-65. https://doi.org/10.11299/metallomicsresearch.MR202104
403. Yokel RA (2000) The toxicology of Al in the brain: a review. Neurotoxicology 21(5):813–28
404. Yu G, Wu L, Su Q et al (2024) Neurotoxic effects of heavy metal pollutants in the environment: focusing on epigenetic mechanisms. Environ Pollut 345:123563. https://doi.org/10.1016/j.envpol.2024.123563
405. Venkatesan D, Muthukumar S, Iyer M et al (2024) Heavy metals toxicity on epigenetic modifications in the pathogenesis of Alzheimer's disease (AD. J Biochem Mol Toxicol 38(6):e23741. https://doi.org/10.1002/jbt.23741
406. Wei Y, Zhou YF, Xiao L et al (2024) Associations of heavy metals with cognitive function: an epigenome-wide view of DNA methylation and mediation analysis. Ann Neurol 96(1):87–98. https://doi.org/10.1002/ana.26942
407. Renzetti S, Cagna G, Calza S et al (2021) The effects of the exposure to neurotoxic elements on Italian school children behavior. Sci Rep 11:9898. https://doi.org/10.1038/s41598-021-88969-z
408. Lucchini RG, Guazzetti S, Renzetti S et al (2019) Neurocognitive impact of metal exposure and social stressors among schoolchildren in Taranto, Italy. Environ Health 18(1):67. https://doi.org/10.1186/s12940-019-0505-3
409. Stratmann M, Özel F, Marinopoulou M et al (2024) Prenatal exposure to endocrine disrupting chemicals and the association with behavioural difficulties in 7-year-old children in the SELMA study. J Expo Sci Environ Epidemiol. doi: https://doi.org/10.1038/s41370-024-00739-x
410. Vasconcellos ACS, Hallwass G, Bezerra JG et al (2021) Health risk assessment of Hg exposure from fish consumption in Munduruku indigenous communities in the Brazilian Amazon. Int J Environ Res Public Health 18(15):7940. https://doi.org/10.3390/ijerph18157940 27
411. Qing Y, Li Y, Yang J et al (2022) Risk assessment of Hg through dietary exposure in China. Environ Pollut 1(312):120026. https://doi.org/10.1016/j.envpol.2022.120026
412. Weinhouse C, Gallis JA, Ortiz E et al (2021) A population-based Hg exposure assessment near an artisanal and small-scale gold mining site in the Peruvian Amazon. J Expo Sci Environ Epidemiol 31(1):126–136. https://doi.org/10.1038/s41370-020-0234-2
413. Mufakhir FR, Yoga GP, Darusman T et al (2024) Hg risk assessment scenarios: exposure from fish dietary behaviors of Katingan River Basin community. Int J Environ Health Res 34(9):3317–3333. https://doi.org/10.1080/09603123.2024.2303980
414. Deng C, Xie H, Ye X et al (2016) Hg risk assessment combining internal and external exposure methods for a population living near a municipal solid waste incinerator. Environ Pollut 219:1060–1068. https://doi.org/10.1016/j.envpol.2016.09.006
415. Santos-Lima CD, Mourão DS, Carvalho CF et al (2020) Neuropsychological effects of Hg exposure in children and adolescents of the Amazon Region, Brazil. Neurotoxicology 79:48–57.https://doi.org/10.1016/j.neuro.2020.04.004
416. Vlasak T, Dujlovic T, Barth A (2023) Mn exposure and cognitive performance: a meta-analytical approach. Env Pollut 332:121884. https://doi.org/10.1016/j.envpol.2023.121884
417. Lopes de Andrade V, Marreilha Dos Santos AP, Aschner M (2021) Neurotoxicity of metal mixtures. Adv Neurotoxicol 5:329–364. https://doi.org/10.1016/bs.ant.2020.12.003
418. Bakulski KM, Seo YA, Hickman RC et al (2020) Heavy metals exposure and Alzheimer's disease and related dementias. J Alzheimers Dis 76(4):1215–1242. https://doi.org/10.3233/JAD-200282
419. Isam F, Shohag S, Akhter S et al (2022) Exposure of metal toxicity in Alzheimer's disease: an extensive review. Front Pharmacol 29(13):903099. https://doi.org/10.3389/fphar.2022.903099
420. Biswas S, Anjum A, Banna HU et al (2019) Mn attenuates the effects of As on neurobehavioral and biochemical changes in mice co-exposed to As and Mn. Env Sci Pollut Res 26:29257–29266. https://doi.org/10.1007/s11356-019-06112-y

421. U.S. Environmental Protection Agency (2007) Framework for Metals Risk Assessment. EPA 120/R-07/001. Office of the Science Advisor, Risk Assessment Forum https://www.epa.gov/sites/default/files/2013-09/documents/metals-risk-assessment-final.pdf. Accessed 13 March 2025
422. Newman MC, Diamond GL, Menzie C et al (2004) Issue Paper on Metal Exposure Assessment. U.S. Environmental Protection Agency. https://www.epa.gov/sites/default/files/2014-11/documents/exposure_final.pdf
423. Dorne JCM, Kass GEN, Bordajandi LR et al (2010) Metal Ions in Toxicology In: Sigel A, Sigel H, Sigel RKO (eds) The Royal Society of Chemistry, vol 8. pp 27–60. https://doi.org/10.1039/9781849732116-00403
424. Ogwu MC, Izah SC, Sawyer WE et al (2025) Environmental risk assessment of trace metal pollution: a statistical perspective. Environ Geochem Health 47:94. https://doi.org/10.1007/s10653-025-02405-z
425. Hu XC, Dai M, Sun JM et al (2023) The utility of machine learning models for predicting chemical contaminants in drinking water: promise, challenges, and opportunities. Curr Env Health Rep 10(1):45–60. https://doi.org/10.1007/s40572-022-00389-x
426. Schaumlöffel D (2016) The position of metallomics within other omics fields. In: Michalke B (ed) https://doi.org/10.1002/9783527694907.ch1
427. Wild CP (2005) Complementing the genome with an "exposome": the outstanding challenge of environmental exposure measurement in molecular epidemiology. Cancer Epidemiol Biomarkers Prev 14(8):1847–50. https://doi.org/10.1158/1055-9965.EPI-05-0456
428. Sarigiannis DA (2019) The exposome: a new tool for improved health risk assessment. In: Roig B, Weiss K, Thireau V (eds) Management of emerging public health issues and risks. Academic Press, Cambridge, MA, USA. https://doi.org/10.1016/C2016-0-00995-6
429. Sillé FCM, Karakitsios S, Kleensang A et al (2020) The exposome—a new approach for risk assessment. ALTEX Altern Anim Exp 37(1): 3–23. https://doi.org/10.14573/altex.2001051
430. Rodríguez-Moro G, Ramírez-Acosta S, Arias-Borrego A et al (2018) Environmental Metallomics. In: Arruda M (ed) Metallomics. Advances in Experimental Medicine and Biology, vol 1055. Springer, Cham. https://doi.org/10.1007/978-3-319-90143-5_3
431. Błażewicz A, Grabrucker AM (2023) Metal profiles in autism spectrum disorders: a crosstalk between toxic and essential metals. Int J Mol Sci 24:308. https://doi.org/10.3390/ijms24010308
432. Liu M, Deng P, Li G et al (2024) Neurotoxicity of combined exposure to the heavy metals (Lead and As) in Zebrafish (*Danio rerio*). Toxics 12:282. https://doi.org/10.3390/toxics12040282
433. Woodruff TJ, Zota AR, Schwartz JM (2011) Environmental chemicals in pregnant women in the United States: NHANES 2003-2004. Environ Health Perspect 119(6):878–85. https://doi.org/10.1289/ehp.1002727
434. Zhang X, Li Z (2024) Assessing chronic gestational exposure to environmental chemicals in pregnant women: ädvancing the co-LEADK model. Env Res 247:118160. https://doi.org/10.1016/j.envres.2024.118160
435. Kippler M, Engström K, Jurkovic Mlakar S et al (2013) Sex-specific effects of early life Cd exposure on DNA methylation and implications for birth weight. Epigenetics 8(5):494–503. https://doi.org/10.4161/epi.24401
436. Kim N, Filipovic D, Bhattacharya S et al (2024) Epigenetic toxicity of heavy metals—implications for embryonic stem cells. Env Int 193:109084. https://doi.org/10.1016/j.envint.2024.109084
437. Bitto A, Pizzino G, Irrera N et al (2014) Epigenetic modifications due to heavy metals exposure in children living in polluted areas. Curr Genomics 15(6):464–8. https://doi.org/10.2174/138920291506150106153336
438. Mei Z, Liu G, Zhao B et al (2023) Emerging roles of epigenetics in Lead-induced neurotoxicity. Env Int 181:108253. https://doi.org/10.1016/j.envint.2023.108253
439. Amadi CN, Orish CN, Frazzoli C et al (2022) Association of autism with toxic metals: a systematic review of case-control studies. Pharmacol Biochem Behav 212:173313. https://doi.org/10.1016/j.lea db.2021.173313
440. Gull A, Dar AA, Sharma M (2018) Effects of heavy metals on the health of pregnant women and fetus: a review. Int J Theor Appl Sci 10(1):2249–3247. 1–7
441. Issah I, Duah MS, Arko-Mensah J et al (2024) Exposure to metal mixtures and adverse pregnancy and birth outcomes: a systematic review. Sci Total Environ 15(908):168380. https://doi.org/10.1016/j.scitotenv.2023.168380
442. Rice D, Barone S (2000) Critical periods of vulnerability for the developing nervous system: evidence from humans and animal models. Environ Health Perspect 108:511–533. https://doi.org/10.1289/ehp.00108s3511

443. van der Zalm AJ, Barroso J, Browne P et al (2022) A framework for establishing scientific confidence in new approach methodologies. Arch Toxicol 96:2865–2879. https://doi.org/10.1007/s00204-022-03365-4
444. Casanova R, Barnard RT, Gaussoin SA et al (2018) Using high-dimensional machine learning methods to estimate an anatomical risk factor for Alzheimer's disease across imaging databases. NeuroImage 183:401–411. https://doi.org/10.1016/j.neuroimage.2018.08.040
445. Servais B, Mahmoudi N, Gautam V et al (2024) Engineering brain-on-a-chip platforms. Nat Rev Bioeng 2:691–709. https://doi.org/10.1038/s44222-024-00184-3
446. Gorman BL, Torti SV, Torti FM et al (2024) Mass spectrometry imaging of metals in tissues and cells: Methods and biological applications. Biochim Biophys Acta 1868(2):130329. https://doi.org/10.1016/j.bbagen.2023.130329
447. Brooks J, Everett J, Lermyte F et al (2020) Analysis of neuronal iron deposits in Parkinson's disease brain tissue by synchrotron X-ray spectromicroscopy. J Trace Elem Med Biol 62:126555. https://doi.org/10.1016/j.jtemb.2020.126555
448. Lee H, Lee MJ, Kim EJ et al (2021) Iron accumulation in the oculomotor nerve of the progressive supranuclear palsy brain. Sci Rep 3 11(1):2950. https://doi.org/10.1038/s41598-021-82469-w
449. Chen P, Miah MR, Aschner M (2016) Metals and neurodegeneration. F1000Res 17(5):Rev-366. https://doi.org/10.12688/f1000research.7431.1
450. Miyazaki I, Asanuma M (2023) Multifunctional metallothioneins as a target for neuroprotection in Parkinson's disease. Antioxidants (Basel) 12(4):894. https://doi.org/10.3390/antiox12040894
451. Carmona A, Roudeau S, Ortega R (2021) Molecular mechanisms of environmental metal neurotoxicity: a focus on the interactions of metals with synapse structure and function. Toxics 9(9):198. https://doi.org/10.3390/toxics9090198

Epilogue: The Inorganic Soul of the Brain

As we close this volume, it is worth reflecting on the scales involved in neurometallomics. From a single metal ion, we can zoom out to the staggering complexity of the human "metallome." From the fundamental physics of a metal–protein interaction, we can speculate about the roles of metals in brain development and function. For decades, metals such as iron, copper, and zinc were often relegated to the status of "trace elements" or passive cofactors. However, as the preceding chapters have demonstrated, these elements are far from secondary; they are the dynamic, inorganic regulators of the central nervous system.

We began this book by establishing the Foundations of Brain Metallomics (Chaps. 1–3), then moved quickly into the high-tech arsenal required to "see" the invisible. The evolution of Omics tools, fluorescent probes, and the precision of ICP-MS and Atomic Absorption Spectroscopy (Chaps. 4–7) has transformed the landscape of metals in the brain from a "black box" of bulk tissue into a vibrant, high-resolution map of elemental flux.

The breakthrough workflows described in Chap. 8, specifically Cryo-Correlative Light and Synchrotron X-ray Fluorescence Imaging, represent the current pinnacle of our field. We are no longer just measuring how much metal is present; we are seeing where it sits in relation to a single protein, inside a single neuron, at the very moment of signaling.

Finally, Chaps. 9–11 bridge the gap between detection methods and their applicability to understanding brain physiology and pathology. For example, measuring metals and the metallome, we now understand that the "crosstalk" between metals and amyloidogenic proteins is not merely a biochemical curiosity. It is of key relevance to Alzheimer's, Parkinson's, and other neurodegenerative diseases. Besides, by investigating the metallome in neuropsychiatric and

Andreas M. Grabrucker (ed.), *Brain Metallomics*, Neuromethods, vol. 237,
https://doi.org/10.1007/978-1-0716-5468-2
© The Editor(s) (if applicable) and The Author(s), under exclusive license to Springer Science+Business Media, LLC, part of Springer Nature 2026

neurodevelopmental disorders, we begin to see that metal dyshomeostasis is a common motif. Thus, it is fair to say that the functions of metals and the metallome are key to understanding the entire brain physiology, from the earliest stages of brain development to aging.

Importantly, as Chap. 12 reminded us, the brain does not exist in a vacuum. Our metallome is a living record of our environment, reflecting the air we breathe, the food we eat, and the water we drink. Understanding this interaction is perhaps one of our greatest tools for future public health advocacy.

The future of Brain Metallomics likely lies at the intersection of artificial intelligence and single-cell resolution. As our datasets grow to include thousands of elemental signatures across millions of cells, we will need more than human brains to identify the patterns. However, if successful, we can move toward a "Personalized Metallomics," in which a patient's elemental profile could dictate their therapeutic path.

We hope this textbook has served as more than just a collection of data. We hope it has been a catalyst. The "metallome" of the brain is still giving up its secrets, and advances in all areas, from detection to understanding function, will reveal crucial new insights.

Andreas M. Grabrucker

FSC
www.fsc.org
MIX
Papier | Fördert gute Waldnutzung
FSC® C083411

Kolibri 360 GmbH
Ferdinand-Jühlke-Straße 7
99095 Erfurt, Deutschland
produktsicherheit@kolibri360.de